16. $\displaystyle\int \cosh ax\,dx = \frac{1}{a}\sinh ax + c.$

17. $\displaystyle\int \frac{dx}{a^2 + x^2} = \frac{1}{a}\tan^{-1}\frac{x}{a} + c.$

18. $\displaystyle\int \frac{dx}{\sqrt{a^2 - x^2}} = \sin^{-1}\frac{x}{a} + c.$

19. $\displaystyle\int \frac{dx}{\sqrt{x^2 \pm a^2}} = \ln\left| x + \sqrt{x^2 \pm a^2}\right| + c.$

20. $\displaystyle\int (ax + b)^n\,dx = \frac{(ax + b)^{n+1}}{a(n + 1)} + c, \quad n \neq -1.$

21. $\displaystyle\int \frac{dx}{ax + b} = \frac{1}{a}\ln|ax + b| + c.$

22. $\displaystyle\int \frac{dx}{x^2 - a^2} = \frac{1}{2a}\ln\left|\frac{x - a}{x + a}\right| + c.$

23. $\displaystyle\int \frac{dx}{x(ax + b)} = \frac{1}{b}\ln\left|\frac{x}{ax + b}\right| + c.$

24. $\displaystyle\int \frac{x\,dx}{ax + b} = \frac{x}{a} - \frac{b}{a^2}\ln|ax + b| + c.$

25. $\displaystyle\int \frac{x\,dx}{(ax + b)^2} = \frac{1}{a^2}\left(\ln|ax + b| + \frac{b}{ax + b}\right) + c.$

26. $\displaystyle\int x(ax + b)^n\,dx = \frac{(ax + b)^{n+1}}{a^2}\left(\frac{ax + b}{n + 2} - \frac{b}{n + 1}\right) + c, \quad n \neq -1, -2.$

27. $\displaystyle\int \frac{x\,dx}{\sqrt{ax + b}} = \frac{2(ax - ab)}{3a^2}\sqrt{ax + b} + c.$

28. $\displaystyle\int \frac{dx}{x\sqrt{ax + b}} = \begin{cases} \dfrac{1}{\sqrt{b}}\ln\left|\dfrac{\sqrt{ax + b} - \sqrt{b}}{\sqrt{ax + b} + \sqrt{b}}\right| + c, & b > 0 \\[2ex] \dfrac{2}{\sqrt{-b}}\tan^{-1}\sqrt{\dfrac{ax + b}{-b}} + c, & b < 0. \end{cases}$

29. $\displaystyle\int \frac{dx}{x^2\sqrt{ax + b}} = -\frac{\sqrt{ax + b}}{bx} - \frac{a}{2b}\int \frac{dx}{x\sqrt{ax + b}} + c, \quad b \neq 0.$

30. $\displaystyle\int \frac{dx}{ax^2 + bx + c} = \begin{cases} \dfrac{2}{\sqrt{4ac - b^2}}\tan^{-1}\dfrac{2ax + b}{\sqrt{4ac - b^2}}, & b^2 - 4ac < 0. \\[2ex] \dfrac{1}{\sqrt{b^2 - 4ac}}\ln\left|\dfrac{2ax + b - \sqrt{b^2 - 4ac}}{2ax + b + \sqrt{b^2 - 4ac}}\right|, & b^2 - 4ac > 0. \end{cases}$

Duane M. Broline

An Introduction to Differential Equations

with Difference Equations, Fourier Series, and Partial Differential Equations

N. Finizio and G. Ladas
University of Rhode Island

Wadsworth Publishing Company
Belmont, California

A division of Wadsworth, Inc.

ABOUT THE COVER: "Dirichlet Problem" (Miles Color Art A25) is the work of Professor E. P. Miles, Jr., and associates of Florida State University using an InteColor 80501 computer. Programming by Eric Chamberlain and photograph by John Owen. The function graphed is the discrete limiting position for the solution by relaxation of the heat distribution in an insulated rectangular plate with fixed temperature at boundary positions. This is an end-position photograph following intermediate positions displayed as the solution converges from an assumed initial average temperature to the ultimate (harmonic function) steady-state temperature induced by the constantly maintained boundary conditions.

Mathematics Editor: Richard Jones
Signing Representative: Richard Giggey

Printed in the United States of America
1 2 3 4 5 6 7 8 9 10—86 85 84 83 82

Library of Congress Cataloging in Publication Data

Finizio, N.
An introduction to ordinary differential equations, with difference equations, Fourier series, and partial differential equations.

Includes index.
1. Differential equations. I. Ladas, G.
II. Title.
QA372.F55 515.3′52 81-1971
ISBN 0-534-00960-3 AACR2

To my mother,

Eva Finizio

and to my father,

Vito Finizio

To my mother,

Ageliki Ladas

and in loving memory of my father,

Efthimios Ladas

List of Applications

Each entry in this list is coded according to whether the application appears in detail (A), little or no detail (B) or is an exercise or example associated with an application (C).

Preface

This book is designed for an introductory, one-semester or one-year course in differential equations, both ordinary and partial. Its prerequisite is elementary calculus.

Perusal of the table of contents and the list of applications shows that the book contains the theory, techniques, and applications covered in the traditional introductory courses in differential equations. A major feature of this text is the quantity and variety of applications of current interest in physical, biological, and social sciences. We have furnished a wealth of applications from such diverse fields as astronomy, bioengineering, biology, botany, chemistry, ecology, economics, electric circuits, finance, geometry, mechanics, medicine, meteorology, pharmacology, physics, psychology, seismology, sociology, and statistics.

Our experience gained in teaching differential equations at the elementary, intermediate, and graduate levels at the University of Rhode Island convinced us of the need for a book at the elementary level which emphasizes to the students the relevance of the various equations to which they are exposed in the course. That is to say, that the various types of differential equations encountered are not merely the product of some mathematician's imagination but rather that the equations occur in the course of scientific investigations of real-world phenomena.

The goal of this book, then, is to make elementary differential equations more useful, more meaningful, and more exciting to the student. To accomplish this, we strive to demonstrate that differential equations are very much "alive" in present-day applications. This approach has indeed had a satisfying effect in the courses we have taught recently.

During the preparation and class testing of this text we continuously kept in mind both the student and the teacher. We have tried to make the presentation direct, yet informal. Definitions and theorems are stated precisely and rigorously, but theory and rigor have been minimized in favor of comprehension of technique. The general approach is to use a larger number of routine examples to illustrate the new concepts, definitions, methods of solution, and theorems. Thus, it is intended that the material will be easily accessible to the student. Hopefully the presence of modern applications in addition to the traditional applications of geometry, physics, and chemistry will be refreshing to the teacher.

Numerous routine exercises in each section will help to test and strengthen the student's understanding of the new methods under discussion. There are over 1600 exercises in the text with answers to odd-numbered exercises provided. Some thought-provoking exercises from *The American Mathematical*

Monthly, Mathematics Magazine, and The William Lowell Putnam Mathematics Competition are inserted in many sections, with references to the source. These should challenge the students and help to train them in searching the literature. Review exercises appear at the end of every chapter. These exercises should serve to help the student review the material presented in the chapter. Some of the review exercises are problems that have been taken directly from physics and engineering textbooks. The inclusion of such problems should further emphasize that differential equations are very much present in applications and that the student is quite apt to encounter them in areas other than mathematics.

Every type of differential equation studied and every method presented is illustrated by real-life applications which are incorporated in the same section (or chapter) with the specific equation or method. Thus, the student will see immediately the importance of each type of differential equation that they learn how to solve. We feel that these "modern" applications, even if the student only glances at some of them, will help to stimulate interest and enthusiasm toward the subject of differential equations specifically and mathematics in general.

Many of the applications are integrated into the main development of ideas, thus blending theory, technique, and application. Frequently, the mathematical model underlying the application is developed in great detail. It would be impossible in a text of this nature to have such development for every application cited. Therefore, some of the models are only sketched, and in some applications the model alone is presented. In practically all cases, references are given for the source of the model. Additionally, a large number of applications appear in the exercises; these applications are also suitably referenced. Consequently, applications are widespread throughout the book, and although they vary in depth and difficulty, they should be diverse and interesting enough to whet the appetite of every reader. As a general statement, every application that appears, even those with little or no detail, is intended to illustrate the relevance of differential equations outside of their intrinsic value as mathematical topics. It is intended that the instructor will probably present only a few of the applications, while the rest can demonstrate to the reader the relevance of differential equations in real-life situations.

The first eight chapters of this book are reproduced from our text *Ordinary Differential Equations with Modern Applications*, Second Edition, Wadsworth Publishing Co., 1981.

The additional chapters 9, 10, and 11 treat difference equations, Fourier series, and partial differential equations, respectively. Each of these chapters provides a thorough introduction to its respective topic. We feel that the chapters on difference equations and partial differential equations are more extensive than one usually finds at this level. Our purpose for including these topics is to allow more course options for users of this book.

We are grateful to Katherine MacDougall, who so skillfully typed the manuscript of this text.

A special word of gratitude goes to Professors Gerald Bradley, John Haddock, Thomas Hallam, Ken Kalmanson, Gordon McLeod, and David Wend, who painstakingly reviewed portions of this book and offered numerous valuable suggestions for its improvement.

Thanks are also due to Dr. Clement McCalla, Dr. Lynnell Stern, and to our students Carl Bender, Thomas Buonanno, Michael Fascitelli, and especially Neal Jamnik, Brian McCartin, and Nagaraj Rao who proofread parts of the material and doublechecked the solutions to some of the exercises.

Special thanks are due to Richard Jones, the Mathematics Editor of Wadsworth Publishing Company for his continuous support, advice, and active interest in the development of this project.

N. Finizio
G. Ladas

Contents

6 BOUNDARY VALUE PROBLEMS

7 NUMERICAL SOLUTIONS OF DIFFERENTIAL EQUATIONS

8 NONLINEAR DIFFERENTIAL EQUATIONS AND SYSTEMS

9 DIFFERENCE EQUATIONS

10 FOURIER SERIES

11 AN INTRODUCTION TO PARTIAL DIFFERENTIAL EQUATIONS

CHAPTER 1

Elementary Methods— First-Order Differential Equations

1.1 INTRODUCTION AND DEFINITIONS

Differential equations are equations that involve derivatives of some unknown function(s). Although such equations should probably be called "derivative equations," the term "differential equations" (*aequatio differentialis*) initiated by Leibniz in 1676 is universally used. For example,

$$y' + xy = 3 \tag{1}$$

$$y'' + 5y' + 6y = \cos x \tag{2}$$

$$y'' = (1 + y'^2)(x^2 + y^2) \tag{3}$$

$$\frac{\partial^2 u}{\partial t^2} - \frac{\partial^2 u}{\partial x^2} = 0 \tag{4}$$

are differential equations. In Eqs. (1)–(3) the unknown function is represented by y and is assumed to be a function of the single independent variable x, that is, $y = y(x)$. The argument x in $y(x)$ (and its derivatives) is usually suppressed for notational simplicity. The terms y' and y'' in Eqs. (1)–(3) are the first and second derivatives, respectively, of the function $y(x)$ with respect to x. In Eq. (4) the unknown function u is assumed to be a function of the two independent variables t and x, that is, $u = u(t, x)$, $\partial^2u/\partial t^2$ and $\partial^2u/\partial x^2$ are the second partial derivatives of the function $u(t, x)$ with respect to t and x, respectively. Equation (4) involves partial derivatives and is a *partial differential equation.* Equations (1)–(3) involve ordinary derivatives and are *ordinary differential equations.*

In this book we are primarily interested in studying ordinary differential equations, however, an introduction to difference equations and an introduction to partial differential equations are presented in Chapters 9 and 11, respectively.

DEFINITION 1

An ordinary differential equation of order n is an equation that is, or can be put, in the form

$$y^{(n)} = F(x,y,y', \ldots, y^{(n-1)}), \tag{5}$$

where $y,y', \ldots, y^{(n)}$ are all evaluated at x.

The independent variable x belongs to some interval I (I may be finite or infinite), the function F is given, and the function $y = y(x)$ is unknown. For the most part the functions F and y will be real valued. Thus, Eq. (1) is an ordinary differential equation of order 1 and Eqs. (2) and (3) are ordinary differential equations of order 2.

DEFINITION 2

A solution of the ordinary differential equation (5) *is a function* $y(x)$ *defined over a subinterval* $J \subset I$ *which satisfies Eq.* (5) *identically over the interval* J.

Clearly, any solution $y(x)$ of Eq. (5) should have the following properties:

1. y should have derivatives at least up to order n in the interval J.
2. For every x in J the point $(x, y(x), y'(x), \ldots, y^{(n-1)}(x))$ should lie in the domain of definition of the function F, that is, F should be defined at this point.
3. $y^{(n)}(x) = F(x, y(x), y'(x), \ldots, y^{(n-1)}(x))$ for every x in J.

As an illustration we note that the function $y(x) = e^x$ is a solution of the second-order ordinary differential equation $y'' - y = 0$. In fact.

$$y''(x) - y(x) = (e^x)'' - e^x = e^x - e^x = 0.$$

Clearly, e^x is a solution of $y'' - y = 0$ valid for all x in the interval $(-\infty, +\infty)$. As another example, the function $y(x) = \cos x$ is a solution of $y'' + y = 0$ over the interval $(-\infty, +\infty)$. Indeed,

$$y''(x) + y(x) = (\cos x)'' + \cos x = -\cos x + \cos x = 0.$$

In each of the illustrations the solution is valid on the whole real line $(-\infty, +\infty)$. On the other hand, $y = \sqrt{x}$ is a solution of the first-order ordinary differential equation $y' = 1/2y$ valid only in the interval $(0, +\infty)$ and $y = \sqrt{x(1-x)}$ is a solution of the ordinary differential equation $y' = (1 - 2x)/2y$ valid only in the interval $(0, 1)$.

As we have seen, $y = e^x$ is a solution of the ordinary differential equation $y'' - y = 0$. We further observe that $y = e^{-x}$ is also a solution and moreover $y = c_1 e^x + c_2 e^{-x}$ is a solution of this equation for arbitrary values of the constants c_1 and c_2. It will be shown in Chapter 2 that $y = c_1 e^x + c_2 e^{-x}$ is the "general solution" of the ordinary differential equation $y'' - y = 0$. By the general solution we mean a solution with the property that any solution of $y'' - y = 0$ can be obtained from the function $c_1 e^x + c_2 e^{-x}$ for some special values of the constants c_1 and c_2. Also, in Chapter 2 we will show that the general solution of the ordinary differential equation $y'' + y = 0$ is given by $y(x) = c_1 \cos x + c_2 \sin x$ for arbitrary values of the constants c_1 and c_2.

In this chapter we present elementary methods for finding the solutions of some first-order ordinary differential equations, that is, equations of the form

$$y' = F(x, y), \tag{6}$$

together with some interesting applications.

The differential of a function $y = y(x)$ is by definition given by $dy = y'dx$.

With this in mind, the differential equation (6) sometimes will be written in the differential form $dy = F(x,y)dx$ or in an algebraically equivalent form. For example, the differential equation

$$y' = \frac{3x^2}{x^3 + 1}(y + 1)$$

can be written in the form

$$dy = \left[\frac{3x^2}{x^3 + 1}(y + 1)\right]dx \quad \text{or} \quad y' - \frac{3x^2}{x^3 + 1}y = \frac{3x^2}{x^3 + 1}$$

There are several types of first-order ordinary differential equations whose solutions can be found explicitly or implicitly by integrations. Of all tractable types of first-order ordinary differential equations, two deserve special attention: differential equations with *variables separable,* that is, equations that can be put into the form

$$y' = \frac{P(x)}{Q(y)} \quad \text{or} \quad P(x)dx = Q(y)dy,$$

and *linear equations,* that is, equations that can be put into the form

$$y' + a(x)y = b(x).$$

Both appear frequently in applications, and many other types of differential equations are reducible to one or the other of these types by means of a simple transformation.

APPLICATIONS 1.1.1

Differential equations appear frequently in mathematical models that attempt to describe real-life situations. Many natural laws and hypotheses can be translated via mathematical language into equations involving derivatives. For example, derivatives appear in physics as velocities and accelerations, in geometry as slopes, in biology as rates of growth of populations, in psychology as rates of learning, in chemistry as reaction rates, in economics as rates of change of the cost of living, and in finance as rates of growth of investments.

It is the case with many mathematical models that in order to obtain a differential equation that describes a real-life problem, we usually assume that the actual situation is governed by very simple laws—which is to say that we often make idealistic assumptions. Once the model is constructed in the form of a differential equation, the next step is to solve the differential equation and utilize the solution to make predictions concerning the behavior of the real problem. In case these predictions are not in reasonable agreement with reality, the scientist must reconsider the assumptions that led to the model and attempt to construct a model closer to reality.

First-order ordinary differential equations are very useful in applications. Let the function $y = y(x)$ represent an unknown quantity that we want to study. We know from calculus that the first derivative $y' = dy/dx$ represents the rate of change of y per unit change in x. If this rate of change is known (say, by experience or by a physical law) to be equal to a function $F(x, y)$, then the quantity y satisfies the first-order ordinary differential equation $y' = F(x, y)$. We next give some specific illustrations.

Biology

■ It has long been observed that some large colonies of bacteria tend to grow at a rate proportional to the number of bacteria present. For such a colony, let $N = N(t)$ be the number of bacteria present at any time t. Then, if k is the constant of proportionality, the function $N = N(t)$ satisfies the first-order ordinary differential equation[1]

$$\dot{N} = kN. \tag{7}$$

This equation is called the *Malthusian law* of population growth. T. R. Malthus observed in 1798 that the population of Europe seemed to be doubling at regular time intervals, and so he concluded that the rate of population increase is proportional to the population present. In Eq. (7) $\dot{N}$ stands for dN/dt. (As is customary, derivatives with respect to x will be denoted by primes and derivatives with respect to t by dots.) In this instance it is the time that is the independent variable. Equation (7) is a separable differential equation and its solution $N(t) = N(0)e^{kt}$ is computed in Example 3 of Section 1.3. Here $N(0)$ is the number of bacteria present initially, that is, at time $t = 0$. The solution $N(t)$ can be represented graphically as in Figure 1.1.

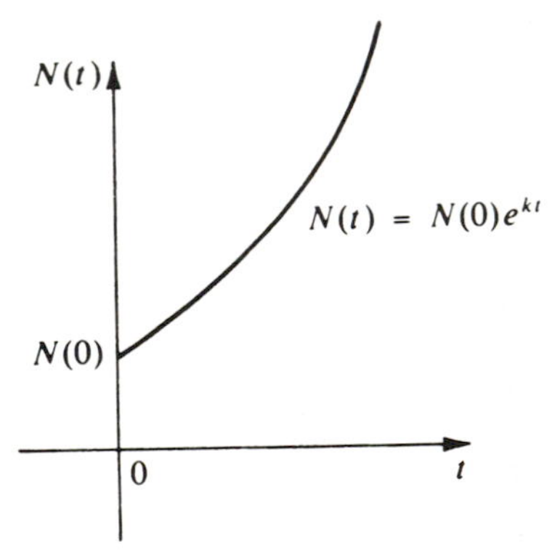

Figure 1.1

It should be emphasized that Eq. (7) is a mathematical model describing a colony of bacteria that grows according to a very simple, perhaps oversimplified, law: It grows at a rate proportional to the number of bacteria present at any time t. On the other hand, assuming this very simple law of growth leads us to a very simple differential equation. The solution, $N(t) = N(0)e^{kt}$, of Eq. (7) provides us with an approximation to the actual size of this colony of bacteria.

[1]Since the function $N(t)$ takes on only integral values, it is not continuous and so not differentiable. However, if the number of bacteria is very large, we can assume that it can be approximated by a differentiable function $N(t)$, since the changes in the size of the population occur over short time intervals.

Clearly, a more realistic mathematical model for the growth of this colony of bacteria is obtained if we take into account such realistic factors as overcrowding, limitations of food, and the like. Of course, the differential equation will then become more complex. It goes without saying that a mathematical model that is impossible to handle mathematically is useless, and consequently some simplifications and modifications of real-life laws are often necessary in order to derive a mathematically tractable model.

Pharmacology
Drug Dosages

■ It is well known in pharmacology[2] that penicillin and many other drugs administered to patients disappear from their bodies according to the following simple rule: If $y(t)$ is the amount of the drug in a human body at time t, then the rate of change $\dot{y}(t)$ of the drug is proportional to the amount present. That is, $y(t)$ satisfies the separable differential equation

$$\dot{y} = -ky, \tag{8}$$

where $k > 0$ is the constant of proportionality. The negative sign in (8) is due to the fact that $y(t)$ decreases as t increases, and hence the derivative of $y(t)$ with respect to t is negative. For each drug the constant k is known experimentally.

The solution of the differential equation (8) is (see Example 3 of Section 1.3)

$$y(t) = y_0 e^{-kt}, \tag{9}$$

where $y_0 = y(0)$ is the initial amount (initial dose) of the drug. As we see from Eq. (9) (see also Figure 1.2), the amount of the drug in the patient's body tends

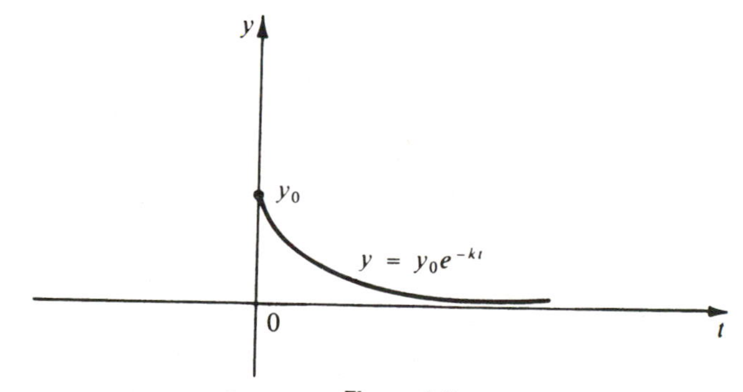

Figure 1.2

to zero as $t \to \infty$. However, in many cases it is necessary to maintain (approximately) a constant concentration (and therefore approximately a constant amount) of the drug in the patient's body for a long time. To achieve this it is necessary to give the patient an initial booster dose y_0 of the drug and then at equal intervals of time, say every τ hours, give the patient a dose D of the drug.

[2]This model, as well as other mathematical models in medicine, is discussed by J. S. Rustagi in *Int. J. Math. Educ. Sci. Technol.* 2 (1971): 193–203.

Equation (9) indicates the amount of the drug in the patient's body at any time t; hence, it is simple to determine the amount of the dose D. In fact, at time τ, and before we administer the dose D, the amount of the drug present in the body is

$$y(\tau) = y_0 e^{-k\tau}.$$

If we want to maintain the initial amount y_0 of the drug in the body at the times $\tau, 2\tau, 3\tau, \ldots$, the dose D should satisfy the equation

$$y_0 e^{-k\tau} + D = y_0.$$

Hence, the desired dose is given by the equation

$$D = y_0(1 - e^{-k\tau}). \tag{10}$$

Operations Research

■ Southwick and Zionts[3] developed an optimal control-theory approach to the education-investment decision which led them to the first-order linear (also separable) differential equation

$$\dot{x} = 1 - kx, \tag{11}$$

where x denotes the education of an individual at time t and the constant k is the rate at which education is being made obsolete or forgotten.

Psychology

■ In learning theory the separable first-order differential equation

$$\dot{p}(t) = a(t)G(p(t)) \tag{12}$$

is a basic model of the instructor/learner interaction.[4] Here G is known as the characteristic learning function and depends on the characteristics of the learner and of the material to be learned, $p(t)$ is the state of the learner at time t, and $a(t)$ is a measure of the intensity of instruction [the larger the value of $a(t)$ the greater the learning rate of the learner, but also, the greater the cost of the instruction].

Mechanics

■ Newton's second law of motion, which states that "the time rate of change of momentum of a body is proportional to the resultant force acting on the body and is in the direction of this resultant force," implies immediately that the motion of any body is described by an ordinary differential equation. Recall that the momentum of a body is the product mv of its mass and its velocity v. If F is the resultant force acting on the body, then

$$\frac{d}{dt}(mv) = kF, \tag{13}$$

[3]L. Southwick and S. Zionts, *Operations Res.* 22 (1974): 1156–1174.

[4]V. G. Chant, *J. Math. Psychol.* 11 (1974): 132–158.

where k is a constant of proportionality. Equation (13) is an ordinary differential equation in v whose particular form depends on m and F. The mass m can be constant or a function of t. Also, F can be constant, a function of t, or even a function of t and v.

Electric Circuits

■ Kirchhoff's voltage law states that, "the algebraic sum of all voltage drops around an electric circuit is zero." This law applied to the RL-series circuit in Figure 1.3 gives rise to the first-order linear differential equation (see also Section 1.4)

$$L\dot{I} + RI = V(t), \tag{14}$$

where $I = I(t)$ is the current in the circuit at time t.

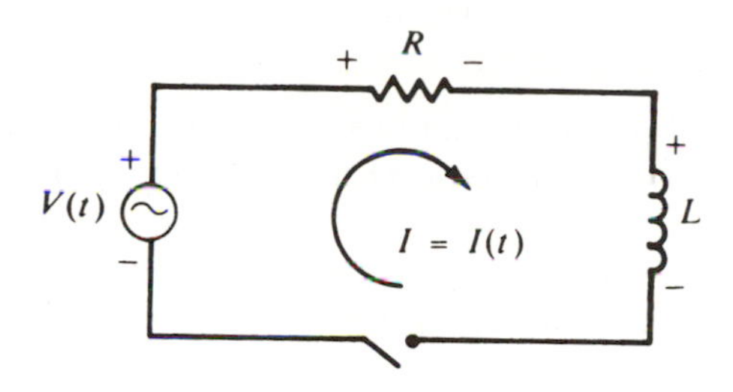

Figure 1.3

Orthogonal Trajectories

■ Consider the one-parameter family of curves given by the equation

$$F(x,y) = c. \tag{15}$$

Computing the differential of Eq. (15), we obtain

$$F_x dx + F_y dy = 0,$$

where F_x and F_y are the partial derivatives of F with respect to x and y, respectively. Thus,

$$\frac{dy}{dx} = -\frac{F_x}{F_y} \tag{16}$$

gives the slope of each curve of the family (15). We want to compute another family of curves such that each member of the new family cuts each member of the family (15) at right angles; that is, we want to compute the *orthogonal trajectories* of the family (15). In view of Eq. (16), the slope of the orthogonal trajectories of the family (15) is given by [the negative reciprocal of (16)]

$$\frac{dy}{dx} = \frac{F_y}{F_x}. \tag{17}$$

The general solution of Eq. (17) gives the orthogonal trajectories of the family (15).

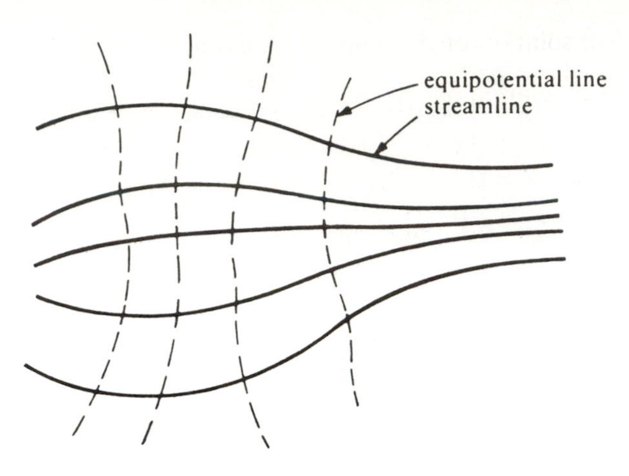

Figure 1.4

There are many physical interpretations and uses of orthogonal trajectories:

1. In *electrostatic fields* the *lines of force* are orthogonal to the *lines of constant potential.*

2. In *two-dimensional flows of fluids* the lines of motion of the flow—called *streamlines*—are orthogonal to the *equipotential* lines of the flow (see Figure 1.4).

3. In *meterology* the orthogonal trajectories of the *isobars* (curves connecting all points that report the same barometric pressure) give the direction of the wind from high- to low-pressure areas.

We have given only a few of the many applications of first-order ordinary differential equations. Many others are developed in detail in subsequent sections. Additionally, we present a number of applications in the exercises. The emphasis there is twofold: first, to expose the reader to the diversity of models incorporating ordinary differential equations; and second, to sharpen the reader's skill at solving differential equations while simultaneously emphasizing the fact that many differential equations are not products of the instructor's imagination but rather are extracted from real-life models.

EXERCISES

In Exercises 1 through 8, answer true or false.

1. $y = e^x + 3e^{-x}$ is a solution of the differential equation $y'' - y = 0$.

2. $y = 5 \sin x + 2 \cos x$ is a solution of the differential equation $y'' + y = 0$.

3. $y = \sin 2x$ is a solution of the differential equation $y'' - 4y = 0$.

4. $y = \cos 2x$ is a solution of the differential equation $y'' + 4y = 0$.

5. $y = (1/2x)\, e^{x^2}$ is a solution of the differential equation $xy' + y = xe^{x^2}$.

6. $y = e^x$ is a solution of the differential equation $y' + y = 0$.

7. $y = e^{-x}$ is a solution of the differential equation $y' - y = 0$.

8. $y = 2 \ln x + 4$ is a solution of the differential equation $x^2y'' - xy' + y = 2 \ln x$.

In Exercises 9 through 14, a differential equation and a function are listed. Show that the function is a solution of the differential equation. If more than one function is listed, show that both functions are solutions of the differential equation.

9. $y'' + y' = 2;\ 2x,\ 2x - 3$

10. $y'' - (\tan x)y' - \dfrac{\tan x}{x} y = \dfrac{1}{x^2} y^3;\ x \sec x$

11. $y''' - 5y'' + 6y' = 0;\ e^{3x}$

12. $y''' - 5y'' + 6y' = 0;\ e^{2x}, 1$

13. $m\ddot{s} = \frac{1}{2}gt^2;\ \dfrac{g}{24\,m} t^4 \left(\ddot{s} = \dfrac{d^2s}{dt^2}\right)$

14. $y'' - 2y' + y = \dfrac{1}{x}(y - y');\ e^x \ln x$

15. Show that the functions $y_1(x) = e^{-x}$ and $y_2(x) = xe^{-x}$ are solutions of the ordinary differential equation $y'' + 2y' + y = 0$.

16. Show that the function $y(x) = c_1e^{-x} + c_2xe^{-x} + 1$ is a solution of the ordinary differential equation $y'' + 2y' + y = 1$ for any values of the constants c_1 and c_2.

17. Show that the functions $y_1(x) \equiv 0$ and $y_2(x) = x^2/4, x \geq 0$, are solutions of the ordinary differential equation

$$y' = y^{1/2}.$$

18. Show that the function

$$y(x) = \begin{cases} 0 & \text{for } x \leq c \\ \dfrac{(x - c)^2}{4} & \text{for } x > c, \end{cases}$$

for any real number c, is a solution of the differential equation

$$y' = y^{1/2}.$$

(*Warning:* Don't forget to show that the solution is differentiable everywhere and in particular at $x = c$.)

19. Show that the function

$$x(t) = \frac{1}{k} + \frac{kx_0 - 1}{k} e^{-k(t-t_0)},$$

where x_0 and t_0 are constants, is a solution of Eq. (11). Show also that this solution passes through the point (t_0, x_0); that is, $x(t_0) = x_0$.

20. Show that $N(t) = ce^{kt}$ for any constant c is a solution of Eq. (7). What is the meaning of the constant c?

21. Find the differential equation of the orthogonal trajectories of the family of straight lines $y = cx$. Using your geometric intuition, guess the solution of the resulting differential equation.

In Exercises 22 through 28, answer true or false.

22. $y = xe^x$ is a solution of the differential equation $y' - 2y = e^x(1 - x)$.

23. $y = x^2$ is a solution of the differential equation $y''' = 0$.

24. $y = x + 1$ is a solution of the differential equation $yy' - y^2 = x^2$.

25. $y = \sin^2 x$ is a solution of the differential equation $y'' + y = \cos^2 x$.

26. $y = -1 + e^{-x}$ is a solution of the differential equation $y'' = y'(y' + y)$.

27. $y = -\sqrt{\dfrac{4 - x^3}{3x}}$ is a solution of the differential equation $y' = \dfrac{x^2 + y^2}{2xy}$.

28. $y = -\sqrt{\dfrac{4 - x^3}{3x}}$ is a solution of the differential equation $y' = -\dfrac{x^2 + y^2}{2xy}$.

29. Verify that each member of the one-parameter family of curves

$$x^2 + y^2 = 2cx$$

cuts every member of the one-parameter family of curves

$$x^2 + y^2 = 2ky$$

at right angles and vice versa. (*Hint:* Show that slopes are negative reciprocals.)

30. Prove that if the family of solutions of the differential equation

$$y' + p(x)y = q(x), \qquad p(x)q(x) \neq 0$$

is cut by the line $x = k$, the tangents to each member of the family at the points of intersection are concurrent.[5] [*Hint:* The equation of the tangent line to the solution at the point (k, c) is $y - c = [q(k) - cp(k)](x - k)$, and this equation passes through the point $[k + 1/p(k), q(k)/p(k).]$

[5]From the William Lowell Putnam Mathematical Competition, 1954. See *Amer. Math. Monthly* 61 (1954): 545.

1.2 EXISTENCE AND UNIQUENESS

We saw in the last section that first-order differential equations occur in many diverse mathematical models. Naturally, the models are useful if the resulting differential equation can be solved explicitly or at least if we can predict some of the properties of its solutions. Before we attempt to discover any ingenious technique to solve a differential equation, it would be very useful to know whether the differential equation has any solutions at all. That is, do there *exist* solutions to the differential equation? For example, the differential equation $(y')^2 + y^2 + 1 = 0$ has no real solution since the left-hand side is always positive.

The general form of a first-order ordinary differential equation is

$$y' = F(x,y). \tag{1}$$

For the sake of motivation, assume that Eq. (1) represents the motion of a particle whose velocity at time x is given by $F(x,y)$, where y is the position (the state) of the particle at time x. Since we imagine that we are observing the motion of the particle, it is plausible that if we know its position y_0 at time x_0, that is, if

$$y(x_0) = y_0 \tag{2}$$

we should be able to find its position at a later time x. Starting at time x_0 the particle will move and will move in a unique way. That is, the differential equation (1) *has* a solution that satisfies the condition (2), and moreover it has *only one* solution, that is, the motion that we are observing. The condition (2) is called an *initial condition* and Eq. (1) together with the initial condition (2) is called an *initial value problem (IVP)*. The term "initial" has been adopted from physics. In Eq. (2) y_0 is the initial position of the particle at the initial time x_0.

On the basis of the motivation above we may expect that an IVP which is a reasonable mathematical model of a real-life situation should have a solution (existence), and in fact it should have only one solution (uniqueness). We shall now state (see Appendix D for a proof)[6] a basic existence and uniqueness theorem for the IVP (1)–(2) which is independent of any physical considerations and which covers a wide class of first-order differential equations.

THEOREM 1

Consider the IVP

$$y' = F(x,y), \qquad y(x_0) = y_0.$$

[6]This theorem and many others that appear in this text are truly important for an over-all appreciation of differential equations. Often the proofs of these theorems require mathematical sophistication which is customarily beyond the scope of an elementary treatment. The reader can rest assured that these theorems are accurately presented and that their proofs have been satisfactorily scrutinized. Such proofs can be found in almost any advanced text on differential equations.

Assume that the functions F and $\partial F/\partial y$ are continuous in some rectangle

$$\mathcal{R} = \left\{(x,y): \quad \begin{array}{l} |x - x_0| \leq A \\ |y - y_0| \leq B \end{array}\right\}, \quad A > 0, B > 0$$

about the point (x_0,y_0). Then there is a positive number $h \leq A$ such that the IVP has one and only one solution in the interval $|x - x_0| \leq h$.

Thus, in the notation of Definition 1 of Section 1.1

$$I = \{x: |x - x_0| \leq A\} \quad \text{and} \quad J = \{x: |x - x_0| \leq h\}.$$

Let us illustrate this theorem by a few examples.

EXAMPLE 1 Show that the IVP

$$y' = x^2 + y^2 \tag{3}$$

$$y(0) = 0 \tag{4}$$

has a unique solution in some interval of the form $-\text{h} \leq x \leq h$.

Proof Here $F(x,y) = x^2 + y^2$ and $\partial F/\partial y = 2y$ are continuous in any rectangle $\mathcal{R}$ about (0, 0). By Theorem 1 there exists a positive number h such that the IVP (3)–(4) has a unique solution in the interval $|x - 0| \leq h$; that is, $-h \leq x \leq h$.

EXAMPLE 2 Assume that the coefficients $a(x)$ and $b(x)$ of the first-order linear differential equation

$$y' + a(x)y = b(x) \tag{5}$$

are continuous in some open interval I. Show that Eq. (5) has a unique solution through any point (x_0, y_0) where $x_0 \in I$.

Solution Choose a number A such that the interval $|x - x_0| \leq A$ is contained in the interval I. Then $F(x, y) = b(x) - a(x)y$ and $F_y(x, y) = -a(x)$ are continuous in

$$\mathcal{R} = \left\{(x, y): \quad \begin{array}{c} |x - x_0| \leq A \\ -\infty < y < +\infty \end{array}\right\}$$

By Theorem 1, Eq. (5) has a unique solution satisfying the initial condition $y(x_0) = y_0$ [in other words, through the point (x_0, y_0)], for any real number y_0.

REMARK 1 For linear equations such as Eq. (5), one can prove that its solutions exist throughout any interval about x_0 in which the coefficients $a(x)$ and $b(x)$ are continuous. For example, the IVP

$$y' + \frac{1}{x-2}y = \ln x$$
$$y(1) = 3$$

has a unique solution in the interval (0, 2), the IVP

$$y' + \frac{1}{x-2}y = \ln x$$
$$y(3) = -1$$

has a unique solution in the interval $(2, \infty)$, and the IVP

$$y' + \frac{1}{x-2}y = \ln |x|$$
$$y(-1) = 5$$

has a unique solution in the inverval $(-\infty, 0)$.

EXAMPLE 3 The functions $y_1(x) \equiv 0$ and $y_2(x) = x^2/4$ are two different solutions of the IVP

$$y' = y^{1/2} \tag{6}$$
$$y(0) = 0. \tag{7}$$

Is this in violation of Theorem 1?

Solution Here $F(x, y) = y^{1/2}$ and $F_y(x, y) = \frac{1}{2}y^{-1/2}$. Since F_y is not continuous at the point (0, 0), one of the hypotheses of Theorem 1 is violated, and consequently there is no guarantee that the IVP (6)–(7) has a unique solution. In fact, in addition to y_1 and y_2, the IVP (6)–(7) has infinitely many solutions through the point (0, 0). For each $c > 0$, these solutions are given by

$$y(x) = \begin{cases} 0, & x \leq c \\ \dfrac{(x-c)^2}{4}, & c < x \end{cases}$$

and their graphs are given in Figure 1.5.

REMARK 2 As we have seen in Example 3 for the IVP (6)–(7), we have existence of solutions but not uniqueness. Uniqueness of solutions of initial value problems is very important. When we know that an IVP has a unique solution we can apply any method (including guessing) to find its solution. For example, we know that the IVP

$$y' + xy = 0 \tag{8}$$
$$y(0) = 0 \tag{9}$$

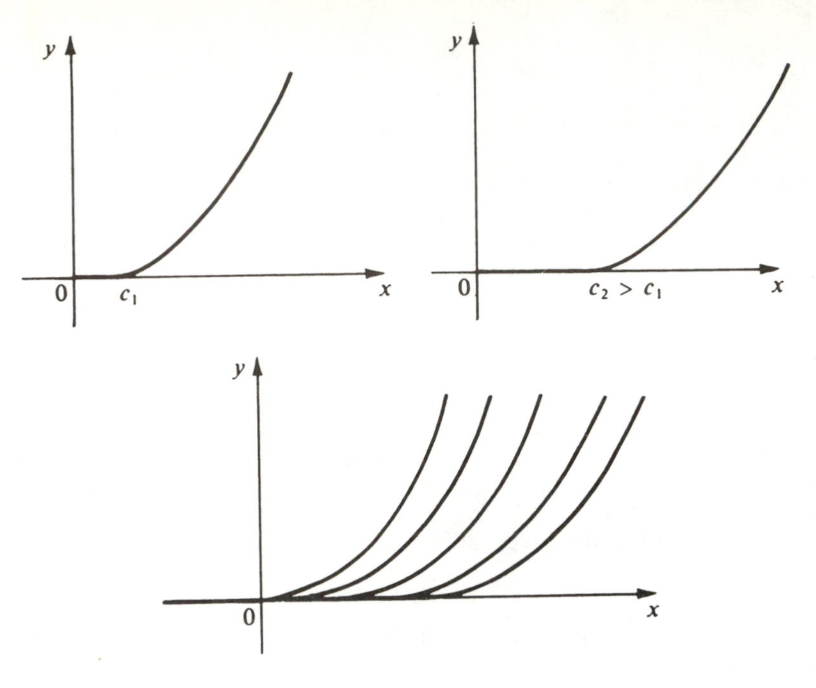

Figure 1.5

has a unique solution. Just by inspection we see that y = 0 satisfies the IVP. Thus, $y(x) = 0$ is the only solution of the IVP (8)–(9) and no further work is required in solving this IVP.

Theorem 1 gives conditions under which a first-order differential equation possesses a unique solution. Knowing that a solution exists, we desire to develop methods for determining this solution either exactly or approximately. The following method, called the method of *isoclines,* is useful in finding an approximate solution.

Consider again the first-order differential equation

$$y' = F(x, y). \tag{10}$$

For a better understanding of the differential equation (10) and its solutions, it is useful to interpret y' as the slope of the tangent line drawn to the solution at the point (x, y). Then at each point (x_0, y_0) in the domain D of F, the differential equation (10) assigns a slope y' equal to $F(x_0, y_0)$. Hence, if we know that (10) has a solution through a given point (x_0, y_0), then (without knowing the solution explicitly) we know immediately the equation of the tangent line to this solution at the point (x_0, y_0). In particular, its slope is $F(x_0, y_0)$. A short segment along this tangent about (x_0, y_0) will give us, graphically, an approximation to the true solution of (10) near the point (x_0, y_0). Thus for graphical approximations of the solutions of (10) it is useful to draw a short line segment at each point $(a, b) \in D$ with slope $F(a, b)$. Such line segments are called *linear elements.* The totality of all linear elements is called the *direction field* of the differential equation (10). The construction of the direction field can

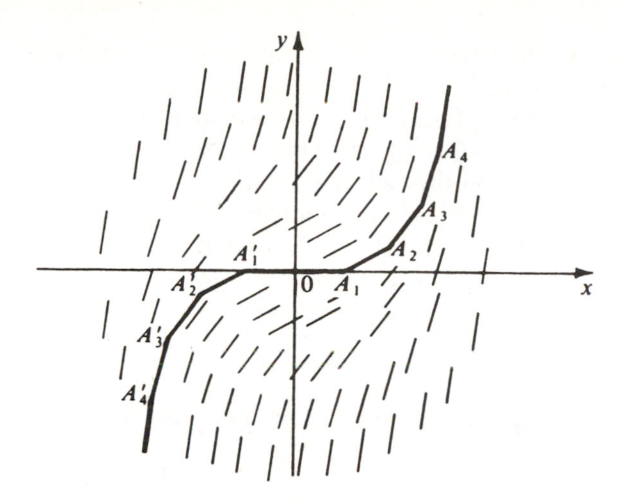

Figure 1.6

be carried out more efficiently by setting $F(x, y) = c$ and realizing that for any constant c all points on the (level) curve $F(x, y) = c$ have the same slope c. This is the *method of isoclines* (equal slopes).

For example, the isoclines of the differential equation

$$y' = x^2 + y^2 \tag{11}$$

are the circles $x^2 + y^2 = c$, centered at the origin (0, 0) with radius $\sqrt{c}$ (see Figure 1.6). At each point on the circle $x^2 + y^2 = c$ the differential equation defines the slope c. Therefore, the linear elements of the direction field for all these points are parallel, and each one of them has slope c. In Figure 1.6 we have drawn a few isoclines (for $c = 1, 4, 9, 16$) and linear elements which give a good idea of the direction field of the differential equation (11).

Now suppose that we want to find a graphical approximation of the solution of the differential equation (10) which goes through the point (x_0, y_0). First we draw a few isoclines $C_1, C_2, C_3, \ldots$ corresponding to the values $c = c_1, c_2, c_3, \ldots$ [C_i is the curve $F(x, y) = c_i$; see Figure 1.7]. Through the point (x_0, y_0) we

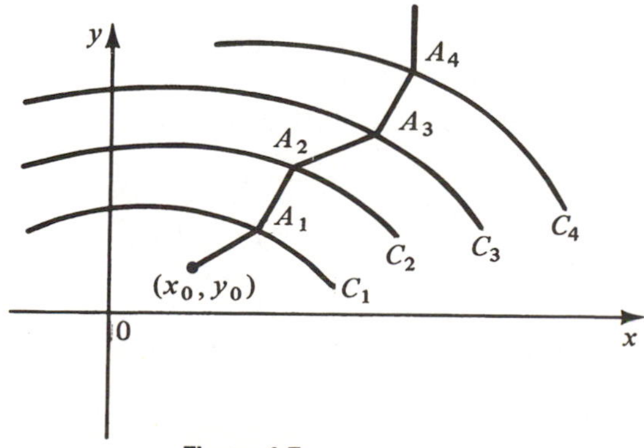

Figure 1.7

draw the linear element with slope $F(x_0, y_0)$ and extend it until it meets the isocline C_1 at a point A_1. Through A_1 we draw the linear element with slope c_1 until it meets the isocline C_2 at a point A_2. Proceeding in this fashion we construct a polygonal curve through (x_0, y_0) which is an approximation of the true solution of the differential equation through that point.

In Figure 1.6 we have drawn an approximate solution of the differential equation (11) through the point (0, 0). Note that, from Example 1, the differential equation (11) has a unique solution through (0, 0).

EXERCISES

In Exercises 1 through 6, check whether the hypotheses of Theorem 1 are satisfied.

1. $y' + xy = 3$, $y(0) = 0$

2. $xy' + y = 3$, $y(0) = 1$

3. $y' = \dfrac{x - y}{x + y}$, $y(0) = -1$

4. $y' = y^{1/2}$, $y(1) = 1$

5. $y' = y^{2/3}$, $y(0) = 0$

6. $y' = \dfrac{x - y}{x + y}$, $y(1) = -1$

7. Review Remark 1 of this section and apply it to the differential equation

$$xy' + \frac{1}{2x + 3}y = \ln|x - 2|$$

for its solutions through the points $(-3, 0)$, $(-1, 5)$, $(1, -7)$, and $(3, 0)$.

8. **Astronomy** In an article[7] concerning the accumulation processes in the primitive solar nebula, the following first-order differential equation was obtained:

$$\frac{dx}{dt} = \frac{ax^{5/6}}{(b - Bt)^{3/2}},$$

where a, b, and B are constants. Find all points (t_0, x_0) through which this differential equation has unique solutions.

9. Show that the only solution of the IVP

$$xy' - y = 1$$
$$y(2) = 3$$

is $y(x) = 2x - 1$.

In Exercises 10 through 13, answer true or false.

[7] A. G. W. Cameron, *Icarus* 18 (1973): 407–450.

10. The unique solution of the IVP

$$y' - xy = 1 - x^2$$
$$y(0) = 0$$

is $y(x) = -x$.

11. The only solution of the IVP

$$y' - xy = 1 - x^2$$
$$y(0) = 0$$

is $y(x) = x$.

12. The unique solution of the IVP

$$xy' + y^2 = 1$$
$$y(-2) = 1$$

is $y(x) = 1$.

13. The differential equation $y' = y^{3/4}$ has a unique solution through the point (0, 0) but not through the point (2, 0).

Use the method of isoclines to find graphical approximations to the solutions of the following IVPs.

14. $y' = y - x$, $y(0) = 0$

15. $y' = xy$, $y(1) = 2$

16. $y' = x^2 + y^2$, $y(0) = 1$

1.3 VARIABLES SEPARABLE

A separable differential equation is characterized by the fact that the two variables of the equation together with their respective differentials can be placed on opposite sides of the equation. In such equations the equality sign "separates" one variable from the other. Algebraic manipulations enable us to write separable differential equations in the form $y' = P(x)/Q(y)$ or, more explicitly,

$$Q(y)dy = P(x)dx. \tag{1}$$

To obtain the general solution of a separable differential equation we first separate the two variables as in Eq. (1) and then integrate both sides, to obtain

$$\int P(x)dx = \int Q(y)dy + c, \tag{2}$$

where c is an arbitrary constant of integration. After performing the integrations in Eq. (2), it is desirable to solve the resulting expression for the dependent variable y, thereby obtaining y explicitly in terms of x. If this is not possible or convenient, then Eq. (2) gives implicitly the general solution of Eq. (1).

The following are examples of differential equations that are separable:

$$x\,dx - (5y^4 + 3)\,dy = 0 \tag{3}$$

$$2x(y^2 + y)\,dx + (x^2 - 1)\,y\,dy = 0 \tag{4}$$

$$e^{x-y}y' = \sin x. \tag{5}$$

In fact, they can be brought, respectively, into the forms (3′), (4′), and (5′):

$$x\,dx = (5y^4 + 3)dy \tag{3′}$$

$$\frac{2x}{x^2 - 1}\,dx = -\frac{1}{y + 1}\,dy, \qquad \text{for } x \neq \pm 1, y \neq 0, y \neq -1 \tag{4′}$$

$$e^{-y}\,dy = e^{-x}\sin x\,dx. \tag{5′}$$

EXAMPLE 1 Find the general solution of Eq. (3).

Solution As we have seen, Eq. (3) is separable because it can be written in the form (3′), where the variables x and y are separated. Integrating both sides of Eq. (3′), we obtain

$$\int x\,dx = \int (5y^4 + 3)\,dy + c.$$

Thus,

$$\tfrac{1}{2}x^2 = y^5 + 3y + c$$

is an implicit representation of the general solution of Eq. (3).

EXAMPLE 2 Solve the differential equation (4).

Solution The variables of Eq. (4) can be separated as in Eq. (4′). Integrating both sides of Eq. (4′) yields

$$\ln|x^2 - 1| = -\ln|y + 1| + c.$$

Thus,[8]

$$\ln|x^2 - 1| + \ln|y + 1| = c \Rightarrow \ln|(x^2 - 1)(y + 1)| = c.$$

Taking exponentials of both sides and using the fact that $e^{\ln p} = p$, we have

$$|(x^2 - 1)(y + 1)| = e^c \Rightarrow (x^2 - 1)(y + 1) = \pm e^c.$$

Since c is an arbitrary constant, clearly $\pm e^c$ is again an arbitrary constant, different than zero, and for economy in notation we can still denote it by c with $c \neq 0$. In what follows we shall use this convention frequently.

[8]The symbol ⇒ stands for "imply" (or "implies"), "it follows," "then," and will be used when appropriate.

$$(x^2 - 1)(y + 1) = c \Rightarrow y = -1 + \frac{c}{x^2 - 1}, \qquad c \neq 0.$$

Now, do not forget that Eq. (4′) is obtained from Eq. (4) by dividing through by the expression $(x^2 - 1)(y^2 + y)$. Thus, we should ensure that this expression is not zero. Hence, we require that $x \neq \pm 1$, $y \neq -1$, and $y \neq 0$. We must finally examine what happens when $x = \pm 1$ and when $y = 0$ or $y = -1$. Going back to the original Eq. (4) we see that the four lines $x = \pm 1$, $y = 0$, and $y = -1$ also satisfy the differential equation (4). If we relax the restriction $c \neq 0$, the curve $y = -1$ will be contained in the formula $y = -1 + [c/(x^2 - 1)]$ for $c = 0$. However, the curves $x = \pm 1$ and $y = 0$ are not contained in the same formula, no matter what the value of c is. Sometimes such curves are called *singular solutions* and the one-parameter family of solutions

$$y = -1 + \frac{c}{x^2 - 1},$$

where c is an arbitrary constant (parameter), is called the *general solution.*

The following initial value problems are frequently encountered in applications of exponential growth and decay, and readers are advised to familiarize themselves with the solutions.

EXAMPLE 3 Prove that the solution of the IVP

$$\dot{y} = ky, \qquad k \text{ is a constant}$$
$$y(0) = y_0$$

is $y(t) = y_0 e^{kt}$. Similarly, show that the solution of the IVP

$$\dot{y} = -ky, \qquad k \text{ is a constant}$$
$$y(0) = y_0$$

is $y(t) = y_0 e^{-kt}$.

Proof Clearly, the solution to the second IVP follows from the solution to the first IVP by replacing the constant k by $-k$. Thus, it suffices to solve the first IVP. Since $\dot{y} = dy/dt$, separating the variables y and t we obtain, for $y \neq 0$,

$$\frac{dy}{y} = k\, dt.$$

Integrating both sides, we have

$$\ln|y| = kt + c \Rightarrow |y| = e^{kt+c} = e^c e^{kt} \Rightarrow y = \pm e^c e^{kt} \Rightarrow y = ce^{kt}, \qquad c \neq 0.$$

Since $y = 0$ is also a solution of $\dot{y} = ky$, it follows that the general solution of $\dot{y} = ky$ is $y = ce^{kt}$, where c is an arbitrary constant. Using the initial condition, we find that $y_0 = c$, that is, $y(t) = y_0 e^{kt}$, and the proof is complete. The solutions of the above IVPs for $k > 0$ and $k < 0$ are represented in Figure 1.8.

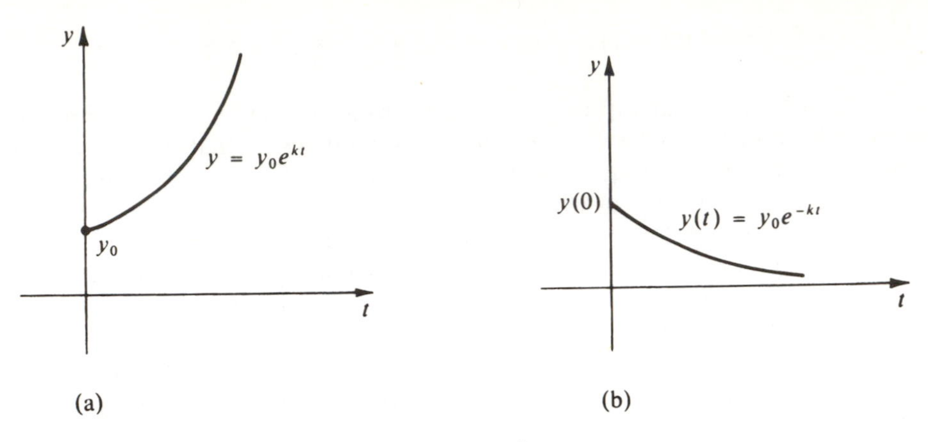

Figure 1.8

APPLICATIONS 1.3.1

One frequently encounters separable differential equations in applications. We shall mention a few instances here. Review also the applications in Section 1.1.1 that involved separable differential equations.

Psychology

■ In 1930, Thurstone[9] obtained a differential equation as a mathematical model describing the state of a learner, or the *learning curve* of a learner, while learning a specific task or body of knowledge. If $y(t)$ denotes the state of a learner at time t, then Thurstone's equation is the following separable differential equation:

$$\frac{dy}{y^{3/2}(1-y)^{3/2}} = \frac{2k}{\sqrt{m}}\, dt. \tag{6}$$

Here k and m are positive constants that depend on the individual learner and the complexity of the task, respectively. For the solution of the differential equation (6), see Exercise 16.

Biology

■ Assume that a colony of bacteria increases at a rate proportional to the number present. If the number of bacteria doubles in 5 hours, how long will it take for the bacteria to triple?

Solution Let $N(t)$ be the number of bacteria present at time t. Then the assumption that this colony of bacteria increases at a rate proportional to the number present can be mathematically written as

$$\frac{dN}{dt} = kN, \tag{7}$$

[9] L. L. Thurstone, *J. Gen. Psychol.* 3 (1930): 469–493.

where k is the constant of proportionality. From Example 3 we have

$$N(t) = N(0)e^{kt}, \tag{8}$$

where $N(0)$ is the initial number of bacteria in this colony. Since the number of bacteria doubles in 5 hours, we have

$$\begin{aligned} N(5) &= 2N(0) \\ \Rightarrow \quad N(0)e^{5k} &= 2N(0) \Rightarrow e^{5k} = 2 \\ \Rightarrow \quad k &= \tfrac{1}{5}\ln 2. \end{aligned}$$

The time t that is required for this colony to triple must satisfy the equation

$$\begin{aligned} N(t) &= 3N(0) \\ \Rightarrow \quad N(0)e^{kt} &= 3N(0) \Rightarrow e^{kt} = 3 \\ \Rightarrow \quad t &= \frac{1}{k}\ln 3 = 5\frac{\ln 3}{\ln 2} = 5\frac{1.09}{0.69} = 7.89 \text{ hours.} \end{aligned}$$

■ The sum of \$5000 is invested at the rate of 8% per year compounded continuously. What will the amount be after 25 years?

Finance

Solution Let $y(t)$ be the amount of money (capital plus interest) at time t. Then the rate of change of the money at time t is given by

$$\frac{dy}{dt} = \frac{8}{100}y. \tag{9}$$

Equation (9) is clearly a separable differential equation. Its solution (by Example 3) is

$$y(t) = y(0)e^{(8/100)t}.$$

Since $y(0) = 5000$ (the initial amount invested), we find that

$$y(25) = 5000e^{(8/100)\cdot 25} = 5000e^{2} = \$36{,}945.28.$$

■ In Section 1.1.1 we observed that according to Newton's second law of motion a moving body of mass m and velocity v is governed by the differential equation

Mechanics

$$\frac{d}{dt}(mv) = kF, \tag{10}$$

where F is the resultant force acting on the body and k is a constant of proportionality. If it happens that F is a function of the velocity v and does not depend explicitly on time, and if m is a constant, then Eq. (10) is a separable differential equation.

In the cgs (centimeter-gram-second) system and in the English system, whose units are given in Table 1.1, the constant k is equal to 1.

	Cgs system	English system
Distance	centimeter	foot
Mass	gram	slug
Time	second	second
Force	dyne	pound
Velocity	cm/sec	ft/sec
Acceleration	cm/sec^2	ft/sec^2

Table 1.1 Table of Units

EXAMPLE 4 From a great height above the earth, a body of mass $m = 2$ kilograms (kg) is thrown downward with initial velocity $v_0 = 10^5$ centimeters per second (cm/sec). In addition to its weight, air resistance is acting upon this body, which is numerically (in dynes) equal to twice its speed at any time. Find the velocity of the body after $t = 10^3$ sec. (Take $g = 900$ cm/sec^2.)

Solution Let $v(t)$ be the velocity of the body at time t. The resultant force acting upon this falling body is

$$F = (\text{weight}) - (\text{air resistance}) = (mg - 2v) \text{ dynes}.$$

Thus, by Newton's second law of motion, Eq (10),

$$\frac{d}{dt}(mv) = mg - 2v$$

$$\Rightarrow \quad m\frac{dv}{dt} = mg - 2v$$

$$\Rightarrow \quad \frac{m}{mg - 2v}\,dv = dt. \tag{11}$$

Also,

$$v(0) = 10^5. \tag{12}$$

The solution of the IVP (11)–(12) is $v(t) = \frac{1}{2}[mg - (mg - 2v_0)e^{-2t/m}]$, and so

$$v(10^3) = \left(9 - \frac{8}{e}\right) 10^5 \text{ cm/sec} = \left(9 - \frac{8}{e}\right) \text{km/sec} = 6.05 \text{ km/sec}.$$

Medicine

■ Here we shall derive a separable differential equation that serves as a mathematical model of the dye-dilution procedure for measuring cardiac output. From the practical point of view this is a reasonable model, in spite of the many simplifications that we introduce.

In the dye-dilution method an amount of dye of mass D_0 milligrams (mg) is injected into a vein so near the heart that we shall assume that the heart contains D_0 milligrams of dye at time $t = 0$. The dye is mixed with the blood passing through the heart, and at each stroke of the heart the diluted mixture flows out and is replaced by blood from the veins. To derive our model we shall assume that the mixture of blood and dye inside the heart is uniform and flows out at a constant rate of r liters per minute. We also assume that the heart is a container of constant volume V liters.

With these data and assumptions, we can now formulate an initial value problem for the amount $D(t)$ of dye in the heart at any time t. This is a special case of mixture problems that we will study in the next section. Since $dD(t)/dt$ represents the rate of change of the dye in the heart at time t, we have the equation

$$\frac{dD(t)}{dt} = \text{rate in} - \text{rate out},$$

where "rate in" is the rate at which dye runs into the heart at time t, and "rate out" is the rate at which dye runs out of the heart at time t. Here

$$\text{rate in} = 0$$

as no dye runs into the heart, and

$$\text{rate out} = \frac{D(t)}{V} r,$$

where $D(t)/V$ is the concentration of dye in the heart (in other words, the amount of dye per liter of diluted mixture). Thus, the differential equation describing this mixture problem is

$$\frac{dD(t)}{dt} = -\frac{r}{V} D(t). \tag{13}$$

We also have the initial condition

$$D(0) = D_0, \tag{14}$$

where D_0 is the (initial) amount of dye injected at time $t = 0$. The solution of the IVP (13)–(14) is

$$D(t) = D_0 e^{-(r/V)t}.$$

The quantity (concentration) $D(t)/V$ is useful in measuring the cardiac output.

Radioactive Dating
Archaeology, Paleontology, Geology, Arts

■ It is known, from experiments, that radioactive substances decay at a rate proportional to the amount of the substance that is present. Thus, if $N(t)$ is the number of atoms of the substance present at any time t, then

$$\dot{N}(t) = -kN(t), \tag{15}$$

where the positive constant k is called the *decay* constant of the substance. [The negative sign is due to the fact that $N(t) > 0$ and decreasing, so it has negative "derivative" $\dot{N}(t)$.] If

$$N(0) = N_0 \tag{16}$$

is the (initial) number of atoms of the substance at time $t = 0$, the solution of the IVP (15)–(16) is given by

$$N(t) = N_0 e^{-kt}. \tag{17}$$

If we know the values of k, $N(t)$, and N_0, then from Eq. (17) we can find the time t that it took the radioactive substance to decay from its initial value N_0 to its present value $N(t)$. In fact, from Eq. (17) we obtain $e^{-kt} = N(t)/N_0$ and (taking logarithms of both sides)

$$t = -\frac{1}{k}\ln\frac{N(t)}{N_0} = \frac{1}{k}\ln\frac{N_0}{N(t)}. \tag{18}$$

Equation (17) is the basic idea behind radioactive dating when we try to determine the age of materials of archaeology, rocks, fossils, old paintings, and so on. The value of $N(t)$ is easily computed from the present amount of the substance. To find the decay constant k, we utilize the *half-life* of the radioactive substance, that is, the time required for half of a given sample of a radioactive substance to decay. Half-life is easily measured in the laboratory, and tables are available that give the half-life of many radioactive substances. Using Eq. (17), we see that if T is the half-life of a radioactive substance, then

$$\frac{1}{2} = e^{-kT} \Rightarrow k = \frac{\ln 2}{T};$$

that is, the decay constant of a radioactive substance is obtained by dividing the natural logarithm of 2 by the half-life of the substance. Although the value N_0 is in general unknown, we can utilize some additional information known about some radioactive substances to compute their age. For example, it is known that each decaying uranium 238 atom gives rise to a single lead 206 atom. Thus, the number P of lead 206 atoms found in a uranium mineral containing $N(t)$ atoms of uranium 238 is given by

$$P = N_0 - N(t) = N(t)(e^{kt} - 1),$$

and solving for t we find the age of the mineral to be

$$t = \frac{1}{k}\ln\left[1 + \frac{P}{N(t)}\right].$$

Those with some interest in radioactive dating of old paintings should read the work of Coreman,[10] where it is proved beyond any scientific doubt that the paintings sold by Van Meegeren are faked Vermeers and De Hooghs. (Vermeer and De Hooghs are famous seventeenth-century Dutch painters.)

[10]P. Coreman, *Van Meegeren's Faked Vermeers and De Hooghs* (Amsterdam: Meulenhoff, 1949). See also the articles in *Science* 155 (1967): 1238–1241 and *Science* 160 (1968): 413–415.

Medicine

■ Certain diseases[11] that affect humans can be considered to impart immunity for life. That is, an individual who has had the disease and survived is protected from ever having that disease again. This is the case (rather it is very close to being the case, since there are exceptions to all rules in these matters) for such common diseases as measles, mumps, and chicken pox, as well as for smallpox, which is now believed to be completely eradicated.

In trying to assess the effect of such a disease, we can consider one cohort group, that is, all the individuals born in one specific year, and define $N(t)$ to be the number who have survived to age t, and $S(t)$ the number who have not had the disease and are still susceptible to it at age t.

If p is the probability of a susceptible's getting the disease ($0 < p < 1$), and $1/m$ the proportion of those who die due to the disease, then the following relation can be derived:

$$dS(t)/dt = -pS(t) + (S(t)/N(t))dN(t)/dt + pS^2(t)/mN(t). \tag{19}$$

This equation was first derived by Daniel Bernoulli (in 1760) in his work on the effects of smallpox. The equation is actually less formidable than it looks. If we multiply both sides by N/S^2 and regroup terms, we get

$$(1/S)dN/dt - (N/S^2)dS/dt = pN/S - p/m. \tag{20}$$

Recall that $(d/dt)(N/S) = (1/S)dN/dt - (N/S^2)dS/dt$, and substitute in the left-hand side of this last equation to get

$$d(N/S)/dt = pN/S - p/m.$$

Since N/S occurs as a variable, define $y = N/S$ and rewrite the equation as

$$dy/dt = py - p/m. \tag{21}$$

Equation (21) is a differential equation with separable variables, so we can write this relation for its solution:

$$\int_0^a \frac{dy/dt}{py - p/m}\,dt = a.$$

Or

$$\int_{y(0)}^{y(a)} \frac{dy}{py - p/m} = a.$$

So,

$$\frac{1}{p}\log\left|\frac{y(a) - 1/m}{y(0) - 1/m}\right| = a.$$

From this we get

$$\frac{y(a) - 1/m}{y(0) - 1/m} = e^{ap}.$$

[11]The authors wish to thank Stavros Busenberg for bringing this model to their attention.

Solving for $my(a)$, we have

$$my(a) = 1 + (my(0) - 1)e^{ap}. \tag{22}$$

Since, at birth, every member of the cohort group is susceptible, $S(0) = N(0)$, yielding $y(0) = 1$. Equation (22) now gives $my(a) = 1 + (m - 1)e^{ap}$. Finally, recalling that $y(a) = N(a)/S(a)$, we get

$$S(a) = mN(a)/[1 + (m - 1)e^{ap}]. \tag{23}$$

This expression gives the number of susceptibles in terms of the number of survivors to age a of the cohort group, and the two basic constants m and p. The values of these parameters in the right-hand side of the equation can be determined from census records and statistics on the disease.

EXAMPLE 5 Using the population data that were available to him, Bernoulli estimated $p = 1/8$ and $m = 8$ for the case of smallpox in Paris of the 1760s. Using these constants in Eq. (23), the number of susceptibles at age a among a cohort group with $N(a)$ survivors of that age would be

$$S(a) = 8N(a)/(1 + 7e^{a/8}).$$

So, by age 24 ($a = 24$), $S(a)/N(a)$ would be about 0.056. In other words, only one in about eighteen of those twenty-four years old would not have had smallpox.

REMARK 1 Formula (23) was derived on the basis of what one believes is a reasonable, but certainly not foolproof, relation (19). To see that (19) is reasonable, let us examine what its terms mean. The left-hand side, $dS(t)/dt$, is the rate at which the number of susceptibles is changing (decreasing). The terms on the right can be collected into two groups. The first, $-pS(t)$, is the rate at which we expect susceptibles at age t to be infected with the disease. This term includes those who will die from the disease. Now we need an expression for the rate of removal of susceptibles due to death from all *other* causes. This is given by the second group of terms $[dN(t)/dt + pS(t)/m](S(t)/N(t))$, where $dN(t)/dt$ is the rate of change of the whole cohort group due to death for all reasons, $pS(t)/m$ is the rate of death due to the particular disease, and $S(t)/N(t)$ is the proportion of susceptibles among the cohort group. Thus, $[dN(t)/dt + pS(t)/m](S(t)/N(t))$ is indeed the rate of decrease of susceptibles due to death for all reasons other than the disease. So. Eq. (19) is an equality between two different ways of expressing the rate of change of susceptibles.

All these assumptions leading to Eq. (19) are reasonable on the intuitive level and lead to the mathematical formulation, or model, of the situation we are examining. Once we have Eq. (19), we can use differential equations to get relation (23).

There are special cases where we can have reasonable expectations for the ratio $S(a)/N(a)$, and we should make sure that (23) fulfills these expectations. For example, if the mortality rate of the disease is very high ($m \to 1$), we could expect just about all survivors $N(a)$ to also be susceptible (that is, only people who have not had the disease survive). Here we would expect that $S(a)/N(a)$

$\to 1$ as $m \to 1$. In the opposite circumstance of very low mortality ($m \to \infty$), we can easily verify that $S(a)/N(a) = e^{-ap}$. Is this what you would expect in this case?

EXERCISES

In Exercises 1 through 14, some of the differential equations are separable and some are not. Solve those that are separable.

1. $(1 + x)y\,dx + x\,dy = 0$

2. $(x^2 + y^2)dx - 2xy\,dy = 0.$

3. $y' = y^{1/2}$

4. $xy' + y = 3$

5. $y' + xy = 3$

6. $\dfrac{dp}{dt} = Ap - Bp^2;\ A, B \neq 0$

7. $xy' - \dfrac{y}{\ln x} = xy^2$

8. $yy' = y + x^2$

9. $y' = x - xy - y + 1$

10. $3xy\,dx + (x^2 + 4)dy = 0$

11. $\cos x \sin y\,dy - (\cos y \cos x + \cos x)dx = 0$

12. $yy' = y^2x^3 + y^2x$

13. $y^4dx + (x^2 - 3y)dy = 0$

14. $(1 + y^2)\cos x\,dx = (1 + \sin^2 x)2y\,dy$

15. Compute and graph a curve passing through the point $(0, 1)$ and whose slope at any point (x, y) on the curve is equal to the ordinate y.

16. Solve Thurstone's differential equation (6). (*Hint:* Set $y = \sin^2\theta$.)

17. Biology Volterra[12] obtained differential equations of the forms

$$\frac{dx}{dt} = x(a_1 + a_2y)$$

$$\frac{dy}{dt} = y(b_1 + b_2x)$$

as a mathematical model describing the competition between two species coexisting in a given environment. From this system one obtains (by the chain rule, in other words, $dy/dx = dy/dt \cdot dt/dx$) the separable differential equation

$$\frac{dy}{dx} = \frac{y(b_1 + b_2x)}{x(a_1 + a_2y)}.$$

Solve this differential equation.

[12]Vito Volterra, *Lecons sur la théorie mathématique de la lutte pour la vie* (Paris: Gauthier-Villars, 1931).

18. Assume that the rate at which a radioactive substance decays is proportional to the amount present. In a certain sample 50% of the substance disappears in a period of 1600 years. (This is true, for example, for radium 226.)

(a) Write the differential equation that describes the decay of the substance.
(b) Compute the decay constant of the substance.
(c) What percentage of the original sample will disappear in 800 years?
(d) In how many years will only one fifth of the original amount remain?

19. Solve the differential equation

$$\frac{dx}{dt} = \frac{ax^{5/6}}{(b - Bt)^{3/2}}$$

given in the astronomy application in Exercise 8 of Section 1.2.

20. Chemistry Solve the differential equation

$$\frac{dx}{dt} = k(a - x)(b - x), \quad k, a, b > 0$$

which occurs in *chemical reactions*. What will the value of x be after a very long time, that is, as $t \to \infty$?

21. Botany Solve the differential equation

$$\frac{dI}{dw} = 0.088(2.4 - I),$$

which seems to fit the data collected in a botanical experiment,[13] and find the value of I as $w \to \infty$.

22. Compute the *orthogonal trajectories* of the one-parameter family of curves $y = cx$.

23. A colony of bacteria increases at a rate proportional to the number present. If the number of bacteria triples in 4 hours, how long will it take for the bacteria to be 27 times the initial number?

24. Biology Assume that a population grows according to Verhurst's *logistic law* of population growth

$$\frac{dN}{dt} = AN - BN^2,$$

where $N = N(t)$ is the population at time t, and the constants A and B are the *vital coefficients* of the population. What will the size of this population be at any time t? What will the population be after a very long time, that is, as $t \to \infty$?

[13] *Australian J. Bot.* 4–5 (1956–57): 159.

25. From the surface of the earth a body of mass 2 slugs is thrown upward with initial velocity of $v_0 = -30$ ft/sec (the convention adopted here is that the positive direction is downward). In addition to its weight, air resistance is acting upon this body which is numerically (in pounds) equal to three times its speed at any time. For how long will the body be moving upward?

26. A tank contains 50 gallons of brine in which 5 pounds of salt is dissolved. Pure water runs into the tank at the rate of 3 gallons per minute. The mixture, kept uniform by stirring, runs out of the tank at the same rate as the inflow. How much salt is in the tank after 15 minutes?

27. The sum of \$15,000 is invested at the rate of 6% per year compounded continuously. In how many years will the money double?

Statistics The solutions of the separable differential equation

$$y' = \frac{A - x}{B + Cx + Dx^2}y$$

give most of the important distributions of statistics for appropriate choices of the constants A, B, C, and D. Solve the differential equation in the following cases.

28. $C = D = 0$, $B > 0$, and A arbitrary (normal distribution).

29. $A = B = D = 0$ and $C > 0$ (exponential distribution).

30. $B = D = 0$, $C > 0$, and $A > -C$ (gamma distribution).

31. $B = 0$, $C = -D$, $(A - 1)/C < 1$, and $A/C > -1$ (beta distribution).

32. In the medicine application dealing with cardiac output assume that $V =$ 500 milliliters ($\frac{1}{2}$ liter), $r = 120$ liters per minute, and $D_0 = 2$ milligrams. Find the amount of dye in the heart after: 1 second, 2 seconds, $2\frac{1}{2}$ seconds, and 10 seconds.

33. How long will it take for \$2000 to grow to \$8000 if this sum is invested at 5.5% annual interest compounded continuously?

34. A sum of \$1000 is invested at i% annual interest compounded continuously. Find i if it is desired that the money be doubled at the end of 10 years.

35. Repeat Exercise 34 if it is desired that the money be tripled at the end of 10 years.

36. A person borrows \$5000 at 18% annual interest compounded continuously. How much will this person owe the lender at the end of 1 year?

37. How long does it take to become a millionaire if \$1000 is invested at 8% annual interest compounded continuously? What if the initial investment is \$10,000?

1.4 FIRST-ORDER LINEAR DIFFERENTIAL EQUATIONS

A first-order linear differential equation is an equation of the form

$$a_1(x)y' + a_0(x)y = f(x).$$

We always assume that the coefficients $a_1(x)$, $a_0(x)$, and the function $f(x)$ are continuous functions in some interval I and that the leading coefficient $a_1(x) \neq 0$ for all x in I. If we divide both sides of the differential equation by $a_1(x)$ and set $a(x) = a_0(x)/a_1(x)$ and $b(x) = f(x)/a_1(x)$, we obtain the equivalent differential equation

$$y' + a(x)y = b(x), \tag{1}$$

where $a(x)$ and $b(x)$ are continuous functions of x in the interval I. The general solution of Eq. (1) can be found explicitly by observing that the change of variables

$$w = ye^{\int a(x)dx} \tag{2}$$

transforms Eq. (1) into a separable differential equation. In fact (recall that $(d/dx)[\int a(x)dx] = a(x)$),

$$\begin{aligned} w' &= y' e^{\int a(x)dx} + ya(x)e^{\int a(x)dx} \\ &= [y' + a(x)y]e^{\int a(x)dx} \\ &= b(x)e^{\int a(x)dx}. \end{aligned} \tag{3}$$

This is a separable differential equation with general solution

$$\begin{aligned} w(x) &= c + \int b(x)e^{\int a(x)dx}dx \\ \Rightarrow \quad ye^{\int a(x)dx} &= c + \int b(x)e^{\int a(x)dx}\,dx \\ \Rightarrow \quad y &= e^{-\int a(x)dx}\left[c + \int b(x)e^{\int a(x)dx}\,dx\right]. \end{aligned}$$

We summarize our results in the form of a theorem.

THEOREM 1

If $a(x)$ and $b(x)$ are continuous functions in some interval I, then the general solution of the differential equation

$$y' + a(x)y = b(x)$$

is given by the formula

$$y(x) = e^{-\int a(x)dx}\left[c + \int b(x)e^{\int a(x)dx}\,dx\right]. \tag{4}$$

REMARK 1 If you read the method above carefully once more, you will realize that the general solution of a first-order linear differential equation is found in three steps as follows.

Step 1 Multiplying both sides of Eq. (1) by $e^{\int a(x)dx}$, we obtain

$$[ye^{\int a(x)dx}]' = b(x)e^{\int a(x)dx}$$

[This is a restatement of Eq. (3).] The term $e^{\int a(x)dx}$ is called an *integrating factor*.

Step 2 Integrating both sides, we find

$$ye^{\int a(x)dx} = c + \int b(x)e^{\int a(x)dx}\,dx.$$

Step 3 Dividing both sides by $e^{\int a(x)dx}$ yields

$$y(x) = e^{-\int a(x)dx}[c + \int b(x)e^{\int a(x)dx}\,dx],$$

which is the general solution of Eq. (1).

REMARK 2 Before you apply the method described above, make sure that the coefficient of y' is 1. In general, you have to divide by the coefficient of y' before you apply the method. That is the case in Exercises 1, 2 and 5, and Example 2.

REMARK 3 It will be helpful in computations if you recall that for any positive function $p(x)$ the following identities hold:

$$e^{\ln p(x)} = p(x).$$
$$e^{-\ln p(x)} = \frac{1}{p(x)}.$$

REMARK 4 From Remark 1 of Section 1.2 the solutions of the differential equation (1) exist throughout the interval I, where the functions $a(x)$ and $b(x)$ are continuous. This, together with Example 2 of Section 1.2, implies that the IVP

$$y' + a(x)y = b(x)$$
$$y(x_0) = y_0,$$

where $x_0 \in I$, has a unique solution which exists throughout the interval I. The reader can verify by direct substitution (see Exercise 18) that the unique solution of the IVP is given by the formula

$$y(x) = y_0 e^{-\int_{x_0}^{x} a(s)ds} + \int_{x_0}^{x} b(s)e^{-\int_{s}^{x} a(t)dt}\,ds. \qquad (5)$$

EXAMPLE 1 Find the general solution of the differential equation

$$\frac{dy}{dx} + (\tan x)\, y = \sin x$$

in the interval $(0, \pi/2)$.

Solution This is a first-order linear differential equation with

$$a(x) = \tan x \quad \text{and} \quad b(x) = \sin x,$$

both continuous in the interval $(0, \pi/2)$. Multiplying both sides by

$$e^{\int \tan x dx} = e^{-\ln \cos x} = \frac{1}{\cos x},$$

we obtain

$$\left(y \frac{1}{\cos x}\right)' = \frac{\sin x}{\cos x}.$$

Integrating both sides, we find

$$y \frac{1}{\cos x} = c - \ln \cos x.$$

Dividing both sides by $1/\cos x$ (in other words, multiplying both sides by $\cos x$) yields the solution

$$y(x) = (\cos x)(c - \ln \cos x), \quad 0 < x < \frac{\pi}{2}.$$

EXAMPLE 2 Solve the IVP

$$xy' - 2y = -x^2$$
$$y(1) = 0.$$

Solution The differential equation can be written in the form $y' - (2/x)y = -x$ with $x \neq 0$. Here $a(x) = -2/x$ and $b(x) = -x$. Therefore, the interval I, where $a(x)$ and $b(x)$ are continuous, is either $(-\infty, 0)$ or $(0, \infty)$. Since $x_0 = 1$ lies in the interval $(0, \infty)$, it follows from Remark 4 that the IVP has a unique solution that exists in the entire interval $(0, \infty)$. Although we could use formula (5) to immediately obtain the solution of the IVP, it is perhaps more instructive to proceed directly by using the steps described in the text. Multiplying both sides by the integrating factor,

$$e^{-\int (2/x) dx} = e^{-2 \ln x} = e^{-\ln x^2} = \frac{1}{x^2},$$

we have

$$\left(y \frac{1}{x^2}\right)' = -\frac{1}{x}.$$

Integrating both sides, we obtain

$$y\frac{1}{x^2} = c - \ln x.$$

So, the general solution is

$$y(x) = x^2(c - \ln x).$$

Using the initial condition $y(1) = 0$, we find

$$0 = 1^2(c - \ln 1)$$

$$\Rightarrow \quad c = 0,$$

and therefore the solution of the IVP is

$$y(x) = -x^2 \ln x.$$

The graph of the solution is given in Figure 1.9.

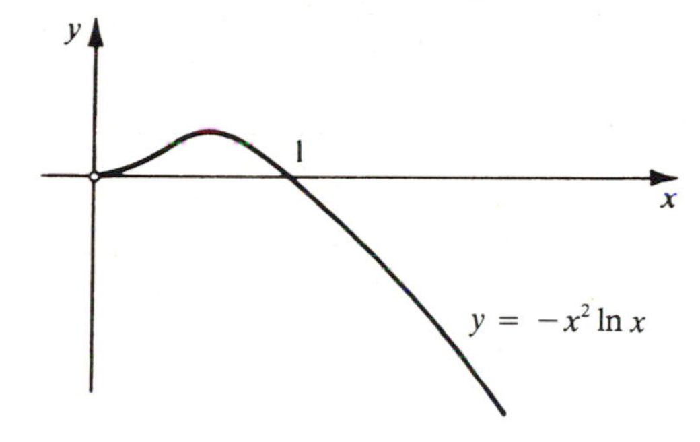

Figure 1.9

APPLICATIONS 1.4.1

In the literature there are many occurrences of the applications of linear differential equations, some of which we illustrate below.

Circuit Theory

■ Consider the RL-series circuit of Figure 1.10, which contains a resistance R, an inductance L, and a generator that supplies a voltage $V(t)$ when the switch

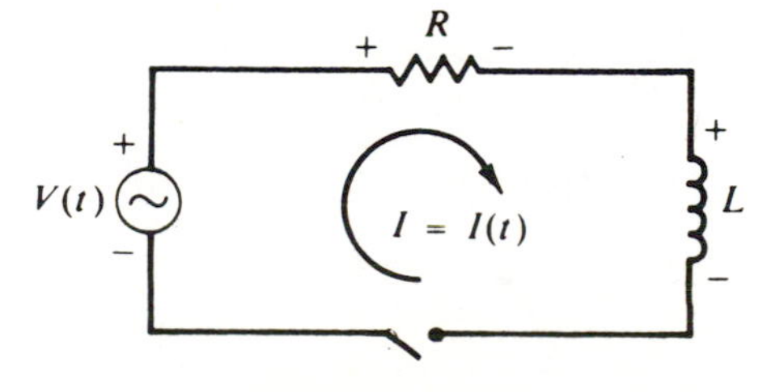

Figure 1.10

is closed. The current $I = I(t)$ in the circuit satisfies the linear first-order differential equation

$$L\frac{dI}{dt} + RI = V(t). \tag{6}$$

This differential equation is obtained from Kirchhoff's voltage law (see the electric circuits application in Section 1.1.1) and the following facts:

1. Across the resistor the voltage drop is RI.
2. Across the inductor the voltage drop is $L(dI/dt)$.
3. $V(t)$ is the only voltage increase in this circuit.

The reader can verify that the general solution of Eq. (6) is given by (see Exercise 19)

$$I(t) = e^{-(R/L)t}\left[c + \frac{1}{L}\int V(t)e^{(R/L)t}\,dt\right]. \tag{7}$$

Table 1.2 lists the units and conventional symbols of circuit elements that will be used in this book in connection with circuit-theory applications.

Quantity	Unit	Symbol
Resistance	ohm	R
Inductance	henry	L
Capacitance	farad	C
Voltage	volt	V
Current	ampere	I
Charge	coulomb	Q
Time	seconds	sec

Symbols of circuit elements

V	Generator or battery
R	Resistor
L	Inductor
C	Capacitor
	Switch

Table 1.2 Table of Units

Geometry

■ With the geometric interpretation of the concept of derivative in mind, it follows that the solutions of the differential equation

$$y' = F(x, y)$$

are curves $y = y(x)$ whose slope at any point (x, y) is equal to $F(x, y)$.

EXAMPLE 3 Find a curve in the xy plane, passing through the point (1, 2), whose slope at the point (x, y) is given by $2 - y/x$.

Solution The differential equation that describes the curve is

$$y' = 2 - \frac{y}{x}. \tag{8}$$

Since the curve passes through the point (1, 2), we should have

$$y(1) = 2. \tag{9}$$

Let us solve this IVP (8)–(9). The differential equation can be written in the form

$$y' + \frac{1}{x}y = 2$$

and so is a linear first-order differential equation. Multiplying both sides by $e^{\int(1/x)dx} = e^{\ln x} = x$, we obtain

$$(xy)' = 2x$$

$$\Rightarrow \quad xy = c + x^2.$$

From $y(1) = 2$ it follows that $c = 1$, and so $y = x + (1/x)$ is the curve with the desired properties (Figure 1.11).

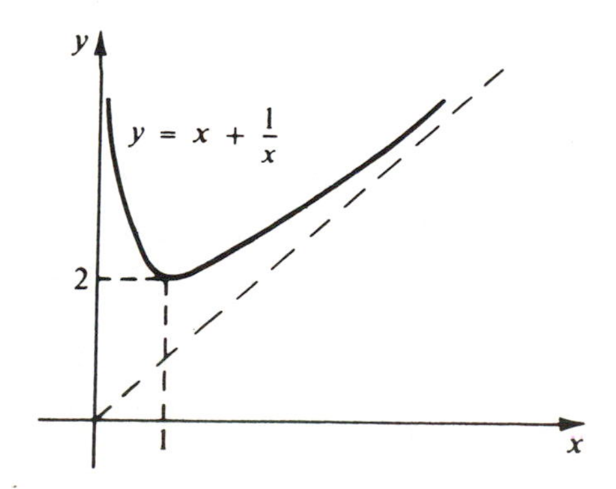

Figure 1.11

Economics ■ Assume that the rate of change of the price P of a commodity is proportional to the difference $D - S$ of the demand D and supply S in the market at any time t. Thus,

$$\frac{dP}{dt} = \alpha(D - S), \tag{10}$$

where the constant α is called the "adjustment" constant. In the simple case where D and S are given as linear functions of the price, that is,

$$D = a - bP, \quad S = -c + dP \qquad (a, b, c, d \text{ positive constants}),$$

Eq. (10) is a linear first-order differential equation. Let us denote by $\bar{P}$ the "equilibrium" price, that is, the price at which

$$a - b\bar{P} = D = S = -c + d\bar{P}.$$

It is left to the reader (see Exercise 26) to verify that the solutions of Eq. (10) is given by

$$P(t) = [P(0) - \bar{P}]\, e^{-\alpha(b+d)t} + \bar{P}. \tag{11}$$

Observe from Eq. (11) that as $t \to \infty$, the price of the commodity $P(t) \to \bar{P}$.

Mixtures ■ First-order linear differential equations arise as mathematical models in rate problems involving mixtures. Let us illustrate this by means of a typical example.

EXAMPLE 4 A large tank contains 81 gallons of brine in which 20 pounds of salt is dissolved. Brine containing 3 pounds of dissolved salt per gallon runs into the tank at the rate of 5 gallons per minute. The mixture, kept uniform by stirring, runs out of the tank at the rate of 2 gallons per minute. How much salt is in the tank at the end of 37 minutes?

Solution Let $y(t)$ be the amount of salt in the tank at any time t. Then $y'(t)$ is the rate of change of the salt in the tank at time t. Clearly,

$$y'(t) = \text{rate in} - \text{rate out},$$

where "rate in" is the rate at which salt runs into the tank at time t, and "rate out" is the rate at which salt runs out of the tank at time t. Here

$$\text{rate in} = (3 \text{ lb/gal})(5 \text{ gal/min}) = 15 \text{ lb/min},$$

that is, 15 pounds of salt per minute flows into the tank. To compute the "rate out," we should first find the *concentration* of salt at time t, that is, the amount of salt per gallon of brine at time t. Since

$$\begin{aligned}\text{concentration} &= \frac{\text{pounds of salt in the tank at time } t}{\text{gallons of brine in the tank at time } t}\\ &= \frac{y(t)}{81 + (5-2)t},\end{aligned}$$

we have

$$\text{rate out} = \left[\frac{y(t)}{81 + 3t}\ \text{lb/gal}\right](2\ \text{gal/min}) = \frac{2y(t)}{81 + 3t}\ \text{lb/min}.$$

Thus, the differential equation describing this mixture problem is

$$y' = 15 - \frac{2y}{81 + 3t}. \tag{12}$$

We also have the initial condition

$$y(0) = 20, \tag{13}$$

which expresses the fact that initially, that is, at $t = 0$, there is 20 pounds of salt in the tank. We must now solve the IVP (12)–(13). Equation (12) is linear and can be written in the form

$$y' + \frac{2}{81 + 3t}y = 15.$$

Multiplying both sides by

$$\exp\left(\int \frac{2}{81 + 3t}\,dt\right) = \exp\left(\frac{2}{3}\int \frac{dt}{27 + t}\right) = \exp\left[\frac{2}{3}\ln(27 + t)\right] = (27 + t)^{2/3},$$

we obtain

$$[y(t)(27 + t)^{2/3}]' = 15(27 + t)^{2/3}$$

Integrating both sides with respect to t, we obtain

$$y(t)(27 + t)^{2/3} = 15(27 + t)^{5/3}\cdot\frac{3}{5} + c,$$

and so

$$y(t) = 9(27 + t) + c(27 + t)^{-2/3}$$

Using the initial condition (13), we find

$$20 = 9(27) + c(27)^{-2/3},$$

and so $c = -2007$. Thus,

$$y(t) = 9(27 + t) - 2007(27 + t)^{-2/3},$$

and the amount of salt in the tank at the end of 37 minutes is

$$y(37) = 9(27 + 37) - 2007(27 + 37)^{-2/3} = 450.6\ \text{lb}.$$

EXERCISES

In Exercises 1 through 14, some of the differential equations are linear and some are not. Determine those that are linear and solve them.

1. $xy' + y = 3$

2. $xy' + y = 3x$

3. $(x^2 + y^2)dx - 2xy\,dy = 0$

4. $(1 + x)y\,dx + x\,dy = 0$

5. $xy' - y = 2x^2$

6. $y^2 dy + y \tan x\, dx = \sin^3 x\, dx$

7. $y' - \dfrac{3}{x-1}y = (x-1)^4$

8. $xy' + 6y = 3x + 1$

9. $y' + \dfrac{1}{\sin x}y - y^2 = 0$

10. $e^x\, dx + x^3\, dy + 4x^2 y\, dx = 0$

11. $xy' + y = x^5$

12. $y' - \dfrac{x}{1+x^2} = -\dfrac{x}{1+x^2}y$

13. $yy' - 7y = 6x$

14. $y\dfrac{dx}{dy} + x = y$

15. Find a curve in the xy plane that passes through the point $(1, 1 + e)$ and whose slope at any point (x, y) is given by

$$2 - \frac{y}{x} + \frac{e^x}{x}.$$

16. Compute the orthogonal trajectories of the family of concentric circles

$$x^2 + y^2 = R^2.$$

17. Find the solution of the IVP

$$y' - y = b(x)$$
$$y(0) = 1,$$

where

$$b(x) = \begin{cases} 0, & x < 0 \\ x, & x \geq 0. \end{cases}$$

18. Verify formula (5) and apply it to the IVPs of Examples 2 and 3.

19. Verify Eq. (7) of this section.

20. In an RL-series circuit assume that the voltage $V(t)$ is a constant V_0. Show that the current at any time t is given by

$$I(t) = \frac{V_0}{R} + \left(I_0 - \frac{V_0}{R}\right)e^{-(R/L)t},$$

where I_0 is the current at time $t = 0$.

21. In an RL-series circuit, $L = 4$ henries, $R = 5$ ohms, $V = 8$ volts, and $I(0) = 0$ amperes. Find the current at the end of 0.1 seconds. What will the current be after a very long time?

22. Assume that the voltage $V(t)$ in Eq. (6) is given by $V_0 \sin \omega t$, where V_0 and ω are given constants. Find the solution of the differential equation (6) subject to the initial condition $I(0) = I_0$.

23. In an RL-series circuit [see Eq. (6)] it is given that $L = 3$ henries; $R = 6$ ohms, $V(t) = 3 \sin t$, and $I(0) = 10$ amperes. Compute the value of the current at any time t.

24. A large tank contains 40 gallons of brine in which 10 pounds of salt is dissolved. Brine containing 2 pounds of dissolved salt per gallon runs into the tank at the rate of 4 gallons per minute. The mixture, kept uniform by stirring, runs out of the tank at the rate of 3 gallons per minute.
(a) How much salt is in the tank at any time t?
(b) Find the amount of salt in the tank at the end of 1 hour.

25. A tank initially contains 10 gallons of pure water. Starting at time $t = 0$ brine containing 3 pounds of salt per gallon flows into the tank at the rate of 2 gal/min. The mixture is kept uniform by stirring and the well-stirred mixture flows out of the tank at the same rate as the inflow. How much salt is in the tank after 5 minutes? How much salt is in the tank after a very long time?

26. Verify Eq. (11) of this section.

27. The *Bernoulli equation* is the differential equation

$$y' + a(x)y = b(x)y^n, \qquad n \neq 0, 1.$$

Show that the transformation $w = y^{1-n}$ reduces the Bernoulli differential equation to the linear differential equation

$$w' + (1 - n)a(x)w = (1 - n)b(x).$$

In Exercises 28 through 31, solve the Bernoulli differential equation (see Exercise 27).

28. $y' - \dfrac{1}{x}y = -\dfrac{1}{2y}$

29. $y' + \dfrac{1}{x}y = -2xy^2$

30. $y' - 2xy = 4xy^{1/2}$

31. $xy' - \dfrac{y}{2 \ln x} = y^2$

In Exercises 32 through 37 solve the initial value problem.

32. $y' - xy = (1 - x^2)e^{(1/2)x^2}$
$y(0) = 0$

33. $xy' + y = 2x$
$y(2) = 2$

34. $xy' - \dfrac{y}{\ln x} = 0$
$y(e) = -1$

35. $(1 + x)^2y' + 2xy = -2x$
$y(0) = -1$

36. $(1 - x)y' + xy = x(x - 1)^2$
$y(5) = 24$

37. $(x - 1)y' - 3y = (x - 1)^5$
$y(-1) = 16$

38. Biophysics In a study related to the biophysical limitations associated with deep diving,[14] Bradner and Mackay, obtained the following first-order linear ordinary differential equation

$$y' - Ay = B + be^{-ax},$$

where a, b, A, and B are constants. Show that the general solution of this differential equation is given by

$$y(x) = -\frac{B}{A} - \frac{b}{a + A}e^{-ax} + ce^{Ax},$$

where c is an arbitrary constant.

39. Show that the general solution of the differential equation

$$y' - 2xy = x^2$$

is of the form

$$y = Y_0(x) + Ae^{x^2},$$

where $Y_0(x)$ is a solution of the differential equation that obeys the inequalities

$$-\frac{x}{2} - \frac{1}{4x} \le Y_0(x) \le -\frac{x}{2}, \qquad x \ge 2.$$

[*Hint:* Find the general solution of the differential equation and observe that the improper integral $\int_0^\infty t^2 e^{-t^2}\,dt$ converges.[15]

40. Physics Newton's law of cooling states that the time rate of change of the temperature $T = T(t)$ of a body at time t is proportional to the difference $T - T_m$ in temperature between the body and its surrounding medium. That is,

$$\frac{dT}{dt} = -k(T - T_m),$$

where $k > 0$ is a constant of proportionality. This model gives a good approximation of the true physical situation provided that the temperature difference is small, generally less than about 36°F. Show that the general solution of this differential equation is

$$T(t) = T_m + ce^{-kt}.$$

41. A body of temperature 70°F is placed (at time $t = 0$) outdoors, where the temperature is 40°F. After 3 minutes the temperature of the body has fallen to 60°F.

(a) How long will it take the body to cool to 50°F?

(b) What is the temperature of the body after 5 minutes?

[14] H. Bradner and R. S. Mackay, *Bull. Math. Biophys.* 25 (1963): 251–72.

[15] *Amer. Math. Monthly* 63 (1956): 414.

42. A *Riccati equation* is a differential equation of the form

$$y' = f(x) + g(x)y + h(x)y^2,$$

where the functions f, g, and h are continuous in some interval I. Show that if $y_1(x)$ is a particular solution of the Riccati differential equation, then the transformation

$$y(x) = y_1(x) + \frac{1}{w(x)}$$

reduces it to the linear first-order differential equation

$$w' + [g(x) + 2y_1(x)h(x)]w = -h(x).$$

Mass Behavior In a research article on the theory of the propagation of a single act in a large population (in other words, an act that is performed at most once in the lifetime of an individual, such as suicide), Rapoport[16] derived the following Riccati differential equation:

$$y = (1 - y)[x(t) + by]. \tag{14}$$

Here $y = y(t)$ is the fraction of the population who performed the act at time t, $x(t)$ is the external influence or stimulus and the product by is the imitation component. Solve this Riccati differential equation in each of the following cases. [*Hint:* $y = 1$ is a particular solution.[17]]

43. $x(t) = \dfrac{1}{t} - \dfrac{1}{10}, b = \dfrac{1}{10}.$

44. $x(t) = -\dfrac{1}{t\ln t} - 0.03, b = 0.03.$

1.5 EXACT DIFFERENTIAL EQUATIONS

A differential equation of the form

$$M(x, y)dx + N(x, y)dy = 0 \tag{1}$$

is called *exact* if there is a function $f(x, y)$ whose total differential is equal to $M(x, y)dx + N(x, y)dy$, that is (suppressing the arguments x and y)

$$df = M\,dx + N\,dy. \tag{2}$$

[16] A. Rapoport, *Bull. Math. Biophys.* 14 (1952): 159–69.

[17] For a detailed study of the differential equation (14), see the paper by J. Z. Hearon, *Bull. Math. Biophys.* 17 (1965): 7–13.

(Recall that the total differential of a function f is given by $df = f_x dx + f_y dy$, provided that the partial derivatives of the function f with respect to x and y exist.)

If Eq. (1) is exact, then [because of (2) and (1)] it is equivalent to

$$df = 0.$$

Thus, the function $f(x, y)$ is a constant and the general solution of Eq. (1) is given by

$$f(x, y) = c. \tag{3}$$

For example, the differential equation

$$(2x - y)dx + (-x + 4y)dy = 0 \tag{4}$$

is exact because

$$\begin{aligned} d(x^2 - xy + 2y^2) &= \frac{\partial}{\partial x}(x^2 - xy + 2y^2)dx + \frac{\partial}{\partial y}(x^2 - xy + 2y^2)dy \\ &= (2x - y)dx + (-x + 4y)dy. \end{aligned}$$

Thus, the general solution of Eq. (4) is given (implicitly) by

$$x^2 - xy + 2y^2 = c. \tag{5}$$

The following two questions are now in our minds.

QUESTION 1 Is there a systematic way to check if the differential equation (1) is exact?

QUESTION 2 If we know that the differential equation (1) is exact, is there a systematic way to solve it?

Answer 1 The differential equation (1) is exact if and only if

$$M_y = N_x. \tag{6}$$

That is, if the partial derivative of M with respect to y is equal to the partial derivative of N with respect to x, then Eq. (1) is exact; conversely, if Eq. (1) is exact, then (6) holds.

Answer 2 Choose any point (x_0, y_0) in the region where the functions M, N and their partial derivatives M_y, N_x are continuous. Then

$$f(x, y) = \int_{x_0}^{x} M(x, y_0)dx + \int_{y_0}^{y} N(x, y)dy = c \tag{7}$$

gives (implicitly) the general solution of the differential equation (1).

To justify the answers above, it suffices to prove that

$$M_y = N_x \qquad \text{implies that} \qquad df = M\,dx + N\,dy,$$

where f is the sum of the two integrals in (7) and conversely if Eq. (1) is exact, then (6) holds. In fact, from Eq. (7) we have

$$f_x = M(x, y_0) + \int_{y_0}^{y} N_x(x, y)dy = M(x, y_0) + \int_{y_0}^{y} M_y(x, y)dy$$
$$= M(x,y_0) + M(x,y)\Big|_{y_0}^{y} = M(x,y)$$

and similarly $f_y = N(x,y)$. Therefore, $df = f_x dx + f_y dy = M\,dx + N\,dy$, which proves the "if" part of Answer 1. To prove the converse, observe that, since Eq. (1) is exact, there exists a function f such that $df = M\,dx + N\,dy \Rightarrow f_x = M$ and $f_y = N \Rightarrow f_{xy} = M_y$ and $f_{yx} = N_x$, and since $f_{yx} = f_{xy}$, we have $M_y = N_x$.

REMARK 1 The point (x_0, y_0) in Eq. (7) may be chosen judiciously for the purpose of simplifying $M(x, y_0)$ in the first integral of (7) and the evaluations of both integrals at the lower points of integrations.

REMARK 2 The precise form of Eq. (7) should be

$$f(x, y) = \int_{x_0}^{x} M(t, y_0)dt + \int_{y_0}^{y} N(x, s)ds = c, \tag{7'}$$

and the assumption should be made that the segments from (x_0, y_0) to (x, y_0) and from (x, y_0) to (x, y) lie in the region where the functions M, N and M_y, N_x exist and are continuous.

REMARK 3 As shown, Eq. (7) is useful to establish the existence of the function f and also to verify the test for exactness, that is Eq. (6). Students sometimes find the presence of the "arbitrary" point (x_0, y_0) and the judicious selection of values for x_0 and y_0 confusing. Consequently alternative systematic methods for the determination of f are given in Examples 3 and 4.

EXAMPLE 1 Solve the differential equation

$$(3x^2 + 4xy^2)dx + (2y - 3y^2 + 4x^2y)dy = 0.$$

Solution Here

$$M(x, y) = 3x^2 + 4xy^2 \quad \text{and} \quad N(x, y) = 2y - 3y^2 + 4x^2y$$
$$\Rightarrow \quad M_y = 8xy \quad \text{and} \quad N_x = 8xy$$
$$\Rightarrow \quad M_y = N_x.$$

Thus, the differential equation is exact and from Eq. (7), with $(x_0, y_0) = (0, 0)$, its general solution is given implicitly by

$$\int_0^x 3x^2\,dx + \int_0^y (2y - 3y^2 + 4x^2y)dy = c.$$

$$\Rightarrow \qquad x^3 + y^2 - y^3 + 2x^2y^2 = c.$$

EXAMPLE 2 Solve the IVP

$$\left[\cos x \ln(2y - 8) + \frac{1}{x}\right]dx + \frac{\sin x}{y - 4}dy = 0 \tag{8}$$

$$y(1) = \frac{9}{2}. \tag{9}$$

Solution Here

$$M(x, y) = \cos x \ln(2y - 8) + \frac{1}{x} \quad \text{and} \quad N(x, y) = \frac{\sin x}{y - 4}$$

$$\Rightarrow \quad M_y = \cos x \frac{2}{2y - 8} = \frac{\cos x}{y - 4} \quad \text{and} \quad N_x = \frac{\cos x}{y - 4}$$

$$\Rightarrow \quad M_y = N_x.$$

Thus, the differential equation is exact and from Eq. (7) its solutions are given by

$$f(x, y) = \int_{x_0}^{x}\left[\cos x \ln(2y_0 - 8) + \frac{1}{x}\right]dx + \int_{y_0}^{y}\frac{\sin x}{y - 4}dy = c,$$

where x_0 is any number $\neq 0$ and $y_0 > 4$. Let us make the judicious choice $x_0 = 1$ and $y_0 = \frac{9}{2}$. Then

$$f(x, y) = \ln x + \sin x \ln(y - 4) + \sin x \ln 2 = c.$$

Using the initial condition (9), we find that $c = 0$. Thus, the solution of the IVP is implicitly given by

$$\ln x + \sin x \ln 2(y - 4) = 0.$$

Solving with respect to y, we find that

$$y = 4 + \tfrac{1}{2}e^{-\ln x/\sin x}$$

REMARK 4 In the differential equation above, the largest interval of existence of the solution through $(1, \frac{9}{2})$ is $(0, \pi)$ (by the existence and uniqueness theorem, Theorem 1 of Section 1.2), and so we assume that $x > 0$.

REMARK 5 Sometimes it is easier to solve an exact differential equation by regrouping its terms into two groups of terms—one being the group of terms of the form $p(x)dx$ and $q(y)dy$ and the other the remaining terms—and recognizing that each group is (in fact) a total differential of a function.

EXAMPLE 3 Solve the differential equation

$$(3x^2 + 4xy^2)dx + (2y - 3y^2 + 4x^2y)dy = 0.$$

Solution From Example 1 we know that this equation is exact. Regrouping terms in accordance with Remark 5, we have

$$\begin{aligned}[3x^2\,dx + (2y - 3y^2)dy] + (4xy^2\,dx + 4x^2y\,dy) &= 0\\ \Rightarrow \quad d(x^3 + y^2 - y^3) + d(2x^2y^2) &= 0\\ \Rightarrow \quad d(x^3 + y^2 - y^3 + 2x^2y^2) &= 0\\ \Rightarrow \quad x^3 + y^2 - y^3 + 2x^2y^2 &= c.\end{aligned}$$

EXAMPLE 4 Solve the differential equation

$$(3x^2 + 4xy^2)dx + (2y - 3y^2 + 4x^2y)dy = 0.$$

Solution From Example 1 we know that this equation is exact, and therefore there exists a function f such that $f_x = M$ and $f_y = N$. Consider

$$f_x = M = 3x^2 + 4xy^2.$$

Integrating with respect to x we obtain

$$f = x^3 + 2x^2y^2 + h(y).$$

The "constant" of integration is a function of y due to the partial integration with respect to x. To determine $h(y)$ we utilize the fact that $f_y = N$. Taking the partial derivative with respect to y, this last expression yields

$$f_y = 4x^2y + h'(y).$$

Setting this equal to N ($N = 2y - 3y^2 + 4x^2y$), we have

$$h'(y) = 2y - 3y^2 \Rightarrow h(y) = y^2 - y^3.$$

Thus,

$$f(x, y) = x^3 + 2x^2y^2 + y^2 - y^3,$$

and the general solution is given (implicitly) by

$$x^3 + 2x^2y^2 + y^2 - y^3 = c.$$

Note that the solution for $h(y)$ omitted a constant of integration which normally would appear. This omission is permissible since otherwise it would be combined with the constant c when we set $f = c$.

As an alternative construction we could have integrated the expression $f_y = N$ with respect to y. In this case the "constant" of integration would be a function of x, say $g(x)$, and $g(x)$ would be determined by setting $f_x = M$.

APPLICATION 1.5.1

Geometry

■ Compute the orthogonal trajectories of the one-parameter family of curves

$$x^2 + 2xy - y^2 + 4x - 4y = c. \tag{10}$$

Solution First, we should compute the slope of the curves (10). Differentiating with respect to x, we obtain

$$2x + 2y + 2xy' - 2yy' + 4 - 4y' = 0,$$

and so

$$y' = -\frac{x + y + 2}{x - y - 2}.$$

Thus, the slope of the orthogonal trajectories of (10) is

$$y' = \frac{x - y - 2}{x + y + 2}. \tag{11}$$

Equation (11) is the differential equation of the orthogonal trajectories of the curves (10). This equation is not separable or linear, but after writing it in the form

$$(-x + y + 2)dx + (x + y + 2)dy = 0, \tag{12}$$

we recognize it as an exact differential equation. Regrouping the terms in Eq. (12) in the form

$$(-x\,dx + y\,dy + 2dx + 2dy) + (y\,dx + x\,dy) = 0, \tag{13}$$

we recognize that Eq. (13) is equivalent to (the total differential)

$$\begin{aligned} &\tfrac{1}{2}d(-x^2 + y^2 + 4x + 4y + 2xy) = 0 \\ \Rightarrow \quad &-\tfrac{1}{2}d(x^2 - y^2 - 4x - 4y - 2xy) = 0. \end{aligned}$$

Thus, the general solution of the differential equation (13) is

$$x^2 - 2xy - y^2 - 4x - 4y = \text{const.} = k. \tag{14}$$

The curves given by Eq. (14) are the orthogonal trajectories of (10). We recall from analytic geometry that an equation of the form $Ax^2 + Bxy + Cy^2 + Dx + Ey + F = 0$ represents a hyperbola, parabola, or ellipse depending on whether $B^2 - 4AC$ is positive, zero, or negative.[18] For the family (10), $B^2 - 4AC = (2)^2 - 4(1)(-1) = 8$, and for the family (14), $B^2 - 4AC = (-2)^2 - 4(1)(-1) = 8$. Thus, both of these families consist of hyperbolas. The standard forms for hyperbolas are $\alpha x^2 - \beta y^2 = \alpha\beta$, $\beta y^2 - \alpha x^2 = \alpha\beta$, and $xy = \gamma$ with $\alpha > 0$, $\beta > 0$, and γ arbitrary. Equations (10) and (14) can be written in these forms by the transformations of variables known as rotation of axes and translation of axes (in that order). In general, the rotation angle θ is

[18] G. B. Thomas, Jr., *Calculus and Analytic Geometry,* Alternate Edition (Reading, Mass.: Addison–Wesley, 1972), p. 508.

chosen so as to eliminate the cross-product term and hence must satisfy the condition

$$\cot 2\theta = \frac{A - C}{B}.$$

For the family (10), we have

$$\cot 2\theta = \frac{1 - (-1)}{2} = 1$$

$$\Rightarrow \quad 2\theta = 45°$$

$$\theta = 22\tfrac{1}{2}°.$$

If the variables corresponding to the rotated x and y axes are denoted by u and v, respectively, we have

$$x = u\cos\theta - v\sin\theta \tag{15}$$

$$y = u\sin\theta + v\cos\theta. \tag{16}$$

Now

$$\sin\theta = \sqrt{\frac{1 - \cos 2\theta}{2}} = \sqrt{\frac{1 - 1/\sqrt{2}}{2}} = \sqrt{\frac{\sqrt{2} - 1}{2\sqrt{2}}}$$

$$\cos\theta = \sqrt{\frac{1 + \cos 2\theta}{2}} = \sqrt{\frac{1 + 1/\sqrt{2}}{2}} = \sqrt{\frac{\sqrt{2} + 1}{2\sqrt{2}}}.$$

Substituting (15) and (16) into Eqs. (10) and (14), we obtain, respectively,

$$u^2 - v^2 - 2u + 2v = \frac{c}{\sqrt{2}} \tag{17}$$

$$uv - u - v = \frac{k}{\sqrt{2}}. \tag{18}$$

By completing the square Eq. (17) can be written in the form

$$(u - 1)^2 - (v - 1)^2 = \frac{c}{\sqrt{2}}. \tag{19}$$

By factoring, Eq. (18) can be written in the form

$$(u - 1)(v - 1) = \frac{k}{\sqrt{2}} + 1. \tag{20}$$

The translation of axes $s = u - 1$, $t = v - 1$ enables us to write (19) and (20) in the respective standard forms

$$s^2 - t^2 = \frac{c}{\sqrt{2}} \tag{21}$$

$$st = \frac{k}{\sqrt{2}} + 1. \tag{22}$$

These families are illustrated in Figure 1.12. F_1 refers to the family (21) and F_2 refers to the family (22).

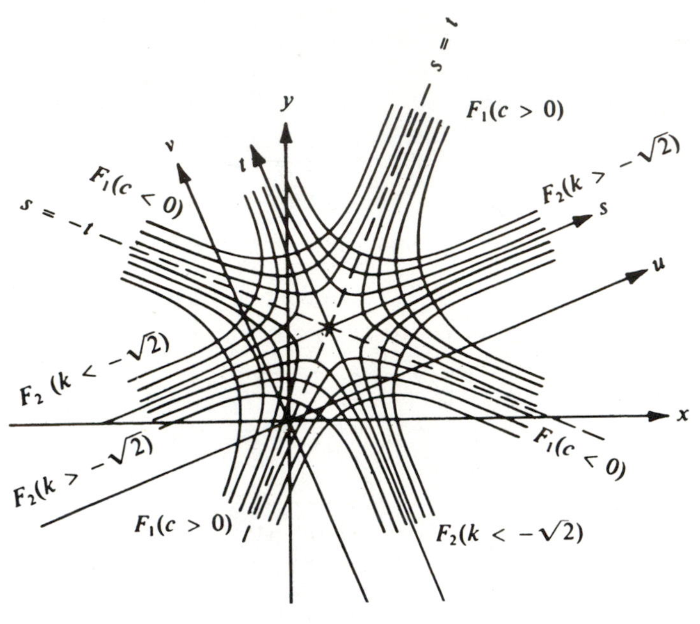

Figure 1.12

EXERCISES

In Exercises 1 through 12, some of the differential equations are exact and some are not. Determine those that are exact and solve them.

1. $(x - y)dx + (-x + y + 2)dy = 0$

2. $(x + y)dx + (x - y)dy = 0$

3. $y' = \dfrac{y - x + 1}{-x + y + 3}$

4. $(x^2 + y^2)dx - 2xy\,dy = 0$

5. $(x^2 + y^2)dx + 2xy\,dy = 0$

6. $y' = y^{1/2}$

7. $ye^{xy}\,dx + (xe^{xy} + 1)\,dy = 0$

8. $\cos y\,dx + \sin x\,dy = 0$

9. $(y + \cos x)dx + (x + \sin y)dy = 0$

10. $(3x^2y + y^2)dx = (-x^3 - 2xy)dy$

11. $(e^x \cos y - x^2)dx + (e^y \sin x + y^2)dy = 0$

12. $(2x - y \sin xy)dx + (6y^2 - x \sin xy)dy = 0$

13. Compute the orthogonal trajectories of the one-parameter family of curves $x^3 - 3xy^2 + x + 1 = c$.

14. If $u(x, y)$ is a harmonic function, that is, $u_{xx} + u_{yy} = 0$, show that the orthogonal trajectories of the one-parameter family of curves $u(x, y) = c$ satisfy an exact differential equation.

15. Consider the one-parameter family of curves

$$Ax^2 + Bxy + Cy^2 + Dx + Ey = c, \tag{23}$$

where c is the parameter and A, B, C, D, and E are arbitrary (but fixed) coefficients. Show that the differential equation for the orthogonal trajectories of this family is

$$(Bx + 2Cy + E)dx + (-2Ax - By - D)dy = 0. \tag{24}$$

16. (a) Under what conditions will the differential equation (24) be exact? (b) Using the condition obtained in part (a), find the equation of orthogonal trajectories and show that they and the family (23) are hyperbolas.

Solve the following initial value problems:

17. $(x - y)\,dx + (-x + y + 2)dy = 0$
$y(1) = 1$

18. $(x + y)\,dx + (x - y)\,dy = 0$
$y(0) = 2$

19. $(x^2 + y^2)dx + 2xy\,dy = 0$
$y(1) = -1$

20. $y' = \dfrac{y - x + 1}{-x + y + 3}$
$y(1) = 2$

21. Integrating Factors If the differential equation

$$M(x, y)\,dx + N(x, y)\,dy = 0 \tag{25}$$

is not exact, that is, $M_y \neq N_x$, we can sometimes find a (nonzero) function μ that depends on x or y or both x and y such that the differential equation

$$\mu M\,dx + \mu N\,dy = 0 \tag{26}$$

is exact, that is, $(\mu M)_y = (\mu N)_x$. The function μ is then called an *integrating factor* of the differential equation (25). Since (26) is exact, we can solve it, and its solutions will also satisfy the differential equation (25).

Show that a function $\mu = \mu(x, y)$ is an integrating factor of the differential equation (25) if and only if it satisfies the partial differential equation

$$N\mu_x - M\mu_y = (M_y - N_x)\mu. \tag{27}$$

22. In general, it is very difficult to solve the partial differential equation (27) without some restrictions on the functions M and N of Eq. (25). In this and the following exercise the restrictions imposed on M and N reduce Eq. (27) into a first-order linear differential equation whose solutions can be found explicitly. If it happens that the expression

$$\frac{1}{N}(M_y - N_x)$$

is a function of x alone, it is always possible to choose μ as a function of x only. Show that with these assumptions the function

$$\mu(x) = e^{\int (1/N)(M_y - N_x)dx}$$

is an integrating factor of the differential equation $M\,dx + N\,dy = 0$.

23. If it happens that the expression

$$\frac{1}{M}(M_y - N_x)$$

is a function of y alone, it is always possible to choose μ as a function of y only. Show that with these assumptions the function

$$\mu(y) = e^{-\int (1/M)(M_y - N_x)dy}$$

is an integrating factor of the differential equation $M\,dx + N\,dy = 0$.

For each of the following differential equations, find an integrating factor and then use it to solve the differential equation. [*Hint:* Use Exercise 22 or 23]

24. $y\,dx - x\,dy = 0$

25. $(x^2 - 2y)dx + x\,dy = 0$

26. $y\,dx + (2x - y^2)dy = 0$

27. $(y - 2x)dx - x\,dy = 0$

28. $y\,dx - (x - 2y)dy = 0$

29. $(x^4 + y^4)dx - xy^3dy = 0$

30. $(x^2 - y^2 + x)dx + 2xy\,dy = 0$

31. Verify that $\mu(x) = e^{\int a(x)dx}$ is an integrating factor of the first-order linear differential equation

$$y' + a(x)y = b(x),$$

and then use it to find its solution. [*Hint:* Write the differential equation in the equivalent form $[a(x)y - b(x)]dx + dy = 0$].

Verify that each of the following functions is an integrating factor of the differential equation $y\,dx - x\,dy = 0$ and then use the function to solve the equation.

32. $\mu(y) = \frac{1}{y^2}$ for $y \neq 0$

33. $\mu(x, y) = \frac{1}{xy}$ for $x \neq 0$ and $y \neq 0$

34. $\mu(x, y) = \frac{1}{x^2 + y^2}$ for $x \neq 0$ or $y \neq 0$

35. $\mu(x) = \frac{1}{x^2}$ for $x \neq 0$

36. $\mu(x, y) = \frac{1}{x^2 - y^2}$ for $x \neq \pm y$

37. $\mu(x, y) = \frac{1}{(x - y)^2}$ for $x \neq y$

38. $\mu(x, y) = \frac{1}{(x + y)^2}$ for $x \neq -y$

39. Verify that the function $\mu(x, y) = 1/(x^2 + y^2)$ is an integrating factor of the differential equation

$$(2x^2 + 2y^2 + x)dx + (x^2 + y^2 + y)dy = 0,$$

and then use it to find its solution.

1.6 HOMOGENEOUS EQUATIONS

A homogeneous differential equation is an equation of the form

$$y' = \frac{g(x, y)}{h(x, y)}, \tag{1}$$

where the functions g and h are homogeneous of the same degree. Recall that a function $g(x, y)$ is called *homogeneous of degree n* if $g(\alpha x, \alpha y) = \alpha^n g(x, y)$. For example, the functions x/y, $\sin x/y$, $x - y$, $x + y + \sqrt{x^2 + y^2}$, $x^2 + \sqrt{x^4 + y^4}$, $x^3 + y^3 + xy^2$ are homogeneous of degree zero, zero, 1, 1, 2, and 3, respectively. The following are examples of homogeneous differential equations:

$$y' = \frac{x - y}{x + y}, \qquad y' = \frac{x^2 + y^2}{x^2 + 3xy - y^2}, \qquad y' = \sin\frac{x}{y},$$

$$y' = \frac{x + \sqrt{x^2 + y^2}}{x - \sqrt{x^2 + y^2}}.$$

To solve the homogeneous differential equation (1), set $y = wx$ and observe that this substitution reduces Eq. (1) to a separable differential equation. In fact,

$$(wx)' = \frac{g(x, wx)}{h(x, wx)} \Rightarrow w'x + w = \frac{x^n g(1, w)}{x^n h(1, w)},$$

where n is the degree of homogeneity of g and h. Thus,

$$w'x = \frac{g(1, w)}{h(1, w)} - w,$$

which is easily recognized as a separable differential equation.

EXAMPLE 1 Solve the differential equation

$$y' = \frac{x^3 + y^3}{xy^2}. \tag{2}$$

Solution Here $g(x, y) = x^3 + y^3$ and $h(x, y) = xy^2$ are homogeneous of degree 3. $[g(\alpha x, \alpha y) = (\alpha x)^3 + (\alpha y)^3 = \alpha^3(x^3 + y^3) = \alpha^3 g(x, y)$, and $h(\alpha x, \alpha y) =$

$(\alpha x)(\alpha y)^2 = \alpha^3 xy^2 = \alpha^3 h(x, y)$.] Thus, Eq. (2) is a homogeneous differential equation. Set $y = wx$. Then, from Eq. (2),

$$(wx)' = \frac{x^3 + w^3x^3}{xw^2x^2} \Rightarrow w'x + w = \frac{1 + w^3}{w^2} = \frac{1}{w^2} + w$$

$$\Rightarrow \qquad w'x = \frac{1}{w^2} \Rightarrow w^2\,dw = \frac{1}{x}\,dx \quad \text{(separable)}$$

$$\Rightarrow \qquad \frac{1}{3}w^3 = \ln|x| + c \Rightarrow \frac{1}{3}\frac{y^3}{x^3} = \ln|x| + c.$$

Thus,

$$y = x(3\ln|x| + c)^{1/3}.$$

EXAMPLE 2 Solve the IVP

$$(x^2 + y^2)\,dx + 2xy\,dy = 0 \tag{3}$$

$$y(1) = -1. \tag{4}$$

Solution The differential equation (3) can be written in the form

$$y' = -\frac{x^2 + y^2}{2xy}, \tag{5}$$

which is clearly homogeneous. [The reader can verify that the hypotheses of the existence theorem, Theorem 1 of Section 1.2, are satisfied for the IVP (3)–(4). Also, we are looking for a solution through the point $(1, -1)$, and so the division by $2xy$ in Eq. (5) is permissible in a small rectangle about the point $(1, -1)$.] Setting $y = wx$ in Eq. (5), we obtain

$$(wx)' = -\frac{x^2 + w^2x^2}{2xwx} \Rightarrow w'x + w = -\frac{1 + w^2}{2w} \Rightarrow w'x = -\frac{1 + 3w^2}{2w}$$

$$\Rightarrow \qquad \frac{2w}{1 + 3w^2}\,dw + \frac{dx}{x} = 0 \Rightarrow \frac{1}{3}\ln(1 + 3w^2) + \ln|x| = c$$

$$\Rightarrow \qquad \ln[(1 + 3w^2)|x|^3] = c \Rightarrow \left(1 + 3\frac{y^2}{x^2}\right)x^3 = c.$$

Using the initial condition (4), we find that $c = 4$

$$\Rightarrow \qquad \left(1 + 3\frac{y^2}{x^2}\right)x^3 = 4 \Rightarrow y^2 = \frac{4 - x^3}{3x} \Rightarrow y = \pm\sqrt{\frac{4 - x^3}{3x}}.$$

Clearly, one of the two signs should be rejected. In this example the right sign is the negative one, because $y(1) = -1$. So the solution is

$$y = -\sqrt{\frac{4 - x^3}{3x}}.$$

APPLICATION 1.6.1

Geometry

■ Compute the orthogonal trajectories of the one-parameter family of circles $x^2 + y^2 = 2cx$.

Solution By taking derivatives of both sides, we find that $2x + 2yy' = 2c$. Since $2c = (x^2 + y^2)/x$, it follows that

$$2x + 2yy' = \frac{x^2 + y^2}{x} \Rightarrow y' = \frac{y^2 - x^2}{2xy}.$$

Thus, the slope of each curve in this family at the point (x, y) is

$$y' = \frac{y^2 - x^2}{2xy}.$$

The differential equation of the orthogonal trajectories is, therefore,

$$y' = \frac{2xy}{x^2 - y^2}. \tag{6}$$

Equation (6) is a homogeneous differential equation. (It is not separable, it is not linear, it is not exact.) Setting $y = wx$ in (6), we obtain

$$(wx)' = \frac{2x^2w}{x^2 - w^2x^2} \Rightarrow w'x + w = \frac{2w}{1 - w^2} \Rightarrow w'x = \frac{w + w^3}{1 - w^2}$$

$$\Rightarrow \quad \frac{(1 - w^2)dw}{w(1 + w^2)} = \frac{dx}{x} \Rightarrow \left(\frac{1}{w} - \frac{2w}{1 + w^2}\right) dw = \frac{dx}{x}.$$

Integrating both sides and replacing w by y/x, we find the one-parameter family of circles $x^2 + y^2 = 2ky$. Clearly, every circle in the family $x^2 + y^2 = 2cx$ has center at the point $(c, 0)$ of the x axis and radius $|c|$, while every circle in the family of orthogonal trajectories $x^2 + y^2 = 2ky$ has center at the point $(0, k)$ of the y axis and radius $|k|$ (see Figure 1.13).

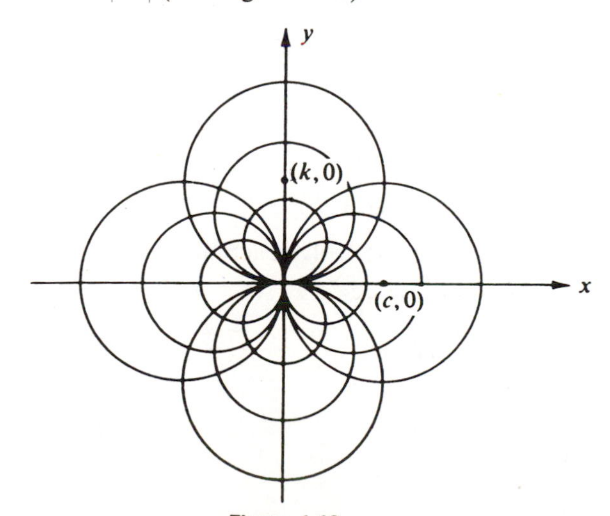

Figure 1.13

EXERCISES

In Exercises 1 through 12, some differential equations are homogeneous and some are not. Solve those that are homogeneous.

1. $(5x - y)dx + 3x\,dy = 0$

2. $xy' + y = 3$

3. $(x^2 + y^2)dx - 2xy\,dy = 0$

4. $(x^2 + y^2 + 1)dx - 2xy\,dy = 0$

5. $-x^2y\,dx + (x^3 + y^3)dy = 0$

6. $(2x - 3y)dx + (7y^2 + x^2)dy = 0$

7. $3y\,dx + (7x - y)\,dy = 0$

8. $e^{y/x} + y' - \dfrac{y}{x} = 0$

9. $xy\,dx - (x^2 - y^2)dy = 0$

10. $(xy + 1)\,dx + y^2\,dy = 0$

11. $(x - y)\,dx + (2x + y)\,dy = 0$

12. $y' = \dfrac{x}{y} + \dfrac{y}{x}$

13. Compute the orthogonal trajectories of the one-parameter family of curves

$$x^2 + 3xy + y^2 = c.$$

14. Find a curve in the xy plane passing through the point $(1, 2)$ and whose slope at any point (x, y) is given by $-(x + 2y)/y$.

15. Show that any differential equation of the form

$$y' = \frac{ax + by + c}{Ax + By + C}, \qquad \text{with } aB - bA \neq 0,$$

can be reduced to a homogeneous differential equation by means of the transformation

$$x = X + x_0$$
$$y = Y + y_0,$$

where (x_0, y_0) is the solution of the system

$$ax + by + c = 0$$
$$Ax + By + C = 0.$$

Using the method of Exercise 15, solve each of the following differential equations.

16. $y' = \dfrac{x - y}{x + y + 2}$

17. $y' = \dfrac{2x + y - 4}{x - y + 1}$

18. $y' = \dfrac{3x - 2y + 7}{2x + 3y + 9}$

19. $y' = \dfrac{5x - y - 2}{x + y + 4}$

20. $y' = \dfrac{x - y + 5}{2x - y - 3}$

21. $y' = \dfrac{-x + y + 1}{3x - y - 1}$

22. $y' = \dfrac{y}{x - y + 1}$

23. $y' = \dfrac{2x}{x - y + 1}$

Solve the following initial value problems.

24. $y' = -\dfrac{x + 2y}{y}$
$y(1) = 1$

25. $(x^2 + y^2)dx - 2xy\,dy = 0$
$y(1) = 1$

26. $y' = \sqrt{\dfrac{x + y}{2x}}$
$y(1) = 2$

27. $y' = \dfrac{2x + y - 4}{x - y + 1}$
$y(2) = 2$

1.7 EQUATIONS REDUCIBLE TO FIRST ORDER

In this section we study two types of higher-order ordinary differential equations that can be reduced to first-order equations by means of simple transformations.

A. Equations of the form

$$y^{(n)} = F(x, y^{(n-1)}), \tag{1}$$

containing only two consecutive derivatives $y^{(n)}$ and $y^{(n-1)}$ can be reduced to first order by means of the transformation

$$w = y^{(n-1)} \tag{2}$$

In fact, differentiating both sides of Eq. (2) with respect to x we find $w' = y^{(n)}$, and using (1) we obtain

$$w' = F(x, w).$$

EXAMPLE 1 Compute the general solution of the differential equation

$$y''' - \frac{1}{x}y'' = 0. \tag{3}$$

Solution Setting $w = y''$, Eq. (3) becomes

$$w' - \frac{1}{x}w = 0. \tag{4}$$

Equation (4) is separable (and linear) with general solution

$$w(x) = c_1 x.$$

Thus, $y'' = c_1 x$. Integrating with respect to x, we obtain $y' = \frac{1}{2}c_1 x^2 + c_2$.

Another integration yields $y = \frac{1}{6} c_1x^3 + c_2x + c_3$. As c_1 is an arbitrary constant, the general solution of Eq. (3) is

$$y(x) = c_1x^3 + c_2x + c_3. \tag{5}$$

REMARK 1 The general solution of an ordinary differential equation of order n contains n arbitrary constants. For example, Eq. (3) is of order 3, and as we have seen, its general solution (5) contains the three arbitrary constants c_1, c_2, and c_3.

B. Second-order differential equations of the form

$$y'' = F(y, y') \tag{6}$$

(they should not contain x) can be reduced to first order by means of the transformation

$$w = y'. \tag{7}$$

In fact, from (7) we obtain (using the chain rule)

$$y'' = \frac{dw}{dx} = \frac{dw}{dy} \cdot \frac{dy}{dx} = \frac{dw}{dy} w.$$

Thus, Eq. (6) becomes

$$w\frac{dw}{dy} = F(y, w),$$

which is first order with independent variable y and unknown function w. Sometimes this latter equation can be solved by one of our earlier methods.

EXAMPLE 2 Solve the IVP

$$y'' = y'(y' + y) \tag{8}$$

$$y(0) = 0$$

$$y'(0) = -1.$$

Solution Here the differential equation (8) does not contain x and therefore can be reduced to a first-order differential equation by means of the transformation $w = y'$. In fact, Eq. (8) becomes

$$\frac{dw}{dy} - w = y, \tag{9}$$

which is linear. Multiplying both sides of (9) by e^{-y} we obtain $(d/dy)(we^{-y}) = ye^{-y}$. Integrating with respect to y we find $we^{-y} = -ye^{-y} - e^{-y} + c_1$, and so

$$y'(x) = w = -y - 1 + c_1e^y.$$

Using the initial conditions, we find that $c_1 = 0$. Thus,

$$y' = -y - 1$$

and $y(x) = -1 + ce^{-x}$. Using the initial condition $y(0) = 0$, we find that $c = 1$, and the solution of Eq. (8) is

$$y(x) = -1 + e^{-x}.$$

APPLICATION 1.7.1

Chemistry
Diffusion between Two Compartments

■ Assume that two compartments A_1 and A_2, of volumes V_1 and V_2, respectively, are separated by a barrier. Through the barrier a solute can diffuse from one compartment to the other at a rate proportional to the difference $c_1 - c_2$ in concentration of the two compartments, from the higher concentration to the lower. Find the concentration in each compartment at any time t.

Solution Let $y_1(t)$ and $y_2(t)$ be the amount of the solute in the compartments A_1 and A_2, respectively, at time t. Then

$$c_1(t) = \frac{y_1(t)}{V_1} \quad \text{and} \quad c_2(t) = \frac{y_2(t)}{V_2}$$

are the concentrations in compartments A_1 and A_2, respectively. The differential equations describing the diffusion are

$$\dot{y}_1(t) = k(c_2 - c_1) \quad \text{and} \quad \dot{y}_2 = k(c_1 - c_2),$$

where k is a constant of proportionality ($k > 0$). Dividing by the volumes of the compartments, we get

$$\dot{c}_1 = \frac{k}{V_1}(c_2 - c_1) \quad \text{and} \quad \dot{c}_2 = \frac{k}{V_2}(c_1 - c_2) = -\frac{\dot{y}_1}{V_2} = -\frac{V_1}{V_2}\dot{c}_1. \tag{10}$$

Differentiating both sides of the first equation and using the second equation, we find that

$$\ddot{c}_1 = \frac{k}{V_1}(\dot{c}_2 - \dot{c}_1) = -k\left(\frac{1}{V_1} + \frac{1}{V_2}\right)\dot{c}_1$$

or

$$\ddot{c}_1 + k\left(\frac{1}{V_1} + \frac{1}{V_2}\right)\dot{c}_1 = 0. \tag{11}$$

Equation (11) is a differential equation of form A and can be reduced to a first-order differential equation. We leave the computational details for Exercises 5, 6, and 7.

EXERCISES

Compute the general solution of the following differential equations.

1. $y'' + y' = 3$

2. $y^{(5)} - y^{(4)} = 0$

Solve the following initial value problems.

3. $y'' = \dfrac{1 + y'^2}{2y}$

$y(0) = 1$

$y'(0) = -1$

4. $y'' + y = 0$

$y(0) = 0$

$y'(0) = 1$

In Exercises 5 through 7, let $c_1(0)$ and $c_2(0)$ be the initial concentrations in the two compartments in the chemistry application. Show that:

5. $c_1(t) = c_1(0) - \dfrac{V_2}{V_1 + V_2}[c_2(0) - c_1(0)]\left\{\exp\left[-k\left(\dfrac{1}{V_1} + \dfrac{1}{V_2}\right)t\right] - 1\right\}$

6. $c_2(t) = c_2(0) - \dfrac{V_1}{V_1 + V_2}[c_1(0) - c_2(0)]\left\{\exp\left[-k\left(\dfrac{1}{V_1} + \dfrac{1}{V_2}\right)t\right] - 1\right\}$

7. $c_1(\infty) = c_2(\infty) = \dfrac{V_1c_1(0) + V_2c_2(0)}{V_1 + V_2}$. That is, after a long time, the concentrations in the two compartments are equal, and so equilibrium would be reached.

Find the general solution of the following differential equations.

8. $xy'' + y' = 3$

9. $y^{(5)} - \dfrac{1}{x}y^{(4)} = 0$

10. $y''' + y'' = 1$

11. $y'' - y = 0$

12. $y'' = 2y' + 2y'y$

13. $y'' - 2y^{-3}y' = 0$

Astrophysics Differential equations of the form

$$y'' = (1 + y'^2)\,f(x, y, y')$$

have been obtained in connection with the study of orbits of satellites.[19] Solve this differential equation in each of the following cases.

14. $f(x, y, y') = \dfrac{1}{y}$

15. $f(x, y, y') = y'$

16. $f(x, y, y') = 1$

REVIEW EXERCISES

In Exercises 1 through 7, solve the initial value problem.

1. $(1 + x)y\,dx + x\,dy = 0$

$y(1) = 2$

2. $y' = 1 + y^2, \quad \dfrac{-\pi}{2} < x < \dfrac{\pi}{2}$

$y\left(\dfrac{\pi}{4}\right) = 1$

[19] See *Notices AMS* (Jan. 1975): A142.

3. $(1 + x^2)y' - 2xy = (x^2 + 1)^2$
$y(0) = 0$

4. $2xyy' - 2y^2 + x = 0$
$y(1) = 1$

5. $(2x^2 + y^2)dx + 2xy\,dy = 0$
$y(1) = -1$

6. $y' = \dfrac{y^2 - x^2 - x}{2xy}$
$y(1) = 1$

7. $xy'' + y' = 2x$
$y(1) = 2$
$y'(1) = 0$

In Exercises 8 through 21, solve the differential equation.

8. $\dfrac{dx}{dt} = (x - 2)(x - 3)$

9. $\dfrac{dx}{dt} = \dfrac{x - 2}{x - 3}$

10. $\dfrac{dx}{dt} = x - x^2$

11. $\dfrac{dx}{dt} = \dfrac{\cos t}{\cos x}$

12. $y' + y = \dfrac{x}{y}$ (Bernoulli)

13. $y' = (1 + x + x^2) - (1 + 2x)y + y^2$
(Riccati with particular solution $y_1(x) = x$)

14. $\dfrac{dx}{dt} = \dfrac{3x^{5/6}}{(1 - t)^{3/2}}$

15. $(3x - 2y)y\,dx = x^2dy$

16. $(x^2 + 2y)\,dx - x\,dy = 0$

17. $xy' - y = x^2e^x$

18. $xy'' - y' = x^2e^x$

19. $(x^4 + y^4)\,dx + (4x - y)y^3dy = 0$

20. $(x + y + 1)\,dx + (3x + y + 1)\,dy = 0$

21. $(x - 2y)\,dx + (y - 2x)\,dy = 0$

22. Show that the IVP

$$y' = \frac{x^2 + 2y}{x}$$
$$y(1) = 0$$

has a unique solution in some interval of the form $1 - h < x < 1 + h$ with $h > 0$.

23. Find the largest interval of existence of the solution of the IVP in Exercise 22.

24. Compute the orthogonal trajectories of the one-parameter family of curves $x^2 + y^2 = 2cy$.

25. Find the solution of the IVP

$$y' + \frac{1}{x}y = b(x)$$

$$y(1) = 0,$$

where

$$b(x) = \begin{cases} 1, & x \leq 1 \\ \dfrac{1}{x}, & x > 1. \end{cases}$$

26. Compute the orthogonal trajectories of the one-parameter family of curves $x^2 + b^2y^2 = 1$.

27. Circuits In the RL-series circuit shown in Figure 1.14, assume that $V(t) = V_0 \sin \omega t$ and $I(0) = 0$. Show that

$$I(t) = \frac{V_0 \omega L}{R^2 + \omega^2 L^2} e^{-Rt/L} + \frac{V_0}{(R^2 + \omega^2 L^2)^{1/2}} \sin(\omega t - \phi),$$

where ϕ is an angle defined by

$$\sin \phi = \frac{\omega L}{(R^2 + \omega^2 L^2)^{1/2}}, \quad \cos \phi = \frac{R}{(R^2 + \omega^2 L^2)^{1/2}}.$$

Note that the current $I(t)$ is the sum of two terms; as $t \to \infty$, the first term dies out and the second term dominates. For this reason the first term is called the *transient* and the second term the *steady state.*

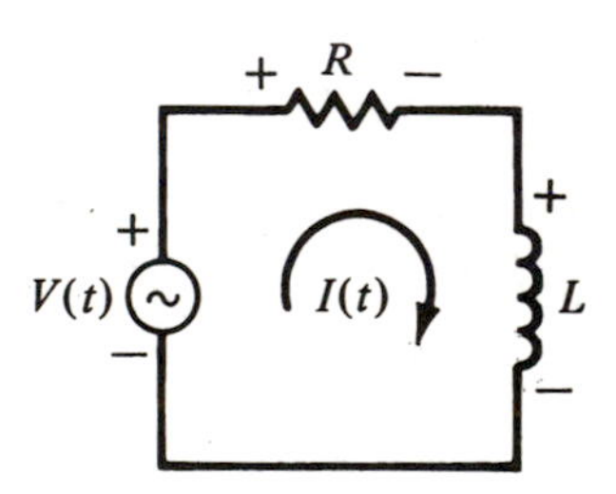

Figure 1.14

28. A colony of bacteria increases at a rate proportional to the number present. If the number of bacteria doubles in 3 hours, how long will it take for the bacteria to triple?

29. A colony of bacteria increases at a rate proportional to the number present. In 1 hour their number increases from 2000 to 5000. How long does it take this colony of bacteria to double?

30. The sum of $4000 is invested at the rate of 12% compounded continuously. What will the amount be after 6 years?

31. It is known that the half-life of radiocarbon is 5568 years. How old is a wooden archaeological specimen which has lost 15% of its original radiocarbon?

32. Find a curve in the xy plane, passing through the origin and whose slope at any point (x, y) is given by $y + e^x$.

33. A large tank contains 30 gallons of brine in which 10 pounds of salt is dissolved. Brine containing 2 pounds of dissolved salt per gallon runs into the tank at the rate of 3 gallons per minute. The mixture, kept uniform by stirring, runs out of the tank at the rate of 2 gallons per minute.
(a) How much salt is in the tank at any time t?
(b) How much salt will be in the tank after 10 minutes?

34. In the circuit[20] shown in Figure 1.15, calculate and sketch the current i for $t \geq 0$, knowing that $i(0) = -10$ mA and given $R = 500\ \Omega$, $L = 10$ mH, and $e = 8$ volts. [*Note:* 1mA $= 10^{-3}$ A and 1mH $= 10^{-3}$ H]

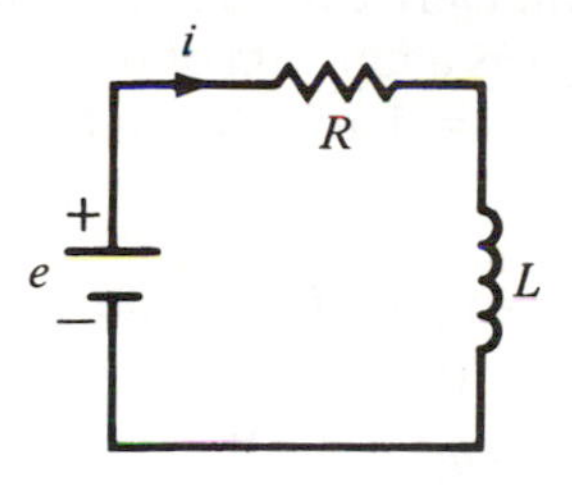

Figure 1.15

35. Radioactive ^{14}C is produced[21] in the earth's atmosphere as a result of bombardment with cosmic rays from outer space. The radioactive carbon is taken up by living things, with the result that they contain 10 disintegrations per minute of ^{14}C per g of carbon. On their death, this ^{14}C decays with a half-life of 5568 years, and no fresh ^{14}C is absorbed. Carbon derived from wood taken from an Egyptian tomb proved to have a ^{14}C content of 7.62 disintegrations per min. per g. What was the age of the tomb?

36. In Figure 1.16, find and sketch[22] the voltage v_2 across C_2 as a function of time after the switch is closed. Initial voltage across C_1 and C_2 is zero. [*Note:* 1μf $= 10^{-6}$ f; 1 kilohm $= 10^3\ \Omega$.]

[20]This is Exercise 12a in C. A. Desoer and E. S. Kuh, *Basic Circuit Theory* (New York: McGraw-Hill Book Co., 1969), p. 169. Reprinted by permission of McGraw-Hill Book Company.
[21]This is Exercise 9 in B. G. Harvey, *Introduction to Nuclear Physics and Chemistry,* 2nd ed. (Englewood Cliffs, N.J.: Prentice-Hall, 1969), p. 54. Reprinted by permission of Prentice-Hall, Inc.
[22]This is Exercise 5–33 in J. L. Shearer, A. T. Murphy, and H. H. Richardson, *Introduction to System Dynamics* (Reading, Mass.: Addison-Wesley, 1971), p. 143. © 1971 Addison-Wesley Publishing Company, Inc. Reprinted with permission of Addison-Wesley Publishing Company.

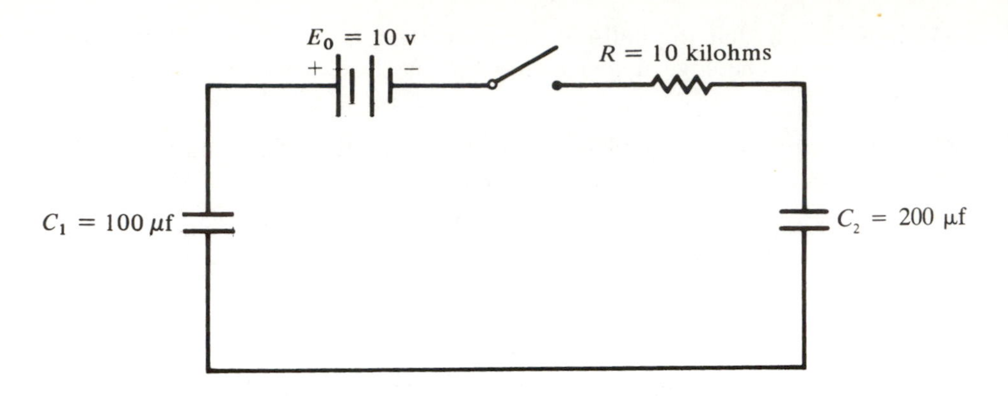

Figure 1.16

37. The capacitor C in the circuit[23] shown in Figure 1.17 is charged to 10 volts when the switch is thrown. Derive the differential equation for the capacitor voltage. Solve the voltage as a function of time and sketch a plot of it, given that $R = 1$ kilohm and $C = 1$ μf. (*Note:* $1\mu\text{f} = 10^{-6}$ f; 1 kilohm $= 10^3$ Ω.)

38. Show[24] that the general solution of the differential equation

$$C\frac{dv}{dt} + \frac{1}{R}v = A_1 \cos(\omega t + \phi_1)$$

is of the form

$$v(t) = K_1 e^{-(1/RC)t} + A_2 \cos(\omega t + \phi_2),$$

where K_1 is an arbitrary constant,

$$A_2 = \frac{A_1}{\sqrt{(1/R)^2 + (\omega C)^2}},$$

and

$$\phi_2 = \phi_1 - \tan^{-1} \omega RC.$$

39. A yacht[25] having a displacement (weight) of 15,000 lb is drifting toward a collision with a pier at a speed of 2 ft/sec. Over what minimum distance can a 150 lb crewman hope to stop the yacht by exerting an average force equal to his weight?

[23]This is Exercise 5–28 in Shearer et al., *System Dynamics,* p. 141. © 1971 Addison-Wesley Publishing Company, Inc. Reprinted with permission of Addison-Wesley Publishing Company.
[24]This exercise is developed from material found in Desoer and Kuh, *Basic Circuit Theory,* pp. 121–22. Used by permission of McGraw-Hill Book Company.
[25]This is Exercise 2-1.3 in J. Norwood, Jr., *Intermediate Classical Mechanics* (Englewood Cliffs, N.J.: Prentice-Hall, 1979), p. 49. Reprinted by permission of Prentice-Hall, Inc.

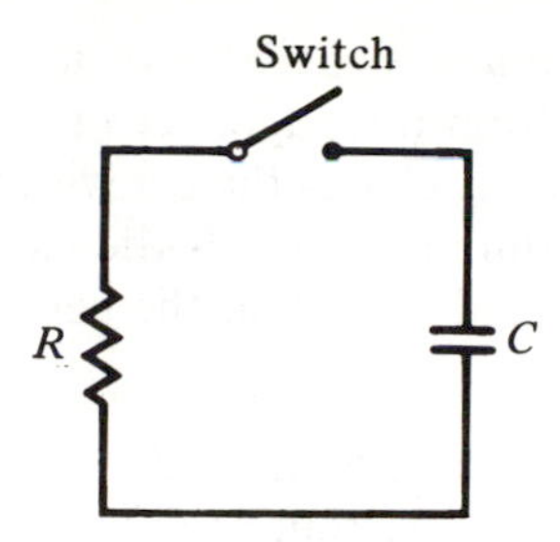

Figure 1.17

40. A particle[26] is moving on a straight line with constant acceleration f. If x is the distance from some fixed point on the straight line, show that the equation $d^2x/dt^2 = f$ can be written as

$$v\frac{dv}{dx} = f,$$

where v is the particle's speed. If $v = u$ at $x = 0$, integrate this equation and show that

$$v^2 = u^2 + 2f \cdot x.$$

41. A ball[27] is thrown vertically downwards from the top of a tall building. Assuming a model with constant gravity and air resistance proportional to its speed, show that if the building is sufficiently tall, the ball's velocity on hitting the ground is approximately independent of its initial speed.

42. A hawk, eagle, and sparrow are in the air.[28] The eagle is 50 feet above the sparrow and the hawk is 100 feet below the sparrow. The sparrow flies straight forward in a horizontal line. Both hawk and eagle fly directly towards the sparrow. The hawk flies twice as fast as the sparrow. The hawk and eagle reach the sparrow at the same time. How far does each fly and at what rate does the eagle fly?

43. Solve the differential equations[29]

$$y' = \frac{x^2 - y^2 + 1}{x^2 - y^2 - 1}, \qquad y' = -\frac{y(2x + y - 1)}{x(x + 2y - 1)}.$$

[*Hint:* The first has $(x + y)^{-1}$ as an integrating factor; the second is exact.]

[26]This is Problem 2.4 in D. N. Burghes and A. M. Downs, *Modern Introduction to Classical Mechanics and Control* (West Sussex, England: Ellis Horwood, 1975), p. 31. Reprinted by permission of Ellis Horwood Limited, Publishers.

[27]This is Example 5.1 in Burghes and Downs, *Classical Mechanics*, p. 109. Reprinted by permission of Ellis Horwood Limited, Publishers.

[28]From *Amer. Math. Monthly* 40 (1933): 436–37.

[29]From *Amer. Math. Monthly* 50 (1943): 572.

44. A cat[30] is running along a straight edge of a garden. A dog, sitting in the garden at a distance b from the edge, sees the cat when it is at its nearest point. The dog immediately chases the cat with twice the cat's speed in such a way that it is always running towards the cat. Find the time that elapses before the cat is caught and show that the cat runs a distance $(2/3)b$ before being caught.

45. In an RL-series circuit, $L = 2$ henries, $R = 4$ ohms, $V(t) = e^{-t}$ volts, and $I(0) = 5$ amperes. Find the current at any time t.

[30]This is Exercise 2.17 in Burghes and Downs, *Classical Mechanics,* p. 34. Reprinted by permission of Ellis Horwood Limited, Publishers.

Linear Differential Equations

2.1 INTRODUCTION AND DEFINITIONS

A linear differential equation of order n is a differential equation of the form

$$a_n(x)\, y^{(n)} + a_{n-1}(x)\, y^{(n-1)} + \cdots + a_1(x)\, y' + a_0(x)\, y = f(x). \tag{1}$$

We always assume that the coefficients $a_n(x), a_{n-1}(x), \ldots, a_0(x)$ and the function $f(x)$ are continuous functions in some interval I and that the leading coefficient $a_n(x) \neq 0$ for all $x \in I$. The interval I is called the *interval of definition of the differential equation.* When the function f is identically zero, we say that Eq. (1) is *homogeneous.* When $f(x)$ is not identically zero, Eq. (1) is called *nonhomogeneous.* If all the coefficients $a_n(x), a_{n-1}(x), \ldots, a_0(x)$ are constants, we speak of Eq. (1) as a linear differential equation with *constant coefficients;* otherwise, it is a linear differential equation with *variable coefficients.* The following are examples of linear differential equations:

$$xy' - 2y = x^3, \quad x \neq 0 \tag{2}$$

$$y'' + 2y' + 3y = \cos x \tag{3}$$

$$y^{(4)} - y = 0. \tag{4}$$

Equation (2) is a nonhomogeneous linear differential equation of order 1 with variable coefficients. Equation (3) is a nonhomogeneous linear differential equation of order 2 with constant coefficients. Equation (4) is a homogeneous linear differential equation of order 4 with constant coefficients. The term *linear* refers to the fact that each expression in the differential equation is of degree *one* or degree *zero* in the variables $y, y', \ldots, y^{(n)}$. The following are nonlinear differential equations:

$$y'' + y^2 = \sin x$$

$$y''' + yy' = x$$

$$y'' + \sin y = 0.$$

The first differential equation is nonlinear because of the term y^2, the second because of the term yy', and the third because of the term $\sin y$ [recall that $\sin y = y - (y^3/3!) + (y^5/5!) - \cdots$].

Our main concern in this chapter, aside from applications, is to develop the elements of the theory of solutions of linear differential equations and to discuss methods for obtaining their general solution.

Of course, we already know how to find the general solution of any linear differential equation of order 1 with variable or constant coefficients. This was done in Section 1.4 of Chapter 1. For example, the general solution of Eq. (2) is easily found to be

$$y(x) = x^3 + cx^2, \tag{5}$$

where c is an arbitrary constant.

Furthermore (subject to some algebraic difficulties only), we shall learn in Sections 2.4, 2.5, and 2.11 how to find the general solution of any linear differential equation with constant coefficients, no matter what the order of the differential equation is.

However, matters are not so simple when we try to solve linear differential equations with variable coefficients of order 2 or higher. In fact, there is *no way* to find explicitly the general solution of a second-order (or higher) differential equation with *variable coefficients* unless the differential equation is of a very special form (for example, of the Euler type, Section 2.7) or unless we know (or can guess) one of its solutions (see Section 2.8). A practical way out of this difficulty is to approximate the solutions of differential equations with variable coefficients. For example, the variable coefficients of a differential equation can be approximated by constant coefficients, in "small" intervals, and hopefully the solution of the differential equation with constant coefficients is an "approximation" to the solution of the original differential equation with variable coefficients. Another way, which will give a better approximation, is to look for a power-series solution of the differential equation with variable coefficients (see Section 2.9 and Chapter 5). A direct substitution of the power-series solution into the differential equation will enable us to compute as many coefficients of the power series as we please in our approximation. Finally, the reader should be told that in more advanced topics on differential equations, called "qualitative theory of differential equations," one can discover many properties of the solutions of differential equations with variable coefficients directly from the properties of these coefficients and without the luxury of knowing explicitly the solutions of the differential equation.

For the sake of motivation, let us consider the second-order linear differential equation

$$\ddot{y} + a^2 y = 0, \tag{6}$$

together with the initial conditions

$$y(0) = y_0, \qquad \dot{y}(0) = v_0. \tag{7}$$

As is customary, dots stand for derivatives with respect to time t. It will be shown in the applications that the *initial value problem (IVP)* (6)–(7) describes the motion of a vibrating spring [Eq. (6) is Eq. (10) with $a^2 = K/m$]. It is easy to verify that the function $y_1(t) = \cos at$ when substituted into Eq. (6) reduces the differential equation to an identity. In fact, $y_1 = \cos at \Rightarrow \dot{y}_1 = -a \sin at \Rightarrow \ddot{y}_1 = -a^2 \cos at$, and substitution of the results into Eq. (6) yields $-a^2 \cos at + a^2 \cos at = 0$, or $0 = 0$. Hence, y_1 is a solution. In a similar fashion we can verify that the function $y_2(t) = \sin at$ is also a solution of the differential equation (6). Furthermore, the sum of the two solutions is also a solution, and any constant multiple of any one of these solutions is again a solution. Thus, for any constants c_1 and c_2, the linear combination

$$y(t) = c_1y_1(t) + c_2y_2(t) \tag{8}$$

of two solutions is also a solution.

The preceding comments serve to distinguish linear homogeneous differential equations from other types of differential equations. For linear homogeneous differential equations we have the principle that if y_1 and y_2 are solutions and if c_1 and c_2 are arbitrary constants, then the *linear combination* $c_1y_1 + c_2y_2$ is also a solution. This result and its proof are presented as Theorem 1 of Section 2.2.

Note that the functions $y_1 = \cos at$ and $y_2 = \sin at$, although they satisfy the differential equation (6), *do not*, in general, satisfy the initial conditions, since $\dot{y}_1(0) = 0 (\neq v_0)$ and $y_2(0) = 0\ (\neq y_0)$. It seems reasonable, however, to suspect that if the constants c_1 and c_2 are chosen properly, then $y(t)$ as given in (8) will be a solution of the initial value problem (6)–(7). Now $y(t)$ satisfies the differential equation (6) and will satisfy the initial conditions provided that c_1 and c_2 satisfy the system of algebraic equations

$$y_0 = y(0) = c_1y_1(0) + c_2y_2(0)$$
$$v_0 = \dot{y}(0) = c_1\dot{y}_1(0) + c_2\dot{y}_2(0).$$

Since $y_1(0) = 1$, $\dot{y}_1(0) = 0$, $y_2(0) = 0$, and $\dot{y}_2(0) = a$, this system takes the form

$$y_0 = c_1$$
$$v_0 = ac_2.$$

Hence, a solution to the initial value problem is

$$y(t) = y_0 \cos at + \frac{v_0}{a} \sin at. \tag{9}$$

In Section 2.3 we establish that Eq. (9) is the *only* solution of the IVP (6)–(7).

It would be natural to ask what motivated us to try cos *at* as a solution. Or we might ask whether there are simple solutions other than cos *at* and sin *at*. [The reader can verify that e^{iat} and e^{-iat}, where $i^2 = -1$, are solutions of Eq. (6).] How many "different" solutions are there to the differential equation (6)? Is (9) the only solution to the initial value problem, or would some other choice of y_1 and y_2 produce different values for c_1 and c_2? [For example, if $y_1 = e^{iat}$, $y_2 = e^{-iat}$, $c_1 = \frac{1}{2}y_0 - (i/2a)v_0$, $c_2 = \frac{1}{2}y_0 + (i/2a)v_0$, then $c_1y_1 + c_2y_2$ also solves the initial value problem (6)–(7).]

The answers to these questions will be obtained as we proceed through the remainder of this chapter.

APPLICATIONS 2.1.1

Linear differential equations occur in many mathematical models of real-life situations. Newton's second law of motion, for example, involves a second derivative (the acceleration), and consequently linear differential equations of second order have been of primary interest in motion problems.

Mechanics
The Vibrating Spring

■ A light coil spring of natural length L is suspended from a point on the ceiling [Figure 2.1(a)].

We attach a mass m to the lower end of the spring and allow the spring to come to an equilibrium position [Figure 2.1(b)]. Assume that at the equilibrium position the length of the spring is $L + l$. Naturally the stretch l is caused by the weight mg of the mass m. According to Hooke's law, the force needed to produce a stretch of length l is proportional to l. That is, $mg = Kl$, with the constant of proportionality, K, known as the *spring constant*. Next we displace the mass vertically (up or down) to a fixed (initial) position, and at that point we give it a vertical (initial) velocity (directed up or down), setting the mass in a vertical motion [Figure 2.1(c)]. It is this motion that we wish to investigate. For convenience, we choose the origin 0 at the equilibrium point, and the positive direction for the vertical (y) axis is downward.

Let $y = y(t)$ be the position of the mass at time t. Then by Newton's second law of motion, we have

$$m\ddot{y} = F,$$

where the double dots denote second derivative with respect to time (in other words, the acceleration of the mass). There are two forces acting upon the mass: one is the force of gravity, equal to mg, and the other is the restoring force of the spring, which by Hooke's law is $K(l + y)$. Thus (since $mg = Kl$),

$$F = mg - K(l + y) = -Ky.$$

The differential equation that describes the motion of the mass m is

$$m\ddot{y} = -Ky,$$

or

$$\ddot{y} + \frac{K}{m}y = 0. \tag{10}$$

Recall that the mass was set to motion by first displacing it to an initial

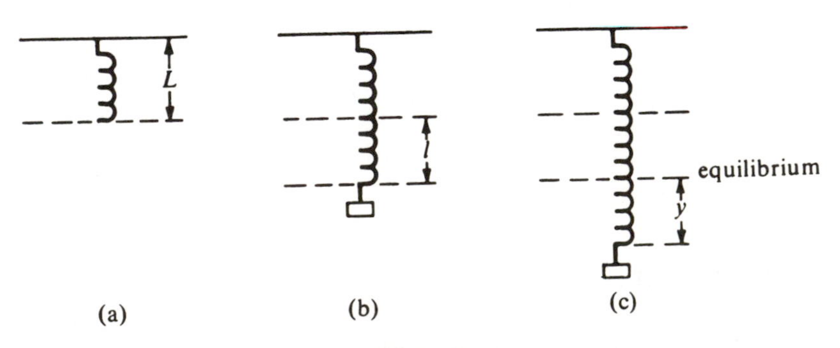

Figure 2.1

position, say y_0, and then giving it an initial velocity v_0. Thus, together with the differential equation (10), we have the two initial conditions

$$y(0) = y_0, \qquad \dot{y}(0) = v_0. \tag{11}$$

In Eq. (11) y_0 is positive if the initial position is below the equilibrium and negative if it is above. Also, v_0 is positive if the initial velocity was directed downward and negative if it was directed upward.

Equation (10) we recognize to be a second-order, homogeneous, linear differential equation with constant coefficients. The differential equation (10), together with the initial conditions (11), constitute an initial value problem.

In Table 2.1 we list the units used in the two most common systems of measurement.

	Cgs system	English system
Distance	centimeter (cm)	foot (ft)
Mass	gram (g)	slug
Time	seconds (sec)	seconds (sec)
Force	dyne (dyn)	pound (lb)
Velocity	cm/sec	ft/sec
Acceleration	cm/sec^2	ft/sec^2
Spring constant	dyn/cm	lb/ft
Gravity (g)	980 cm/sec^2	32 ft/sec^2

Table 2.1 Units of Measurement

EXAMPLE 1 A spring is stretched 12 cm by a force equal to 360 dyn. A mass of 300 g is attached to the end of the spring and released 3 cm above the point of equilibrium with initial velocity 5 cm/sec directed downward. Find the position of the mass at any time t.

Solution Using Hooke's law we compute the spring constant:

$$K = \tfrac{360}{12}\text{ dyn/cm} = 30 \text{ dyn/cm}.$$

The IVP describing the motion of the mass is [from Eqs. (10) and (11)]

$$\begin{aligned} \ddot{y} + \tfrac{30}{300}y &= 0 \\ y(0) &= -3 \\ \dot{y}(0) &= 5. \end{aligned}$$

From Eq. (9) the solution of this IVP is

$$y(t) = -3\cos\frac{t}{\sqrt{10}} + 5\sqrt{10}\sin\frac{t}{\sqrt{10}}.$$

The reader is urged to read Exercise 40 of Section 2.4, where the vibrating spring is discussed in greater detail.

Diffusion

■ Suppose that a gas diffuses into a liquid in a long and narrow pipe. Suppose that this process takes place for such a long period of time that the concentration $y(x)$ of gas in the pipe depends only on the distance x from some initial point 0 (and is independent of time). Assume that the gas reacts chemically with the liquid and that the amount $H(x)$ of gas that disappears by this reaction is proportional to $y(x)$. That is,

$$H(x) = ky(x). \tag{12}$$

We want to compute the concentration $y(x)$ of gas in the liquid, at any point x, given that we know the concentration at the points 0 and l, say

$$y(0) = A \qquad \text{and} \qquad y(l) = 0. \tag{13}$$

To solve this problem we consider a small section S of the pipe between the point x and $x + \Delta x$ (Figure 2.2). Then the amount $G(x)$ of gas diffusing into S at the point x minus the amount $G(x + \Delta x)$ diffusing out of S at the point $x + \Delta x$ must be equal to the amount of gas disappearing in S by the chemical reaction. To compute $G(x)$ and $G(x + \Delta x)$, we use Fick's law of diffusion, which states that the amount of gas that passes through a unit area per unit time is proportional to the rate of change of the concentration. That is,

$$G(x) = -Dy'(x), \tag{14}$$

where $D > 0$ is the *diffusion coefficient* and the negative sign justifies the fact that the gas diffuses from regions of high concentration to regions of low concentration. Similarly,

$$G(x + \Delta x) = -Dy'(x + \Delta x). \tag{15}$$

Equation (12) gives the amount of gas that disappears at the point x because of the chemical reaction of the gas with the liquid. Therefore, the total amount of gas that disappears by the chemical reaction in the section S is given by

$$\int_x^{x+\Delta x} ky(s)ds. \tag{16}$$

Thus,

$$[-Dy'(x)] - [-Dy'(x + \Delta x)] = \int_x^{x+\Delta x} ky(s)ds. \tag{17}$$

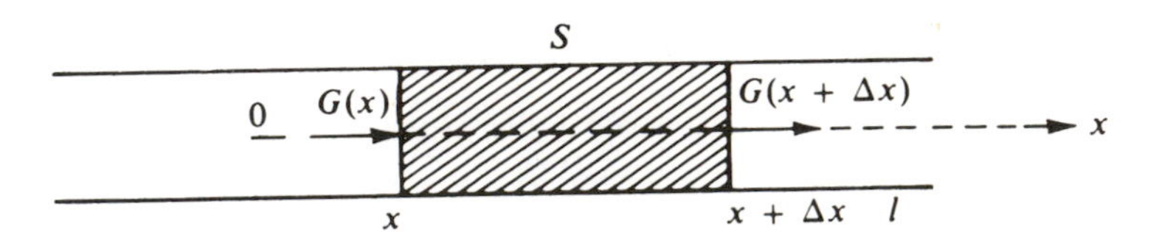

Figure 2.2

A good approximation to the integral in (16) is given by $ky(x + \Delta x/2) \cdot \Delta x$. Thus after some rearrangement Eq. (17) becomes

$$\frac{y'(x + \Delta x) - y'(x)}{\Delta x} = \frac{k}{D} y\left(x + \frac{\Delta x}{2}\right), \tag{18}$$

where $y[x + (\Delta x/2)]$ is the value of y at the midpoint between the limits of integration and Δx the length of the interval of integration. Taking limits on both sides of Eq. (18) as $\Delta x \to 0$, we obtain the second-order linear differential equation

$$y'' - \frac{k}{D} y = 0. \tag{19}$$

The problem consisting of the differential equation (19) and the "boundary" conditions (13) is referred to as a *boundary value problem (BVP)*. Boundary value problems for ordinary differential equations are discussed in Chapter 6 and boundary value problems for partial differential equations are discussed in Chapter 11.

EXERCISES

In Exercises 1 through 19, classify the differential equation as being either linear or nonlinear. Furthermore, classify the linear ones as homogeneous or nonhomogeneous, with constant coefficients or with variable coefficients, and state its order. (Note that derivatives are denoted by primes, lowercase roman numerals, or by arabic numbers in parentheses.)

1. $y'' + xy = 0$

2. $y''' + x^2 y = e^x$

3. $y'' + yy^{iv} = 5$

4. $y'' + \cos y = 0$

5. $y^{(5)} - 2y^{(4)} + y = 2x^2 + 3$

6. $(1 + x^2)y'' - xy' + y = 0$

7. $(\sin x)y' + e^{x^2} y = 1$

8. $2y''' + 3y'' - 4y' + xy = 0$

9. $x^2 y'' - e^x y' - 2 = 0$

10. $y' + y^{1/2} = 3x$

11. $y'' + xy = x$

12. $2y - 3xy'' + 4y' = 0$

13. $y''' - 2 = 0$

14. $e^x (y')^2 + 3y = 0$

15. $y'' + 5y' - 6y = 0$

16. $yy' = 3$

17. $xy''' + 4xy'' - xy = 1,\ x > 0$

18. $7y' - xy = 0$

19. $7y'' + 3xy' - 4y = 0$

20. A spring is stretched 2 cm by a force equal to 16 dyn. A mass of 32 g is attached to the end of the spring. The system is then set to motion by pulling the mass 4 cm above the point of equilibrium and releasing it, at time $t = 0$, with an initial velocity of 2 cm/sec directed downward. Derive an IVP that describes the motion of the mass. Use Eq. (9) to find the solution of this IVP.

21. A spring is stretched 1.5 in. by a 4-lb weight. The system is then set to motion by pushing the mass (note that mg = weight and that $g = 32$ ft/sec^2) 6 in. below the point of equilibrium and releasing it, at time $t = 0$, with an initial velocity of 4 ft/sec directed upward. Use Eq. (9) to find the position of the mass at any time t.

The solution of the vibrating spring is given by Eq. (9),

$$y(t) = y_0 \cos \sqrt{\frac{K}{m}}\,t + \frac{v_0\sqrt{m}}{\sqrt{K}} \sin \sqrt{\frac{K}{m}}\,t.$$

The graph of $y(t)$ is periodic of period $2\pi \sqrt{m}/\sqrt{K}$; consequently, the motion of the mass repeats itself after every time interval of length $2\pi \sqrt{m}/\sqrt{K}$. For this reason we say that the motion is *oscillatory* or that *vibrations* occur. If we recall a technique from trigonometry [that an expression of the form $\alpha \cos t + \beta \sin t$ can be written as $A \cos (t - \phi)$, where $A = \sqrt{\alpha^2 + \beta^2}$ and $\phi = \cos^{-1}(\alpha/\sqrt{\alpha^2 + \beta^2}) = \sin^{-1}(\beta/\sqrt{\alpha^2 + \beta^2})$], the solution can be written in the form

$$y(t) = A \cos\left(\sqrt{\frac{K}{m}}\,t - \phi\right),$$

with $A = \sqrt{y_0^2 + mv_0^2/K}$ and $\phi = \cos^{-1}(y_0/A)$. Hence, we say that the motion is oscillatory of *amplitude* A, period $2\pi\sqrt{m}/\sqrt{K}$, and *phase* ϕ.

In each of Exercises 22, 23, and 24, evaluate the amplitude and the period of the oscillation and sketch the graph of $y(t)$.

22. $y_0 = 5$ in., $K = m$, $v_0 = 0$ in./sec

23. $y_0 = 0$ cm, $K = 5m$, $v_0 = 10\sqrt{5}$ cm/sec

24. $y_0 = 3$ cm, $K = m$, $v_0 = 4$ cm/sec

25. A function f is said to be bounded[1] as $x \to \infty$ if there exists a positive number M such that $\lim_{x\to\infty} f(x) \leq M$. Show that all solutions of the differential equation $y'' + e^x y = 0$ are bounded as $x \to \infty$. *Hint:* Multiply the differential equation by $e^{-x}y'$ and integrate from 0 to T, obtaining

$$y^2(T) + 2\int_0^T e^{-x}y'y''dx = y^2(0).$$

Note that for some ρ such that $0 \leq \rho \leq T$, we must have

$$2\int_0^T e^{-x}y'y''\,dx = 2\int_0^\rho y'y''\,dx = [y'(\rho)]^2 - [y'(0)]^2.$$

Hence,

$$y^2(T) + [y'(\rho)]^2 = y^2(0) + [y'(0)]^2.$$

[1]Problem 25 and its method of solution are taken from the 1966 William Lowell Putnam Mathematical Competition, *Amer. Math. Monthly* 74, no. 7 (1967).

2.2 LINEAR INDEPENDENCE AND WRONSKIANS

In Section 2.1 we posed certain questions concerning the number of solutions to a linear differential equation and to the manner of finding solutions to linear differential equations. In this section and the next we concentrate on answering the question: *How many solutions can a linear differential equation have?* To gain some insight into this question, we consider an example.

EXAMPLE 1 Solve the second-order homogeneous linear differential equation

$$y'' = 0. \tag{1}$$

Solution Because of the simplicity of this equation, we can integrate it directly to obtain solutions. Integrating Eq. (1) with respect to x, we obtain

$$y' = c_1, \tag{2}$$

where c_1 is a constant of integration. Integrating Eq. (2) with respect to x, we have

$$y = c_1x + c_2, \tag{3}$$

where c_2 is another constant of integration. Regardless of the values given to the constants c_1 and c_2, the expression for y given in (3) is a solution of the differential equation (1). Thus, there are an infinity of solutions to the differential equation (1). Notice that $y_1 = x$ and $y_2 = 1$ are solutions of Eq. (1) and Eq. (3) shows that they play a special role in that *every* solution can be expressed as a linear combination of these two solutions. That is, if y is a solution of Eq. (1), then there exist constants c_1 and c_2 such that $y = c_1y_1 + c_2y_2$. Of course, if $y = y_1$, then $c_1 = 1$ and $c_2 = 0$, and if $y = y_2$, then $c_1 = 0$ and $c_2 = 1$. Since it is possible to express every solution in terms of the solutions y_1 and y_2, we speak of y_1 and y_2 as constituting a *fundamental set of solutions* for Eq. (1). Equation (3) is referred to as the *general solution* of Eq. (1). For this differential equation the general solution contains two arbitrary constants, c_1 and c_2. Consequently, if we try to formulate conditions that reduce the infinity of solutions to a single solution, we need two conditions which serve to determine c_1 and c_2. For example, suppose that we want a solution of Eq. (1) that satisfies the initial conditions $y(1) = 1$, $y'(1) = 2$. From Eq. (3) we have

$$1 = y(1) = c_1(1) + c_2,$$

and, differentiating Eq. (3),

$$2 = y'(1) = c_1.$$

Hence, $c_1 = 2$ and $c_2 = -1$, and the solution we seek is

$$y = 2x - 1.$$

Intuitively, we suspect that to solve an nth-order equation requires n integrations, and as a consequence the general solution of such a differential equation involves n constants of integration. Furthermore, we suspect that in this case there should be a set of n special solutions having the property that every solution can be expressed as a linear combination of these special solutions. In what follows these notions will be made more precise.

DEFINITION 1

A collection of m functions $f_1, f_2, \ldots, f_m$, each defined and continuous on the interval $a \le x \le b$ is said to be linearly dependent on $a \le x \le b$ if there exist constants $\alpha_1, \alpha_2, \ldots, \alpha_m$, not all of which are zero, such that

$$\alpha_1 f_1 + \alpha_2 f_2 + \cdots + \alpha_m f_m = 0 \tag{4}$$

for every x in the interval $a \le x \le b$. Otherwise, the functions are said to be linearly independent on this interval.

EXAMPLE 2 (a) Show that the functions $f_1(x) = 3x + (12/5)$ and $f_2(x) = 5x + 4$ are linearly dependent on the interval $(-\infty, \infty)$.

(b) Show that the functions $f_1(x) = x$ and $f_2(x) = x^2$ are linearly independent on the interval $-1 \le x \le 1$.

Solution (a) To show that the functions f_1 and f_2 are linearly dependent we must show that there exist constants α_1 and α_2, not both zero, such that

$$\alpha_1\left(3x + \frac{12}{5}\right) + \alpha_2(5x + 4) = 0$$

for all x in the interval $-\infty < x < \infty$. This is possible if we choose for example, $\alpha_1 = 5$ and $\alpha_2 = -3$.

(b) Otherwise f_1 and f_2 are linearly dependent on the interval $-1 \le x \le 1$. That is, there exist constants α_1 and α_2, not both zero, such that

$$\alpha_1 x + \alpha_2 x^2 = 0$$

for all x in the interval $-1 \le x \le 1$. But for $x = 1$ and $x = -1$ we find

$$\alpha_1 + \alpha_2 = 0 \qquad \text{and} \qquad -\alpha_1 + \alpha_2 = 0.$$

The only solution of this system is $\alpha_1 = \alpha_2 = 0$. This contradicts our assumption that f_1 and f_2 are linearly dependent. Hence, f_1 and f_2 are linearly independent on the interval $-1 \le x \le 1$.

If the functions $f_1, f_2, \ldots, f_n$ are solutions to a linear homogeneous differential equation, there is a simple test that serves to determine whether they are linearly independent or not.

Before we state this test, we need the following definition.

DEFINITION 2

Let $f_1, f_2, \ldots, f_n$ be n functions which together with their first $n - 1$ derivatives are continuous on the interval $a \le x \le b$. The Wronskian of $f_1, f_2, \ldots, f_n$, *evaluated at x, is denoted by* $W(f_1, f_2, \ldots, f_n; x)$ *and is defined to be the determinant*[2]

$$W(f_1, f_2, \ldots, f_n; x) = \begin{vmatrix} f_1 & f_2 & \cdots & f_n \\ f_1' & f_2' & \cdots & f_n' \\ f_1'' & f_2'' & \cdots & f_n'' \\ \cdot & \cdot & \cdot & \cdot \\ \cdot & \cdot & \cdot & \cdot \\ \cdot & \cdot & \cdot & \cdot \\ f_1^{(n-1)} & f_2^{(n-1)} & \cdots & f_n^{(n-1)} \end{vmatrix}$$

Each of the functions appearing in this determinant is to be evaluated at x.

EXAMPLE 3 Given $f_1(x) = x^2$, $f_2(x) = \cos x$, find $W(f_1, f_2; x)$.

Solution From Definition 2 and the functions given we compute

$$W(x^2, \cos x; x) = \begin{vmatrix} x^2 & \cos x \\ 2x & -\sin x \end{vmatrix} = -x^2 \sin x - 2x \cos x.$$

In the introduction to this chapter we mentioned that linear homogeneous differential equations have the property that linear combinations of solutions are also solutions. This principle, which distinguishes linear from nonlinear differential equations, is embodied in the following theorem.

THEOREM 1

If each of the functions $y_1, y_2, \ldots, y_m$ *are solutions to the same linear homogeneous differential equation,*

$$a_n(x)y^{(n)} + a_{n-1}(x)y^{(n-1)} + \cdots + a_1(x)y' + a_0(x)y = 0,$$

then for every choice of the constants $c_1, c_2, \ldots, c_m$ *the linear combination*

$$c_1y_1 + c_2y_2 + \cdots + c_my_m$$

is also a solution.

Proof[3] The proof depends on two basic properties of differentiation,

$$(f_1 + f_2 + \cdots + f_k)' = f_1' + f_2' + \cdots + f_k'$$

[2]For a brief discussion of determinants, their properties, and their application to linear systems of algebraic equations, see Appendix A.

[3]In Chapter 1 we expressed the philosophy that difficult or sophisticated proofs would be omitted in this text. The proof of this theorem is neither difficult nor sophisticated. Nevertheless, we choose not to give a proof under the most general circumstances but rather demonstrate a proof for a specific version of the theorem, with the implication that the general proof follows along similar lines of reasoning. This format will be followed in many places in the text. That is, the theorem will be stated in its general form and the proof given for a specific case.

and

$$(cf)' = cf',$$

where c is a constant.

We present the details for the case $m = 2$, $n = 3$. We assume then that y_1 and y_2 are solutions of the differential equation

$$a_3(x)y''' + a_2(x)y'' + a_1(x)y' + a_0(x)y = 0.$$

Substituting $c_1y_1 + c_2y_2$ into this differential equation, we obtain

$$a_3(x)(c_1y_1 + c_2y_2)''' + a_2(x)(c_1y_1 + c_2y_2)'' + a_1(x)(c_1y_1 + c_2y_2)' + a_0(x)(c_1y_1 + c_2y_2) \stackrel{?}{=} 0$$

$$a_3(x)(c_1y_1''' + c_2y_2''') + a_2(x)(c_1y_1'' + c_2y_2'') + a_1(x)(c_1y_1' + c_2y_2') + a_0(x)(c_1y_1 + c_2y_2) \stackrel{?}{=} 0$$

$$c_1[a_3(x)y_1''' + a_2(x)y_1'' + a_1(x)y_1' + a_0(x)y_1] + c_2[a_3(x)y_2''' + a_2(x)y_2'' + a_1(x)y_2' + a_0(x)y_2] \stackrel{?}{=} 0$$

$$c_1(0) + c_2(0) \stackrel{?}{=} 0$$

$$0 + 0 = 0.$$

Suppose that we are given n functions $y_1, y_2, \ldots, y_n$, each of which is a solution to the same nth-order linear homogeneous differential equation,

$$a_n(x)y^{(n)} + a_{n-1}(x)y^{(n-1)} + \cdots + a_1(x)y' + a_0(x)y = 0. \tag{5}$$

We wish to determine whether or not these functions are linearly dependent. These functions will be linearly dependent if we can find n constants $\alpha_1, \alpha_2, \ldots, \alpha_n$ not all zero such that

$$\alpha_1y_1 + \alpha_2y_2 + \cdots + \alpha_ny_n = 0. \tag{6}$$

Since each y_i is a solution of Eq. (5), the derivatives $y_i, y_i', y_i'', \ldots, y_i^{(n-1)}$ all exist. Equation (6) can be differentiated $n - 1$ times to obtain the system of equations

$$\begin{aligned}
\alpha_1y_1 + \alpha_2y_2 + \cdots + \alpha_ny_n &= 0\\
\alpha_1y_1' + \alpha_2y_2' + \cdots + \alpha_ny_n' &= 0\\
\alpha_1y_1'' + \alpha_2y_2'' + \cdots + \alpha_ny_n'' &= 0\\
\cdots\cdots\cdots\cdots\cdots\cdots\cdots\cdots&\cdots\\
\alpha_1y_1^{(n-1)} + \alpha_2y_2^{(n-1)} + \cdots + \alpha_ny_n^{(n-1)} &= 0.
\end{aligned} \tag{7}$$

We recognize Eqs. (7) as a system of homogeneous linear algebraic equations for the unknowns $\alpha_1, \alpha_2, \ldots, \alpha_n$. Such a system of equations has solutions[4] other than the trivial solution $\alpha_1 = 0, \alpha_2 = 0, \ldots, \alpha_n = 0$ if and only if the determinant of its coefficients is zero. That is, the functions $y_1, y_2, \ldots, y_n$ are linearly dependent if $W(y_1, y_2, \ldots, y_n; x) = 0$ for *every* x in the interval of definition of the differential equation. It can be shown that the converse is also

[4]See Appendix A.

true: if the functions are linearly dependent, then their Wronskian is zero for some x. Therefore we have the following criterion for testing linear independence (and linear dependence).

THEOREM 2

n solutions $y_1, y_2, \ldots, y_n$ of the nth-order differential equation

$$a_n(x)y^{(n)} + a_{n-1}(x)y^{(n-1)} + \cdots + a_1(x)y' + a_0(x)y = 0$$

defined on an interval $a \le x \le b$ are linearly independent if and only if $W(y_1, y_2, \ldots, y_n; x) \neq 0$ for every x in the interval $a \le x \le b$.

For example, the solutions $y_1(x) = \cos x$ and $y_2(x) = \sin x$ of the differential equation $y'' + y = 0$ are linearly independent because their Wronskian

$$\begin{vmatrix} \cos x & \sin x \\ -\sin x & \cos x \end{vmatrix} = \cos^2 x + \sin^2 x = 1$$

is never zero. Also, the solutions e^x and e^{-x} of the differential equation $y'' - y = 0$ are linearly independent because their Wronskian

$$\begin{vmatrix} e^x & e^{-x} \\ e^x & -e^{-x} \end{vmatrix} = -e^0 - e^0 = -2$$

is never zero.

REMARK 1 In Example 2 we demonstrated that the functions x and x^2 are linearly independent on the interval $-1 \le x \le 1$. However, their Wronskian

$$\begin{vmatrix} x & x^2 \\ 1 & 2x \end{vmatrix} = x^2$$

vanishes at the point $x = 0$ of the interval $-1 \le x \le 1$. Does this contradict Theorem 2? No. Rather, there must not exist a second-order linear differential equation with interval of definition $-1 \le x \le 1$ that has x and x^2 as solutions. The reader can verify that $y_1 = x$ and $y_2 = x^2$ are solutions of the second-order linear differential equation

$$x^2y'' - 2xy' + 2y = 0.$$

However, any interval of definition I for this differential equation must exclude $x = 0$, since we have assumed that $a_2(x) \neq 0$ in I. Thus, we conclude that Theorem 2 is not contradicted by this example.[5]

An interesting property of the Wronskian is that it satisfies a first-order linear differential equation. Thus, it follows from Section 1.2 that the Wronskian is either identically zero or never vanishes. Equation (8), which is called *Abel's*

[5]For a discussion of the connection between the Wronskian having the value zero and linear dependence of functions, the reader is referred to the article by G. H. Meisters, "Local Linear Dependence and the Vanishing of the Wronskian," *Amer. Math. Monthly* 68, no. 9 (1961): 847–56.

formula is also the basis for determining a second linearly independent solution of a second-order linear differential equation when one solution is already known (see Section 2.8). For simplicity we show the details for the second-order linear differential equation, but we state the theorem for the more general case.

THEOREM 3

If $y_1, y_2, \ldots, y_n$ are solutions of the differential equation

$$a_n(x)y^{(n)} + a_{n-1}(x)\,y^{n-1} + \cdots + a_1(x)y' + a_0(x)y = 0,$$

where each $a_i(x)$ is defined and continuous on $a \le x \le b$ and $a_n(x) \neq 0$ on $a \le x \le b$, then either $W(y_1, \ldots, y_n; x)$ vanishes identically on $a \le x \le b$ or $W(y_1, \ldots, y_n; x)$ is never zero on $a \le x \le b$.

Proof The proof is for the case $n = 2$. For brevity we set $W = W(y_1, y_2; x)$, $p(x) = a_1(x)/a_2(x)$, and $q(x) = a_0(x)/a_2(x)$. Note that both p and q are continuous functions since $a_2(x) \neq 0$. Now

$$W = \begin{vmatrix} y_1 & y_2 \\ y_1' & y_2' \end{vmatrix} = y_1y_2' - y_1'y_2.$$

Therefore,

$$\begin{aligned} W' &= y_1y_2'' + y_1'y_2' - (y_1'y_2' + y_1''y_2) = y_1y_2'' - y_1''y_2 \\ &= y_1(-py_2' - qy_2) - (-py_1' - qy_1)y_2 = -p(y_1y_2' - y_1'y_2) \\ &= -pW. \end{aligned}$$

Using Eq. (5) of Section 1.4, with $a(x) = p(x)$ and $b(x) = 0$, we find that the differential equation $W' = -pW$ has the solution

$$W(x) = W(x_0)\exp\left[-\int_{x_0}^{x} p(s)dx\right] \tag{8}$$

Since the exponential function is never zero, it follows from Eq. (8) that if W is zero at a point x_0 on the interval $a \le x \le b$, then it is identically zero; and if W is different from zero at a point x_0, then it is never zero.

DEFINITION 3

Suppose that $y_1, y_2, \ldots, y_n$ are n solutions of the differential equation

$$a_n(x)y^{(n)} + a_{n-1}(x)y^{(n-1)} + \cdots + a_1(x)y' + a_0(x)y = 0.$$

Suppose also that these functions are linearly independent on the interval of definition of this differential equation. We then say that these functions form a fundamental set (or fundamental system) of solutions for the differential equation.

For example, the functions $\cos x$ and $\sin x$ constitute a fundamental set of solutions for the differential equation $y'' + y = 0$. Also, the functions e^x and

e^{-x} form a fundamental set of solutions for the differential equation $y'' - y = 0$.

Discovering a fundamental set of solutions for a homogeneous linear differential equation is of great importance because, as we shall prove in Section 2.3, any solution of a homogeneous linear differential equation is a linear combination of the solutions in the fundamental set. More precisely, if $y_1, y_2, \ldots, y_n$ form a fundamental set of solutions for a linear homogeneous differential equation of order n, then the expression

$$y = c_1y_1 + c_2y_2 + \cdots + c_ny_n,$$

where $c_1, c_2, \ldots, c_n$ are arbitrary constants, is the general solution of the differential equation.

REMARK 2 Our use of Wronskians is primarily as a test to determine whether or not a given collection of solutions to a differential equation are linearly independent. Eggan and Insell[6] present a different use of the Wronskian concept. Their results are embodied in the following theorem.

THEOREM 4

If f is a real-valued function defined on the real line R and if there exists a positive integer n such that f has $2n$ continuous derivatives on R and $W(f, f', \ldots, f^{(n)}) = 0$ on R, then f is a solution of an nth-order homogeneous linear differential equation with constant coefficients, not all of which are zero.

EXERCISES

In Exercises 1 through 15, compute the Wronskian of the set of functions.

1. $1, x, x^2$

2. $\sin x^2, \cos x^2$

3. $\sin 3x, \cos 3x$

4. $1 + x, 2 - 3x, 4 - x$

5. $\ln x, e^x$

6. $x^{1/2}, x^{3/2}$

7. e^{3x}, e^{2x}

8. $6x^2, 14x^2, 2x$

9. $x^2, x^2 \ln x$

10. e^{7x}, e^{7x}

11. e^{7x}, xe^{7x}

12. e^x, xe^x, x^2e^x

13. $x^{\lambda_1}, x^{\lambda_2}; \lambda_1 \neq \lambda_2$

14. $e^{\lambda_1 x}, e^{\lambda_2 x}; \lambda_1 \neq \lambda_2$

15. $f(x), cf(x)$; c a constant

In Exercises 16 through 23, integrate the differential equation directly to find the general solution.

16. $y' = e^{2x}$

17. $y''' = 0$

18. $y''' = x^3$

19. $y'' = \sin x$

[6]L. C. Eggan and A. J. Insell, *Amer. Math. Monthly* 80, no. 3 (1973).

20. $y'' = 3x$ **21.** $y^{(4)} = 0$ **22.** $y''' = x^2$ **23.** $y^{(5)} = 0$

In Exercises 24 through 33, show that the functions form a fundamental set of solutions for the differential equation. Write down the general solution of the differential equation.

24. $\sin 3x$, $\cos 3x$; $y'' + 9y = 0$

25. $e^{2x}, e^{-2x}; y'' - 4y = 0$

26. $e^{iat}, e^{-iat}; \ddot{y} + a^2y = 0$ $(i^2 = -1)$, $a \neq 0$

27. $\frac{1}{x}; xy' + y = 0, x < 0$

28. e^{3x}, e^x; $y'' - 4y' + 3y = 0$

29. $e^{-2x}, xe^{-2x}; y'' + 4y' + 4y = 0$

30. $e^x \cos 2x$, $e^x \sin 2x$; $y'' - 2y' + 5y = 0$

31. $\cos x + \sin x$, $\cos x - \sin x$; $y'' + y = 0$

32. $\frac{1}{x}\cos(\ln x), \frac{1}{x}\sin(\ln x)$; $x^2y'' + 3xy' + 2y = 0$, $x > 0$

33. 1, $\cos x$, $\sin x$; $y''' + y' = 0$

In Exercises 34 through 37, a second-order differential equation is given. Write an expression for the Wronskian of two linearly independent solutions without finding the solutions themselves. [*Hint:* Refer to Eq. (8).]

34. $2y'' - 3y' + y = 0$ **35.** $xy'' - 3y' - 5y = 0$, $x \neq 0$

36. $y'' + (\cos x)y' + e^xy = 0$ **37.** $y'' + y = 0$

In Exercises 38 through 41, the functions are linearly independent solutions of the differential equation. In each case show that the corresponding Wronskian satisfies Eq. (8).

38. $y'' + 4y = 0$, $y_1 = \sin 2x$, $y_2 = \cos 2x$

39. $y'' - 7y' + 6y = 0$, $y_1 = e^x$, $y_2 = e^{6x}$

40. $xy'' + y' = 0$, $y_1 = \ln x$, $y_2 = 1$, $x > 0$

41. $3y'' + 48y' + 192y = 0$, $y_1 = e^{-8x}, y_2 = xe^{-8x}$

42. Astronomy Kopal[7] obtained a differential equation of the form

$$v\frac{d^2\psi}{dv^2} + 4\frac{d\psi}{dv} = 0, \qquad v > 0.$$

Show that $\psi = 1$ and $\psi = v^{-3}$ are linearly independent solutions of this differential equation and obtain the general solution.

[7] Z. Kopal, "Stress History of the Moon and of Terrestial Planets," *Icarus* 2 (1963): 381.

43. Flows Barton and Raynor, in their study of peristaltic flow in tubes,[8] obtained the following linear homogeneous differential equation with variable coefficients.

$$\frac{d^2p}{dv^2} + \frac{1}{v}\frac{dp}{dv} = 0, \qquad v > 0.$$

Show that $p(v) = 1$ and $p(v) = \ln v$ are linearly independent solutions of this differential equation, and obtain the general solution.

2.3 EXISTENCE AND UNIQUENESS OF SOLUTIONS

As we pointed out in Chapter 1, it is reassuring to know that *there is* (existence) a solution to the differential equation that one hopes to solve. The situation is even better if, in addition, there is *only one* solution (uniqueness) to the differential equation. The main result of this section is Theorem 1 below, which states conditions on the coefficient functions that guarantee the existence and uniqueness of solutions to the IVP containing an nth-order linear differential equation. This *existence and uniqueness theorem* is, from one point of view, the most important theorem associated with nth-order linear differential equations. Although we choose not to present the proof, it is possible to prove this theorem using arguments based on a knowledge of calculus only.[9]

THEOREM 1 EXISTENCE AND UNIQUENESS

Let $a_0(x)$, $a_1(x)$, . . . , $a_n(x)$ and $f(x)$ be defined and continuous on $a \le x \le b$ with $a_n(x) \neq 0$ for $a \le x \le b$. Let x_0 be such that $a \le x_0 \le b$ and let β_0, β_1, . . . , β_{n-1} be any constants. Then there exists a unique function y satisfying the initial value problem

$$a_n(x)y^{(n)} + a_{n-1}(x)y^{(n-1)} + \cdots + a_1(x)y' + a_0(x)y = f(x) \tag{1}$$

$$y(x_0) = \beta_0, y'(x_0) = \beta_1, \ldots, y^{(n-1)}(x_0) = \beta_{n-1}. \tag{2}$$

Furthermore, the solution y is defined in the entire interval $a \le x \le b$.

Using Theorem 1 it is easy to *construct a fundamental set* of solutions for the *homogeneous* differential equation

$$a_n(x)y^{(n)} + a_{n-1}(x)y^{(n-1)} + \cdots + a_1(x)y' + a_0(x)y = 0. \tag{3}$$

We define y_1 to be the unique solution of Eq. (3) satisfying the initial conditions $y(x_0) = 1$, $y'(x_0) = 0$, $y''(x_0) = 0$, . . . , $y^{(n-1)}(x_0) = 0$. Define y_2 to be the unique

[8] C. Barton and S. Raynor, *Bull. Math. Biophys.* 30 (1968): 663–80.

[9] See E. A. Coddington, *An Introduction to Ordinary Differential Equations* (Englewood Cliffs, N.J.: Prentice-Hall, 1961), or D. Willett, "The Existence-Uniqueness Theorem for an nth-Order Linear Ordinary Differential Equation," *Amer. Math. Monthly* 5, no. 2 (1968).

solution of Eq. (3) satisfying the initial conditions $y(x_0) = 0$, $y'(x_0) = 1$, $y''(x_0) = 0, \ldots, y^{(n-1)}(x_0) = 0$. Define y_3 to be the unique solution of Eq. (3) satisfying the initial conditions $y(x_0) = 0$, $y'(x_0) = 0$, $y''(x_0) = 1, \ldots, y^{(n-1)}(x_0) = 0$, and so on. Define y_n to be the unique solution of Eq. (3) satisfying the initial conditions $y(x_0) = 0$, $y'(x_0) = 0, \ldots, y^{(n-1)}(x_0) = 1$. By Theorem 1 each of these functions exists and is uniquely determined. It remains to show that these functions are linearly independent. In this case

$$W(y_1, y_2, \ldots, y_n; x_0) = \begin{vmatrix} 1 & 0 & 0 & \cdots & 0 \\ 0 & 1 & 0 & \cdots & 0 \\ 0 & 0 & 1 & \cdots & 0 \\ \cdot & \cdot & \cdot & & \cdot \\ \cdot & \cdot & \cdot & & \cdot \\ \cdot & \cdot & \cdot & & \cdot \\ 0 & 0 & 0 & \cdots & 1 \end{vmatrix} = 1 \neq 0.$$

Thus, these functions constitute a fundamental set of solutions.

The observations of the preceding paragraph emphasize the fact that for any linear differential equation whose coefficients satisfy the hypotheses of Theorem 1, there always exists a fundamental set of solutions. These solutions, although guaranteed by the theory, are not always easy to construct in practice. Nevertheless, there are some forms of linear differential equations for which systematic construction of a fundamental set of solutions can be described. These forms of equations are discussed in Sections 2.4 through 2.12.

REMARK 1 The existence and uniqueness theorem for the IVP (1)–(2) is something that should be expected to be true when the IVP is a "good" mathematical model of a real-life situation. For example, in the case of the vibrating spring discussed in Section 2.1.1, the existence and uniqueness of the IVP (10)–(11) of that section simply means that the mass m moves (existence) and, in fact, moves in only one way (uniqueness).

THEOREM 2

If $y_1, y_2, \ldots, y_n$ constitute a fundamental set of solutions of Eq. (3), *then given any other solution, y, satisfying the initial conditions $y(x_0) = \beta_0$, $y'(x_0) = \beta_1, \ldots, y^{(n-1)}(x_0) = \beta_{n-1}$, there exist unique constants $c_1, c_2, \ldots, c_n$ such that*

$$y = c_1y_1 + c_2y_2 + \cdots + c_ny_n.$$

That is, this other solution, y, can be expressed *uniquely* as a linear combination of the fundamental set of solutions.

Proof Set $Y = c_1y_1 + c_2y_2 + \cdots + c_ny_n$. Thus, by Theorem 1 of Section 2.2, it follows that the function Y is a solution of the differential equation (3). We calculate $Y', Y'', \ldots, Y^{(n-1)}$ and then evaluate Y and each derivative at $x = x_0$. This process leads to the system

$$
\begin{aligned}
c_1y_1(x_0) + c_2y_2(x_0) + \cdots + c_ny_n(x_0) &= \beta_0 \\
c_1y_1'(x_0) + c_2y_2'(x_0) + \cdots + c_ny_n'(x_0) &= \beta_1 \qquad (4) \\
\cdots\cdots\cdots\cdots\cdots\cdots\cdots\cdots\cdots\cdots\cdots\cdots \\
c_1y_1^{(n-1)}(x_0) + c_2y_2^{(n-1)}(x_0) + \cdots + c_ny_n^{(n-1)}(x_0) &= \beta_{n-1}.
\end{aligned}
$$

Equations (4) constitute a nonhomogeneous system of linear algebraic equations in the unknowns $c_1, c_2, \ldots, c_n$ whose determinant of coefficients is $W(y_1, \ldots, y_n; x_0)$.

Since these functions form a fundamental set of solutions, their Wronskian is not zero. By Cramer's rule (for details, see Appendix A) there is a unique solution for the unknowns c_i. Now the functions Y and y are both solutions of the differential equation (3), with the same initial conditions, and, by the uniqueness part of Theorem 1, $Y = y$. The proof is complete.

On the basis of Theorem 2, the following result is well justified.

COROLLARY 1

If $y_1, y_2, \ldots, y_n$ form a fundamental set of solutions for Eq. (3), *then the expression*

$$y = c_1y_1 + c_2y_2 + \cdots + c_ny_n,$$

where the c_i are arbitrary constants is the general solution of Eq. (3).

For example, $y = c_1 \cos x + c_2 \sin x$ is the general solution of the differential equation $y'' + y = 0$, and $y = c_1e^x + c_2e^{-x}$ is the general solution of the differential equation $y'' - y = 0$.

At this point we realize that the problem of finding all solutions of Eq. (3) boils down to finding a fundamental set of solutions. Sections 2.4 through 2.8 are devoted to discussing methods associated with the problem of finding a fundamental set of solutions.

EXAMPLE 1

Consider the IVP

$$y'' - y' - 2y = 0 \qquad (5)$$

$$y(0) = 1, y'(0) = 8. \qquad (6)$$

(a) Show that the functions $y_1(x) = e^{-x}$ and $y_2(x) = e^{2x}$ form a fundamental set of solutions for Eq. (5).

(b) Find the general solution of Eq. (5).

(c) Find the unique solution of the IVP (5)–(6).

Solution (a) By direct substitution of y_1 and y_2 into Eq. (5), we find

$$(e^{-x})'' - (e^{-x})' - 2(e^{-x}) = e^{-x} + e^{-x} - 2e^{-x} = 0$$

and

$$(e^{2x})'' - (e^{2x})' - 2(e^{2x}) = 4e^{2x} - 2e^{2x} - 2e^{2x} = 0.$$

Therefore y_1 and y_2 are solutions of (5). Their Wronskian

$$\begin{vmatrix} e^{-x} & e^{2x} \\ -e^{-x} & 2e^{2x} \end{vmatrix} = 2e^{x} + e^{x} = 3e^{x}$$

is never zero. Hence y_1 and y_2 are linearly independent solutions of Eq. (5). Since Eq. (5) is second order, it follows that y_1 and y_2 form a fundamental set of solutions for the equation.

(b) By Corollary 1 the general solution of Eq. (5) is

$$y(x) = c_1e^{-x} + c_2e^{2x}. \tag{7}$$

(c) To find the unique solution of the IVP (5)–(6), we must use the initial conditions (6) in (7) and

$$y'(x) = -c_1e^{-x} + 2c_2e^{2x} \tag{8}$$

Setting $x = 0$ in (7) and (8) we find,

$$\left.\begin{aligned} c_1 + c_2 &= 1 \\ -c_1 + 2c_2 &= 8 \end{aligned}\right\} \Rightarrow c_1 = -2, c_2 = 3.$$

The unique solution of the IVP (5)–(6) is therefore

$$y(x) = -2e^{-x} + 3e^{2x}. \tag{9}$$

If we want to check that (9) is indeed *the* solution of the IVP (5)–(6), we must check three things: First, (9) is a solution of Eq (5). Second, $y(0) = 1$. Third, $y'(0) = 8$. In fact,

$$(-2e^{-x} + 3e^{2x})'' - (-2e^{-x} + 3e^{2x})' - 2(-2e^{-x} + 3e^{2x}) =$$
$$(-2e^{-x} + 12e^{2x}) - (2e^{-x} + 6e^{2x}) - 2(-2e^{-x} + 3e^{2x}) =$$
$$-2e^{x} + 12e^{2x} - 2e^{-x} - 6e^{2x} + 4e^{-x} - 6e^{2x} = 0;$$
$$y(0) = -2e^0 + 3e^0 = 1; y'(x) = 2e^{-x} + 6e^{2x} \Rightarrow y'(0) = 2e^0 + 6e^0 = 8.$$

EXAMPLE 2 In Section 2.1 we encountered the differential equation

$$\ddot{y} + a^2y = 0, \qquad a \neq 0. \tag{10}$$

Show that $y_1 = \cos at$ and $y_2 = \sin at$ form a fundamental set of solutions for Eq. (10), and find its general solution.

Solution In Section 2.1 we demonstrated that y_1 and y_2 are solutions of Eq. (10). To show that they are linearly independent, we compute their Wronskian:

$$W(y_1, y_2; t) = \begin{vmatrix} \cos at & \sin at \\ -a\sin at & a\cos at \end{vmatrix} = a\cos^2 at + a\sin^2 at$$
$$= a(\cos^2 at + \sin^2 at) \quad = a \neq 0.$$

Since $W(y_1, y_2; t) \neq 0$ and y_1 and y_2 are solutions, we conclude that y_1 and y_2 constitute a fundamental system of solutions for Eq. (10), and every solution of Eq. (10) can be expressed as a linear combination of y_1 and y_2. That is, if $y(t)$ is *any* solution of Eq. (10), then $y = c_1y_1 + c_2y_2$ for some choice of constants c_1, c_2. Since e^{iat} and e^{-iat} are also solutions of Eq. (10), these functions must be expressible as linear combinations of y_1 and y_2. These linear combinations are readily obtained if one uses the *Euler identity* of trigonometry, that is, $e^{i\theta} = \cos\theta + i\sin\theta$ ($\Rightarrow e^{-i\theta} = \cos\theta - i\sin\theta$). Thus, $e^{iat} = c_1y_1 + c_2y_2$, with $c_1 = 1$, $c_2 = i$; and $e^{-iat} = c_1y_1 + c_2y_2$, with $c_1 = 1$, $c_2 = -i$. The functions e^{iat} and e^{-iat} also form a fundamental set of solutions for Eq. (10). (See Exercise 21.)

EXERCISES

For each of the initial value problems in Exercises 1 through 10, indicate whether or not the existence and uniqueness theorem, Theorem 1, applies.

1. $y'' + xy = 0$, $y(0) = 1$, $y'(0) = 0$, $-5 \le x \le 5$
2. $xy'' + y = 0$, $y(0) = 1$, $y'(0) = 0$, $-5 \le x \le 5$
3. $xy'' + y = 0$, $y(1) = 1$, $y'(1) = 1$, $1 \le x \le 5$
4. $(1 - x)y'' - xy' + e^xy = 0$, $y(0) = 0$, $y'(0) = 4$, $-\frac{1}{2} \le x \le \frac{1}{2}$
5. $(\sin x)y'' + xy' + y = 2$, $y\left(\frac{3\pi}{4}\right) = 1$, $y'\left(\frac{3\pi}{4}\right) = 0$, $\frac{3\pi}{4} \le x \le \frac{5\pi}{4}$
6. $x^2y'' + (x\cos x)y' - 3y = 1$, $y\left(\frac{3\pi}{4}\right) = 1$, $y'\left(\frac{3\pi}{4}\right) = 1$, $\frac{3\pi}{4} \le x \le \frac{5\pi}{4}$
7. $(x^3 - 1)y''' - 3y'' + 4xy = 0$, $y(-1) = 0$, $y'(-1) = 2$, $y''(-1) = 0$, $-1 \le x \le 0$
8. $y''' + yy' = 2$, $y(0) = 0$, $y'(0) = -1$, $0 \le x \le 3$
9. $3y'' + y' - 2y = 0$, $y(0) = 2$, $y'(0) = 0$, $0 \le x < \infty$
10. $(\cos x)y'' + 3y = 1$, $y(1) = 0$, $y'(0) = 0$, $1 \le x < \infty$
11. Solve the IVP

$$y'' + (\sin x)\, y' + e^xy = 0$$
$$y(0) = 0, \qquad y'(0) = 0.$$

(*Hint:* Note that $y = 0$ is a solution and then use Theorem 1.)

For each of Exercises 12 through 17, determine an interval in which there exists a unique solution determined by the conditions $y(x_0) = y_0$, $y'(x_0) = y_1$, where x_0 is to be a point of the interval.

12. $(x - 1)y'' + 3y' = 0$

13. $2xy'' - 7(\cos x)y' + y = e^{-x}$

14. $y'' + 4(\tan x)y' - xy = 0$

15. $(\cos x)y'' + y = \sin x$

16. $(x^2 - 4)y'' + 3x^3y' + \dfrac{4}{x-1}y = 0$

17. $(x^2 + 1)y'' + (2x - 1)\,y' + 3y = 0$

18. As we explained in Section 2.1.1, the IVP

$$m\ddot{y} + Ky = 0 \tag{11}$$

$$y(0) = y_0,\ \dot{y}(0) = v_0 \tag{12}$$

describes the motion of a spring of mass m and spring constant K. Initially, at time $t = 0$, the mass is located at the position y_0 and has initial velocity v_0.

(a) Show that

$$y_1(t) = \cos\sqrt{\frac{K}{m}}\,t, \qquad y_2(t) = \sin\sqrt{\frac{K}{m}}\,t$$

form a fundamental set of solutions of the differential equation (11).

(b) Find the general solution of the differential equation.

(c) Find the unique solution of the IVP (11)–(12).

19. Give a physical interpretation of the IVP

$$\ddot{y} + 4y = 0$$

$$y(0) = \dot{y}(0) = 0.$$

Find the solution to the IVP first by physical intuition and then by applying Theorem 1.

20. **Mechanics** A projectile of mass m leaves the earth with initial velocity v_0 in a direction that makes an angle θ with the horizontal. Assume that the motion takes place on the xy plane, that the projectile was fired at the origin 0, and that the x axis is the horizontal (Figure 2.3). Use Newton's second law of motion to show that the position $(x(t), y(t))$ of the mass m at time t satisfies the two initial value problems

$$\begin{aligned} m\ddot{x} &= 0 \\ x(0) &= 0 \\ \dot{x}(0) &= v_0\cos\theta \end{aligned} \qquad \text{and} \qquad \begin{aligned} m\ddot{y} &= -mg \\ y(0) &= 0 \\ \dot{y}(0) &= v_0\sin\theta. \end{aligned}$$

Show (by direct integrations) that the solutions of these initial value problems are given by $x(t) = (v_0\cos\theta)t$ and $y(t) = -\frac{1}{2}gt^2 + (v_0\sin\theta)t$.

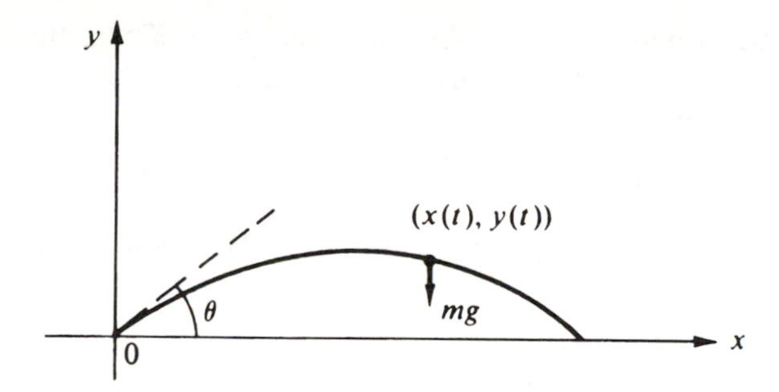

Figure 2.3

21. Show that the functions $y_1 = e^{iat}$ and $y_2 = e^{-iat}$ form a fundamental set of solutions for the differential equation

$$\ddot{y} + a^2y = 0, \qquad a \neq 0.$$

22. Show that the functions $y_1 = \cosh at$ and $y_2 = \sinh at$ form a fundamental set of solutions for the differential equation

$$\ddot{y} - a^2y = 0, \qquad a \neq 0.$$

23. Find the general solution of the differential equation

$$(x - a)(x - b)y'' + 2(2x - a - b)y' + 2y = 0$$

[*Hint:* Compute[10] the second derivative of the expression $(x - a)(x - b)y$.]

2.4 HOMOGENEOUS DIFFERENTIAL EQUATIONS WITH CONSTANT COEFFICIENTS—THE CHARACTERISTIC EQUATION

The linear homogeneous differential equation with constant coefficients has the form

$$a_ny^{(n)} + a_{n-1}y^{(n-1)} + \cdots + a_1y' + a_0y = 0, \tag{1}$$

where each coefficient a_i is a constant and the leading coefficient $a_n \neq 0$. To find the general solution of Eq. (1), we need to find n linearly independent solutions.

As we examine Eq. (1) carefully (especially in the case $n = 1$), we see that in a sense any solution of this equation must have the property that derivatives of this solution merely produce constant multiples of themselves. This observation motivates us to try $e^{\lambda x}$, with λ a constant, as a solution of Eq. (1). If $y = e^{\lambda x}$, then $y' = \lambda e^{\lambda x}$, $y'' = \lambda^2e^{\lambda x}, \ldots, y^{(n)} = \lambda^ne^{\lambda x}$. Substitution of $y = e^{\lambda x}$ in Eq. (1) yields

$$a_n\lambda^ne^{\lambda x} + a_{n-1}\lambda^{n-1}e^{\lambda x} + \cdots + a_1\lambda e^{\lambda x} + a_0e^{\lambda x} = 0$$

or

$$e^{\lambda x}(a_n\lambda^n + a_{n-1}\lambda^{n-1} + \cdots + a_1\lambda + a_0) = 0. \tag{2}$$

[10]See also *Math. Magazine* 44 (1971): 18, 56, 297.

The exponential function never takes the value zero. Thus, the only way that $y = e^{\lambda x}$ can be a solution of the differential equation is that λ be a root of the polynomial equation

$$a_n\lambda^n + a_{n-1}\lambda^{n-1} + \cdots + a_1\lambda + a_0 = 0. \tag{3}$$

We conclude, then, that if λ is a root of Eq. (3), $y = e^{\lambda x}$ is a solution of the differential equation (1).

DEFINITION 1

The polynomial $f(\lambda) \equiv a_n\lambda^n + a_{n-1}\lambda^{n-1} + \cdots + a_1\lambda + a_0$ *is called the characteristic polynomial for Eq.* (1), *and the equation* $f(\lambda) = 0$ *is called the characteristic equation for Eq.* (1). *The roots of the characteristic equation are called characteristic roots.*

In Section 2.5 we will demonstrate that knowledge of the roots of the characteristic equation is all that is needed to construct the general solution of homogeneous linear differential equations with constant coefficients.

Consequently, the problem of solving homogeneous linear differential equations with constant coefficients can be reduced to the problem of finding the roots of the characteristic equation for the differential equation. The characteristic polynomial is of degree n and the *fundamental theorem of algebra* states that every polynomial of degree n ($n \geq 1$) has at least one root and, therefore, counting multiplicities, exactly n roots. However, if n is large, indeed, when $n \geq 5$, the problem of finding the roots of the characteristic equation may be far from trivial. Some elementary methods for finding roots of polynomial equations are given in Appendix C.

REMARK 1 If $n = 1$, the characteristic equation is $a_1\lambda + a_0 = 0$ and has the root $\lambda = -a_0/a_1$. If $n = 2$, the characteristic equation is the quadratic equation $a_2\lambda^2 + a_1\lambda + a_0 = 0$ and its two roots are given by the quadratic formula

$$\lambda = \frac{-a_1 \pm \sqrt{a_1^2 - 4a_2a_0}}{2a_2}. \tag{4}$$

If $n = 3$, the characteristic equation is a cubic equation, and the formulas for its solutions (called *Cardan's formulas*) were discovered by Scipione del Ferro in 1515. If $n = 4$, the characteristic equation is a quartic equation, and its solution was discovered by Ludovico Ferrari in the sixteenth century.[11] For $n \geq 5$, the characteristic equation cannot be solved (by radicals) in general unless it is of some special form. This fact was proved by Abel and Galois early in the nineteenth century.

In spite of the possibility of algebraic difficulties surrounding the solution of the characteristic equation, we can develop some information about linearly independent solutions to Eq. (1).

[11] For the solutions of the cubic and quartic equations, the interested reader is referred to the text by E. A. Grove and G. Ladas, *Introduction to Complex Variables* (Boston: Houghton Mifflin, 1974).

We emphasize once more that if λ is a characteristic root, then $e^{\lambda x}$ is a solution of the linear homogeneous differential equation with constant coefficients. In the special case that λ is a complex number, it is conventional to write the corresponding solution in an alternative form. First; to say that λ is a complex number means that λ is of the form $\alpha + i\beta$, where α and β are real numbers and $i^2 = -1$. The real number α is called the *real part* of λ, and the real number β is called the *imaginary part* of λ. We then write

$$\begin{aligned} e^{\lambda x} &= e^{(\alpha + i\beta)x} = e^{\alpha x + i\beta x} = e^{\alpha x}e^{i\beta x} \\ &= e^{\alpha x}(\cos \beta x + i \sin \beta x). \end{aligned} \tag{5}$$

The last form follows from the Euler identity, $e^{i\theta} = \cos \theta + i \sin \theta$ ($\Rightarrow e^{-i\theta} = \cos \theta - i \sin \theta$). Suppose now that the coefficients a_i of the differential equation are real numbers. Then the only way that a characteristic root can be a complex number is that the characteristic polynomial contain a quadratic factor of the form $b_2\lambda^2 + b_1\lambda + b_0$ for which $b_1^2 < 4b_2b_0$. In this case the complex characteristic roots occur in *conjugate pairs* ($\alpha - i\beta$ is called the *conjugate* of the complex number $\alpha + i\beta$). Thus, two characteristic roots would be $\alpha + i\beta$ and $\alpha - i\beta$, where [refer to Eq. (4) and recall that $\sqrt{-K} = i\sqrt{K}$]

$$\alpha = -\frac{b_1}{2b_2}, \qquad \beta = \frac{\sqrt{4b_0b_2 - b_1^2}}{2b_2}$$

In addition to the solution (5) there would also be the solution $e^{\alpha x}(\cos \beta x - i \sin \beta x)$. For further discussion of this case, see Example 2 below, Exercises 21 through 29, and Remark 1 in Section 2.5.

EXAMPLE 1 Solve the differential equation $y'' - y = 0$.

Solution Set $y = e^{\lambda x}$; then (since $a_2 = 1$, $a_1 = 0$, $a_0 = -1$) λ must satisfy the characteristic equation

$$\lambda^2 - 1 = 0 \Rightarrow (\lambda + 1)(\lambda - 1) = 0 \Rightarrow \lambda = \pm 1.$$

We have two solutions, $y_1 = e^x$ and $y_2 = e^{-x}$. These two solutions are linearly independent, since their Wronskian has the value $-2 (\neq 0)$. Thus, any solution y of this differential equation is of the form

$$y = c_1e^x + c_2e^{-x},$$

where c_1 and c_2 are arbitrary constants.

EXAMPLE 2 Solve the differential equation of the vibrating spring example of Section 2.1.1:

$$\ddot{y} + a^2y = 0,$$

with $a^2 = K/m$.

Solution Set $y = e^{\lambda t}$; then (since $a_2 = 1$, $a_1 = 0$, $a_0 = a^2$) λ must satisfy the equation

$$\lambda^2 + a^2 = 0, \qquad \lambda = \pm ia.$$

We have two solutions, $y_1 = e^{iat}$ and $y_2 = e^{-iat}$. These two solutions form a fundamental set of solutions, since their Wronskian has the value $-2ia \neq 0$. Hence, any solution to this differential equation is of the form

$$y(t) = c_1 e^{iat} + c_2 e^{-iat},$$

or (using the Euler identity $e^{i\theta} = \cos\theta + i\sin\theta$) we find

$$y(t) = \alpha_1 \cos at + \alpha_2 \sin at,$$

with $\alpha_1 = c_1 + c_2$ and $\alpha_2 = i(c_1 - c_2)$.

EXERCISES

In Exercises 1 through 10, write out the characteristic equation for the differential equation.

1. $4y''' - 2y'' + 6y' - 7y = 0$

2. $2y^{iv} + 3y''' - y'' + 2y' - y = 0$

3. $y''' - y' + 2y = 0$

4. $y'' - y' + 6y = 0$

5. $5y''' - 5y'' + y' - 2y = 0$

6. $6y''' - 4iy'' + (3 + i)y' - 2y = 0$

7. $3y''' + 4y'' = 0$

8. $6y^{iv} - 3y''' + y'' - 7y' - 6y = 0$

9. $3y^{v} - 2y^{iv} + y'' - 2y' = 0$

10. $2y''' - 2y'' - y' + 3y = 0$

In Exercises 11 through 20, determine whether the associated characteristic equation has complex roots.

11. $y'' - 5y' + 6y = 0$

12. $2y'' + 3y' + y = 0$

13. $2y'' + 3y' - 2y = 0$

14. $y'' + 9y = 0$

15. $3y'' - 5y' + 3y = 0$

16. $y'' + y' + y = 0$

17. $2y'' - 4y' - y = 0$

18. $4y'' - 3y' + y = 0$

19. $y'' + 3y' + 4y = 0$

20. $2y'' + y = 0$

In Exercises 21 through 29, the second-order homogeneous linear differential equation has characteristic roots which occur in conjugate pairs. In each case determine the characteristic roots λ_1 and λ_2. Write the general solution in the form $y = c_1 e^{\lambda_1 x} + c_2 e^{\lambda_2 x}$. Next, following the method of Example 2, write the general solution so that the complex unit i is absorbed into the constants.

21. $y'' + 16y = 0$

22. $2y'' + 14y' + 25y = 0$

23. $y'' + y' + y = 0$

24. $y'' + 9y = 0$

25. $4y'' - 8y' + 5y = 0$

26. $2y'' - 6y' + 5y = 0$

27. $y'' + 4y = 0$

28. $y'' - 3y' + \frac{5}{2}y = 0$

29. $y'' + 25y = 0$

30. The definitions of the hyperbolic cosine and hyperbolic sine are

$$\cosh x = \frac{e^x + e^{-x}}{2}$$

and

$$\sinh x = \frac{e^x - e^{-x}}{2}.$$

(a) Show that these definitions can be manipulated algebraically to yield $e^x = \cosh x + \sinh x$ and $e^{-x} = \cosh x - \sinh x$.

(b) Show that the general solution of the differential equation $y'' - y = 0$ can be written in the form $y = c_1 \cosh x + c_2 \sinh x$.

In Exercises 31 through 37 write two equivalent versions of the general solution. [*Hint:* See Example 1 and Exercise 30.]

31. $y'' - 4y = 0$

32. $y'' - 16y = 0$

33. $y'' - 25y = 0$

34. $y'' - 49y = 0$

35. $y'' - 81y = 0$

36. $y'' - 121y = 0$

37. $y'' - 9y = 0.$

38. Apply the method of Example 2 to write the general solution of

$$y^{iv} + 13y'' + 36y = 0$$

in terms of trigonometric functions. [*Hint:* $\gamma^2 + 13\gamma + 36 = (\gamma + 4)(\gamma + 9)$.]

39. Economics Second-order linear differential equations with constant coefficients of the form

$$\ddot{D} - \beta D = 0$$

occur in the Domar burden-of-debt model, where the income grows at a constant relative rate β $(0 < \beta < 1)$ and D is the total public debt.[12] Find the characteristic roots associated with this differential equation. Write the general solution of this differential equation in the form $y(t) = c_1 e^{\lambda_1 t} + c_2 e^{\lambda_2 t}$ and also in terms of the hyperbolic sine and hyperbolic cosine.

40. The Vibrating Spring with Damping In practice, a vibrating spring is most often subjected to frictional forces and other forces (for example, air resistance) which act to retard (dampen) the motion and eventually cause the system to come to rest. Thus, it is realistic to assume that the spring is subjected to a *damping force*. Generally speaking, the damping force is difficult to formulate precisely; however, experiments have shown that the magnitude of the damping force is approximately proportional to the velocity of the mass, provided that the velocity of the mass is small. Naturally,

[12]See E. D. Domar, "The Burden of the Debt and the National Income," *Amer. Econ. Rev.* (Dec. 1944): 798–827.

the damping force, as described, acts in a direction opposite to that of the mass. Thus, the damping force is negative when dy/dt is positive and positive when dy/dt is negative. Consequently, we can express the damping force as $-\alpha(dy/dt)$, where $\alpha > 0$ is called the *damping constant.*

Using Newton's second law, show that the equation of motion in this case is

$$m\ddot{y} + \alpha\dot{y} + Ky = 0. \tag{6}$$

Write the characteristic equation for this differential equation. Describe conditions on α, K, and m which guarantee that the characteristic roots will not be complex numbers. What conditions will guarantee that the characteristic roots are complex?

2.5 HOMOGENEOUS DIFFERENTIAL EQUATIONS WITH CONSTANT COEFFICIENTS—THE GENERAL SOLUTION

In Sections 2.2 and 2.3 we demonstrated that for linear differential equations, the basic problem is to find a fundamental set of solutions. In Section 2.4 we demonstrated that linear homogeneous differential equations with constant coefficients will have solutions of the form $e^{\lambda x}$ provided that λ is a root of the characteristic equation associated with the differential equation. The loose end that still persists is to relate the solutions $e^{\lambda x}$ with a fundamental set of solutions of the differential equation. This we accomplish in the present section via theorems and a number of illustrative examples.

THEOREM 1

If all the roots of the characteristic equation are distinct, say $\lambda = \lambda_i$, $i = 1, 2, \ldots, n$, then the n functions $y_i = e^{\lambda_i x}$, $i = 1, 2, \ldots, n$ constitute a fundamental set of solutions.

Proof The proof is for $n = 2$.

$$\begin{aligned} W(y_1, y_2; x) &= \begin{vmatrix} e^{\lambda_1 x} & e^{\lambda_2 x} \\ \lambda_1 e^{\lambda_1 x} & \lambda_2 e^{\lambda_2 x} \end{vmatrix} \\ &= \lambda_2 e^{(\lambda_1+\lambda_2)x} - \lambda_1 e^{(\lambda_1+\lambda_2)x} \\ &= (\lambda_2 - \lambda_1)e^{(\lambda_1+\lambda_2)x} \neq 0. \end{aligned}$$

Thus, y_1 and y_2 form a fundamental set of solutions. The proof for the general case would follow along similar lines of reasoning. However, the evaluation of the corresponding Wronskian would not be trivial. In fact, the Wronskian reduces to a determinant commonly called the *Vandermonde determinant.*[13]

[13]For the statement and evaluation of the van der Monde determinant, see C. W. Curtis, *Linear Algebra,* 2nd ed. (Boston: Allyn and Bacon, 1968), p. 140.

EXAMPLE 1 Find the general solution of the differential equation

$$y''' - 6y'' + 11y' - 6y = 0.$$

Solution For this differential equation the characteristic equation is

$$\lambda^3 - 6\lambda^2 + 11\lambda - 6 = 0.$$

Note that the coefficients in this polynomial add up to zero. Whenever this happens it is always true that $\lambda = 1$ is a root. Thus we obtain (after long division or synthetic division)

$$\lambda^3 - 6\lambda^2 + 11\lambda - 6 = (\lambda - 1)(\lambda^2 - 5\lambda + 6) = (\lambda - 1)(\lambda - 2)(\lambda - 3).$$

Hence, $\lambda = 1, 2, 3$ and the fundamental set of solutions is $y_1 = e^x$, $y_2 = e^{2x}$, $y_3 = e^{3x}$. The general solution is

$$y = c_1e^x + c_2e^{2x} + c_3e^{3x}.$$

If the roots of the characteristic equation are not all distinct, that is, if the characteristic equation has at least one repeated root, then we cannot generate a fundamental system of solutions by considering just the exponentials $e^{\lambda_i x}$. To try to obtain a feeling for the form of the general solution in this case, we first consider some special examples.

EXAMPLE 2 Solve the differential equation

$$y'' - 2y' + y = 0. \tag{1}$$

Solution For the differential equation (1) the characteristic polynomial is $\lambda^2 - 2\lambda + 1 = (\lambda - 1)^2$. Thus $\lambda = 1$ is a root of multiplicity 2. We introduce a new unknown function v by the relation $v = y' - y$. Then $v' = y'' - y'$, which, using the differential equation (1), takes the form

$$v' = (2y' - y) - y' = y' - y = v \Rightarrow v' = v \Rightarrow v = c_1e^x.$$

Therefore,

$$v = y' - y = c_1e^x.$$

This last differential equation is a first-order linear differential equation and can be solved by the methods of Section 1.4 to yield

$$y = e^x(\textstyle\int e^{-x}c_1e^x\,dx + c_2) = e^x(c_1x + c_2) = c_1(xe^x) + c_2e^x.$$

That is,

$$y = c_1y_1 + c_2y_2,$$

where $y_1 = xe^x$ and $y_2 = e^x$. The reader can verify that both y_1 and y_2 are solutions of Eq. (1). Furthermore,

$$W(y_1, y_2; x) = \begin{vmatrix} xe^x & e^x \\ (x+1)e^x & e^x \end{vmatrix} = e^x \cdot e^x \begin{vmatrix} x & 1 \\ x+1 & 1 \end{vmatrix} = e^{2x}(-1) \neq 0.$$

Our fundamental set of solutions is therefore $y_1 = xe^x$ and $y_2 = e^x$, and the general solution is

$$y = c_1xe^x + c_2e^x.$$

In this example we obtained a simpler problem by making the substitution $v = y' - y$. For an alternative approach, see Exercises 46 and 47.

EXAMPLE 3 Solve the differential equation

$$y''' - 3y'' + 3y' - y = 0. \tag{2}$$

Solution The characteristic polynomial for the differential equation (2) is $\lambda^3 - 3\lambda^2 + 3\lambda - 1 = (\lambda - 1)^3$. Hence, $\lambda = 1$ is a root of the characteristic equation of multiplicity 3. *Taking a clue from Example 2, we guess that the fundamental system of solutions for Eq. (2) is $y_1 = e^x$, $y_2 = xe^x$, and $y_3 = x^2e^x$.* Certainly, y_1 is a solution of Eq. (2). As for y_2, we have

$$y_2 = xe^x \Rightarrow y_2' = (x + 1)e^x \Rightarrow y_2'' = (x + 2)e^x \Rightarrow y_2''' = (x + 3)e^x.$$

Substituting these results in Eq. (2) yields

$$\begin{aligned}(x + 3)e^x - 3(x + 2)e^x + 3(x + 1)e^x - xe^x &= e^x(x + 3 - 3x - 6 + 3x + 3 - x)\\ &= e^x(0) = 0.\end{aligned}$$

Hence, y_2 is also a solution. Similarly, for y_3 we obtain

$$\begin{aligned}y_3 = x^2e^x \Rightarrow y_3' = (x^2 + 2x)e^x \Rightarrow y_3'' &= (x^2 + 4x + 2)e^x\\ \Rightarrow y_3''' &= (x^2 + 6x + 6)e^x.\end{aligned}$$

Substitution of these results in Eq. (2) yields

$$e^x[x^2 + 6x + 6 - 3(x^2 + 4x + 2) + 3(x^2 + 2x) - x^2] = e^x(0) = 0.$$

Therefore, y_3 is a solution. To show that these solutions are linearly independent, we compute their Wronskian.

$$\begin{aligned}W(y_1, y_2, y_3; x) &= \begin{vmatrix} e^x & xe^x & x^2e^x \\ e^x & (x + 1)e^x & (x^2 + 2x)e^x \\ e^x & (x + 2)e^x & (x^2 + 4x + 2)e^x \end{vmatrix}\\ &= e^x \cdot e^x \cdot e^x \begin{vmatrix} 1 & x & x^2 \\ 1 & x + 1 & x^2 + 2x \\ 1 & x + 2 & x^2 + 4x + 2 \end{vmatrix}\\ &= e^{3x}(2) \neq 0.\end{aligned}$$

We conclude, then, that y_1, y_2, and y_3 form a fundamental set of solutions. The general solution is

$$y = c_1e^x + c_2xe^x + c_3x^2e^x.$$

The results embodied in Examples 2 and 3 are capable of generalization, and this generalization is stated without proof in the following theorem.[14]

THEOREM 2

If $\lambda = \lambda_i$ is a root of multiplicity m of the characteristic equation associated with the differential equation $a_n y^{(n)} + a_{n-1} y^{(n-1)} + \cdots + a_1 y' + a_0 y = 0$, then there are m linearly independent solutions associated with $\lambda = \lambda_i$. These solutions are as follows: $e^{\lambda_i x}, xe^{\lambda_i x}, x^2 e^{\lambda_i x}, \ldots, x^{m-1} e^{\lambda_i x}$.

REMARK 1 When the coefficients a_i of the differential equation are all real numbers, we know from the theory of polynomial equations that if the characteristic equation has a complex root, such as $\lambda = \alpha + i\beta$, of multiplicity m, then the characteristic equation also has the complex conjugate $\alpha - i\beta$ as a root again of multiplicity m. Applying Theorem 2 first to the root $\alpha + i\beta$ and then to $\alpha - i\beta$ we find $2m$ complex valued solutions. It can be shown however, that the real and imaginary parts of the solutions

$$e^{(\alpha + i\beta)x}, xe^{(\alpha + i\beta)x}, \ldots, x^{m-1} e^{(\alpha + i\beta)x}$$

and

$$e^{(\alpha - i\beta)x}, xe^{(\alpha - i\beta)x}, \ldots, x^{m-1} e^{(\alpha - i\beta)x},$$

namely the $2m$ functions

$$e^{\alpha x} \cos \beta x, e^{\alpha x} \sin \beta x, xe^{\alpha x} \cos \beta x, xe^{\alpha x} \sin \beta x, \ldots, x^{m-1} e^{\alpha x} \cos \beta x, x^{m-1} e^{\alpha x} \sin \beta x,$$

are linearly independent solutions. From this and the previous results it follows that *the general solution of a linear differential equation with real and constant coefficients can always be expressed as a linear combination of n real valued functions.*

COROLLARY 1

Consider the second-order linear differential equation with real coefficients

$$a_2 y'' + a_1 y' + a_0 y = 0, \ a_2 \neq 0. \tag{3}$$

Assume that λ_1 and λ_2 are roots of the characteristic equation $a_2 \lambda^2 + a_1 \lambda + a_0 = 0$. Then the form of the general solution $y(x)$ of Eq. (3) is described by the following cases:

CASE 1 *Real and Distinct Roots* $\lambda_1 = \lambda_2$

$$y(x) = c_1 e^{\lambda_1 x} + c_2 e^{\lambda_2 x},$$

where c_1 and c_2 are arbitrary constants.

[14]For a proof, see Coddington, *Ordinary Differential Equation* (Englewood Cliffs, N.J. Prentice-Hall, 1961)

CASE 2 *Equal Roots* $\lambda_1 = \lambda_2 = \lambda$

$$y(x) = c_1 e^{\lambda x} + c_2 x e^{\lambda x}.$$

CASE 3 *Complex Conjugate Roots* $\lambda_{1,2} = k \pm i\ell$

$$y(x) = c_1 e^{kx} \cos \ell x + c_2 e^{kx} \sin \ell x.$$

EXAMPLE 4 Solve the differential equation

$$y' - (2 + 3i)y = 0. \tag{4}$$

Solution The characteristic equation for the differential equation (4) is $\lambda - (2 + 3i) = 0$, hence the solution is

$$y = c_1 e^{(2+3i)x} = c_1(e^{2x} \cos 3x + ie^{2x} \sin 3x).$$

Note that the functions $e^{2x} \cos 3x$ and $e^{2x} \sin 3x$ are not, by themselves, solutions of the differential equation (4), rather it is the specific combination $e^{2x} \cos 3x + ie^{2x} \sin 3x$ that is a solution. Naturally, the difference between this example and Case 3 of Corollary 1 is that the constants in the differential equation (4) are *complex* and that the complex characteristic root $\lambda = 2 + 3i$ *does not* occur in a conjugate pair of characteristic roots.

EXAMPLE 5 Solve the differential equation

$$y^{(5)} - 3y^{(4)} + 4y''' - 4y'' + 3y' - y = 0. \tag{5}$$

Solution The characteristic equation for the differential equation (5) is

$$\lambda^5 - 3\lambda^4 + 4\lambda^3 - 4\lambda^2 + 3\lambda - 1 = 0.$$

Since the coefficients add to zero, it follows that $\lambda = 1$ is a root and we can write (see Appendix C)

$$\begin{aligned} \lambda^5 - 3\lambda^4 + 4\lambda^3 - 4\lambda^2 + 3\lambda - 1 &= (\lambda - 1)(\lambda^4 - 2\lambda^3 + 2\lambda^2 - 2\lambda + 1) \\ &= (\lambda - 1)(\lambda - 1)(\lambda^3 - \lambda^2 + \lambda - 1) \\ &= (\lambda - 1)(\lambda - 1)(\lambda - 1)(\lambda^2 + 1). \end{aligned}$$

Thus, $\lambda = 1, 1, 1, i, -i$, and the general solution has the form

$$y = c_1 e^x + c_2 x e^x + c_3 x^2 e^x + c_4 \cos x + c_5 \sin x.$$

EXAMPLE 6 The characteristic equation of a certain differential equation has the following roots: 1 as a simple root, -2 as a triple root, $2 \pm 3i$ as simple roots, and $-1 \pm 2i$ as double roots; that is, the characteristic polynomial can be factored as

$$a_{10}(\lambda - 1)(\lambda + 2)^3(\lambda^2 - 4\lambda + 13)(\lambda^2 + 2\lambda + 5)^2,$$

where a_{10} is the coefficient of $y^{(10)}$ in the original differential equation. Write down the general solution of this differential equation.

Solution It follows from Theorems 1 and 2 and Remark 1 that the general solution of the differential equation in question is

$$y = c_1e^x + c_2e^{-2x} + c_3xe^{-2x} + c_4x^2e^{-2x} + c_5e^{2x}\cos 3x + c_6e^{2x}\sin 3x$$
$$+ c_7e^{-x}\cos 2x + c_8e^{-x}\sin 2x + c_9xe^{-x}\cos 2x + c_{10}xe^{-x}\sin 2x.$$

APPLICATION 2.5.1

It is suggested that the reader review Exercise 40 of Section 2.4 before studying this application.

Mechanics
The Vibrating Spring with Damping

■ The motion of the vibrating spring subjected to damping (retarding) forces is governed by the differential equation [see Eq. (6), Exercise 40, Section 2.4]

$$\ddot{y} + \frac{\alpha}{m}\dot{y} + \frac{K}{m}y = 0,$$

where $\alpha > 0$ is the damping constant, $K > 0$ is the spring constant, and $m > 0$ is the mass. To avoid fractions, it is convenient to set $2a = \alpha/m$ and $b^2 = K/m$, $b > 0$. Thus, the equation of motion has the form

$$\ddot{y} + 2a\dot{y} + b^2y = 0. \tag{6}$$

The characteristic equation associated with Eq. (6) is

$$\lambda^2 + 2a\lambda + b^2 = 0.$$

Consequently, the characteristic roots are

$$\lambda_1 = -a + \sqrt{a^2 - b^2}, \qquad \lambda_2 = -a - \sqrt{a^2 - b^2}.$$

We realize that the form of the general solution of Eq. (6) depends very heavily upon the nature of the characteristic roots. The three possible cases are discussed separately.

CASE 1 $b^2 > a^2 (\Rightarrow b > a)$ The general solution of Eq. (6) is

$$y(t) = e^{-at}(c_1\cos\sqrt{b^2 - a^2}\,t + c_2\sin\sqrt{b^2 - a^2}\,t).$$

Graphically, the solution looks as shown in Figure 2.4. We observe that the motion [that is $y(t)$] oscillates about the t axis, and for this reason we term the motion *oscillatory*. We note, however, that the "amplitude" of the motion[15] decreases as time goes on. This situation is referred to as being *underdamped*

[15]As in Exercise 21, Section 2.1, the solution in this case can be written in the form

$$y(t) = Ae^{-at}\cos\left[\sqrt{b^2 - a^2}t - \phi\right]$$
$$A = \sqrt{c_1^2 + c_2^2} \quad \text{and} \quad \phi = \cos^{-1}\left[c_1/(c_1^2 + c_2^2)^{1/2}\right].$$

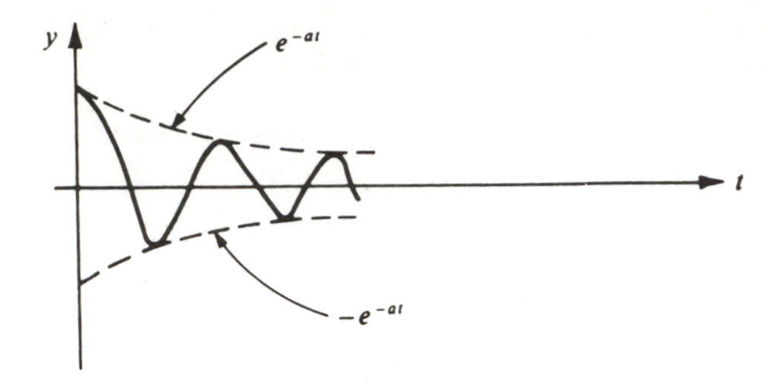

Figure 2.4

in the sense that the amount of damping (or the damping constant) is less than the spring stiffness, and hence there is not enough damping to overcome the oscillations.

CASE 2 $b^2 = a^2$ ($\Rightarrow b = a$) The general solution of Eq. (6) is

$$y(t) = e^{-at}(c_1 + c_2 t).$$

Since $y(t)$ contains no sine or cosine terms, we note that there are no vibrations (oscillations). However, in reference to Case 1, we realize that even the slightest decrease in the damping will produce vibrations. For this reason, this case is called *critically damped.*

CASE 3 $b^2 < a^2$ ($\Rightarrow b < a$) Here the general solution of Eq. (6) is

$$y(t) = c_1 e^{\lambda_1 t} + c_2 e^{\lambda_2 t},$$

with both λ_1 and λ_2 negative real numbers. In this case there are no oscillations and the graph of the solution approaches the t axis very quickly as time goes on. That is, the damping force is so strong that it causes the system to "slow down" very quickly. This case is referred to as being *overdamped.*

EXAMPLE 7 It is known that the motion of a certain damped vibrating spring is governed by the differential equation

$$\ddot{y} + 4\dot{y} + 4y = 0.$$

Classify this case as being underdamped, critically damped, or overdamped.

Solution Comparing this differential equation with Eq. (6), we note that $2a = 4$ and $b^2 = 4$; hence $a = b = 2$. This is Case 2, and therefore the motion is classified as critically damped.

EXERCISES

In Exercises 1 through 10, write the general solution of the differential equation.

1. $2y'' + 3y' + y = 0$

2. $8y'' - 6y' + y = 0$

3. $y'' + y' + y = 0$

4. $9y'' - 6y' + y = 0$

5. $y'' + 6y = 0$

6. $y'' - 9y = 0$

7. $y' - 3y = 0$

8. $y''' - 7y'' + 5y' + y = 0$

9. $y^{iv} + y''' - 3y'' - y' + 2y = 0$

10. $y''' - 6y'' + 12y' - 8y = 0$

In Exercises 11 through 19, the factored form of the characteristic equation for certain differential equations are given. In each case state the order of the differential equation and give the form of the general solution.

11. $(\lambda - 3)(\lambda + 2)^2(\lambda^2 - 2\lambda + 5)(\lambda^2 + 4\lambda + 5)^2$

12. $(\lambda + 1)(\lambda - 1)(\lambda^2 + 1)(\lambda^2 + \lambda + 1)$

13. $(2\lambda - 3)(5\lambda + 4)(\lambda^2 + 4)^2(\lambda^2 - \lambda - 1)$

14. $(\lambda - 3)^3(\lambda + 2)^4$

15. $(\lambda^2 + 4)(\lambda^2 + 1)^2$

16. $(\lambda + 1)(\lambda + 3)(\lambda + 2)^2(\lambda^2 - 2\lambda + 5)^2(\lambda^2 + 9)$

17. $(\lambda - 7)(2\lambda - 7)(\lambda^2 + \lambda + 1)(\lambda + 4)^3$

18. $(\lambda^2 + 3)(\lambda^2 + 1)^3(\lambda + 1)^3$

19. $(\lambda + 2)(\lambda - 5)^2(\lambda - 4)^3(\lambda^2 + 4)^3$

Solve the initial value problems in Exercises 20 and 21.

20. $y'' - 2y' + y = 0$
$y(0) = 1$
$y'(0) = -1$

21. $y'' + y' + y = 0$
$y(0) = 1$
$y'(0) = \sqrt{3}$

22. Solve the differential equation

$$y'' - iy' + 12y = 0.$$

Do the roots of the characteristic equation occur in conjugate pairs? Determine whether or not the real or imaginary part of either of the fundamental solutions satisfies the differential equation.

In Exercises 23 through 29, find the solution of the initial value problem.

23. $y'' + 3y = 0,\quad y(0) = 3,\quad y'(0) = -6\sqrt{3}$

24. $y'' - 4y = 0,\quad y(0) = 0,\quad y'(0) = 4$

25. $y''' - y'' = 0,\quad y(0) = 1,\quad y'(0) = 3,\quad y''(0) = 2$

26. $y''' - 4y'' + 4y' = 0,\quad y(0) = 2,\quad y'(0) = -1\quad y''(0) = 0$

27. $y''' - y'' - y' + y = 0, \quad y(0) = 0, \quad y'(0) = 5 \quad y''(0) = 2$

28. $y'' + 4y = 0, \quad y\left(\frac{\pi}{4}\right) = 1, \quad y'\left(\frac{\pi}{4}\right) = 0$

29. $y''' - 2y'' + 2y' = 0, \quad y(0) = -2, \quad y'(0) = 0, \quad y''(0) = 4$

Find the solution of each boundary value problem in Exercises 30 and 31.

30. $y'' + 3y' + 2y = 0, \quad y(0) = 0, \quad y(-1) = e$

31. $y'' + y = 0, \quad y(0) = 3, \quad y\left(\frac{\pi}{2}\right) = 2$

32. A spring is stretched 1.5 cm by a force of 4 dyn. A mass of 4 g is attached to the end of the spring. The system is then set into motion by pulling the mass 6 cm above the point of equilibrium and releasing it, at time $t = 0$, with an initial velocity of $5\sqrt{6}$ cm/sec directed downward. Find the position of the mass at any time t.

33. In Exercise 32, assume that, in addition, air resistance acts upon the mass. This air resistance force equals $\frac{20}{3}$ times the velocity of the mass at time t. Find the position of the mass as a function of time. Discuss the behavior of the motion as $t \to \infty$. [*Hint:* See Exercise 40, Section 2.4.]

Assume that each of the differential equations in Exercises 34 through 43 governs the motion of a certain damped vibrating spring. Classify the resulting motion as being underdamped, critically damped, or overdamped.

34. $\ddot{y} + 6\dot{y} + 12y = 0$

35. $\ddot{y} + 20y + 64y = 0$

36. $\ddot{y} + 9\dot{y} + 4y = 0$

37. $5\ddot{y} + 10\dot{y} + 20y = 0$

38. $\ddot{y} + 2\dot{y} + y = 0$

39. $6\ddot{y} + 4\dot{y} + y = 0$

40. $\ddot{y} + 5\dot{y} + y = 0$

41. $\ddot{y} + 8\dot{y} + 16y = 0$

42. $4\ddot{y} + 8\dot{y} + 4y = 0$

43. $\ddot{y} + \dot{y} + y = 0$

44. Show[16] that the graphs in the xy plane of all solutions of the system of differential equations

$$\ddot{x} + \dot{y} + 6x = 0, \qquad \ddot{y} - \dot{x} + 6y = 0$$

which satisfy $\dot{x}(0) = \dot{y}(0) = 0$ are hypocycloids. (A hypocycloid is the path described by a fixed point on the circumference of a circle which rolls on the inside of a given fixed circle.) The problem that is proposed here is the following: Verify that the substitution $z = x + iy$ enables one to collapse the given system into the single differential equation

$$\ddot{z} - i\dot{z} + 6z = 0$$

with initial condition $\dot{z}(0) = 0$. Solve this differential equation.

[16]Problem 44 appeared on the 1971 William Lowell Putnam Mathematical Competition, *Amer. Math. Monthly* 80, no. 2 (1973).

45. Find[17] all twice differentiable functions f such that for all x

$$f'(x) = f(-x).$$

Solve this problem. (*Hint:* Differentiate the expression, use the expression itself, and the result will be a differential equation that you can solve.)

46. One can motivate the form of the general solution in the case of equal roots of the characteristic equation in the following manner.[18]

(a) Show that the solution of the IVP

$$y'' - 2ry' + \left(r^2 - \frac{\alpha^2}{4}\right) y = 0$$

$$y(0) = 0, \qquad y'(0) = 1$$

is

$$y_\alpha(x) = \frac{1}{\alpha}\left[e^{(r+\alpha/2)x} - e^{(r-\alpha/2)x}\right].$$

(b) Show that $\lim_{\alpha\to 0} y_\alpha(x) = xe^{rx}$.

47. (a) Show that the solution of the IVP

$$y'' - 2(r + \beta)y' + r^2 y = 0$$

$$y(0) = 0, \qquad y'(0) = 1$$

is

$$y_\beta(x) = \frac{1}{2\sqrt{\beta(2r + \beta)}}\left[e^{[r+\beta+\sqrt{\beta(2r+\beta)}]x} - e^{[r+\beta-\sqrt{\beta(2r+\beta)}]x}\right].$$

(b) Show that $\lim_{\beta\to 0} y_\beta(x) = xe^{rx}$.

It is often possible to extract information about solutions to a differential equation without explicitly solving the differential equation. Perhaps one of the best examples of this idea[19] is the trigonometric differential equation $y'' + y = 0$.

Define $S(x)$ to be *the* solution of the IVP

$$y'' + y = 0$$

$$S(0) = 0, \qquad S'(0) = 1$$

Define $C(x)$ to be *the* solution of the IVP

$$y'' + y = 0$$

$$C(0) = 1, \qquad C'(0) = 0.$$

[17]Problem 45 appears in *Math. Magazine* 47, no. 5 (1974).

[18]See R. S. Baslaw and H. M. Hastings, "On the Critically Damped Oscillator," *Math Magazine* 48, no. 2 (1975). For further reference, see R. D. Larsson, "General Solutions of Linear Ordinary Differential Equations," *Amer. Math. Monthly* 65, no. 7 (1958).

[19]The outline presented in Exercises 48–53 is extracted from D. A. Kearns, "An Analytic Approach to Trigonometric Function," *Amer. Math. Monthly* 65, no. 8 (1958). See also G. Birkhoff and G–C. Rota, *Ordinary Differential Equations* (Boston: Ginn & Company, 1962), p. 53, and W. Leighton, *An Introduction to the Theory of Differential Equations* (New York: McGraw-Hill, 1952), p. 41.

48. Show that $C(x)$ and $S(x)$ are linearly independent.

49. Show that $S'(x) = C(x)$. (*Hint:* Use the differential equation, the initial conditions, and the uniqueness theorem.)

50. Show that $C'(x) = -S(x)$.

51. Show that $[S(x)]^2 + [C(x)]^2 \equiv 1$. (*Hint:* Multiply $S'' + S = 0$ by S'; and integrate.)

52. Show that for arbitrary a,

$$S(x + a) = S(x)C(a) + S(a)C(x).$$

(*Hint:* First show that $S(x + a)$ is a solution, then use the fact that $S(x)$ and $C(x)$ form a fundamental set.)

53. Show that for arbitrary a,

$$C(x + a) = C(x)C(a) - S(x)S(x).$$

2.6 HOMOGENEOUS EQUATIONS WITH VARIABLE COEFFICIENTS—OVERVIEW

Throughout this book we strive to find solutions of differential equations. In Chapter 1 we learned how to solve a few types of differential equations of order 1 (such as separable, linear, exact, and homogeneous). In the last two sections we learned that (subject to some algebraic difficulties only) we can solve any linear homogeneous differential equation with constant coefficients. In Sections 2.11 and 2.12 we will see that the nonhomogeneous linear differential equation with constant coefficients can be handled in a straightforward manner.

Unfortunately, matters are not so simple when we try to solve linear differential equations with variable coefficients of order 2 or higher. In general there is *no way* to solve, *explicitly,* a linear homogeneous differential equation with variable coefficients unless the differential equation is of a very special form; for example, the Euler-type differential equation and differential equations of order 2, for which we know already one solution. The following two sections deal with these two special cases. Once we solve the homogeneous differential equation the nonhomogeneous equation presents no difficulties (at least in theory), as we will see in Sections 2.10, 2.11, and 2.12. Apparently, then, it is difficult if not impossible to solve linear differential equations with variable coefficients. But what do we do if we have such a differential equation and we desire information about its solution(s)? Perhaps the best we can do with differential equations having variable coefficients, at this level, is to try to *approximate* their solutions by assuming that the solutions are power series. The supposition that the solution is a power series is valid only if the coefficient functions $a_i(x)$ satisfy certain differentiability requirements. When these requirements are met, the supposition that the solution is a power series is nothing more than looking at the Taylor-series representation of the solution. A direct substitution of the power-series solution into the differential equation enables us, in many cases, to compute as many coefficients of the power series as we please for our approximation. We have devoted an entire chapter (Chapter 5) to this approach. We include a brief discussion of power series here to round out the treatment within this chapter.

2.7 EULER DIFFERENTIAL EQUATION

An Euler differential equation is a differential equation of the form

$$a_n x^n y^{(n)} + a_{n-1} x^{n-1} y^{(n-1)} + \cdots + a_1 x y' + a_0 y = 0, \tag{1}$$

where $a_n, a_{n-1}, \ldots, a_1, a_0$ are constants and $a_n \neq 0$. Since the leading coefficient $a_n x^n$ should never be zero, the interval of definition of the differential equation (1) is either the open interval $(0, \infty)$ or the open interval $(-\infty, 0)$. That is, the differential equation should be solved for either $x > 0$ or $x < 0$. The Euler differential equation is probably the simplest type of linear differential equation with variable coefficients. The reason for this is that the change of independent variable

$$x = \begin{cases} e^t & \text{if} \quad x > 0 \\ -e^t & \text{if} \quad x < 0 \end{cases}$$

produces a differential equation with constant coefficients. We illustrate this fact for the second-order case.

EXAMPLE 1 Show by means of the change of independent variables above that the Euler differential equation of second order,

$$a_2 x^2 y'' + a_1 x y' + a_0 y = 0, \tag{2}$$

with a_0, a_1, and a_2 given constants, is reduced to the differential equation

$$a_2 \ddot{y} + (a_1 - a_2)\dot{y} + a_0 y = 0, \tag{3}$$

where dots denote derivatives with respect to t.

Proof First, assume that $x > 0$. Then the transformation $x = e^t$ implies that $dx/dt = e^t$, and so $dt/dx = e^{-t}$. By the chain rule for differentiation, we have

$$y' = \frac{dy}{dx} = \frac{dy}{dt} \cdot \frac{dt}{dx} = \dot{y}e^{-t} = \dot{y}\frac{1}{x} \Rightarrow xy' = \dot{y}$$

and

$$\begin{aligned} y'' &= \frac{dy'}{dt} \cdot \frac{dt}{dx} = \frac{d}{dt}(\dot{y}e^{-t})e^{-t} \\ &= (\ddot{y}e^{-t} - \dot{y}e^{-t})e^{-t} = (\ddot{y} - \dot{y})e^{-2t} = (\ddot{y} - \dot{y})\frac{1}{x^2} \Rightarrow x^2 y'' = \ddot{y} - \dot{y}. \end{aligned}$$

Thus, substituting xy' and x^2y'' into the differential equation (2), we obtain

$$\begin{aligned} a_2x^2y'' + a_1xy' + a_0y &= a_2(\ddot{y} - \dot{y}) + a_1\dot{y} + a_0y \\ &= a_2\ddot{y} + (a_1 - a_2)\dot{y} + a_0y, \end{aligned}$$

proving that Eq. (3) holds. Now, if we assume that $x < 0$, we set $x = -e^t$, so $dx/dt = -e^t$ and $dt/dx = -e^{-t}$. Again, by the chain rule we have $y' = -\dot{y}e^{-t}$ and $y'' = (\ddot{y} - \dot{y})e^{-2t}$. Thus,

$$\begin{aligned} a_2x^2y'' + a_1xy' + a_0y &= a_2(-e^t)^2(\ddot{y} - \dot{y})e^{-2t} + a_1(-e^t)(-\dot{y}e^{-t}) + a_0y \\ &= a_2\ddot{y} + (a_1 - a_2)\dot{y} + a_0y, \end{aligned}$$

and Eq. (3) holds once again. The proof is complete.

EXAMPLE 2 Solve the Euler differential equation

$$2x^2y'' - 3xy' - 3y = 0,$$

(a) for $x > 0$, and (b) for $x < 0$.

Solution (a) Using the transformation $x = e^t$, the differential equation becomes [see Eq. (3)]

$$\begin{aligned} 2\ddot{y} + (-3 - 2)\dot{y} - 3y &= 0 \\ 2\ddot{y} - 5\dot{y} - 3y &= 0. \end{aligned}$$

The characteristic equation for this latter differential equation is $2\lambda^2 - 5\lambda - 3 = 0$, and so the characteristic roots are $\lambda_1 = -\frac{1}{2}$, $\lambda_2 = 3$. Thus,

$$y(t) = c_1e^{-(1/2)t} + c_2e^{3t} = c_1(e^t)^{-1/2} + c_2(e^t)^3.$$

But $x = e^t$, so

$$y(x) = c_1x^{-1/2} + c_2x^3.$$

(b) Using the transformation $x = -e^t$, we obtain the same differential equation as in part (a), with the same characteristic roots. Thus,

$$y(t) = c_1e^{-(1/2)t} + c_2e^{3t} = c_1(e^t)^{-1/2} + c_2(e^t)^3.$$

In this case $e^t = -x$, so

$$y(x) = c_1(-x)^{-1/2} + c_2(-x)^3.$$

EXAMPLE 3 Discuss the form of the solution to the Euler differential equation of order 2.

Solution From Example 1 we see that the nature of the solutions depends on the roots λ_1 and λ_2 of the characteristic equation for the differential equation (3),

$$a_2\lambda^2 + (a_1 - a_2)\lambda + a_0 = 0. \tag{4}$$

Thus, we have

$$y(t) = \begin{cases} c_1e^{\lambda_1 t} + c_2e^{\lambda_2 t} & \text{if } \lambda_1, \lambda_2 \text{ real } \lambda_1 \neq \lambda_2 \\ (c_1 + c_2 t)e^{\lambda_1 t} & \text{if } \lambda_1, \lambda_2 \text{ real, } \lambda_1 = \lambda_2 \\ c_1e^{\alpha t}\cos \beta t + c_2 e^{\alpha t} \sin \beta t & \text{if } \lambda_1 = \alpha + i\beta, \lambda_2 = \alpha - i\beta. \end{cases}$$

Now, for $x > 0$, $e^{\lambda t} = (e^t)^\lambda = x^\lambda$ and $t = \ln x$. Thus, the general solution of the Euler differential equation has the form

$$y(x) = \begin{cases} c_1x^{\lambda_1} + c_2x^{\lambda_2} & \text{if } \lambda_1, \lambda_2 \text{ are real, } \lambda_1 \neq \lambda_2 \\ (c_1 + c_2 \ln x)x^{\lambda_1} & \text{if } \lambda_1, \lambda_2 \text{ are real, } \lambda_1 = \lambda_2 \\ c_1x^\alpha \cos(\beta \ln x) + c_2x^\alpha \sin(\beta \ln x) & \text{if } \lambda_1 = \alpha + i\beta, \lambda_2 = \alpha - i\beta. \end{cases}$$

If $x < 0$, then x should be replaced by $-x$ in the formula for $y(x)$. For the last expression we used the relations $x^{\alpha + i\beta} = x^\alpha x^{i\beta} = x^\alpha e^{\ln(x^{i\beta})} = x^\alpha e^{i(\beta \ln x)} = x^\alpha[\cos(\beta \ln x) + i \sin(\beta \ln x)]$.

The results of Examples 1, 2, and 3 associated with the Euler differential equation of order 2 prompt us to conclude that in order to solve any Euler differential equation, we try a substitution of the form $y = x^\lambda$, where λ is to be a root of a polynomial equation.

EXAMPLE 4 Solve the Euler differential equation

$$x^3y''' - x^2y'' - 2xy' - 4y = 0 \qquad \text{for } x > 0. \tag{5}$$

Solution We substitute $y = x^\lambda$ in the differential equation (5) to obtain

$$\lambda(\lambda - 1)(\lambda - 2)x^\lambda - \lambda(\lambda - 1)x^\lambda - 2\lambda x^\lambda - 4x^\lambda = 0.$$

Thus,

$$x^\lambda[\lambda(\lambda - 1)(\lambda - 2) - \lambda(\lambda - 1) - 2\lambda - 4] = 0.$$

Since $x^\lambda \neq 0$, we require that

$$\lambda(\lambda - 1)(\lambda - 2) - \lambda(\lambda - 1) - 2\lambda - 4 = 0$$

or

$$\lambda^3 - 4\lambda^2 + \lambda - 4 = 0$$

$$\Rightarrow (\lambda^2 + 1)(\lambda - 4) = 0 \Rightarrow \lambda = 4, \pm i.$$

Therefore, the solution is

$$y = c_1x^4 + c_2 \cos(\ln x) + c_3 \sin(\ln x).$$

EXAMPLE 5 Substitute $y = x^\lambda$ in Eq. (1) to find the characteristic equation for the general nth-order Euler differential equation.

Solution Equation (1) is the general nth-order Euler differential equation

$$a_nx^ny^{(n)} + a_{n-1}x^{n-1}y^{(n-1)} + \cdots + a_1xy' + a_0y = 0$$

Set $y = x^\lambda \Rightarrow y' = \lambda x^{\lambda-1}$, $y'' = \lambda(\lambda - 1)x^{\lambda-2}, \ldots, y^{(n)} = \lambda(\lambda - 1) \cdots (\lambda - n + 1)x^{\lambda-n}$. Substituting these results into the differential equation, we obtain

$$a_n\lambda(\lambda - 1) \cdots (\lambda - n + 1)x^\lambda + a_{n-1}\lambda(\lambda - 1) \cdots (\lambda - n + 2)x^\lambda + \cdots + a_2\lambda(\lambda - 1)x^\lambda + a_1\lambda x^\lambda + a_0x^\lambda = 0.$$

Since $x^\lambda \neq 0$ we find

$$a_n\lambda(\lambda - 1) \cdots (\lambda - n + 1) + a_{n-1}\lambda(\lambda - 1) \cdots (\lambda - n + 2) + \cdots + a_2\lambda(\lambda - 1) + a_1\lambda + a_0 = 0. \tag{6}$$

Equation (6) is the characteristic equation for the general nth-order Euler differential equation.

EXERCISES

Find the general solution of the Euler equations in Exercises 1 through 15.

1. $5x^2y'' - 3xy' + 3y = 0, x > 0$

2. $x^2y'' - xy' + y = 0, x > 0$

3. $3x^2y'' + 4xy' + y = 0, x < 0$

4. $x^2y'' + 3xy' + 2y = 0, x < 0$

5. $x^2y'' + 4xy' - 4y = 0, x < 0$

6. $(x - 1)^2y'' + 5(x - 1)y' + 4y = 0, x - 1 > 0$ [*Hint:* Try $y = (x - 1)^\lambda$.]

7. $2x^2y'' + xy' - 3y = 0, x < 0$

8. $x^2y'' + 3xy' + 5y = 0, x > 0$

9. $x^3y''' + 3x^2y'' - 2xy' + 2y = 0, x > 0$

10. $x^2y'' + \frac{7}{2}xy' - \frac{3}{2}y = 0, x > 0$

11. $(x + 3)^2y'' + 3(x + 3)y' + 5y = 0, x + 3 < 0$

12. $(x - 2)^2y'' - (x - 2)y' + y = 0, x > 2$

13. $x^2y'' - 6y = 0, x < 0$

14. $x^3y''' + 4x^2y'' - 8xy' + 8y = 0, x < 0$

15. $x^3y''' + 4x^2y'' - 8xy' + 8y = 0, x > 0$

Solve the initial value problems in Exercises 16 through 20.

16. $x^2y'' - xy' + y = 0, y(1) = 1, y'(1) = 0$

17. $x^2y'' - xy' + y = 0, y(-1) = 1, y'(-1) = 0$

18. $x^2y'' + 3xy' + 2y = 0, y(1) = 0, y'(1) = 1$

19. $x^2y'' + \frac{7}{2}xy' - \frac{3}{2}y = 0, y(-4) = 1, y'(-4) = 0$

20. $x^3y''' + 4x^2y'' - 8xy' + 8y = 0$, $y(1) = 0$, $y'(1) = 1$, $y''(1) = 0$

21. Just as in the case of linear differential equations with constant coefficients (see Exercises 46 and 47 of Section 2.5), the form of the solution in the case of equal characteristic roots can be obtained via a limit process.[20]

(a) If λ_1 and λ_2 are roots of the characteristic equation (4), then

$$y(x) = \frac{x^{\lambda_1} - x^{\lambda_2}}{\lambda_1 - \lambda_2}$$

is a solution of Eq. (2). (Why?)

(b) Show that

$$\lim_{\lambda_1 \to \lambda_2} y(x) = x^{\lambda_2} \ln x.$$

22. Astronomy Kopal[21] obtained a differential equation of the form

$$r\frac{d^2\psi}{dr^2} + 4\frac{d\psi}{dr} = 0, \qquad r > 0.$$

Solve the differential equation. [*Hint:* Multiplication of the differential equation by r produces an Euler differential equation.]

23. Flows Barton and Raynor[22] in their study of peristaltic flow in tubes, obtained the following linear homogeneous differential equation with variable coefficients:

$$\frac{d^2p}{dr^2} + \frac{1}{r}\frac{dp}{dr} = 0, \qquad r > 0.$$

Solve this differential equation. [*Hint:* Multiplication by r^2 produces an Euler differential equation.]

In Exercises 24 through 32, write the characteristic equation associated with the Euler differential equation. Find the characteristic roots and obtain the general solution.

24. $3x^3y''' - x^2y'' + 4xy' - 4y = 0$, $x > 0$

25. $x^4y^{iv} - 5x^3y''' + 3x^2y'' - 6xy' + 6y = 0$, $x > 0$

26. $2x^4y^{iv} + 3x^3y''' - 4x^2y'' + 8xy' - 8y = 0$, $x < 0$

27. $2x^3y''' + x^2y'' - 12xy' - 2y = 0$, $x < 0$

28. $x^5y^{v} - 2x^3y''' + 4x^2y'' = 0$, $x < 0$

29. $7x^4y^{iv} - 2x^3y''' + 3x^2y'' - 6xy' + 6y = 0$, $x > 0$

30. $x^5y^{v} + 3x^3y''' - 9x^2y'' + 18xy' - 18y = 0$, $x > 0$

31. $x^6y^{vi} - 12x^4y^{iv} = 0$, $x > 0$

[20] See B. Schweizer, "On the Euler-Cauchy Equation," *Amer. Math. Monthly* 68, no. 6 (1961).
[21] Kopal, "Stress History of the Moon." See footnote 8.
[22] Barton and Raynor, *Bull. Math. Biophys.* See footnote 7.

32. $x^3y''' + 3x^2y'' - 2xy' - 2y = 0,\ x > 0$

Solve each of the differential equations in Exercises 33 through 40.

33. $y'' - \frac{5}{x}y' + \frac{5}{x^2}y = 0, x > 0$

34. $xy''' - \frac{6}{x^2}y = 0, x < 0$

35. $x^2y^{iv} - xy''' = 0,\ x < 0$

36. $3xy'' - 4y' + \frac{5}{x}y = 0, x > 0$

37. $(x - 4)y'' + 4y' - \frac{4}{x - 4}y = 0, x > 4$

38. $(x + 2)y'' - y' + \frac{1}{x + 2}y = 0, x + 2 > 0$

39. $y'' + \frac{5}{x - 1}y' + \frac{4}{(x - 1)^2}y = 0, x < 1$

40. $5y'' - \frac{3}{x - 3}y' + \frac{3}{(x - 3)^2}y = 0, x < 3$

2.8 REDUCTION OF ORDER

As the terminology implies, *reduction of order* is a device whereby the problem of solving a (linear) differential equation of certain order is replaced by a problem of solving a (linear) differential equation of lower order. The procedure associated with reduction of order is somewhat systematic and hinges upon our having specific knowledge of at least one solution to the original differential equation. Perhaps the most interesting application of the reduction of order is that *we can find a second linearly independent solution of a differential equation of order 2 with variable coefficients provided that we know one nontrivial (not identically zero) solution of it.* The method is embodied in the following theorem.

THEOREM 1

If y_1 is a known solution of the linear differential equation

$$a_n(x)y^{(n)} + a_{n-1}(x)y^{(n-1)} + \cdots + a_1(x)y' + a_0(x)y = 0, \tag{1}$$

which has the property that it is never zero in the interval of definition of the differential equation, then the change of dependent variable

$$y = y_1u \tag{2}$$

produces a linear differential equation of order $n - 1$ for u'.

Proof As we have done many times before, we consider a special case rather than confuse the reader with the cumbersome symbolism that the general case requires. Thus, we restrict our attention to the case $n = 2$. Now

$$y = y_1 u$$

$$y' = y_1 u' + y_1' u$$

$$y'' = y_1 u'' + 2y_1' u' + y_1'' u.$$

Substituting these results into Eq. (1) (when $n = 2$) yields

$$a_2(x)[y_1 u'' + 2y_1' u' + y_1'' u] + a_1(x)[y_1 u' + y_1' u] + a_0(x)[y_1 u] = 0$$

$$[a_2(x)y_1]u'' + [2a_2(x)y_1' + a_1(x)y_1]u' + [a_2(x)y_1'' + a_1(x)y_1' + a_0(x)y_1]u = 0$$

Since y_1 is a solution of the differential equation it follows that

$$a_2(x)y_1'' + a_1(x)y_1' + a_0(x)y_1 = 0,$$

and so

$$[a_2(x)y_1]u'' + [2a_2(x)y_1' + a_1(x)y_1]u' = 0. \tag{3}$$

Setting $u' = v$, Eq. (3) becomes a first-order linear differential equation which we can solve by the method shown in Section 1.4.

EXAMPLE 1 Given that $y_1 = x \sin x$ is a solution of the differential equation

$$xy'' - 2y' + \frac{x^2 + 2}{x} y = 0, \tag{4}$$

for $0 < x < \pi$, find another (linearly independent) solution.

Solution We leave it to the reader to verify that y_1 is a solution of the differential equation (4) and proceed directly to Eq. (3). Thus,

$$[x(x \sin x)]u'' + [2x(x \cos x + \sin x) - 2(x \sin x)]u' = 0$$

$$\Rightarrow \qquad (x^2 \sin x)u'' + (2x^2 \cos x)u' = 0$$

Setting $u' = v$ we obtain a first-order linear differential equation, which therefore we can solve explicitly:

$$v = e^{-2 \ln |\sin x|} = \csc^2 x.$$

Hence, $u' = v = \csc^2 x \Rightarrow u = -\cot x$, and so a second solution is

$$y_2 = y_1 u = x \sin x(-\cot x) = -x \cos x.$$

Furthermore, $W(y_1, y_2; x) = x^2 \neq 0$, since $0 < x < \pi$, and so y_1 and y_2 are linearly independent. Therefore, we conclude that the general solution of Eq. (4) is

$$y = c_1 y_1 + c_2 y_2$$

$$= x(c_1 \sin x - c_2 \cos x).$$

EXAMPLE 2 Apply the reduction-of-order technique to the differential equation

$$y''' - 4y'' + 5y' - 2y = 0, \tag{5}$$

given that $y_1 = e^x$ is a solution.

Solution As in Example 1, we leave it to the reader to verify that y_1 is a solution of the differential equation (5). Now

$$\begin{aligned} & y = y_1 u = e^x u \\ \Rightarrow \quad & y' = e^x u + e^x u' \\ \Rightarrow \quad & y'' = e^x u + 2e^x u' + e^x u'' \\ \Rightarrow \quad & y''' = e^x u + 3e^x u' + 3e^x u'' + e^x u'''. \end{aligned}$$

Substitution of these results into Eq. (5) yields (after simplification)

$$e^x u''' - e^x u'' = 0,$$

and setting $v = u'$ we find

$$e^x v'' - e^x v' = 0. \tag{6}$$

Equation (6) has variable coefficients. However, since e^x is never zero, the differential equation is equivalent to $v'' - v' = 0$. This latter differential equation has $v_1 = 1$ and $v_2 = e^x$ as its solutions. Since $u' = v$, it follows that $u_1 = x$ and $u_2 = e^x$. Finally, $y_2 = u_1 y_1 = xe^x$ and $y_3 = u_2 y_1 = e^{2x}$. The general solution to Eq. (5) is

$$\begin{aligned} y &= c_1 y_1 + c_2 y_2 + c_3 y_3 \\ &= c_1 e^x + c_2 x e^x + c_3 e^{2x}. \end{aligned}$$

REMARK 1 Equation (5) has constant coefficients and can be solved by the method of Section 2.5. We solved it by reduction of order in order to illustrate that method with a relatively simple example.

EXAMPLE 3 Use the reduction-of-order method to find the general solution of the differential equation

$$(\tfrac{4}{3}x^3 + 1)\, y''' - 4x^2 y'' + 8xy' - 8y = 0, \qquad x > 0,$$

given that $y_1 = x$ is a solution.

Solution Many of the details of the solution are left as exercises (see Exercises 30 through 34). Substituting $y_1 = x$, $y_1' = 1$, $y_1'' = y_1''' = 0$ into the differential equation yields

$$8x(1) - 8x = 0.$$

Thus, y_1 is indeed a solution. Setting $y = uy_1$ into the differential equation leads to the following differential equation for u:

$$\begin{aligned}[(\tfrac{4}{3}x^3 + 1)y_1]u''' &+ [3(\tfrac{4}{3}x^3 + 1)y_1' - 4x^2y_1]u'' \\ &+ [3(\tfrac{4}{3}x^3 + 1)y_1'' - 8x^2y_1' + 8xy_1]u' \\ &+ [(\tfrac{4}{3}x^3 + 1)y_1''' - 4x^2y_1'' + 8xy_1' - 8y_1]u = 0.\end{aligned} \tag{7}$$

Since y_1 is a solution of the differential equation, the coefficient of u is zero. Set $u' = v$; then the differential equation (7) takes the form

$$\begin{aligned}[(\tfrac{4}{3}x^3 + 1)y_1]v'' + [3(\tfrac{4}{3}x^3 + 1)y_1' - 4x^2y_1]v' \\ + [3(\tfrac{4}{3}x^3 + 1)y_1'' - 8x^2y_1' + 8xy_1]v = 0.\end{aligned}$$

Substituting $y_1 = x$, $y_1' = 1$, $y_1'' = 0$, we have

$$[(\tfrac{4}{3}x^3 + 1)x]v'' + [3(\tfrac{4}{3}x^3 + 1) - 4x^3]v' = 0. \tag{8}$$

This differential equation has the solutions

$$v_1 = \tfrac{4}{3}x - \tfrac{1}{2}x^{-2} \tag{9}$$

$$v_2 = 1. \tag{10}$$

These two solutions give rise to

$$y_2 = u_1y_1 = \tfrac{2}{3}x^3 + \tfrac{1}{2} \tag{11}$$

and

$$y_3 = u_2y_1 = x^2. \tag{12}$$

Now

$$W(y_1, y_2, y_3; x) = -\tfrac{4}{3}x^3 - 1 \neq 0, \tag{13}$$

because $x > 0$. Thus,

$$y = c_1x + c_2(\tfrac{2}{3}x^3 + \tfrac{1}{2}) + c_3x^2.$$

The substitution (2) always reduces the general differential equation (1) to an equation of lower order. In the special case $n = 2$, we can also use another approach. From the proof of Theorem 3 of Section 2.2 [Eq. (8)], we have the relation

$$W(y_1, y_2; x) = C\exp\left[-\int_{x_0}^{x} p(s)\,dx\right],$$

where $p(x) = a_1(x)/a_2(x)$. That is,

$$\begin{aligned}
y_1y_2' - y_1'y_2 &= C\exp\left[-\int_{x_0}^{x} p(s)ds\right] \\
\Rightarrow \frac{y_1y_2' - y_1'y_2}{y_1^2} &= \frac{C\exp\left[-\int_{x_0}^{x} p(s)ds\right]}{y_1^2} \\
\Rightarrow \quad \frac{d}{dx}\left(\frac{y_2}{y_1}\right) &= \frac{C}{y_1^2}\exp\left[-\int_{x_0}^{x} p(s)ds\right] \\
\Rightarrow \quad \frac{y_2}{y_1} &= C\int_{x_0}^{x} \frac{\exp\left[-\int_{x_0}^{t} p(s)ds\right]}{[y_1(t)]^2}dt \\
\Rightarrow \quad y_2 &= Cy_1\int_{x_0}^{x} \frac{\exp\left[-\int_{x_0}^{t} p(s)ds\right]}{[y_1(t)]^2}dt. \qquad (14)
\end{aligned}$$

Equation (14) is called *Abel's formula* and gives a second linearly independent solution of a second-order linear differential equation when one solution is known. Actually, the constant C in Eq. (14) can be omitted since in the general solution we consider $c_1y_1 + c_2y_2$ anyway. Also, the evaluation of the integral at x_0 can be neglected for the same reason.

EXAMPLE 4 Using Abel's formula, find a second linearly independent solution of the differential equation

$$xy'' - 2y' + \frac{x^2+2}{x}y = 0, \qquad 0 < x < \pi,$$

given that $y_1 = x\sin x$ is a solution.

Solution Observe that $a_2(x) = x$, $a_1(x) = -2$ and hence that $p(x) = -2/x$. Substituting into Abel's formula, Eq. (14), we obtain

$$\begin{aligned}
y_2 &= (x\sin x)\int^{x} \frac{(\exp - \int^{t} -2/s\, ds)}{(t\sin t)^2}dt = x\sin x\int^{x} \frac{e^{2\ln t}}{t^2\sin^2 t}dt \\
&= x\sin x\int^{x} \frac{t^2}{t^2\sin^2 t}dt = x\sin x\int^{x} \csc^2 t\, dt = x\sin x(-\cot x) \\
&= -x\cos x.
\end{aligned}$$

APPLICATIONS 2.8.1

Second-order linear ordinary differential equations with variable coefficients occur rather frequently in theoretical physics. They also play an indirect role in such mathematical studies as approximation theory and probability theory. Most often at least one of the linearly independent solutions (very often both) of these equations is an infinite series. It is not uncommon in these circumstances to find one solution and then express the second solution in integral form by means of Abel's formula. How one obtains the first solution is another question altogether. A partial answer to this question is presented in Section 2.9. A more elaborate and detailed treatment of the question is presented in Chapter 5.

Quantum Mechanics

■ In theoretical physics a basic equation of quantum mechanics is the Schrödinger wave equation

$$\frac{ih}{2\pi}\frac{\partial\phi}{\partial t} = -\frac{h^2}{8m\pi^2}\left(\frac{\partial^2\phi}{\partial x^2} + \frac{\partial^2\phi}{\partial y^2} + \frac{\partial^2\phi}{\partial z^2}\right) + V(x, y, z)\phi,$$

where h is a constant known as Planck's constant, V is the potential energy, and m is the mass of the particle whose wave function is ϕ (ϕ is related to the probability that the particle will be found in a differential element of volume at any particular time). The search for acceptable solutions of Schrödinger's equation for the special case of a harmonic oscillator leads to the differential equation

$$y'' - 2xy' + 2py = 0, \tag{15}$$

where p is an arbitrary constant. Equation (15), known as *Hermite's equation,* will be studied in detail in Section 5.4.1.

EXAMPLE 5 Let $p = 2$ in Eq. (15). Given that $y_1 = 4x^2 - 2$ is a solution of Eq. (15) (with $p = 2$), use Abel's formula to obtain a second linearly independent solution.

Solution We leave it to the reader to verify that $y_1 = 4x^2 - 2$ is a solution of $y'' - 2xy' + 4y = 0$. Applying Abel's formula [Eq. (14)] we have

$$y_2 = (4x^2 - 2)\int^x \frac{\exp - \int^t (-2s)\,ds}{(4t^2 - 2)^2}\,dt = (4x^2 - 2)\int^x \frac{e^{t^2}}{(4t^2 - 2)^2}\,dt.$$

Since the integral cannot be evaluated in terms of elementary functions, we content ourselves with having y_2 in integral form.

Mathematical Physics

■ Many theoretical studies in physics utilize *Laplace's equation,*

$$\frac{\partial^2 u}{\partial x^2} + \frac{\partial^2 u}{\partial y^2} + \frac{\partial^2 u}{\partial z^2} = 0.$$

There are various techniques associated with the problem of obtaining solutions to this equation, and there are a number of considerations that govern which technique is most efficient for a given situation. It is not our intention to discuss these considerations but rather to consider a special instance (for more elaboration and a few other considerations see Section 11.8).

If Laplace's equation is expressed in terms of spherical coordinates and then solved by the method of separation of variables for partial differential equations, the result is *Legendre's differential equation* (for details, see Section 5.4.1),

$$(1 - x^2)y'' - 2xy' + p(p + 1)y = 0, \qquad x < 1, \tag{16}$$

where p is a constant.

EXAMPLE 6 If $p = 2$, then $y_1 = \frac{3}{2}x^2 - \frac{1}{2}$ is a solution of Legendre's differential equation. Find an integral form for a second linearly independent solution.

Solution We leave it to the reader to verify that y_1 is a solution of the differential equation $(1 - x^2)y'' - 2xy' + 6y = 0$. Applying Abel's formula, we have

$$y_2 = (\tfrac{3}{2}x^2 - \tfrac{1}{2}) \int^x \frac{\exp\left[-\int^t \frac{-2s}{(1 - s^2)} ds\right]}{(\frac{3}{2}t^2 - \frac{1}{2})^2} dt = (\tfrac{3}{2}x^2 - \tfrac{1}{2}) \int^x \frac{e^{-\ln(1-t^2)}}{(\frac{3}{2}t^2 - \frac{1}{2})^2} dt$$

$$= (\tfrac{3}{2}x^2 - \tfrac{1}{2}) \int^x \frac{dt}{(1 - t^2)(\frac{3}{2}t^2 - \frac{1}{2})^2}. \tag{17}$$

The integral can be evaluated in this case (partial fractions) and is left as an exercise (Exercise 35).

EXERCISES

In Exercises 1 through 12, a differential equation and one of its solutions is given. Apply the reduction-of-order method (as in Examples 1 and 2) to obtain another linearly independent solution.

1. $x^2y'' + xy' - y = 0, x > 0, y_1 = x$

2. $x^2y'' + (2x^2 - x)y' - 2xy = 0, x > 0, y_1 = e^{-2x}$

3. $x^3y'' + (5x^3 - x^2)y' + 2(3x^3 - x^2)y = 0, x > 0, y_1 = e^{-2x}$

4. $x^4y'' + 2x^3y' - y = 0, x > 0, y_1 = e^{1/x}$

5. $x^2y'' + x^2y' - (x + 2)y = 0, x > 0, y_1 = x^{-1}e^{-x}$

6. $xy'' + (x - 1)y' + (3 - 12x)y = 0, x > 0, y_1 = e^{3x}$

7. $x^2(1 - \ln x)y'' + xy' - y = 0, x > e, y_1 = x$

8. $y'' + \frac{2}{x}y' + \frac{9}{x^4}y = 0, x > 0, y_1 = \cos\frac{3}{x}$

9. $x^2y'' + xy' + (x^2 - \frac{1}{4})y = 0, x > 0, y_1 = x^{-1/2}\sin x$

10. $(1 - x^2)y'' - 7xy' + 7y = 0, x > 1, y_1 = x$

11. $y'' + x^{-1/3}y' + (\frac{1}{4}x^{-2/3} - \frac{1}{6}x^{-4/3} - 6x^{-2})y = 0, x > 0, y_1 = x^3e^{-(3/4)x^{2/3}}$

12. $x(x - 2)y'' - 2(x^2 - 3x + 3)y' + (x^2 - 4x + 6)y = 0, x > 2,$
$y_1 = e^x(x^{-1} - x^{-2})$

In Exercises 13 through 19, show that y_1 is a solution of the differential equation. Use Abel's formula to obtain an expression for a second linearly independent solution.

13. $x(1 - 3x \ln x)y'' + (1 + 9x^2 \ln x)y' - (3 + 9x)y = 0,\ x > e,\ y_1 = e^{3x}$

14. $y'' - \left(1 + \dfrac{3}{2x}\right)y' + \dfrac{3}{2x^2}y = 0, \quad x > 0, \quad y_1 = x^{3/2}e^x$

15. $x(1 - 2x \ln x)y'' + (1 + 4x^2 \ln x)y' - (2 + 4x)y = 0,\ x > e,\ y_1 = \ln x$

16. $y'' - 2xy' + 2y = 0$ (Hermite $p = 1$), $y_1 = 2x$

17. $y'' - 2xy' + 6y = 0$ (Hermite $p = 3$), $y_1 = 8x^3 - 12x$

18. $(1 - x^2)y'' - 2xy' + 2y = 0$ (Legendre $p = 1$), $x \ < 1,\ y_1 = x$

19. $(1 - x^2)y'' - 2xy' + 12y = 0$ (Legendre $p = 3$), $|x| < 1,\ y_1 = \frac{5}{2}x^3 - \frac{3}{2}x$

Quantum Mechanics The study of solutions to the Schrödinger wave equation for the hydrogen atom leads to *Laguerre's differential equation,*

$$xy'' + (1 - x)y' + py = 0, \qquad x > 0,$$

where p is a constant. In Exercises 20, 21, and 22, verify that y_1 as given is a solution of Laguerre's differential equation for the given p. Using Abel's formula, obtain an integral form for a second linearly independent solution.

20. $p = 1,\ y_1 = 1 - x$

21. $p = 2,\ y_1 = 2 - 4x + x^2$

22. $p = 3,\ y_1 = 6 - 18x + 9x^2 - x^3$

Approximation Theory One aspect of approximation theory (from an oversimplified point of view) is the problem of approximating a nonpolynomial function by a polynomial. An important class of polynomials for this purpose is the class of Chebyshev polynomials. *Chebyshev polynomials* are solutions of the differential equation (called Chebyshev's differential equation)

$$(1 - x^2)y'' - xy' + p^2y = 0, \qquad x \ < 1,$$

in the special case that p is a nonnegative integer. In Exercises 23, 24, and 25, verify that y_1 is a solution of Chebyshev's differential equation for the given p. In each case use Abel's formula to obtain in integral form a second linearly independent solution.

23. $p = 1,\ y_1 = x$ **24.** $p = 2,\ y_1 = 2x^2 - 1$ **25.** $p = 3,\ y_1 = 4x^3 - 3x$

26. Find the general solution of the differential equation

$$(x^2 + 1)y'' - 2xy' + 2y = 0,$$

given that $y_1(x) = -x$ is a solution.

27. Using Abel's formula, find a second linearly independent solution of the differential equation

$$xy'' + (1 - 2x)y' + (x - 1)y = 0 \quad \text{for} \quad x > 0,$$

given that $y_1(x) = e^x$ is a solution.

28. Solve the IVP

$$xy'' + (1 - 2x)y' + (x - 1)y = 0$$
$$y(1) = 2e$$
$$y'(1) = -3e,$$

given that $y_1(x) = e^x$ is a solution of the differential equation.

29. Show that $y_1 = xe^x$ is a solution of the differential equation

$$y''' - 4y'' + 5y' - 2y = 0.$$

Show that the reduction-of-order method leads to the differential equation

$$xv'' + (3 - x)v' - 2v = 0.$$

30. Verify Eq. (7) of this section.

31. Verify Eq. (8) of this section.

32. Set $w = v'$ in Eq. (8) to obtain

$$[(\tfrac{4}{3}x^3 + 1)x]w' + [3(\tfrac{4}{3}x^3 + 1) - 4x^3]w = 0.$$

Solve this differential equation by separation of variables and then integrate the result to obtain Eq. (9).

33. Verify that $v_2 = 1$ is a solution of the differential equation (8).

34. Verify Eq. (13) of this section.

35. Perform the integrations required to simplify Eq. (17).

36. Let $a_2y'' + a_1y' + a_0y = 0$ be a differential equation with constant coefficients. Assume that y_1 is a nontrivial solution of the differential equation. Show that the reduction-of-order method leads to a (first-order) differential equation with constant coefficients if and only if y_1'/y_1 is equal to a constant.

2.9 SOLUTIONS OF LINEAR HOMOGENEOUS DIFFERENTIAL EQUATIONS BY THE METHOD OF TAYLOR SERIES

In this section we provide a brief introduction to the method of finding Taylor-series representations for solutions of differential equations. This method is perhaps the most widely applicable technique for solving (or approximating the solutions of) differential equations with variable coefficients. There are many ramifications associated with the Taylor-series technique. Discussion of these ramifications is postponed to Chapter 5, where the method is discussed in detail. We include the section here simply to round out our treatment of linear differential equations.

We reiterate the assumption made earlier in this chapter that all the coefficient functions are continuous and that the leading coefficient function $a_n(x)$ does not vanish in the interval of definition of the differential equation.

Consider the initial value problem

$$\begin{gathered} y'' + xy = 0 \\ y(0) = 1, \qquad y'(0) = 0. \end{gathered} \tag{1}$$

Theorem 1 of Section 2.3 guarantees us that this IVP possesses a unique solution. None of the methods that we have discussed can be applied to this differential equation. First, recall from calculus that a Taylor-series representation for a function f is of the form

$$f(x) = f(x_0) + f'(x_0)(x - x_0) + \frac{f''(x_0)}{2!}(x - x_0)^2 + \frac{f'''(x_0)}{3!}(x - x_0)^3 + \cdots$$

or, compactly,

$$f(x) = \sum_{n=0}^{\infty} \frac{f^{(n)}(x_0)}{n!}(x - x_0)^n.$$

We will seek a Taylor expansion for the function y *defined* by the IVP (1). There are essentially two approaches that one can follow, and we illustrate each approach via an example.

EXAMPLE 1 Using a Taylor-series approach, solve the IVP (1).

Solution The Taylor (Maclaurin) expansion has the form

$$y(x) = y(0) + y'(0)x + \frac{y''(0)}{2!}x^2 + \frac{y'''(0)}{3!}x^3 + \cdots \tag{2}$$

Now, $y(0) = 1$, $y'(0) = 0$, and from (1) we have

$$\begin{aligned} y'' &= -xy \\ \Rightarrow \quad y''(0) &= -(0)\cdot(1) = 0. \end{aligned} \tag{3}$$

Differentiating (3), we obtain

$$\begin{aligned} y''' &= -y - xy' \\ \Rightarrow \quad y'''(0) &= -1 - (0)(0) = -1. \end{aligned} \tag{4}$$

Differentiating (4), we have

$$\begin{aligned} y^{\text{iv}} &= -2y' - xy'' \\ \Rightarrow \quad y^{\text{iv}}(0) &= 0. \end{aligned}$$

Similarly,

$$\begin{aligned} y^{(5)} &= -3y'' - xy''' \Rightarrow y^{(5)}(0) = 0 \\ y^{(6)} &= -4y''' - xy^{\text{iv}} \Rightarrow y^{(6)}(0) = 4. \end{aligned}$$

We could proceed in this fashion to generate as many terms of the Taylor expansion as we want. (In general, we cannot find them all in this manner, since there are an infinity of terms; therefore, the number of terms to be computed would have to be prescribed at the onset.) Substituting our results into Eq. (2) yields

$$y = 1 - \frac{1}{3!}x^3 + \frac{4}{6!}x^6 + \cdots .$$

EXAMPLE 2 Using power series, solve the differential equation

$$(2 - x)y'' + y = 0. \tag{5}$$

Solution In this example we are not given the initial values; therefore, we cannot compute the coefficients in the Taylor expansion in the manner of Example 1. Here we assume that the solution y of Eq. (5) has an expansion of the form

$$y = a_0 + a_1x + a_2x^2 + a_3x^3 + a_4x^4 + a_5x^5 + \cdots, \tag{6}$$

where the coefficients a_i are constants to be determined. From (6), we obtain

$$y' = a_1 + 2a_2x + 3a_3x^2 + 4a_4x^3 + 5a_5x^4 + \cdots$$

and

$$y'' = 2a_2 + 6a_3x + 12a_4x^2 + 20a_5x^3 + \cdots.$$

Substituting these results in the differential equation (5) yields

$$(2 - x)(2a_2 + 6a_3x + 12a_4x^2 + 20a_5x^3 + \cdots) + (a_0 + a_1x + a_2x^2 + a_3x^3 + a_4x^4 + a_5x^5 + \cdots) = 0.$$

Doing the multiplication and gathering together terms with like powers of x, we have

$$(4a_2 + a_0) + (12a_3 - 2a_2 + a_1)x + (24a_4 - 6a_3 + a_2)x^2 + (40a_5 - 12a_4 + a_3)x^3 + \cdots = 0. \tag{7}$$

In order that Eq. (7) hold *for every x* in the interval of definition of the differential equation, we must have that the coefficient of each power of x must vanish. Thus, we are led to the *infinite* system of equations

$$\begin{aligned} 4a_2 + a_0 &= 0 \\ 12a_3 - 2a_2 + a_1 &= 0 \\ 24a_4 - 6a_3 + a_2 &= 0 \\ 40a_5 - 12a_4 + a_3 &= 0, \end{aligned} \tag{8}$$

and so on. The first of Eqs. (8) yields $a_2 = -\frac{1}{4}a_0$ and the second yields $a_3 = \frac{1}{6}a_2 - \frac{1}{12}a_1$, which in light of the result for a_2 reduces to $a_3 = -\frac{1}{24}a_0 - \frac{1}{12}a_1$.

Analogous manipulations apply to each subsequent equation, and we are led to the results

$$\begin{aligned} a_2 &= -\tfrac{1}{4}a_0 \\ a_3 &= -\tfrac{1}{24}a_0 - \tfrac{1}{12}a_1 \\ a_4 &= -\tfrac{1}{48}a_1 \\ a_5 &= \tfrac{1}{960}a_0 - \tfrac{1}{240}a_1, \end{aligned} \tag{9}$$

and so on. Thus,

$$\begin{aligned} y &= a_0 + a_1x + (-\tfrac{1}{4}a_0)x^2 + (-\tfrac{1}{24}a_0 - \tfrac{1}{12}a_1)x^3 + (-\tfrac{1}{48}a_1)x^4 \\ &\quad + (\tfrac{1}{960}a_0 - \tfrac{1}{240}a_1)x^5 + \cdots \\ &= a_0(1 - \tfrac{1}{4}x^2 - \tfrac{1}{24}x^3 + \tfrac{1}{960}x^5 + \cdots) \\ &\quad + a_1(x - \tfrac{1}{12}x^3 - \tfrac{1}{48}x^4 - \tfrac{1}{240}x^5 + \cdots). \end{aligned} \tag{10}$$

We note that the solution (10) depends on two arbitrary constants, a_0 and a_1, as it should, since the differential equation (5) is of second order. We also conjecture that the parenthetical expressions in Eq. (10) are the two linearly independent solutions to the differential equation (5). Furthermore, these expressions are infinite series and therefore their convergence must be considered. That is, do the expressions in parentheses in Eq. (10) converge for every x in the interval of definition of the differential equation? These conjectures and observations will be discussed in Chapter 5.

Example 2 illustrates the Taylor-series method most often used. We assume a solution of the form $y = \sum_{n=0}^{\infty} a_n(x - x_0)^n$. We substitute this expression into the differential equation and then rearrange terms so that like powers of $(x - x_0)$ can be combined. The next step is to set the coefficient of each power of $(x - x_0)$ equal to zero, thus obtaining an infinite system of equations [for example, Eqs. (8)]. Finally, we obtain the coefficients a_n from this infinite system.

EXERCISES

In Exercises 1 through 10, use the Taylor-series method of Example 1 to solve the initial value problem. In each exercise, compute the coefficients out to the fifth power of the x term.

1. $y'' + xy' + y = 0,\ y(0) = 1,\ y'(0) = 0$

2. $y''' - (\sin x)y = 0,\ y(0) = 0,\ y'(0) = 1,\ y''(0) = 0$

3. $y'' + y' + e^x y = 0,\ y(0) = 2,\ y'(0) = 1$

4. $y^{\text{iv}} - \ln(1 + x)y = 0,\ y(0) = 1,\ y'(0) = 1,\ y''(0) = 0,\ y'''(0) = 0$

5. $y'' + (3 + x)y = 0,\ y(0) = 1,\ y'(0) = 0$

6. $y'' + x^2y = 0,\ y(0) = 0,\ y'(0) = 2$

7. $y'' - 2xy' + y = 0$, $y(0) = 2$, $y'(0) = 1$
8. $y''' - 3x^2y' + 2xy = 0$, $y(0) = 1$, $y'(0) = 1$, $y''(0) = 0$
9. $y''' + 4y'' + 2y' - x^3y = 0$, $y(0) = 1$, $y'(0) = 1$, $y''(0) = 0$
10. $y'' + (\cos x)y = 0$, $y(0) = 0$, $y'(0) = 1$

In Exercises 11 through 20, use the Taylor-series method of Example 2 to solve the differential equation. In each exercise, compute the coefficients out to the fifth power of the x term.

11. $y'' + y = 0$
12. $y'' + xy = 0$
13. $y'' + x^2y' + xy = 0$
14. $y''' + y = 0$
15. $(x^2 + 2)y'' - 3xy' + 4y = 0$
16. $y'' + x^2y = 0$
17. $(2x^2 + 1)y'' + 3y' - xy = 0$
18. $y' + 3xy = 0$
19. $y''' - 2xy = 0$
20. $(x + 1)y'' + 3y' - 2x^2y = 0$
21. **Electronics** The differential equation

$$(1 + a\cos 2x)y'' + \lambda y = 0, \qquad a \neq -1,$$

has been studied by Campi.[23] The constants a and λ are physical parameters. Recall that

$$\cos t = 1 - \frac{t^2}{2!} + \frac{t^4}{4!} - \cdots,$$

and find a_2, a_3, a_4, and a_5 in terms of a, λ, a_0, and a_1 if the solution y is taken in the form

$$y = a_0 + a_1x + a_2x^2 + a_3x^3 + a_4x^4 + a_5x^5 + \cdots.$$

2.10 NONHOMOGENEOUS DIFFERENTIAL EQUATIONS

In the analysis of the deflection of a beam one encounters the differential equation $EIy^{(4)} = f(x)$ (see Exercises 59–64, Section 2.11), where E and I are physical constants and f is a given function of the independent variable. This equation, although linear, is nonhomogeneous. For nonhomogeneous differential equations, in contrast to homogeneous differential equations, there is no guarantee that $c_1y_1 + c_2y_2$ is a solution merely because y_1 and y_2 happen to be solutions. In fact, it is generally the case for nonhomogeneous differential equations that $c_1y_1 + c_2y_2$ is *not* a solution. As a consequence, it requires a little ingenuity to describe the general solution for nonhomogeneous linear differential equations. However, this is always possible provided we know the general solution of the associated homogeneous differential equation.

[23] E. Campi, "Trigonometric Components of a Frequency-Modulated Wave," *Proc. IRE* (1948).

The general nonhomogeneous linear differential equation has the form

$$a_n(x)y^{(n)} + a_{n-1}(x)y^{(n-1)} + \cdots + a_1(x)y' + a_0(x)y = f(x) \tag{1}$$

or, briefly,

$$\sum_{i=0}^{n} a_i(x)y^{(i)} = f(x). \tag{2}$$

The function f is called the *nonhomogeneous term* for the differential equation (1).

DEFINITION 1

With every nonhomogeneous differential equation (2), *there is an associated homogeneous differential equation defined by*

$$\sum_{i=0}^{n} a_i(x)y^{(i)} = 0. \tag{3}$$

DEFINITION 2

If the n functions $y_1, y_2, \ldots, y_n$ constitute a fundamental system of solutions for the homogeneous differential equation (3), *then the function y_h defined by*

$$y_h = c_1y_1 + c_2y_2 + \cdots + c_ny_n, \tag{4}$$

where the c_i are arbitrary constants, is called the homogeneous solution for Eq. (2). [Note that the homogeneous solution is not an actual solution of Eq. (2). It is the general solution of the associated homogeneous differential equation. In some texts the homogeneous solution is referred to as the *complementary solution.*]

EXAMPLE 1 Find the homogeneous solution of the differential equation

$$y'' - 3y' + 2y = \cos x. \tag{5}$$

Solution The homogeneous equation associated with the differential equation (5) is

$$y'' - 3y' + 2y = 0.$$

Using the methods of Section 2.5, the general solution of this equation is $c_1e^x + c_2e^{2x}$. Thus, the homogeneous solution of the differential equation (5) is

$$y_h = c_1e^x + c_2e^{2x}.$$

EXAMPLE 2 Find the homogeneous solution of the differential equation

$$2x^2y'' - 3xy' - 3y = e^x, \qquad x > 0. \tag{6}$$

Solution The homogeneous equation associated with the differential equation (6) is

$$2x^2y'' - 3xy' - 3y = 0.$$

This latter differential equation is an Euler differential equation, and its solution (given in Example 2 of Section 2.7) is $c_1x^{-1/2} + c_2x^3$. Thus,

$$y_h = c_1x^{-1/2} + c_2x^3.$$

If, somehow or other, we find a function that satisfies Eq. (2), we refer to this function as a *particular solution* of Eq. (2) and denote it by y_p.

THEOREM 1

If $y_1, y_2, \ldots, y_n$ form a fundamental system of solutions for Eq. (3), *and if y_p is any particular solution of Eq.* (2), *then the general solution of Eq.* (2) *can be written in the form*

$$y = y_h + y_p = c_1y_1 + c_2y_2 + \cdots c_ny_n + y_p. \tag{7}$$

Proof It suffices to demonstrate that y as given in the expression (7) satisfies the differential equation (2) and that any solution Y of the differential equation (2) can be written in the form of Eq. (7). Substituting into the differential equation (2), we have

$$\begin{aligned}
\sum_{i=0}^{n} a_i(x)y^{(i)} &= \sum_{i=0}^{n} a_i(x)[c_1y_1 + c_2y_2 + \cdots + c_ny_n + y_p]^{(i)} \\
&= \sum_{i=0}^{n} a_i(x)[c_1y_1^{(i)} + c_2y_2^{(i)} + \cdots + c_ny_n^{(i)} + y_p^{(i)}] \\
&= c_1 \sum_{i=0}^{n} a_i(x)y_1^{(i)} + c_2 \sum_{i=0}^{n} a_i(x)y_2^{(i)} \\
&\quad + \cdots + c_n \sum_{i=0}^{n} a_i(x)y_n^{(i)} + \sum_{i=0}^{n} a_i(x)y_p^{(i)}
\end{aligned}$$

$$\begin{aligned}
\Rightarrow \quad \sum_{i=0}^{n} a_i(x)y^{(i)} &= c_1 \cdot 0 + c_2 \cdot 0 + \cdots + c_n \cdot 0 + f(x) \\
&= f(x).
\end{aligned}$$

That is, the function y as defined by Eq. (7) satisfies the differential equation (2). Now let Y be any solution of the differential equation (2). We should prove that Y can be written in the form of Eq. (7). This is equivalent to proving that the function $Y - y_p$ satisfies the homogeneous differential equation (3). In fact,

$$\sum_{i=0}^{n} a_i(x)(Y - y_p)^{(i)} = \sum_{i=0}^{n} a_i(x)Y^{(i)} - \sum_{i=0}^{n} a_i(x)y_p^{(i)} = f(x) - f(x) = 0.$$

The proof is complete.

From this theorem and the fact that the last few sections provided us with information concerning the fundamental system of solutions of Eq. (3), we realize that we need only concentrate on methods associated with finding the particular solution y_p. There are essentially two popular methods for finding a particular solution: the *method of undetermined coefficients* and the *method of variation of parameters.* (See Sections 2.11 and 2.12.)

EXERCISES

In Exercises 1 through 15, find the homogeneous solution of each nonhomogeneous differential equation.

1. $y'' - 5y' + 6y = x^2 + 3$

2. $y'' + 4y' + 4y = e^x + e^{-2x}$

3. $y'' - y' - 2y = \cos x$

4. $y'' - y = e^x$

5. $y'' + 9y = \cos 3x - \sin 3x$

6. $y''' + y' - 2y = x^3$

7. $y'' - 13y' + 36y = xe^{4x}$

8. $x^2y'' - 2xy' + 2y = \tan x,\ 0 < x < \pi/2$

9. $y'' - 10y' + 25y = x^2e^{5x}$

10. $y''' - y = 3 \ln x,\ x > 0$

11. $y^{iv} - y = x^2$

12. $x^2y'' + 4xy' - 10y = x \ln x,\ x > 0$

13. $3x^2y'' - 2xy' - 8y = 3x + 5,\ x > 0$

14. $y'' + y' = e^{-x}$

15. $xy'' - 2y' + \dfrac{x^2 + 2}{x}y = 4 + \tan x,\quad 0 < x < \pi/2$ (*Hint:* See Example 1, Section 2.8.)

Answer true or false.

16. $3xy''' - 4xy = \cos y$ is a nonhomogeneous differential equation.

17. $y'' + 5y' = \sin x$ has $c_1 + c_2e^{-5x}$ as its homogeneous solution.

18. $y'' + y = x$ has a particular solution $y_p = x$.

19. $y''' - 3xy'' + 4y = x^2$ has a particular solution $y_p = \frac{1}{4}x^2 + \frac{3}{8}x$.

20. $y'' - 3y = \cos x$ has a particular solution $y_p = \cos x$.

21. $3xy''' + 5y'' + 6y' - 3 \cos x = 0$ is a homogeneous differential equation.

22. $y'' + 2y = e^x$ has a particular solution $y_p = \frac{1}{3}e^x$.

23. $y'' - y = e^x$ has a particular solution $y_p = \frac{1}{2}xe^x$.

2.11 THE METHOD OF UNDETERMINED COEFFICIENTS

The *method of undetermined coefficients* is used when we want to compute a particular solution of the nonhomogeneous differential equation

$$a_n y^{(n)} + a_{n-1} y^{(n-1)} + \cdots + a_1 y' + a_0 y = f(x), \tag{1}$$

where the coefficients $a_0, a_1, \ldots, a_n$ are constants and $f(x)$ is a linear combination of functions of the following types:

1. x^α, where α is a positive integer or zero.
2. $e^{\beta x}$, where β is a nonzero constant.
3. $\cos \gamma x$, where γ is a nonzero constant.
4. $\sin \delta x$, where δ is a nonzero constant.
5. A (finite) product of two or more functions of types 1–4.

For example, the function

$$f(x) = 3x^2 - 2 + 5e^{3x} - x(\sin x)e^{2x} + 5\cos 2x + xe^x$$

is a linear combination of functions of types 1–5. On the other hand, the functions $1/x$ and $\log x$ are not of these types.

The key observation that makes the method of undetermined coefficients work is the fact that not only $f(x)$ but also *any derivative* (zeroth derivative included; recall that the zeroth derivative of f is f itself) of any term of $f(x)$ is a linear combination of functions of types 1–5. For example, any derivative of $3x^2$ is a linear combination of the functions x^2, x, and 1, which are all of type 1. We use the symbolism

$$3x^2 \rightarrow \{x^2, x, 1\}$$

to denote that any derivative of the function $3x^2$ on the left is a linear combination of the functions x^2, x, 1 on the right. We also say that the functions x^2, x, 1 *span* the derivatives of the function $3x^2$. As further illustrations, we have

$$2e^{3x} \rightarrow \{e^{3x}\}$$

$$-5 \rightarrow \{1\}$$

$$5\cos 2x \rightarrow \{\cos 2x, \sin 2x\}$$

$$xe^x \rightarrow \{xe^x, e^x\}$$

$$-xe^{2x}\sin x \rightarrow \{xe^{2x}\sin x, xe^{2x}\cos x, e^{2x}\sin x, e^{2x}\cos x\}.$$

Roughly speaking, a particular solution of Eq. (1) is a linear combination of those functions of types 1–5 that span all the derivatives of $f(x)$. The coefficients of this linear combination are the undetermined coefficients (hence the name attached to the method) that are to be determined by substituting the assumed particular solution into the differential equation (1) and equating coefficients of similar terms.

The following example contains the typical features of the method of undetermined coefficients, and the reader is urged to study this example very carefully. (For motivation of the method, see Exercises 65 and 66.)

EXAMPLE 1 Compute a particular solution.of the differential equation

$$y'' - y = -2x^2 + 5 + 2e^x. \tag{2}$$

Solution The method of undetermined coefficients is applicable to this example since the differential equation (2) is of the form of Eq. (1), and $f(x) = -2x^2 + 5 + 2e^x$ is a linear combination of the functions x^2, 1, and e^x (which are of types 1, 1, and 2, respectively). We first compute the functions that span the derivatives of each one of the three terms of the function f. With our previous notation, we have

$$-2x^2 \to \{x^2, x, 1\} \tag{3}$$

$$5 \to \{1\} \tag{4}$$

$$2e^x \to \{e^x\}. \tag{5}$$

Therefore, the derivatives of f are spanned by the functions in the sets $\{x^2, x, 1\}$ and $\{e^x\}$. The set $\{1\}$ in (4) is omitted because it is contained in the larger set (3). Now if *none* of the functions in the sets $\{x^2, x, 1\}$ and $\{e^x\}$ is a solution of the associated homogeneous equation for (2), that is, $y'' - y = 0$, then the particular solution of Eq. (2) is of the form

$$y_p = Ax^2 + Bx + C + De^x,$$

where A, B, C, and D are the undetermined coefficients. On the other hand, if *any* one of the functions in *any* one of the sets $\{x^2, x, 1\}$ and $\{e^x\}$ is a solution of the associated homogeneous equation, then all the elements of that set should be multiplied by the lowest integral power of x, so that the resulting new set does not contain any function that is a solution of the associated homogeneous equation. Now, no function in the set $\{x^2, x, 1\}$ is a solution of $y'' - y = 0$, and so we leave this set as it is. On the other hand, e^x in the set $\{e^x\}$ is a solution of $y'' - y = 0$, and so we must multiply the function in this set by the lowest integral power of x, so that the resulting function is not a solution of $y'' - y = 0$. Since e^x is, but xe^x is not, a solution of $y'' - y = 0$, we should multiply e^x by x, *obtaining* $\{xe^x\}$. Thus, a particular solution of Eq. (2) is a linear combination of the functions in the sets $\{x^2, x, 1\}$ and $\{xe^x\}$. Hence,

$$y_p = Ax^2 + Bx + C + Dxe^x. \tag{6}$$

To obtain the undetermined coefficients A, B, C, and D, we compute

$$y_p' = 2Ax + B + De^x + Dxe^x$$

$$y_p'' = 2A + 2De^x + Dxe^x$$

and substitute these results into the differential equation (2), obtaining

$$(2A + 2De^x + Dxe^x) - (Ax^2 + Bx + C + Dxe^x) = -2x^2 + 5 + 2e^x.$$

Equating coefficients of similar terms, we obtain the following system of equations:

$$2A - C = 5$$
$$2D = 2$$
$$-B = 0$$
$$-A = -2.$$

Thus, $A = 2$, $B = 0$, $C = -1$, $D = 1$, and a particular solution of Eq. (2) is

$$y_p = 2x^2 - 1 + xe^x.$$

The crucial aspect of the method of undetermined coefficients is that we assumed the proper form for a particular solution (such as, xe^x instead of e^x in Example 1). Should we assume an inappropriate form for a particular solution, a contradiction will occur in the resulting system of equations when we attempt to compute the undetermined coefficients. Alternatively, it may happen that the coefficients of unnecessary terms will be found to be equal to zero.

To save space, we suggest that the explanations that lead us to the form (6) for a particular solution of Eq. (2) be abbreviated in accordance with the following format:

$$-2x^2 \rightarrow \{x^2, x, 1\}$$
$$5 \rightarrow \{1\}$$
$$2e^x \rightarrow \{e^x\} \rightarrow \{xe^x\}$$

(e^x is a solution of the associated homogeneous equation and xe^x is not). So

$$y_p = Ax^2 + Bx + C + Dxe^x$$

is the correct form of a particular solution of Eq. (2).

EXAMPLE 2 Find the form of a particular solution of the differential equation

$$y'' + 2y' - 3y = 3x^2e^x + e^{2x} + x \sin x + 2 + 3x.$$

Solution Since e^x is a solution of the associated homogeneous equation but xe^x is not, we have

$$3x^2e^x \rightarrow \{x^2e^x, xe^x, e^x\} \rightarrow \{x^3e^x, x^2e^x, xe^x\}$$
$$e^{2x} \rightarrow \{e^{2x}\}$$
$$x \sin x \rightarrow \{x \sin x, x \cos x, \sin x, \cos x\}$$
$$2 \rightarrow \{1\}$$
$$3x \rightarrow \{x, 1\}.$$

Thus,

$$y_p = Ax^3e^x + Bx^2e^x + Cxe^x + De^{2x} + Ex\sin x$$
$$+ Fx\cos x + G\sin x + H\cos x + Ix + J.$$

EXAMPLE 3 Compute a particular solution of the differential equation

$$y'' - 2y' + y = 2e^x - 3e^{-x}. \tag{7}$$

Solution We have $2e^x \to \{e^x\} \to \{x^2e^x\}$ (e^x and xe^x are solutions of the associated homogeneous equation and x^2e^x is not) and $-3e^{-x} \to \{e^{-x}\}$. Thus,

$$y_p = Ax^2e^x + Be^{-x};$$

hence,

$$y_p' = 2Axe^x + Ax^2e^x - Be^{-x}$$

$$y_p'' = 2Ae^x + 4Axe^x + Ax^2e^x + Be^{-x}.$$

Substituting y_p, y_p', and y_p'' into the differential equation (7), we have

$$(2Ae^x + 4Axe^x + Ax^2e^x + Be^{-x}) - 2(2Axe^x + Ax^2e^x - Be^{-x})$$
$$+ (Ax^2e^x + Be^{-x}) = 2e^x - 3e^{-x},$$
$$\Rightarrow \quad 2Ae^x + 4Be^{-x} = 2e^x - 3e^{-x}.$$

Equating the coefficients of e^x and equating the coefficients of e^{-x}, we obtain $A = 1$ and $B = -\frac{3}{4}$; thus,

$$y_p = x^2e^x - \tfrac{3}{4}e^{-x}.$$

EXAMPLE 4 Compute the general solution of the differential equation

$$y'' - y' - 2y = e^{3x}\cos 2x. \tag{8}$$

Solution The general solution of the nonhomogeneous differential equation is

$$y = y_h + y_p.$$

We first compute the homogeneous solution y_h, which is the general solution of the associated homogeneous equation $y'' - y' - 2y = 0$. Now the characteristic equation is $\lambda^2 - \lambda - 2 = 0$, and therefore the characteristic roots are $\lambda_1 = 2$ and $\lambda_2 = -1$. Hence,

$$y_h = c_1e^{2x} + c_2e^{-x}.$$

As for y_p, we observe that

$$e^{3x}\cos 2x \to \{e^{3x}\cos 2x,\ e^{3x}\sin 2x\}.$$

Since neither of the functions $e^{3x}\cos 2x$ and $e^{3x}\sin 2x$ is a solution of the associated homogeneous equation, it follows that a particular solution of Eq. (8) is of the form

$$y_p = Ae^{3x}\cos 2x + Be^{3x}\sin 2x = e^{3x}(A\cos 2x + B\sin 2x).$$

Substituting this y_p into Eq. (8) and equating coefficients of similar terms, we find that $A = 0$ and $B = \frac{1}{10}$ and so

$$y_p = \tfrac{1}{10}e^{3x}\sin 2x.$$

Therefore, the general solution of the differential equation (8) is

$$y = c_1e^{2x} + c_2e^{-x} + \tfrac{1}{10}e^{3x}\sin 2x.$$

EXAMPLE 5 Solve the initial value problem

$$y''' + y' = 2 + \sin x \tag{9}$$

$$y(0) = 0, \qquad y'(0) = 1, \qquad y''(0) = -1.$$

Solution We first compute the general solution of Eq. (9), which is the sum of the general solution y_h of the associated homogeneous equation $y''' + y' = 0$, and a particular solution y_p of Eq. (9). As for y_h, the characteristic equation is $\lambda^3 + \lambda = 0$, hence the characteristic roots are $\lambda_1 = 0$, $\lambda_2 = i$, $\lambda_3 = -i$. Thus,

$$y_h = c_1 + c_2\cos x + c_3\sin x.$$

From Eq. (9) the nonhomogeneous term $2 + \sin x$ gives rise to the sets

$$2 \to \{1\} \to \{x\}$$

$$\sin x \to \{\sin x, \cos x\} \to \{x\sin x, x\cos x\}.$$

Consequently, we assume y_p to be of the form

$$y_p = Ax + Bx\sin x + Cx\cos x.$$

Therefore,

$$\begin{aligned} y_p' &= A + B\sin x + Bx\cos x + C\cos x - Cx\sin x \\ y_p'' &= 2B\cos x - Bx\sin x - 2C\sin x - Cx\cos x \\ y_p''' &= -3B\sin x - Bx\cos x - 3C\cos x + Cx\sin x. \end{aligned}$$

Substituting y_p' and y_p''' into the differential equation (9), we obtain

$$A - 2B\sin x - 2C\cos x = 2 + \sin x.$$

Therefore, $A = 2$, $B = -\frac{1}{2}$, $C = 0$, and $y_p = 2x - \frac{1}{2}x\sin x$.

So the general solution of Eq. (9) is

$$y = c_1 + c_2\cos x + c_3\sin x + 2x - \tfrac{1}{2}x\sin x. \tag{10}$$

Finally, we use the initial conditions to determine c_1, c_2, and c_3. Differentiating Eq. (10), we have

$$y' = -c_2 \sin x + c_3 \cos x + 2 - \tfrac{1}{2}\sin x - \tfrac{1}{2}x \cos x$$

$$\Rightarrow \quad y'' = -c_2 \cos x - c_3 \sin x - \cos x + \tfrac{1}{2}x \sin x.$$

Now

$$y(0) = 0 \Rightarrow c_1 + c_2 = 0$$

$$y'(0) = 1 \Rightarrow c_3 + 2 = 1$$

$$y''(0) = -1 \Rightarrow -c_2 - 1 = -1.$$

Thus, $c_1 = 0$, $c_2 = 0$, $c_3 = -1$, and the solution of the initial value problem is

$$y(x) = -\sin x + 2x - \tfrac{1}{2}x \sin x.$$

APPLICATIONS 2.11.1

Electric Circuits

■ The electric circuit shown in Figure 2.5 is customarily called an *RLC*-series circuit. Here a generator supplying a voltage of $V(t)$ volts is connected in series with an R-ohm resistor, an inductor of L henries, and a capacitor of C farads. When the switch is closed, a current of $I = I(t)$ amperes flows in the circuit. We would like to find the current I as a function of time. We also want to find the charge $Q = Q(t)$ coulombs in the capacitor at any time t. By definition

$$I = \frac{dQ}{dt} \tag{11}$$

Thus, it suffices to compute Q. It is known from physics that the current I produces a voltage drop across the resistor equal to RI, a voltage drop across the inductor equal to $L(dI/dt)$, and a voltage drop across the capacitor equal to

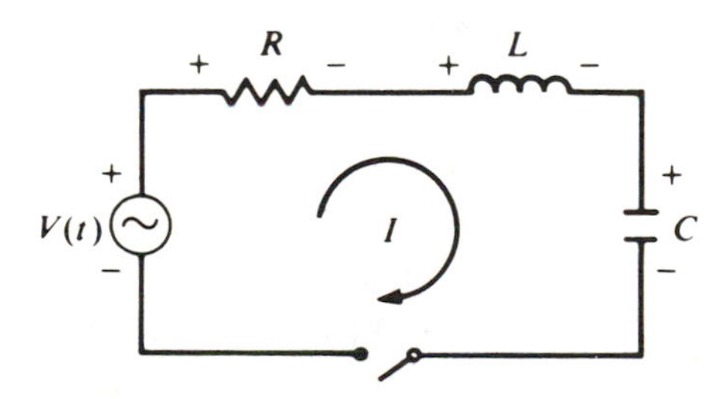

Figure 2.5

$(1/C)Q$. Kirchhoff's voltage law states that the voltage supplied is equal to the sum of the voltage drops in the circuit. Applying this law to the circuit of Figure 2.5 (with the switch closed), we have

$$L\frac{dI}{dt} + RI + \frac{1}{C}Q = V(t). \tag{12}$$

Using the relation (11), Eq. (12) can be written in the form

$$L\frac{d^2Q}{dt^2} + R\frac{dQ}{dt} + \frac{1}{C}Q = V(t). \tag{13}$$

This is a second-order linear nonhomogeneous differential equation. In order to find the charge $Q(t)$ in the capacitor, we must find the general solution of Eq. (13). This general solution involves two arbitrary constants. To determine these two constants, we impose two initial conditions $Q(0) = Q_0$ and $\dot{Q}(0) = 0$. Q_0 is the initial charge on the capacitor and $\dot{Q}(0) = I(0) = 0$, because at time $t = 0$ there is no current flow in the circuit. To find the current $I(t)$, we can use the relation (11) or the differential equation

$$L\frac{d^2I}{dt^2} + R\frac{dI}{dt} + \frac{1}{C}I = \frac{dV(t)}{dt}, \tag{14}$$

which is obtained by differentiating Eq. (12). Initial conditions associated with this differential equation are $I(0) = 0$ and $\dot{I}(0) = (1/L)[V(0) - (1/C)Q(0)]$, the latter condition being obtained from the differential equation (12).

Electromechanical Analogies ■ If we compare Eq. (13) with Eq. (16) of Exercise 57, we observe the analogies between mechanical and electrical systems given in Table 2.2. These analogies are very useful in applications. Often the study of a mechanical system is converted to the study of an equivalent electrical system, which is usually easier to study. For example, this is one of the basic ideas behind the design of analog computers.

Mechanical		Electrical
Mass	$m \leftrightarrow L$	Inductance
Friction (damping)	$\alpha \leftrightarrow R$	Resistance
Spring stiffness	$K \leftrightarrow \frac{1}{C}$	Reciprocal of capacitance
Displacement	$y \leftrightarrow Q$	Charge on capacitor
External force	$F(t) \leftrightarrow V(t)$	Electromotive force
Velocity	$\dot{y} \leftrightarrow I$	Current

Table 2.2 Mechanical–Electrical Analogies

EXERCISES

In Exercises 1 through 6, a nonhomogeneous differential equation is given. Verify that the expression for y_p is the correct form.

1. $y'' + y = x + 2e^{-x}$; $y_p = Ax + B + Ce^{-x}$
2. $y'' - y = e^x + \sin x$; $y_p = Axe^x + B\sin x + C\cos x$
3. $y'' - 4y' + 4y = e^{2x}$; $y_p = Ax^2e^{2x}$
4. $y''' + y' = x$; $y_p = Ax^2 + Bx$
5. $y^{(6)} - 3y^{(4)} = 1$; $y_p = Ax^4$
6. $y'' - y = xe^x$; $y_p = Ax^2e^x + Bxe^x$

In Exercises 7 through 12, write out the assumed form of y_p but do not carry out the calculation of the undetermined coefficients.

7. $y'' - y = x^3 + 3x$
8. $y'' - 3y' + 2y = e^{2x} + \sin x$
9. $y' + y = x^2 + \cos x$
10. $y'' + 4y' + 4y = e^{-2x} + xe^{-2x}$
11. $y''' - y' = x^5 + \cos x$
12. $y''' + y' = xe^x + 3x$.

In Exercises 13 through 26, compute y_p for each differential equation.

13. $y'' + y = x + 2e^{-x}$
14. $y'' - y = e^x + \sin x$
15. $y'' - 4y' + 4y = e^{2x}$
16. $y''' + y' = x$
17. $y^{(6)} - 3y^{(4)} = 1$
18. $y' - y = xe^x$
19. $y''' + 3y'' - 4y = e^{-2x}$
20. $y'' + 4y = 4x^3 - 8x^2 - 14x + 7$
21. $y''' - y' = e^x$
22. $y'' + y = e^x(x + 1)$
23. $y'' - y = x\sin x$
24. $y'' - 3y' + 2y = e^{-x}\cos x$
25. $2y'' + y' - y = e^x(x^2 - 1)$
26. $y'' - 2y' + y = xe^x$

In Exercises 27 through 38, compute the general solution of each nonhomogeneous equation.

27. $y'' + y = \sin x$
28. $y'' + 4y = \sin x$
29. $y'' - y' - 2y = 2xe^{-x} + x^2$
30. $y''' - 3y'' + 3y' - y = x^2 + 5e^x$
31. $y'' - y = 4\cosh x$ [*Hint:* $\cosh x = (e^x + e^{-x})/2$]
32. $y'' = 3$
33. $x^2y'' + 4xy' - 4y = x^3$, $x > 0$
34. $x^2y'' - xy' + y = x^2$, $x > 0$
35. $y'' - 5y' + 6y = e^x\sin x$
36. $y'' - 7y' - 8y = e^x(x^2 + 2)$
37. $y'' - 5y' + 4y = e^{2x}\cos x + e^{2x}\sin x$
38. $y'' + 2y' - 3y = e^{2x}(x + 3)$

In Exercises 39 through 42, solve the initial value problems.

39. $y'' + y = x + 2e^{-x}$
$y(0) = 1$
$y'(0) = -2$

40. $y'' - y = xe^{x}$
$y(0) = 0$
$y'(0) = 1$

41. $y'' - 4y' + 4y = e^{2x}$
$y(0) = 0$
$y'(0) = 0$

42. $y''' + y' = x$
$y(0) = 0$
$y'(0) = 1$
$y''(0) = 0$

In Exercises 43 through 47, answer true or false.

43. A particular solution of $y'' + 3y' + 2y = e^x$ is of the form Axe^x.

44. A particular solution of $y'' - 3y' + 2y = e^x$ is of the form Axe^x.

45. A particular solution of $y'' + y = 1/x$ cannot be found by the method of undetermined coefficients.

46. A particular solution of $y'' + y = \cos x$ is of the form $Ax \cos x + Bx \sin x$.

47. A particular solution of $y'' - 3y = x \ln x$ can be found by the method of undetermined coefficients.

Exercises 48 through 50 refer to the *RLC*-series circuit shown in Figure 2.5. Assuming that initially $Q(0) = I(0) = 0$, find the charge and the current at any time t.

48. $R = 4$ ohms; $L = 1$ henry; $C = 1/13$ farads; $V(t) = 26$ volts.

49. $R = 10$ ohms; $L = 2$ henries; $C = 1/12$ farads; $V(t) = \cos 2t$ volts.

50. $R = 4$ ohms; $L = 2$ henries; $C = 0.5$ farads; $V(t) = 2e^{-t}$ volts.

51. **Learning Theory** In 1962, Brady and Marmasse,[24] in an effort to describe experimental data recorded in a simple avoidance situation using rats, arrived at the following initial value problem:

$$\ddot{u} + \frac{1}{\tau}\dot{u} + bu = \frac{\beta}{\tau}$$

$$u(0) = \alpha, \qquad \dot{u}(0) = \beta,$$

where $u(t)$ represents the value of the learning curve of a rat at time t; α and β are initial values; and b and τ are positive constants characteristic of the experimental situation. Without any additional information about the constants τ, b, and β, one would assume a particular solution of the form $u_p = A$. Show that such a u_p can never satisfy the homogeneous equation if $b \neq 0$. Find u_p.

[24] J. P. Brady and C. Marmasse, *Psychol. Record* 12 (1962): 361–68; see also *Bull. Math. Biophys.* 26 (1964): 77–81.

52. Referring to Exercise 51, show that when $(1/\tau^2) - 4b < 0$, the solutions contain oscillatory terms (linear combinations of sines and cosines). (This phenomenon was observed in the actual experiment with rats.)

53. Solve the nonhomogeneous differential equation of Exercise 51 if $\tau = \frac{1}{5}$, $b = 6$, $\alpha = 1$, and $\beta = 1$. Discuss the behavior of the solution at $t \to \infty$.

54. Medicine: Periodic Relapsing Catatonia In the development of a mathematical theory of thyroid–pituitary interactions, Danziger and Elmergreen[25] obtained the following third-order linear differential equations:

$$\alpha_3\dddot{\theta} + \alpha_2\ddot{\theta} + \alpha_1\dot{\theta} + (1 + K)\theta = KC, \qquad \theta < C \tag{15}$$

and

$$\alpha_3\dddot{\theta} + \alpha_2\ddot{\theta} + \alpha_1\dot{\theta} + \theta = 0, \qquad \theta > C.$$

These equations describe the variation of thyroid hormone with time. Here $\theta = \theta(t)$ is the concentration of thyroid hormone at time t and α_1, α_2, α_3, K, and C are constants. If $K \neq -1$, find θ_p for Eq. (15).

55. Minerals In finding the optimum path for the exploitation of minerals in the simple situation where cost rises linearly with cumulated production, Herfindahl and Kneese[26] presented the following second-order differential equation:

$$\ddot{x} - r\dot{x} - \frac{r\beta}{b}x = -\frac{rK}{b},$$

where r, β, b, and K are nonzero constants. Find x_p for this differential equation.

56. Social Behavior Second-order linear differential equations with constant coefficients of the form

$$\ddot{y} - Ay = B, \qquad A > 0,$$

were obtained by Rashevsky[27] in his study of riots by oppressed groups. Here B is an arbitrary constant. Find the general solution of this differential equation.

57. Forced Motion of the Vibrating Spring In addition to the damping force assumed in Exercise 40, Section 2.4, suppose that an external force equal to $f(t)$ acts upon the mass of the spring. Prove, in this case, that the differential equation of the vibrating spring is given by

$$m\ddot{y} + \alpha\dot{y} + Ky = f(t) \tag{16}$$

Such a motion is called a *forced motion*. When the external force is identically equal to zero, the motion is called a *free motion*.

[25] L. Danziger and G. L. Elmergreen, *Bull. Math. Biophys.* 18 (1956): 1–13. See also N. Rashevsky, *Bull. Math. Biophys.* 30 (1968): 735–49.

[26] O. C. Herfindahl and A. V. Kneese, *Economic Theory of Natural Resources* (Columbus, Ohio: Charles E. Merrill, 1974), p. 182.

[27] N. Rashevsky, *Bull. Math. Biophys.* 30 (1968): 501–18.

58. For the forced motion of Exercise 57, assume that $m = 4$ g, $K = 3$ dyn/cm, and that air resistance acts upon the mass with a force that is 7 times its velocity at time t. The mass is also subjected to an external periodic force equal to $5 \cos t$. The mass is raised to a position 3 cm above the equilibrium position and released with a velocity of 5 cm/sec downward. Thus, the motion of the mass satisfies the IVP

$$\begin{aligned} 4\ddot{y} + 7\dot{y} + 3y &= 5 \cos t \\ y(0) &= -3 \\ \dot{y}(0) &= 5. \end{aligned}$$

Solve this IVP. Describe the limiting motion at $t \to \infty$.

Elasticity: Static Deflection of a Beam In the consideration of the static deflection u of a beam of uniform cross section (for example, an I beam) subjected to a transverse load $f(x)$, there arises the differential equation[28]

$$EI \frac{d^4u}{dx^4} = f(x), \tag{17}$$

where E and I are constants, E being Young's modulus of elasticity and I the moment of inertia of a cross section of the beam about some fixed axis.

In Exercises 59 through 64, assume that $EI = 1$ and solve Eq. (17) for the given f.

59. $f(x) = \cos x$ **60.** $f(x) = e^{-x}$ **61.** $f(x) = \sinh x$

62. $f(x) = 1$ **63.** $f(x) = x^2$ **64.** $f(x) = x^4$

The portion of the method of undetermined coefficients associated with the case wherein $f(x)$ has terms that are solutions of the homogeneous equation may seem unmotivated. Certainly, we did not give any indication where the proper form for y_p came from. To this end, consider the following exercises.[29]

65. (a) Show that the general solution of the differential equation

$$y'' - 2y' + y = e^{ax}, \qquad a \neq 1,$$

is

$$y(x) = c_1 e^x + c_2 x e^x + \frac{e^{ax}}{(a-1)^2}.$$

(b) Given that $y(0) = y_0$ and $y'(0) = y_1$, show that c_1 and c_2 in the solution of part (a) are

$$c_1 = y_0 - \frac{1}{(a-1)^2} \qquad \text{and} \qquad c_2 = y_1 - y_0 - \frac{1}{a-1}$$

hence

$$y(x) = y_0 e^x + (y_1 - y_0) x e^x + \frac{e^{ax} - e^x + (1-a) x e^x}{(a-1)^2}$$

[28] S. Timoshenko, *Vibration Problems in Engineering* (Princeton, N.J.: Van Nostrand, 1928).

[29] These exercises are extracted from A. B. Farnell, "On the Solutions of Linear Differential Equations," *Amer. Math. Monthly* 54, no. 3 (1974).

(c) Show that

$$\lim_{a\to 1} y(x) = y_0e^x + (y_1 - y_0)\, xe^x + \tfrac{1}{2}x^2e^x.$$

66. (a) Show that the general solution of the differential equation

$$y'' + y = \sin ax$$

is

$$y(x) = c_1 \sin x + c_2 \cos x + \frac{\sin ax}{1 - a^2}, \qquad a \neq \pm 1.$$

(b) Given that $y(0) = y_0$ and $y'(0) = y_1$, show that c_1 and c_2 in the solution of part (a) are $c_1 = y_1 - a/(1 - a^2)$ and $c_2 = y_0$; hence

$$y(x) = y_1 \sin x + y_0 \cos x + \frac{\sin ax - a \sin x}{1 - a^2}$$

(c) Show that

$$\lim_{a\to 1} y(x) = y_1 \sin x + y_0 \cos x + \tfrac{1}{2}(\sin x - x \cos x).$$

67. Resonance When an undamped vibrating spring is subjected to a periodic external force of the form $f(t) = M \cos wt$ the differential equation takes the form (see Eq. (16) of Ex. 57)

$$\ddot{y} + w_0^2 y = F \cos wt, \tag{18}$$

where $w_0^2 = K/m$ and $F = M/m$. If $f(t) \equiv 0$ ($\Rightarrow M = 0$) the spring is said to be *free*. For this reason w_0 is called the *natural frequency* of the free vibration.

(a) If $w \neq w_0$, show that $y_p = \dfrac{F}{w_0^2 - w^2} \cos wt$.

(b) If w is "close to" w_0, the solution in part (a) shows that the amplitude of y_p is large and gets even larger if w is chosen closer to w_0. This phenomenon associated with the unbounded growth of the amplitude of y_p as $w \to w_0$ is referred to as *resonance*. That is, resonance is said to occur when the frequency of the external force matches the natural frequency of the system. Find y_p for Eq. (18) in the case $w = w_0$.

2.12 VARIATION OF PARAMETERS

The method of variation of parameters, like the method of undetermined coefficients, is used to compute a particular solution of the nonhomogeneous differential equation

$$a_n(x)y^{(n)} + a_{n-1}(x)y^{(n-1)} + \cdots + a_1(x)y' + a_0(x)y = f(x). \tag{1}$$

The method of undetermined coefficients, however, requires that the coefficients

$a_0, a_1, \ldots,$ and a_n of the differential equation be constants, and that the function f be one of the types described in Section 2.11. When we encounter differential equations for which we cannot apply the method of undetermined coefficients, we use the *variation-of-parameters method.* This method, which in theory "always works," evolves from the following consideration. Suppose that we know the general solution

$$y_h = c_1y_1 + c_2y_2 + \cdots + c_ny_n$$

of the associated homogeneous differential equation

$$a_n(x)y^{(n)} + a_{n-1}(x)y^{(n-1)} + \cdots + a_1(x)y' + a_0(x)y = 0. \tag{2}$$

Is it possible to treat the constants c_i as functions (variables) u_i and to impose appropriate conditions on these functions so that the expression

$$u_1y_1 + u_2y_2 + \cdots + u_ny_n$$

is a solution of the nonhomogeneous equation (1)? Such an approach has been shown to lead to meaningful results, which we present in Theorem 1. This theorem is stated for the general nth-order equation, but for simplicity of presentation we prove it for the case $n = 2$.

THEOREM 1

If $y_1, y_2, \ldots, y_n$ form a fundamental system of solution for Eq. (2), and if the functions $u_1, u_2, \ldots, u_n$ satisfy the system of equations

$$\begin{aligned}
y_1u_1' + y_2u_2' + \cdots + y_nu_n' &= 0,\\
y_1'u_1' + y_2'u_2' + \cdots + y_n'u_n' &= 0,\\
y_1''u_1' + y_2''u_2' + \cdots + y_n''u_n' &= 0,\\
&\cdots\cdots\cdots\cdots\cdots\cdots\\
y_1^{(n-2)}u_1' + y_2^{(n-2)}u_2' + \cdots + y_n^{(n-2)}u_n' &= 0,\\
y_1^{(n-1)}u_1' + y_2^{(n-1)}u_2' + \cdots + y_n^{(n-1)}u_n' &= \frac{f(x)}{a_n(x)},
\end{aligned} \tag{3}$$

then $y = u_1y_1 + u_2y_2 + \cdots u_ny_n$ is a particular solution of Eq. (1).

Proof (Special case $n = 2$) By hypothesis, y_1 and y_2 form a fundamental system of solutions for the equation $a_2(x)y'' + a_1(x)y' + a_0(x)y = 0$, and that u_1 and u_2 satisfy the system of algebraic equations

$$\begin{aligned}
y_1u_1' + y_2u_2' &= 0\\
y_1'u_1' + y_2'u_2' &= \frac{f(x)}{a_2(x)}.
\end{aligned} \tag{4}$$

The determinant of the coefficients of the nonhomogeneous system (4) is the Wronskian of the linearly independent solutions y_1 and y_2. Since this determinant is different from zero, the system has a unique solution u_1', u_2'. (See Appendix

A.) From u_1' and u_2' we determine u_1 and u_2 via integration. All that remains to be shown is that the function

$$y = u_1y_1 + u_2y_2 \tag{5}$$

is a solution of the differential equation. Indeed, from (5) we have

$$y' = u_1y_1' + u_1'y_1 + u_2y_2' + u_2'y_2 = u_1y_1' + u_2y_2'.$$

The last expression is obtained by applying the first of Eqs. (4). Hence,

$$y'' = u_1y_1'' + u_1'y_1' + u_2y_2'' + u_2'y_2' = u_1y_1'' + u_2y_2'' + \frac{f(x)}{a_2(x)}.$$

Once again the last expression follows by applying the second of Eqs. (4). Now

$$\begin{aligned} a_2(x)y'' + a_1(x)y' + a_0(x)y &= a_2(x)\left[u_1y_1'' + u_2y_2'' + \frac{f(x)}{a_2(x)}\right] \\ &\quad + a_1(x)[u_1y_1' + u_2y_2'] + a_0(x)[u_1y_1 + u_2y_2] \\ &= u_1[a_2(x)y_1'' + a_1(x)y_1' + a_0(x)y_1] \\ &\quad + u_2[a_2(x)y_2'' + a_1(x)y_2' + a_0(x)y_2] + f(x) \\ &= f(x). \end{aligned}$$

Therefore, y satisfies Eq. (1) (for $n = 2$) and the proof is complete. Other than the tedious notation, the proof in the general case follows in exactly the same manner.

The reader should observe that Theorem 1 and its proof show us how to construct the particular solution y_p. We first find the fundamental system of solutions for Eq. (2). Using these functions, we then solve Eqs. (3) for $u_1', u_2', \ldots, u_n'$. Next, we integrate each of the $u_1', u_2', \ldots, u_n'$ to find, respectively, $u_1, u_2, \ldots, u_n$. (In these integrations, there is no loss in generality if we set the constant of integration equal to zero, for otherwise the term corresponding to a nonzero constant of integration would be absorbed into the homogeneous solution when we wrote out the general solution). And, finally,

$$u_1y_1 + u_2y_2 + \cdots + u_ny_n$$

is the desired particular solution y_p.

We note that the variation-of-parameters method applies to any nonhomogeneous differential equation no matter what the coefficients and the function f happen to be. The reason for introducing the method of undetermined coefficients is that it is sometimes quicker and easier to apply. The method of undetermined coefficients involves differentiation, while the method of variation of parameters involves integration, and in some cases differentiation is easier than integration.

EXAMPLE 1 Solve the differential equation

$$y'' + y = \csc x.$$

Solution The homogeneous solution is $y_h = c_1 \sin x + c_2 \cos x$. The functions u_1 and u_2 are to be determined from the system of equations [see Eqs. (4)]

$$\sin x\, u_1' + \cos x\, u_2' = 0$$

$$\cos x\, u_1' - \sin x\, u_2' = \csc x.$$

Thus,

$$u_1' = \cos x \csc x \Rightarrow u_1 = \ln|\sin x|$$

$$u_2' = -1 \qquad \Rightarrow u_2 = -x.$$

Therefore, we have $y_p = [\ln|\sin x|] \sin x + (-x) \cos x$, and the general solution is

$$y = y_h + y_p = [c_1 + \ln|\sin x|] \sin x + (c_2 - x) \cos x.$$

EXAMPLE 2 Find y_p for the beam-deflection problem [see Eq. (17), Section 2.11] subjected to the load $f(x) = \ln x$.

Solution The differential equation in question is

$$EI\frac{d^4y}{dx^4} = \ln x,$$

where E and I are constants. Therefore, the homogeneous solution is $y_h = c_1 + c_2x + c_3x^2 + c_4x^3$, and u_1, u_2, u_3, u_4 are to be determined from the system of equations

$$\begin{aligned} u_1' + xu_2' + x^2u_3' + x^3u_4' &= 0 \\ u_2' + 2xu_3' + 3x^2u_4' &= 0 \\ 2u_3' + 6xu_4' &= 0 \\ 6u_4' &= \frac{\ln x}{EI} \end{aligned} \tag{6}$$

Equations (6) can be solved by starting with the last equation and working "upward." Thus,

$$u_4' = \frac{\ln x}{6EI} \qquad \Rightarrow u_4 = \frac{1}{6EI}(x \ln x - x)$$

$$u_3' = \frac{-1}{2EI}(x \ln x) \Rightarrow u_3 = \frac{-1}{2EI}\left(\frac{x^2}{2}\ln x - \frac{x^2}{4}\right)$$

$$u_2' = \frac{1}{2EI}(x^2 \ln x) \Rightarrow u_2 = \frac{1}{2EI}\left(\frac{x^3}{3}\ln x - \frac{x^3}{9}\right)$$

$$u_1' = \frac{-1}{6EI}x^3 \ln x \quad \Rightarrow u_1 = \frac{-1}{6EI}\left(\frac{x^4}{4}\ln x - \frac{x^4}{16}\right)$$

The particular solution y_p is then $u_1\cdot(1) + u_2\cdot(x) + u_3\cdot(x^2) + u_4\cdot(x^3)$ which, after simplification, has the form

$$y_p = \frac{x^4}{288EI}(12\ln x - 25).$$

We state the following theorem without proof.

THEOREM 2

Suppose that $y_1, y_2, \ldots, y_n$ constitutes a fundamental set of solutions for the equation

$$a_n(x)y^{(n)} + a_{n-1}(x)y^{(n-1)} + \cdots + a_1(x)y' + a_0(x)y = 0.$$

Then a particular solution of the equation

$$a_n(x)y^{(n)} + a_{n-1}(x)y^{(n-1)} + \cdots + a_1(x)y' + a_0(x)y = f(x)$$

is given by

$$y_p(x) = \sum_{k=1}^{n} y_k(x)\int_{x_0}^{x} \frac{W_k(y_1, \ldots, y_n; s)}{W(y_1, \ldots, y_n; s)}\cdot\frac{f(s)}{a_n(s)}\,ds$$

where $W(y_1, \ldots, y_n;\ s)$ is the Wronskian of $y_1, \ldots, y_n$ evaluated at s and $W_k(y_1, \ldots, y_n;\ s)$ is the determinant obtained from $W(y_1, \ldots, y_n;\ s)$ by replacing the kth column by $(0, 0, \ldots, 0, 1)$. Furthermore, $y_p(x_0) = y_p'(x_0) = \cdots = y_p^{(n-1)}(x_0) = 0$. (For the important special case $n = 2$, see Exercise 30.)

EXERCISES

In Exercises 1 through 17, use the variation-of-parameters method to compute the general solution of the nonhomogeneous differential equations.

1. $y'' + y = \tan x,\ 0 < x < \pi/2$
2. $y'' + y = \sec x,\ 0 < x < \pi/2$
3. $y'' - 2y' + y = \dfrac{1}{x}e^x,\ x > 0$
4. $y'' + 10y' + 25y = e^{-5x}\ln x/x^2,\ x > 0$
5. $y'' + 6y' + 9y = \dfrac{1}{x^3}e^{-3x}, x > 0$
6. $y'' + y = \csc x \cot x,\ 0 < x < \pi/2$
7. $y'' - 12y' + 36y = e^{6x}\ln x,\ x > 0$
8. $y'' + 4y' + 5y = e^{-2x}\sec x,\ 0 < x < \pi/2$
9. $y'' + y = \sec^3 x,\ 0 < x < \pi/2$
10. $y'' - 4y' + 4y = e^{2x}x^{-4},\ x > 0$

11. $y'' + 2y' + y = x^{-2}e^{-x}\ln x,\ x > 0$

12. $y'' - 2y' + y = e^{2x}/(e^x + 1)^2$

13. $5x^2y'' - 3xy' + 3y = x^{1/2},\ x > 0$

14. $x^2y'' + 4xy' - 4y = x^{1/4}\ln x,\ x > 0$

15. $2x^2y'' + xy' - 3y = x^{-3},\ x > 0$

16. $2x^2y'' + 7xy' - 3y = x^{-2}\ln x,\ x > 0$

17. $x^2y'' + 5xy' + 4y = (x^{-3} + x^{-5})\ln x,\ x > 0$

In Exercises 18 through 25, solve the initial value problems.

18. $y'' + y = \csc x$
$y(\pi/2) = 0$
$y'(\pi/2) = 1$

19. $y'' + y = \tan x$
$y(0) = 1$
$y'(0) = 0$

20. $y'' - 2y' + y = (1/x)e^x$
$y(1) = e$
$y'(1) = 0$

21. $y'' + 6y' + 9y = x^{-3}e^{-3x}$
$y(1) = 4e^{-3}$
$y'(1) = -2e^{-3}$

22. $y'' + y = \sec^3 x$
$y(0) = 1$
$y'(0) = 1$

23. $y'' - 4y' + 4y = x^{-4}e^{2x}$
$y(1) = 0$
$y'(1) = e^2$

24. $y'' - 2y' + y = e^{2x}/(e^x + 1)^2$
$y(0) = 3$
$y'(0) = \frac{5}{2}$

25. $2x^2y'' + xy' - 3y = x^{-3}$
$y(\frac{1}{4}) = 0$
$y'(\frac{1}{4}) = \frac{14}{9}$

26. Find the solution of the IVP

$$(x^2 + 1)y'' - 2xy' + 2y = (x^2 + 1)^2$$
$$y(0) = 0, \qquad y'(0) = 1$$

given that $y_1 = x$ is a solution of the corresponding homogeneous differential equation.

27. Solve the differential equation

$$y^{(5)} - y' - \frac{4}{x}y = 0, \qquad x > 0.$$

[*Hint:* Setting $y = vx$ and then $v^{(4)} - v = w$, we obtain the differential equation $xw' + 5w = 0$. Thus, $v^{(4)} - v = cx^{-5}$, where c is an arbitrary constant. Use the variation-of-parameters method to obtain an integral formula for v (the resulting integrals cannot be expressed in terms of elementary functions).[30]]

[30]The preceding transformations (and a generalization to higher-order equations) are due to M. S. Klamkin, *Math. Magazine* 43 (1970): 272–75.

28. Find the solution of the IVP

$$x(1 - 2x \ln x)y'' + (1 + 4x^2 \ln x)y' - (2 + 4x)y = e^{2x}(1 - 2x \ln x)^2, \qquad \tfrac{1}{2} \le x \le 1$$

$$y(\tfrac{1}{2}) = \frac{e}{2}; \qquad y'(\tfrac{1}{2}) = e(2 + \ln 2)$$

given that $y_1 = \ln x$ is a solution of the corresponding homogeneous differential equation.

29. Seismology In recording an earthquake, the response of the accelerogram is governed by the equation

$$\ddot{x} + 2\eta_0\omega_0\dot{x} + \omega_0^2 x = -a(t),$$

where $x(t)$ is the instrument response recorded on the accelerogram, $a(t)$ the ground acceleration, η_0 the damping ratio of the instrument, and ω_0 the undamped natural frequency of the instrument.[31] Take $\eta_0 = \omega_0 = 1$. Find the general solution of this differential equation when $a(t) = e^{-t}/(t + 1)^2$.

30. If $n = 2$ show that the formula of Theorem 2 for y_p yields

$$y_p(x) = \int_{x_0}^{x} \frac{y_1(s)y_2(x) - y_2(s)y_1(x)}{y_1(s)y_2'(s) - y_1'(s)y_2(s)} \cdot \frac{f(s)}{a_2(s)}\, ds, \qquad \text{with } y_p(x_0) = y_p'(x_0) = 0.$$

REVIEW EXERCISES

In Exercises 1 through 3, classify the differential equation as being either homogeneous or nonhomogeneous, with constant coefficients or variable coefficients; also state its order and indicate verbally a method of solution.

1. $x^2y'' + 8x^3y''' - 46y - 3 = 0$

2. $8y - 2y'' - 7x = 0$

3. $y'' + 8xy' = 8y$

In Exercises 4 and 5, write an expression for the Wronskian of two linearly independent solutions without finding the solutions themselves.

4. $2y'' + \dfrac{7}{x}y' - y = 0$

5. $y'' - 8y' + 8e^x y = 0$

6. Using an appropriate theorem from this chapter, verify that there exists a unique solution to the initial value problem

$$(2 - x^2)y'' + (\cos x)y' + (\tan x)y = 0, \qquad -1 < x < 1,$$

$$y(0) = 0, \qquad y'(0) = 1.$$

[31] *Bull. Seismol. Soc. Amer.* 59 (1969): 1591–98.

What do we know about the interval of existence of the solution?

7. A certain homogeneous linear differential equation with constant coefficients has the characteristic equation

$$(\lambda - 7)^3(\lambda^2 - 3)(\lambda^2 + 2\lambda + 2) = 0.$$

Write the general solution for this differential equation and state its order.

In Exercises 8 through 10, find the first five coefficients in the Taylor expansion of the solution about the point $x_0 = 1$.

8. $x^2y'' - \left(\sin\frac{\pi x}{2}\right)y' + y = 0, \quad y(1) = -1, \quad y'(1) = 1$

9. $(2 + x)y'' + xy' - (3 + x^2)y = 0, \quad y(1) = 0, \quad y'(1) = 3$

10. $(5 - x)y' - (\sec^2 \pi x)y = 0, \quad y(1) = 1$

In Exercises 11 through 26, write the general solution of the given differential equation.

11. $y'' + 17y' + 16y = e^x + 4e^{-x}$

12. $(x^2 - 2x)y'' + (2 - x^2)y' + 2(x - 1)y = 0, \ 0 < x < 1,$ given that $y_1 = x^2$ is a solution.

13. $3x^2y'' - 2xy' - 12y = 0, x > 0$

14. $y'' + y = \csc x + 1, 0 < x < 1$

15. $y''' - 9y'' + 6y' + 16y = 0$

16. $y'' - y = x^3e^x$

17. $(\tan x - \sec^2 x)y'' + \tan x(2 \sec^2 x - 1)y' + \sec^2 x(1 - 2 \tan x)y = 0,$ $0 < x < 1$, given that $y_1 = \tan x$ is a solution.

18. $x^2y'' + 7xy' + 9y = 0, x < 0$

19. $y''' + 27y = x$

20. $9y'' + 48y' + 64y = \sin 2x$

21. $y'' + y = \cot x, 0 < x < \pi$

22. $(x \cos x - \sin x)y'' + (x \sin x)y' - (\sin x)y = 0, \ \pi/4 < x < \pi/2,$ given that $y_1 = \sin x$ is a solution.

23. $y'' + y = x \sin x$

24. $y^{iv} - 2y''' + 2y'' - 2y' + y = 0$

25. $x^2y'' + 9xy' + 17y = 0, x > 0$

26. $3x^2y'' - 2xy' - 12y = x^{-2}, x > 0$

27. A projectile is fired horizontally[32] from a gun which is located 144 ft (44m) above a horizontal plane and has a muzzle speed of 800 ft/sec (240 m/sec). (a) How long does the projectile remain in the air? (b) At what horizontal distance does it strike the ground? (c) What is the magnitude of the vertical component of its velocity as it strikes the ground? Take $g = 32.2$ ft/sec^2.

28. A rifle[33] with a muzzle velocity of 1500 ft/sec shoots a bullet at a small target 150 ft away. How high above the target must the gun be aimed so that the bullet will hit the target? Take $g = 32.2$ ft/sec^2.

29. A batter[34] hits a pitched ball at a height 4.0 ft above ground so that its angle of projection is 45° and its horizontal range is 350 ft. The ball is fair down the left field line where a 24 ft high fence is located 320 ft from home plate. Will the ball clear the fence? Take $g = 32.2$ ft/sec^2.

30. It has been determined experimentally[35] that the magnitude in ft/sec^2 of the deceleration due to air resistance of a projectile is $0.001\ v^2$, where v is expressed in ft/sec. If the projectile is released from rest and keeps pointing downward, determine its velocity after it has fallen 500 ft. [*Hint:* The total acceleration is $g - 0.001v^2$, where $g = 32.2$ ft/sec^2.]

31. The acceleration due to gravity[36] of a particle falling toward the earth is $a = -gR^2/r^2$, where r is the distance from the *center* of the earth to the particle, R is the radius of the earth, and g is the acceleration due to gravity at the surface of the earth. Derive an expression for *escape velocity,* the minimum velocity with which a particle should be projected vertically upward from the surface of the earth if it is not to return to the earth. [*Hint:* $v = 0$ for $r = \infty$.]

32. Automobile A starts[37] from 0 and accelerates at the constant rate of 4 ft/sec^2. A short time later it is passed by truck B, which is travelling in the opposite direction at a constant speed of 45 ft/sec. Knowing that truck B passes point O 25 sec after automobile A started from there, determine when and where the vehicles pass each other.

33. In a vertical takeoff,[38] rocket engines exert a thrust that gives it a constant acceleration of 2.0 g for 40 sec. What velocity is attained at this time and what is its height?

[32]This is Exercise 8 in David Halliday and Robert Resnik, *Physics, Part I,* 3rd ed. (New York: John Wiley & Sons, 1977), p. 68. Copyright © 1977 by John Wiley & Sons. Reprinted by permission of John Wiley & Sons, Inc.

[33]This is Exercise 6, ibid., p. 67. Copyright © 1977 by John Wiley & Sons. Reprinted by permission of John Wiley & Sons, Inc.

[34]This is Exercise 14, ibid., p. 68. Copyright © 1977 by John Wiley & Sons. Reprinted by permission of John Wiley & Sons, Inc.

[35]This is Exercise 11.16 in Ferdinand P. Beer and E. Russell Johnston, Jr., *Vector Mechanics for Engineers: Statics and Dynamics,* 3rd ed. (New York: McGraw-Hill, 1977), p. 446. Reprinted by permission of McGraw-Hill Book Company.

[36]This is Exercise 11.21, ibid., p. 447. Reprinted by permission of McGraw-Hill Book Company.

[37]This is Exercise 11.28, ibid., p. 454. Reprinted by permission of McGraw-Hill Book Company.

[38]This is Exercise 1.3-2 in Eduard C. Pestel and William T. Thompson, *Dynamics* (New York: McGraw-Hill, 1968), p. 17. Reprinted by permission of McGraw-Hill Book Company.

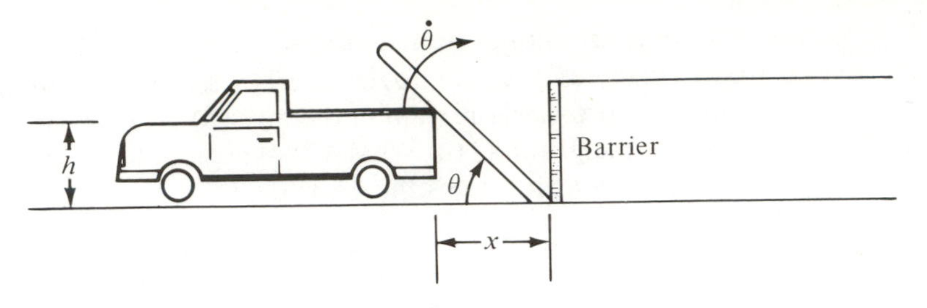

Figure 2.6

34. A train[39] can attain its maximum speed of 60 mph in 4 min., then stop within a distance of 1/2 mile. Assuming that acceleration and deceleration are uniform, find the least time required for the train to go between two stations $8\frac{1}{2}$ miles apart. Draw a velocity-time diagram.

35. It is estimated[40] that an athlete, in putting a shot, releases the shot at a height of 7 ft, with a velocity of 34 ft/sec, at an angle of 40° above the horizontal. Determine the range, the maximum height attained by the shot, and the time of flight.

36. A telephone pole[41] is raised by backing a truck against it, as shown in Figure 2.6. If the angular velocity of the pole is to be a constant, show that the required speed of the truck is $\dot{x} = h\dot{\theta}/\sin^2 \theta$.

37. The displacement[42] of a simple harmonic oscillator is given by $x = a \sin \omega t$. If the values of this displacement x and the velocity $\dot{x}$ are plotted on perpendicular axes, eliminate t to show that the locus of points $(x, \dot{x})$ is an ellipse. Show that this ellipse represents a path of constant energy. [*Hint:* Energy of a simple harmonic oscillator equals $\frac{1}{2}$ (mass)(velocity)^2 + $\frac{1}{2}$ (stiffness)·(displacement)^2. The basic differential equation for the motion is $\ddot{x} + \omega^2 x = 0$, where $\omega^2 = s/m$, s is the stiffness, and m is the mass.]

38. An overdamped spring mass system is displaced a distance A from its equilibrium position and released from rest. Find the position of the mass at any time t. [*Hint:* The differential equation is $m\ddot{y} + \alpha\dot{y} + Ky = 0$.]

39. The equation $m\ddot{x} + sx = F_0 \sin \omega t$ describes[43] the motion of an undamped simple harmonic oscillator driven by a force of frequency ω. Show that the steady-state solution (the particular solution) is given by

$$x = \frac{F_0 \sin \omega t}{m(\omega_0^2 - \omega^2)},$$

[39] This is Exercise 1.3-18, ibid., p. 19. Reprinted by permission of McGraw-Hill Book Company.
[40] This is Exercise 1.4-13, ibid., p. 34. Reprinted by permission of McGraw-Hill Book Company.
[41] This is Exercise 1.4-42, ibid., p. 39. Reprinted by permission of McGraw-Hill Book Company.
[42] This is Exercise 1.11 in H. J. Pain, *The Physics of Vibrations and Waves,* 2nd ed. (New York: John Wiley & Sons, 1976), p. 43. Reprinted by permission of John Wiley & Sons, Inc. Copyright © 1976 by John Wiley & Sons.
[43] This is Exercise 2.3, ibid., p. 68. Copyright © 1976 by John Wiley & Sons. Reprinted by permission of John Wiley & Sons, Inc.

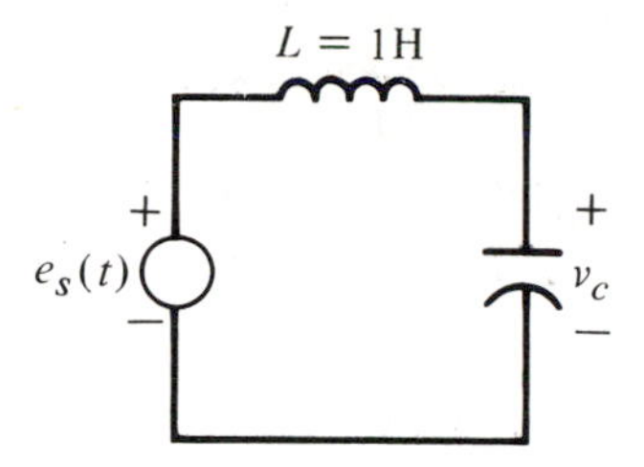

Figure 2.7

where $\omega_0^2 = s/m$. Sketch the behavior of x versus ω and note that the change of sign as ω passes through ω_0^2 defines a phase change of π radians in the displacement. Now show that the general solution for the displacement is given by

$$x = \frac{F_0 \sin \omega t}{m(\omega_0^2 - \omega^2)} + A \cos \omega_0 t + B \sin \omega_0 t,$$

where A and B are constant.

40. Consider[44] the nonlinear time-invariant subharmonic generating circuit shown in Figure 2.7. The inductor is linear and the capacitor has a characteristic

$$v_c = \frac{1}{18} q + \frac{2}{27} q^3.$$

(a) Verify that for an input $e_s = (1/54) \cos t$ volt, a response $q(t) = \cos(t/3)$ coulombs satisfies the differential equation. [*Note:* the charge oscillates at *one-third* of the frequency of the source.]
(b) Calculate the current through the source for the charge found in part (a).

41. The circuit[45] shown in Figure 2.8 is made of linear time-invariant elements. The input is e_s, and the response is v_c. Knowing that $e_s(t) = \sin 2t$ volts, and that at time $t = 0$ the state is $i_L = 2$ amp and $v_c = 1$ volt, calculate the complete response.

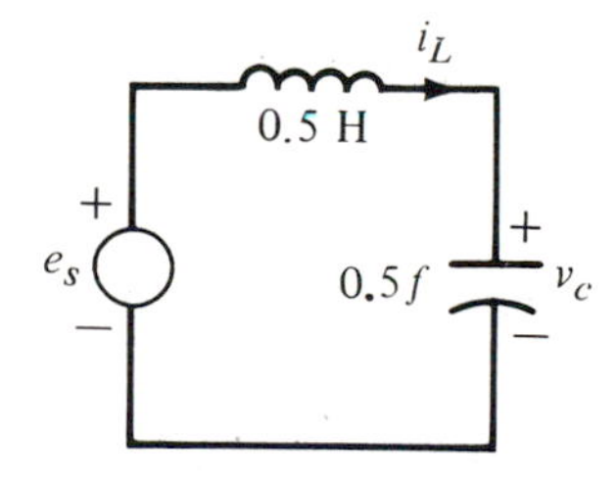

Figure 2.8

[44] This is Exercise 4 in Charles A. Desoer and Ernest S. Kuh, *Basic Circuit Theory* (New York: McGraw-Hill, 1969), p. 332. Reprinted by permission of McGraw-Hill Book Company.
[45] This is Exercise 8, ibid., p. 333. Reprinted by permission of McGraw-Hill Book Company.

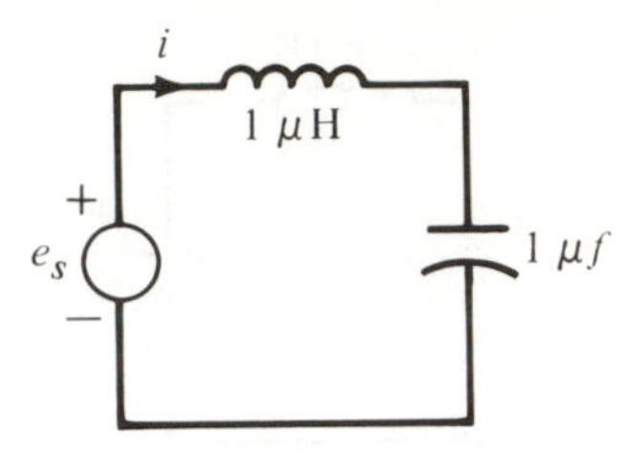

Figure 2.9

42. Let[46] a sinusoidal voltage $e_s(t) = 3 \cos(10^6 t)$ volts be applied at time $t = 0$ to the linear time invariant LC-circuit shown in Figure 2.9. Given $i(0) = 1$ mA and $v(0) = 0$, calculate and sketch $i(t)$ for $t \geq 0$.

43. What[47] is the time of flight of a projectile fired with a muzzle velocity v_0 at an angle α above the horizontal if the launching site is located at height h above the target?

44. Determine[48] the general solution for the motion of an underdamped harmonic oscillator having mass m, damping constant b, and spring constant K when the driving force (the external force) is a constant F_0. Solve the nonhomogeneous equation by assuming a form for the particular solution and then determining the constant coefficients.

45. By differentiation,[49] confirm that $\Phi(\phi) = \dfrac{1}{2\pi}(\cos m\phi + i \sin m\phi)$ is a solution of

$$\frac{1}{\Phi}\frac{d^2\Phi}{d\phi^2} = -m^2.$$

46. If u and v are linearly independent particular solutions of the differential equation

$$y''' + Py'' + Qy' + Ry = 0,$$

prove that the general solution is given by $y = Au + Bv$, where A and B satisfy the equations

$$\frac{A'}{v} = -\frac{B'}{u} = c\,\frac{e^{-\int P dx}}{(u'v - uv')^2}$$

and c is an arbitrary constant.[50]

47. Solve[51] the differential equation

$$(y')^3 - 3y(y')^2 + 4y^3 = a.$$

[*Hint:* Differentiate both sides of the equation with respect to x and factor.]

[46] This is Exercise 11a, ibid., p. 334. Reprinted by permission of McGraw-Hill Book Company.
[47] This is Exercise 2-2.2 in J. Norwood, Jr., *Intermediate Classical Mechanics* (Englewood Cliffs, N.J.: Prentice-Hall, 1979), p. 51. Reprinted by permission of Prentice-Hall, Inc.
[48] This is Exercise 3-3.6, ibid., p. 102. Reprinted by permission of Prentice-Hall, Inc.
[49] This is Exercise 14 in Bernard G. Harvey, *Introduction to Nuclear Physics and Chemistry,* 2nd ed. (Englewood Cliffs, N.J.: Prentice-Hall, 1969), p. 26. Reprinted by permission of Prentice-Hall, Inc.
[50] See *Amer. Math. Monthly* 36 (1929): 112–13.
[51] See *Amer. Math. Monthly* 60 (1953): 264.

CHAPTER 3

Linear Systems

3.1 INTRODUCTION AND BASIC THEORY

The differential equations that we studied in Chapters 1 and 2 are ordinary differential equations involving one unknown function. Consequently, in each case up to now, we had to solve *one* differential equation which involved *one* unknown function. For many reasons, including applications and generalizations, one is interested in studying systems of n differential equations in n unknown functions where n is an integer ≥ 2. Since in this book we are mainly interested in obtaining solutions, we shall restrict our attention to systems of linear differential equations, *linear systems*. *Nonlinear systems*, that is, systems of nonlinear differential equations, are usually (and successfully) studied in more advanced courses, by approximating each nonlinear equation of the system by a linear one. For some results on nonlinear systems, see Chapter 8.

In this chapter we mainly consider systems of *two* linear differential equations in *two* unknown functions of the form

$$\begin{aligned} \dot{x}_1 &= a_{11}(t)x_1 + a_{12}(t)x_2 + f_1(t) \\ \dot{x}_2 &= a_{21}(t)x_1 + a_{22}(t)x_2 + f_2(t), \end{aligned} \tag{1}$$

where the coefficients a_{11}, a_{12}, a_{21}, a_{22}, and the functions f_1, f_2 are all given continuous functions of t on some interval I and x_1, x_2 are unknown functions of t. As usual, the dots over x_1 and x_2 in (1) indicate differentiation with respect to the independent variable t.

REMARK 1 In the treatment of systems of differential equations, it is customary to use the symbols $x_1, x_2, \ldots, x_n$ to denote unknown functions and to use t as the independent variable. This notation has obvious advantages when we are dealing with general systems. However, in special cases alternative notation may be used.

In this section we present a few definitions and state without proof some basic theorems about the solutions of system (1). These definitions and theorems extend easily to systems of n linear differential equations in n unknown functions of the form

$$\begin{aligned} \dot{x}_1 &= a_{11}(t)x_1 + a_{12}(t)x_2 + \cdots + a_{1n}(t)x_n + f_1(t) \\ \dot{x}_2 &= a_{21}(t)x_1 + a_{22}(t)x_2 + \cdots + a_{2n}(t)x_n + f_2(t) \\ &\cdots\cdots\cdots\cdots\cdots\cdots\cdots\cdots\cdots\cdots \\ \dot{x}_n &= a_{n1}(t)x_1 + a_{n2}(t)x_2 + \cdots + a_{nn}(t)x_n + f_n(t), \end{aligned} \tag{2}$$

or, more compactly,

$$\dot{x}_i = \sum_{j=1}^{n} a_{ij}(t)x_j + f_i(t), \qquad i = 1, 2, \ldots, n.$$

In subsequent sections we shall present two elementary methods for finding explicit solutions of system (1) when the coefficients a_{11}, a_{12}, a_{21}, and a_{22} are all known constants. These methods, with varying degrees of technical difficulties, can be extended to solving system (2) when its coefficients a_{ij} are all known constants. For the reader's sake we mention here that the *method of elimination* discussed in Section 3.2 is, perhaps, the easiest method to solve system (1). The *matrix method* outlined in Section 3.3 is recommended for system (2) when $n \geq 3$. This is an elegant method, but its detailed presentation and justification requires a good knowledge of matrix analysis and linear algebra.

DEFINITION 1

A solution of the system (1) *is a pair of functions* $x_1(t)$ *and* $x_2(t)$, *each being differentiable on an interval I and which, when substituted into the two equations in* (1), *makes them identities in t for all t in I.*

For example,

$$x_1(t) = -\tfrac{3}{2}t - \tfrac{5}{4}, x_2(t) = -t - \tfrac{3}{2}$$

is a solution of the system

$$\begin{aligned} \dot{x}_1 &= x_2 + t \\ \dot{x}_2 &= -2x_1 + 3x_2 + 1 \end{aligned} \tag{3}$$

for all t, as the reader can easily verify.

Sometimes it is convenient (and suitable for studying more general systems) to denote a solution of (1) by the column vector

$$\begin{bmatrix} x_1(t) \\ x_2(t) \end{bmatrix}.$$

For example,

$$\begin{bmatrix} -\tfrac{3}{2}t - \tfrac{5}{4} \\ -t - \tfrac{3}{2} \end{bmatrix}$$

is a solution of system (3).

REMARK 2 The algebra of column vectors that is used in this chapter is embodied in the following statements:

(a) *Equality*

$$\begin{bmatrix} x_1(t) \\ x_2(t) \end{bmatrix} = \begin{bmatrix} y_1(t) \\ y_2(t) \end{bmatrix} \quad \text{if and only if} \quad x_1(t) = y_1(t) \text{ and } x_2(t) = y_2(t).$$

(b) *Linear Combination.* For any constants c_1 and c_2,

$$c_1\begin{bmatrix} x_1(t) \\ x_2(t) \end{bmatrix} + c_2\begin{bmatrix} y_1(t) \\ y_2(t) \end{bmatrix} = \begin{bmatrix} c_1x_1(t) + c_2y_1(t) \\ c_1x_2(t) + c_2y_2(t) \end{bmatrix}$$

DEFINITION 2

When both functions f_1 and f_2 in (1) *are equal to zero, the system is called homogeneous. Otherwise it is called nonhomogeneous.*

For example, system (3) is nonhomogeneous. On the other hand, the system

$$\begin{aligned} \dot{x}_1 &= x_2 \\ \dot{x}_2 &= -2x_1 + 3x_2 \end{aligned} \tag{4}$$

is homogeneous. The reader can verify that each of the two vectors

$$\begin{bmatrix} e^t \\ e^t \end{bmatrix} \quad \text{and} \quad \begin{bmatrix} e^{2t} \\ 2e^{2t} \end{bmatrix} \tag{5}$$

is a solution of system (4).

To a great extent, the theory of linear systems resembles the theory of linear differential equations. Let us consider the homogeneous system

$$\begin{aligned} \dot{x}_1 &= a_{11}(t)x_1 + a_{12}(t)x_2 \\ \dot{x}_2 &= a_{21}(t)x_1 + a_{22}(t)x_2, \end{aligned} \tag{6}$$

where the coefficients a_{11}, a_{12}, a_{21}, and a_{22} are all continuous functions on some interval I. As in the case of homogeneous linear differential equations, we have the following.

THEOREM 1

Any linear combination of solutions of (6) *is also a solution of* (6).

For example, each of the two column vectors in (5) is a solution of system (4); therefore, for any constants c_1 and c_2, the linear combination

$$c_1\begin{bmatrix} e^t \\ e^t \end{bmatrix} + c_2\begin{bmatrix} e^{2t} \\ 2e^{2t} \end{bmatrix} = \begin{bmatrix} c_1e^t + c_2e^{2t} \\ c_1e^t + 2c_2e^{2t} \end{bmatrix} \tag{7}$$

is also a solution of (4).

DEFINITION 3

The column vector

$$\begin{bmatrix} 0 \\ 0 \end{bmatrix},$$

that is, $x_1(t) \equiv 0$, $x_2(t) \equiv 0$, is a solution of (6) *for any choice of the coefficients. This solution is called the trivial solution. Any other solution of* (6) *is called a nontrivial solution.*

DEFINITION 4
Two solutions

$$\begin{bmatrix} x_{11}(t) \\ x_{21}(t) \end{bmatrix} \quad \text{and} \quad \begin{bmatrix} x_{12}(t) \\ x_{22}(t) \end{bmatrix} \tag{8}$$

of system (6) *are called linearly independent on an interval I if the statement*

$$c_1 \begin{bmatrix} x_{11}(t) \\ x_{21}(t) \end{bmatrix} + c_2 \begin{bmatrix} x_{12}(t) \\ x_{22}(t) \end{bmatrix} = \begin{bmatrix} 0 \\ 0 \end{bmatrix} \quad \text{for all } t \text{ in } I,$$

that is,

$$c_1 x_{11}(t) + c_2 x_{12}(t) = 0 \text{ and } c_1 x_{21}(t) + c_2 x_{22}(t) = 0$$

for all t in I implies that $c_1 = c_2 = 0$. *Otherwise the solutions* (8) *are called linearly dependent.*

For example, we can demonstrate that the two solutions (5) are linearly independent solutions of the system (4). In fact, if

$$c_1 \begin{bmatrix} e^t \\ e^t \end{bmatrix} + c_2 \begin{bmatrix} e^{2t} \\ 2e^{2t} \end{bmatrix} = \begin{bmatrix} 0 \\ 0 \end{bmatrix},$$

then $c_1e^t + c_2e^{2t} = 0$ and $c_1e^t + 2c_2e^{2t} = 0$. Subtracting the first equation from the second, we find that $c_2e^{2t} = 0$. Thus $c_2 = 0$. Consequently, the first equation becomes $c_1e^t = 0$ and so $c_1 = 0$. Therefore, $c_1 = c_2 = 0$, which establishes our claim. The reader can verify that

$$\begin{bmatrix} e^{2t} \\ 2e^{2t} \end{bmatrix} \quad \text{and} \quad \begin{bmatrix} 3e^{2t} \\ 6e^{2t} \end{bmatrix} \tag{9}$$

are linearly dependent solutions of (4).

The following criterion, reminiscent of the Wronskian determinant, can be used to check the linear dependence or independence of solutions of system (6).

THEOREM 2
The two solutions

$$\begin{bmatrix} x_{11}(t) \\ x_{21}(t) \end{bmatrix} \quad \text{and} \quad \begin{bmatrix} x_{12}(t) \\ x_{22}(t) \end{bmatrix}$$

of the system

$$\dot{x}_1 = a_{11}(t)x_1 + a_{12}(t)x_2$$
$$\dot{x}_2 = a_{21}(t)x_1 + a_{22}(t)x_2$$

are linearly independent on an interval I if and only if the Wronskian determinant

$$\begin{vmatrix} x_{11}(t) & x_{12}(t) \\ x_{21}(t) & x_{22}(t) \end{vmatrix}$$

is never zero on I.

For example, the two solutions (5) of system (4) are linearly independent because

$$\begin{vmatrix} e^t & e^{2t} \\ e^t & 2e^{2t} \end{vmatrix} = 2e^{3t} - e^{3t} = e^{3t} \neq 0 \qquad \text{for all } t.$$

On the other hand, the two solutions (9) of system (4) are linearly dependent because

$$\begin{vmatrix} e^{2t} & 3e^{2t} \\ 2e^{2t} & 6e^{2t} \end{vmatrix} = 6e^{4t} - 6e^{4t} = 0.$$

The following basic *existence and uniqueness theorem* for linear systems is stated here without proof. The reader should recognize the strong resemblance between this theorem and the existence–uniqueness theorem of Chapter 2.

THEOREM 3

Assume that the coefficients a_{11}, a_{12}, a_{21}, a_{22} and the functions f_1 and f_2 of the system

$$\begin{aligned} \dot{x}_1 &= a_{11}(t)x_1 + a_{12}(t)x_2 + f_1(t) \\ \dot{x}_2 &= a_{21}(t)x_1 + a_{22}(t)x_2 + f_2(t) \end{aligned} \tag{1}$$

are all continuous on an interval I. Let t_0 be a point in I and let x_{10} and x_{20} be two given constants. Then the IVP consisting of system (1) *and the initial conditions*

$$x_1(t_0) = x_{10}, \qquad x_2(t_0) = x_{20}$$

has a unique solution

$$\begin{bmatrix} x_1(t) \\ x_2(t) \end{bmatrix}.$$

Furthermore, this unique solution is valid throughout the interval I.

Using this theorem, the reader can verify (by direct substitution) that the unique solution of the IVP

$$\begin{aligned} \dot{x}_1 &= x_2 + t \\ \dot{x}_2 &= -2x_1 + 3x_2 + 1 \\ x_1(0) &= -1 \\ x_2(0) &= -\tfrac{5}{4} \end{aligned} \tag{10}$$

is

$$\begin{aligned} x_1(t) &= \tfrac{1}{4}e^t - \tfrac{3}{2}t - \tfrac{5}{4} \\ x_2(t) &= \tfrac{1}{4}e^t - t - \tfrac{3}{2}. \end{aligned} \tag{11}$$

THEOREM 4

There exist two linearly independent solutions of the system

$$\begin{aligned} \dot{x}_1 &= a_{11}(t)x_1 + a_{12}(t)x_2 \\ \dot{x}_2 &= a_{21}(t)x_1 + a_{22}(t)x_2. \end{aligned} \tag{6}$$

Furthermore, if the two column vectors

$$\begin{bmatrix} x_{11}(t) \\ x_{21}(t) \end{bmatrix} \quad \textit{and} \quad \begin{bmatrix} x_{12}(t) \\ x_{22}(t) \end{bmatrix}$$

are linearly independent solutions of (6), *the general solution of system* (6) *is given by*

$$\begin{bmatrix} x_1(t) \\ x_2(t) \end{bmatrix} = c_1 \begin{bmatrix} x_{11}(t) \\ x_{21}(t) \end{bmatrix} + c_2 \begin{bmatrix} x_{12}(t) \\ x_{22}(t) \end{bmatrix};$$

that is,

$$x_1(t) = c_1 x_{11}(t) + c_2 x_{12}(t)$$

$$x_2(t) = c_1 x_{21}(t) + c_2 x_{22}(t),$$

where c_1 and c_2 are arbitrary constants.

For example, the general solution of (4) is given by

$$\begin{bmatrix} x_1(t) \\ x_2(t) \end{bmatrix} = c_1 \begin{bmatrix} e^t \\ e^t \end{bmatrix} + c_2 \begin{bmatrix} e^{2t} \\ 2e^{2t} \end{bmatrix};$$

that is,

$$x_1(t) = c_1 e^t + c_2 e^{2t} \quad \text{and} \quad x_2(t) = c_1 e^t + 2c_2 e^{2t}. \tag{12}$$

THEOREM 5

If the two solutions

$$\begin{bmatrix} x_{11}(t) \\ x_{21}(t) \end{bmatrix} \quad \textit{and} \quad \begin{bmatrix} x_{12}(t) \\ x_{22}(t) \end{bmatrix}$$

of the system

$$\dot{x}_1 = a_{11}(t)x_1 + a_{12}(t)x_2$$

$$\dot{x}_2 = a_{21}(t)x_1 + a_{22}(t)x_2$$

are linearly independent and if

$$\begin{bmatrix} x_{1p}(t) \\ x_{2p}(t) \end{bmatrix}$$

is a particular solution of the system

$$\begin{aligned} \dot{x}_1 &= a_{11}(t)x_1 + a_{12}(t)x_2 + f_1(t) \\ \dot{x}_2 &= a_{21}(t)x_1 + a_{22}(t)x_2 + f_2(t), \end{aligned} \tag{1}$$

the general solution of (1) *is given by*

$$\begin{bmatrix} x_1(t) \\ x_2(t) \end{bmatrix} = c_1 \begin{bmatrix} x_{11}(t) \\ x_{21}(t) \end{bmatrix} + c_2 \begin{bmatrix} x_{12}(t) \\ x_{22}(t) \end{bmatrix} + \begin{bmatrix} x_{1p}(t) \\ x_{2p}(t) \end{bmatrix};$$

that is,

$$x_1(t) = c_1 x_{11}(t) + c_2 x_{12}(t) + x_{1p}(t)$$

$$x_2(t) = c_1 x_{21}(t) + c_2 x_{22}(t) + x_{2p}(t).$$

For example, the general solution of system (3) is.

$$\begin{bmatrix} x_1(t) \\ x_2(t) \end{bmatrix} = c_1 \begin{bmatrix} e^t \\ e^t \end{bmatrix} + c_2 \begin{bmatrix} e^{2t} \\ 2e^{2t} \end{bmatrix} + \begin{bmatrix} -\frac{3}{2}t - \frac{5}{4} \\ -t - \frac{3}{2} \end{bmatrix};$$

that is,

$$\begin{aligned} x_1(t) &= c_1e^t + c_2e^{2t} - \tfrac{3}{2}t - \tfrac{5}{4} \\ x_2(t) &= c_1e^t + 2c_2e^{2t} - t - \tfrac{3}{2}. \end{aligned} \tag{13}$$

APPLICATIONS 3.1.1

Linear systems of differential equations are utilized in many interesting mathematical models of real-life situations. Here we present a sampling of these applications.

Ecology

■ Consider two species which coexist in a given environment and which interact between themselves in a specific way. For example, one species eats the other, or both species help each other to survive. Let us denote by $N_1(t)$ and $N_2(t)$ the number, at time t, of the first and second species, respectively. If the two species were in isolation from each other, it would be reasonable to assume that they vary at a rate proportional to their number present. That is,

$$\dot{N}_1 = aN_1 \quad \text{and} \quad \dot{N}_2 = dN_2,$$

where the constants of proportionalities a and d are certain constants (positive, negative, or zero, depending on whether the population increases, decreases, or remains constant, respectively).

However, we have assumed that the two species interact between themselves. Let us assume for simplicity that the rate of change of one population due to interaction of the two species is proportional to the size of the other population. Then we obtain the following linear system of differential equations:

$$\begin{aligned} \dot{N}_1 &= aN_1 + bN_2 \\ \dot{N}_2 &= cN_1 + dN_2, \end{aligned}$$

where b and c are certain constants (positive, negative, or zero).

Morphogenesis

■ In Turing's theory of morphogenesis,[1] the state of a cell is represented by the numerical values of a pair of substances x and y which are called *morphogens*. As an illustration, Turing proposed that these morphogens interact chemically according to the kinetic equations

$$\begin{aligned} \dot{x} &= 5x - 6y + 1 \\ \dot{y} &= 6x - 7y + 1. \end{aligned} \tag{14}$$

The general solution of the system (14) is given by

$$\begin{aligned} x &= c_1e^{-t} + c_2te^{-t} + 1 \\ y &= c_1e^{-t} + c_2e^{-t}(t - \tfrac{1}{6}) + 1. \end{aligned} \tag{15}$$

[1]Turing, *Royal Soc. London Phil. Trans.* B237 (1952): 42. See also R. Rosen, *Foundations of Mathematical Biology,* vol. II (New York: Academic Press, 1972), pp. 61–66.

Medicine
Endocrinology

■ Systems of linear differential equations of the form

$$\dot{x}_i = a_{i0} + \sum_{j=1}^{n} a_{ij}\, x_j, \qquad i = 1, 2, \ldots, n, \tag{16}$$

have been used to describe periodic fluctuations of the concentrations of hormones in the bloodstream associated with the menstrual cycle.[2] In 1957, Danziger and Elmergreen[3] proposed mathematical models of endocrine systems that are linear systems of differential equations. For example, reactions of the pituitary and thyroid glands in the regulation of the metabolic rate and reactions of the pituitary and ovary glands in the control of the menstrual cycle can be represented by simple linear systems of the form

$$\begin{aligned} \dot{x}_1 &= -a_{11}x_1 - a_{12}x_2 + a_{10} \\ \dot{x}_2 &= a_{21}x_1 - a_{22}x_2 + a_{20}, \end{aligned} \tag{17}$$

where all the a_{ij} are positive constants with the exception of a_{20}, which is equal to zero.

Recently, Ackerman and colleagues[4] obtained system (17) with $a_{20} > 0$ as a linear approximation of a mathematical model for the detection of diabetes.

Circuit Theory

■ Consider the electric circuits shown in Figures 3.1 and 3.2. Assume that at time t the current flows as indicated in each closed path. Show in each case that the currents satisfy the given *systems* of linear differential equations.

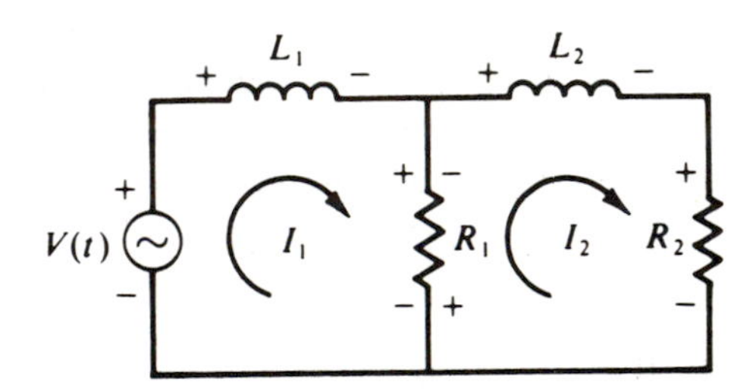

Figure 3.1

$$\begin{aligned} \dot{I}_1 &= -\frac{R_1}{L_1}I_1 + \frac{R_1}{L_1}I_2 + \frac{V(t)}{L_1} \\ \dot{I}_2 &= \frac{R_1}{L_2}I_1 - \frac{R_1 + R_2}{L_2}I_2. \end{aligned} \tag{18}$$

[2] A. Rapoport, *Bull. Math. Biophys.* 14 (1952): 171–83.
[3] L. Danziger and G. L. Elmergreen, *Bull. Math. Biophys.* 19 (1957): 9–18.
[4] E. Ackerman, L. C. Gatewood, J. W. Rosevear, and G. D. Molnal, "Blood Glucose Regulation and Diabetes," in *Concepts and Models of Biomathematics,* ed. F. Heinmets (New York: Marcel Dekker, 1969), chapter 4.

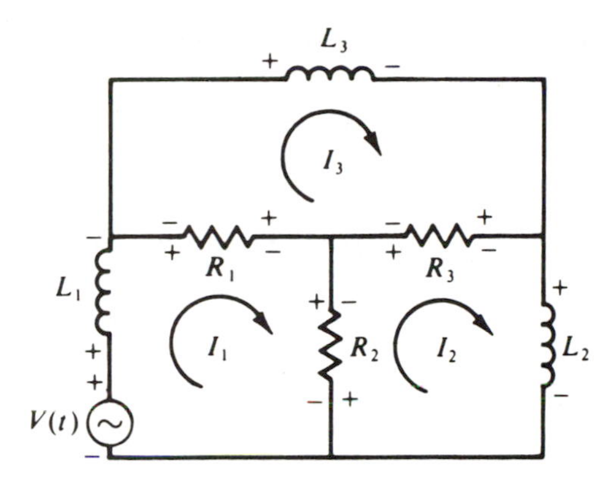

Figure 3.2

$$\begin{aligned}
\dot{I}_1 &= -\frac{R_1 + R_2}{L_1} I_1 + \frac{R_2}{L_1} I_2 + \frac{R_1}{L_1} I_3 + \frac{V(t)}{L_1} \\
\dot{I}_2 &= \frac{R_2}{L_2} I_1 - \frac{R_2 + R_3}{L_2} I_2 + \frac{R_3}{L_2} I_3 \\
\dot{I}_3 &= \frac{R_1}{L_3} I_1 + \frac{R_3}{L_3} I_2 - \frac{R_3 + R_1}{L_3} I_3.
\end{aligned} \tag{19}$$

Solution These systems are obtained by applying Kirchhoff's voltage law (see Section 1.1.1) to each closed path. For example, the first equation in (19) is obtained by applying Kirchhoff's voltage law to the closed path in Figure 3.2, where the current I_1 flows. In fact, we have

$$V(t) - L_1\dot{I}_1 - R_1 I_1 + R_1 I_3 - R_2 I_1 + R_2 I_2 = 0, \tag{20}$$

which after some algebraic manipulations leads to the first differential equation in the system (19). Clearly, the sign in front of any term in Eq. (20) is + or − according to whether the term represents voltage increase or drop, respectively.

Arms Races
The Richardson Model

■ Lewis F. Richardson, in a classic monograph,[5] devised a mathematical model of arms races between two nations. In this model two very simplistic assumptions are made:

1. There are only two nations involved, denoted by A and B.
2. There is only one kind of weapon or missile available.

Let $M_A(t)$ and $M_B(t)$ denote the number of missiles available to countries A and B, respectively. Then $\dot{M}_A(t)$ and $\dot{M}_B(t)$ are the time rate of change of the missile stocks in the two countries. The Richardson model for a two-nation

[5]Lewis F. Richardson, "Generalized Foreign Politics," *British J. Psychol. Monograph Suppl.* 23 (1939).

armament race is given by the following system of nonhomogeneous linear differential equations

$$\dot{M}_A = -a_1 M_A + b_1 M_B + c_1$$

$$\dot{M}_B = a_2 M_A - b_2 M_B + c_2,$$

where the coefficients a_1, b_1, a_2, and b_2 are nonnegative constants. The constants b_1 and a_2 are the "defense coefficients" of the respective nations; the constants a_1 and b_2 are the "fatigue and expense" coefficients, respectively; and the constants c_1 and c_2 are the "grievance coefficients" and indicate the effects of all other factors on the acquisition of missiles.[6]

Economics ■ The economics application described in Section 1.4.1 can be extended to interrelated markets. For simplicity we assume that we have two commodities. Assume that the rates of change of the prices P_1 and P_2 of these commodities are, respectively, proportional to the differences $D_1 - S_1$ and $D_2 - S_2$ of the demand and supply at any time t. Thus, Eq. (10) of Section 1.4.1 is replaced by the system of equations

$$\frac{dP_1}{dt} = \alpha_1(D_1 - S_1)$$

and

$$\frac{dP_2}{dt} = \alpha_2(D_2 - S_2).$$

In the simple case where D_1, D_2, S_1, and S_2 are given as linear functions of the prices P_1 and P_2, the system above takes the form of a linear system in the unknowns P_1 and P_2. That is,[7]

$$\frac{dP_1}{dt} = a_{11}P_1 + a_{12}P_2 + b_1$$

$$\frac{dP_2}{dt} = a_{21}P_1 + a_{22}P_2 + b_2.$$

EXERCISES

1. (a) Show that the column vectors

$$\begin{bmatrix} e^{3t} \\ e^{3t} \end{bmatrix} \quad \text{and} \quad \begin{bmatrix} e^{5t} \\ 3e^{3t} \end{bmatrix}$$

are linearly independent solutions of the linear system

$$\dot{x}_1 = 2x_1 + x_2$$
$$\dot{x}_2 = -3x_1 + 6x_2.$$

[6]For more information on arms races, the interested reader is referred to M. D. Intriligator, *Proceedings of a Conference on the Application of Undergraduate Mathematics in the Engineering, Life, Managerial, and Social Sciences,* ed. P. J. Knopp and G. H. Meyer (Atlanta: Georgia Institute of Technology, 1973), p. 253.

[7]For more details, see R. G. D. Allen, *Mathematical Economics* (London: Macmillan, 1957), p. 480.

(b) Using Theorem 4 of Section 3.1, write down the general solution of the system above.

2. (a) Show that the column vectors

$$\begin{bmatrix} e^t \\ e^t \end{bmatrix} \quad \text{and} \quad \begin{bmatrix} e^{-t} \\ \frac{1}{2}e^{-t} \end{bmatrix}$$

are linearly independent solutions of the linear system

$$\begin{aligned} \dot{x}_1 &= -3x_1 + 4x_2 \\ \dot{x}_2 &= -2x_1 + 3x_2. \end{aligned}$$

(b) Using Theorem 4 of Section 3.1, write down the general solution of the system above.

(c) Using your answer in part (b), solve the IVP

$$\begin{aligned} \dot{x}_1 &= -3x_1 + 4x_2 \\ \dot{x}_2 &= -2x_1 + 3x_2 \\ x_1(0) &= -1, \qquad x_2(0) = 3. \end{aligned}$$

3. Use Theorem 3 of Section 3.1 to prove that the only solution of the IVP

$$\begin{aligned} \dot{x}_1 &= (2\sin t)x_1 + (\ln t)x_2 \\ \dot{x}_2 &= \frac{1}{t-2}x_1 + \frac{e^t}{t+1}x_2 \\ x_1(3) &= 0, \qquad x_2(3) = 0 \end{aligned}$$

is $x_1(t) \equiv 0$ and $x_2(t) \equiv 0$ in the interval $(2, \infty)$.

4. Show that the unique solution of the IVP

$$\begin{aligned} \dot{x} &= 5x - 6y + 1 \\ \dot{y} &= 6x - 7y + 1 \\ x(0) &= 0, \qquad y(0) = 0 \end{aligned}$$

is $x(t) = 1 - e^{-t}, y(t) = 1 - e^{-t}$.

5. Show that the column vectors

$$\begin{bmatrix} e^{-t} \\ e^{-t} \end{bmatrix} \quad \text{and} \quad \begin{bmatrix} te^{-t} \\ (t - \frac{1}{6})e^{-t} \end{bmatrix}$$

are linearly independent solutions of the homogeneous system

$$\begin{aligned} \dot{x} &= 5x - 6y \\ \dot{y} &= 6x - 7y \end{aligned}$$

and that the column vector

$$\begin{bmatrix} 1 \\ 1 \end{bmatrix}$$

is a particular solution of the nonhomogeneous system (14) of Section 3.1.1.

Use Theorem 5 of Section 3.1 to show that Eqs. (15) in Section 3.1.1 give the general solution of system (14).

In Exercises 6 through 11, answer true or false.

6. The column vectors

$$\begin{bmatrix} e^{-t} \\ e^{-3t} \end{bmatrix} \quad \text{and} \quad \begin{bmatrix} e^{-t} \\ -e^{-3t} \end{bmatrix}$$

are linearly independent solutions of the system

$$\dot{x} = -2x + y, \qquad \dot{y} = x - 2y.$$

7. The column vectors

$$\begin{bmatrix} e^{-t} \\ e^{-t} \end{bmatrix} \quad \text{and} \quad \begin{bmatrix} e^{-3t} \\ -e^{-3t} \end{bmatrix}$$

are linearly independent solutions of the system

$$\dot{x} = -2x + y, \qquad \dot{y} = x - 2y.$$

8. The general solution of the system

$$\dot{x} = 3x - 2y, \qquad \dot{y} = 2x - 2y$$

is

$$x = c_1 e^{-t} + c_2 e^{2t}, \qquad y = 2c_1 e^{-t} + \tfrac{1}{2} c_2 e^{2t}.$$

9. The column vectors

$$\begin{bmatrix} \cos t \\ -\sin t \end{bmatrix} \quad \text{and} \quad \begin{bmatrix} \sin t \\ \cos t \end{bmatrix}$$

are linearly independent solutions of the system

$$\dot{x} = x + y, \qquad \dot{y} = x - y.$$

10. The column vector

$$\begin{bmatrix} \cos t \\ -\sin t \end{bmatrix}$$

is the unique solution of the IVP

$$\dot{x} = y, \qquad \dot{y} = -x$$
$$x(0) = 1, \qquad y(0) = 0.$$

11. The system

$$\dot{x} = -x + y, \qquad \dot{y} = -x - y$$

has solutions of the form

$$\begin{bmatrix} Ae^{\lambda t} \\ Be^{\lambda t} \end{bmatrix} \quad \text{with } \lambda = -1 \pm i.$$

12. Consider the electric network shown in Figure 3.3. Find a system of differential equations that describes the currents I_1 and I_2.

13. Use Kirchhoff's voltage law to verify the equations in system (18).

14. Use Kirchhoff's voltage law to verify the equations in system (19).

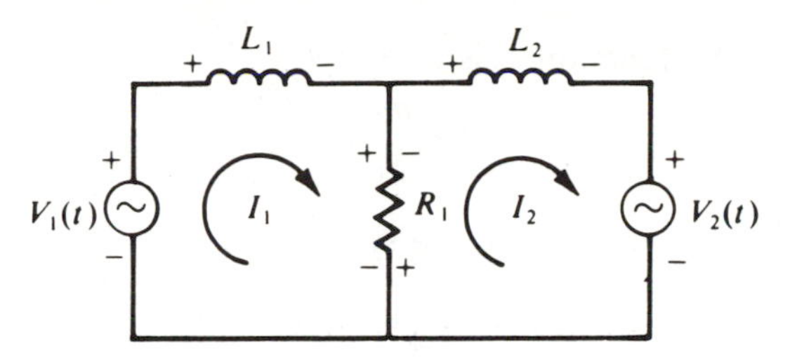

Figure 3.3

15. Professor Umbugio,[8] a mythical mathematics professor, has invented a remarkable scheme for reviewing books. He divides the time he allows himself for reviewing into three fractions, α, β, γ. He devotes the fraction α of his time to a deep study of the title page and jacket. He devotes the fraction β to a spirited search for his name and quotations from his own works. Finally, he spends the fraction γ of his allotted time in a proportionally penetrating perusal of the remaining text. Knowing his characteristic taste for simple and direct methods, we cannot fail to be duly impressed by the differential equations on which he bases his scheme:

$$\frac{dx}{dt} = y - z, \qquad \frac{dy}{dt} = z - x, \qquad \frac{dz}{dt} = x - y. \tag{a}$$

He considers a system of solutions x, y, z which is determined by initial conditions depending on a (small) parameter ϵ independent of t. Therefore, x, y, and z depend on both t and ϵ, and we appropriately use the notation:

$$x = f(t, \epsilon), \qquad y = g(t, \epsilon), \qquad z = h(t, \epsilon). \tag{b}$$

The functions (b) satisfy the equations (a) and the initial conditions

$$f(0, \epsilon) = \tfrac{1}{3} - \epsilon, \qquad g(0, \epsilon) = \tfrac{1}{3}, \qquad h(0, \epsilon) = \tfrac{1}{3} + \epsilon. \tag{c}$$

Professor Umbugio defines his important fractions α, β, and γ by

$$\lim_{\epsilon \to 1} f(2, \epsilon) = \alpha, \qquad \lim_{\epsilon \to 0} g(5, \epsilon) = \beta, \qquad \lim_{\epsilon \to 0} h(279, \epsilon) = \gamma.$$

Deflate the professor! Find α, β, and γ without much numerical computation. [Hint: Observe that $\dot{x} + \dot{y} + \dot{z} = 0$ and $x\dot{x} + y\dot{y} + z\dot{z} = 0 \Rightarrow x + y + z = 1$ and $x^2 + y^2 + z^2 = \frac{1}{3} + 2\epsilon^2 \Rightarrow (x - \frac{1}{3})^2 + (y - \frac{1}{3})^2 + (z - \frac{1}{3})^2 = 2\epsilon^2$. Now take limits as $\epsilon \to 0$.]

16. Reduction of Order Assume that $(y_1(t), y_2(t))$ is a known solution of system (6) with $y_1(t) \neq 0$ for all t in the interval I. Show that the transformation

$$x_1 = y_1 z_1$$
$$x_2 = y_2 z_1 + z_2$$

reduces the system (6) to the single differential equation

$$\dot{z}_2 = \left(a_{22} - \frac{y_2}{y_1} a_{12}\right) z_2.$$

[8]This problem is taken from *Amer. Math. Monthly* 54 (1974): 223.

Furthermore,[9] the function z_1 is given by

$$\dot{z}_1 = \frac{a_{12}}{y_1} z_2.$$

17. Given that $\begin{bmatrix} t \\ 1 \end{bmatrix}$ is a solution of the system

$$\dot{x}_1 = x_1 + (1 - t)x_2$$
$$\dot{x}_2 = \frac{1}{t}x_1 - x_2, \qquad t > 0$$

use reduction of order to compute the general solution of the system.

18. State Definitions 1 through 4 of Section 3.1 for system (2) and the corresponding homogeneous system

$$\dot{x}_i = \sum_{j=1}^{n} a_{ij}(t)x_j, \qquad i = 1, 2, \ldots, n. \tag{21}$$

19. State Theorem 2 of Section 3.1 for n solutions of system (21).

20. State Theorem 3 of Section 3.1 for system (2).

21. State Theorem 4 of Section 3.1 for system (21).

22. State Theorem 5 of Section 3.1 for system (2).

In Exercises 23 through 30, answer true or false.

23. The column vectors

$$\begin{bmatrix} 1 \\ 0 \\ 0 \\ 0 \end{bmatrix}, \quad \begin{bmatrix} e^{-t} \\ -e^{-t} \\ 0 \\ 0 \end{bmatrix}, \quad \begin{bmatrix} e^{t} \\ 0 \\ -e^{t} \\ 0 \end{bmatrix}, \quad \begin{bmatrix} \frac{7}{2}e^{2t} \\ e^{2t} \\ -3e^{2t} \\ 3e^{2t} \end{bmatrix}$$

are linearly independent solutions of the system

$$\dot{x}_1 = x_2 - x_3 + x_4$$
$$\dot{x}_2 = -x_2 + x_4$$
$$\dot{x}_3 = x_3 - x_4$$
$$\dot{x}_4 = 2x_4.$$

24. The general solution of the system in Exercise 23 is

$$x_1 = c_1 + c_2e^{-t} + c_3e^{t} + \tfrac{7}{2}c_4e^{2t}$$
$$x_2 = -c_2e^{-t} + c_4e^{2t}$$
$$x_3 = -c_3e^{t} - 3c_4e^{2t}$$
$$x_4 = 3c_4e^{2t}.$$

[9]For a generalization, see J. W. Evans, *Amer. Math. Monthly* 75 (1968): 637.

25. The column vectors

$$\begin{bmatrix} -\frac{1}{2}e^{-3t} \\ e^{-3t} \\ e^{-3t} \end{bmatrix}, \quad \begin{bmatrix} 4e^{6t} \\ e^{6t} \\ e^{6t} \end{bmatrix}, \quad \text{and} \quad \begin{bmatrix} 0 \\ 0 \\ 0 \end{bmatrix}$$

are solutions of the system

$$\begin{aligned} \dot{x}_1 &= 5x_1 + 2x_2 + 2x_3 \\ \dot{x}_2 &= 2x_1 + 2x_2 - 4x_3 \\ \dot{x}_3 &= 2x_1 - 4x_2 + 2x_3. \end{aligned}$$

26. The general solution of the system in Exercise 25 is

$$\begin{aligned} x_1 &= -\tfrac{1}{2}c_1e^{-3t} + 4c_2e^{6t} \\ x_2 &= c_1e^{-3t} + c_2e^{6t} \\ x_3 &= c_1e^{-3t} + c_2e^{6t}. \end{aligned}$$

27. The column vector

$$\begin{bmatrix} 2e^{6t} \\ e^{6t} \\ 2 \end{bmatrix}$$

is the only solution of the IVP

$$\begin{aligned} \dot{x}_1 &= 5x_1 + 2x_2 + 2x_3 \\ \dot{x}_2 &= 2x_1 + 2x_2 - 4x_3 \\ \dot{x}_3 &= 2x_1 - 4x_2 + 2x_3 \end{aligned}$$

$$x_1(0) = 2, \qquad x_2(0) = 1, \qquad x_3(0) = 0.$$

28. The column vector

$$\begin{bmatrix} 2e^{6t} \\ e^{6t} \\ 0 \end{bmatrix}$$

is the only solution of the IVP in Exercise 27.

29. The column vectors

$$\begin{bmatrix} 2e^{t} \\ e^{t} \\ 3e^{t} \end{bmatrix}, \quad \begin{bmatrix} 3e^{2t} \\ e^{2t} \\ 5e^{2t} \end{bmatrix}, \quad \text{and} \quad \begin{bmatrix} e^{3t} \\ -e^{3t} \\ 2e^{3t} \end{bmatrix}$$

are linearly independent solutions of the system

$$\begin{aligned} \dot{x}_1 &= -10x_1 + x_2 + 7x_3 \\ \dot{x}_2 &= -9x_1 + 4x_2 + 5x_3 \\ \dot{x}_3 &= -17x_1 + x_2 + 12x_3. \end{aligned}$$

30. The unique solution of the IVP

$$\dot{x}_1 = -10x_1 + x_2 + 7x_3$$
$$\dot{x}_2 = -9x_1 + 4x_2 + 5x_3$$
$$\dot{x}_3 = -17x_1 + x_2 + 12x_3$$
$$x_1(0) = 5, \quad x_2(0) = 2, \quad x_3(0) = -2$$

is

$$x_1 = 2e^t + 3e^{2t}, \quad x_2 = e^t + e^{2t}, \quad x_3 = 3e^t - 5e^{2t}.$$

3.2 THE METHOD OF ELIMINATION

The most elementary method for solving a system of linear differential equations in two unknown functions and with constant coefficients is the *method of elimination.* In this method the aim is to reduce the given system to a single differential equation in one unknown function by eliminating the other dependent variable.

EXAMPLE 1 Find the general solution of the homogeneous linear system

$$\begin{aligned} \dot{x}_1 &= x_2 \\ \dot{x}_2 &= -2x_1 + 3x_2. \end{aligned} \tag{1}$$

Solution Differentiating both sides of the first equation with respect to t and using the second equation, we obtain

$$\ddot{x}_1 = \dot{x}_2 = -2x_1 + 3x_2.$$

The first equation in (1) states that $x_2 = \dot{x}_1$, hence

$$\ddot{x}_1 = -2x_1 + 3\dot{x}_1$$
$$\Rightarrow \quad \ddot{x}_1 - 3\dot{x}_1 + 2x_1 = 0. \tag{2}$$

Thus, we have succeeded in eliminating the function x_2 from system (1), obtaining in the process the differential equation (2), which contains one single function and which can be solved. The general solution of Eq. (2) is

$$x_1(t) = c_1e^t + c_2e^{2t}. \tag{3}$$

At this point we use the first equation in system (1) and the value of x_1 given in (3) to find the other unknown function $x_2 = \dot{x}_1$. Thus,

$$x_2(t) = c_1e^t + 2c_2e^{2t}. \tag{4}$$

Equations (3) and (4) give the general solution of system (1).

Other ways to eliminate one of the unknown functions and arrive at a single equation for the remaining unknown are the following: Solve the first equation of the system for x_2 and substitute the value of x_2 into the second equation of the system; or, solve the second equation of the system for x_1 and substitute the value of x_1 into the first equation of the system.

The method of elimination can also be applied efficiently to nonhomogeneous systems, as the following example illustrates.

EXAMPLE 2 Find the general solution of the nonhomogeneous system

$$\begin{aligned} \dot{x}_1 &= x_2 + t \\ \dot{x}_2 &= -2x_1 + 3x_2 + 1. \end{aligned} \tag{5}$$

Solution As in the previous example, differentiating both sides of the first equation with respect to t and using the second equation, we obtain

$$\ddot{x}_1 = \dot{x}_2 + 1 = -2x_1 + 3x_2 + 2.$$

Using the first equation in (5) yields

$$\begin{aligned} \ddot{x}_1 &= -2x_1 + 3(\dot{x}_1 - t) + 2 \\ \Rightarrow \quad \ddot{x}_1 - 3\dot{x}_1 + 2x_1 &= -3t + 2. \end{aligned} \tag{6}$$

Thus, we have succeeded in eliminating the variable x_2 from system (5), obtaining the nonhomogeneous differential equation (6), which contains one single function and can be solved. The solution of the homogeneous equation associated with Eq. (6) is

$$x_{1h} = c_1e^t + c_2e^{2t}. \tag{7}$$

A particular solution of Eq. (6) is of the form $x_{1p} = At + B \Rightarrow \dot{x}_{1p} = A$,

$$\ddot{x}_{1p} = 0 \Rightarrow -3A + 2At + 2B = -3t + 2 \Rightarrow 2A = -3$$

and

$$-3A + 2B = 2 \Rightarrow A = -\tfrac{3}{2} \quad \text{and} \quad B = -\tfrac{5}{4} \Rightarrow x_{1p} = -\tfrac{3}{2}t - \tfrac{5}{4}. \tag{8}$$

Adding (7) and (8), we obtain the general solution of Eq. (6):

$$x_1(t) = c_1e^t + c_2e^{2t} - \tfrac{3}{2}t - \tfrac{5}{4}. \tag{9}$$

Now from the first equation in (5) we have $x_2 = \dot{x}_1 - t$, so

$$x_2(t) = c_1e^t + 2c_2e^{2t} - t - \tfrac{3}{2}. \tag{10}$$

Equations (9) and (10) give the general solution of (5).

EXAMPLE 3 Solve the IVP

$$\dot{x}_1 = -3x_1 + 4x_2 \tag{11}$$

$$\dot{x}_2 = -2x_1 + 3x_2 \tag{12}$$

$$x_1(0) = -1, \qquad x_2(0) = 3. \tag{13}$$

Solution Differentiating both sides of (11) and using (12) and then (11) again, we obtain

$$\begin{aligned}\ddot{x}_1 &= -3\dot{x}_1 + 4\dot{x}_2 = -3\dot{x}_1 + 4(-2x_1 + 3x_2)\\ &= -3\dot{x}_1 - 8x_1 + 12x_2 = -3\dot{x}_1 - 8x_1 + 3(\dot{x}_1 + 3x_1)\end{aligned}$$

$$\Rightarrow \quad \ddot{x}_1 - x_1 = 0,$$

thus,

$$x_1(t) = c_1e^t + c_2e^{-t}. \tag{14}$$

Thus, from Eq. (11) we have

$$4x_2 = \dot{x}_1 + 3x_1 = c_1e^t - c_2e^{-t} + 3c_1e^t + 3c_2e^{-t} = 4c_1e^t + 2c_2e^{-t},$$

and so,

$$x_2(t) = c_1e^t + \tfrac{1}{2}c_2e^{-t}. \tag{15}$$

Using the initial conditions (13) in Eqs. (14) and (15), we obtain

$$\begin{aligned}c_1 + c_2 &= -1\\ c_1 + \tfrac{1}{2}c_2 &= 3.\end{aligned}$$

Thus, $c_1 = 7$, $c_2 = -8$, and the solution of the IVP (11)–(13) is

$$\begin{aligned}x_1(t) &= 7e^t - 8e^{-t}\\ x_2(t) &= 7e^t - 4e^{-t}.\end{aligned}$$

REMARK 1 It can be shown that the general solution of a system of n first-order linear differential equations in n unknown functions contains n arbitrary constants. For $n = 2$, this follows from Theorem 5 of Section 3.1. Sometimes, however, the method of elimination introduces *extraneous* constants. The extraneous constants can be eliminated by substituting the solutions into the system and equating coefficients of similar terms. We illustrate this by the following example.

EXAMPLE 4 Find the general solution of the system

$$\begin{aligned}\dot{x}_1 &= x_2 + 1\\ \dot{x}_2 &= x_1.\end{aligned}$$

Solution Differentiating both sides of the first equation with respect to t and using the second equation, we find $\ddot{x}_1 = \dot{x}_2 = x_1 \Rightarrow \ddot{x}_1 - x_1 = 0 \Rightarrow x_1(t) = c_1e^t + c_2e^{-t}$. Now that we have $x_1(t)$, we can find $x_2(t)$ either from the first equation of the system or from the second. If we use the first equation we find $x_2(t) = c_1e^t - c_2e^{-t} - 1$, and the general solution of the system contains, as expected, two arbitrary constants. However, if we find $x_2(t)$ from the second equation of the system, we have to integrate, which introduces a third arbitrary constant. In this case, $x_1(t) = c_1e^t + c_2e^{-t}$ and $x_2(t) = c_1e^t - c_2e^{-t} + c_3$. If we substitute these functions into the system and equate coefficients of similar terms, we find that c_3 should be -1. Thus the general solution is

$$x_1(t) = c_1e^t + c_2e^{-t}, \qquad x_2(t) = c_1e^t - c_2e^{-t} - 1.$$

APPLICATIONS 3.2.1

Biokinetics

■ Some investigations into the distribution, breakdown, and synthesis of albumin in animals has led to systems of ordinary differential equations.[10] Under normal conditions an animal is assumed to have a constant amount of albumin. Part of this albumin is assumed to be present in the animal's vascular system (plasma) and the remainder in the extravascular fluids (lymph and tissue fluids) (Figure 3.4). The albumin in the plasma (and in the extravascular fluids) is changing in the following manner. Some albumin is transferring to the extravascular fluids (or, respectively, to the plasma), and some albumin is being broken down, and this catabolized protein is replaced by newly synthesized protein. It is not completely known exactly where this breakdown takes place, and we therefore assume that some breakdown occurs in the plasma and some in the extravascular fluids. Thus, in its natural state the study of albumin can be viewed in terms of the compartment model given as Figure 3.4.

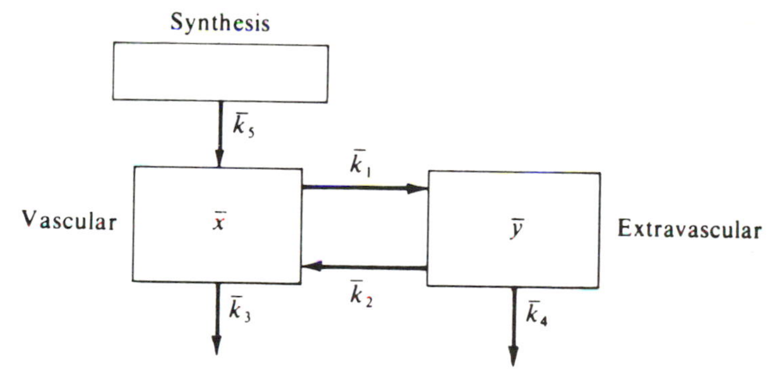

Figure 3.4

In this figure, $\bar{x}$ represents the number of grams of albumin in the plasma; $\bar{y}$ represents the number of grams of albumin in the extravascular fluids; and $\bar{k}_1, \bar{k}_2, \bar{k}_3, \bar{k}_4$, and $\bar{k}_5$ are constants associated with the rate at which the various processes take place. The newly synthesized protein is shown reentering the system via the plasma. $\bar{k}_3$ and $\bar{k}_4$ are associated with the breakdown process, and $\bar{k}_1$ and $\bar{k}_2$ are associated with the rates of interaction. We refer to the model above as being associated with unlabeled albumin.

In order to study the albumin process, some investigators proceed as follows. A certain amount of radioactive albumin, referred to as I^{131}-albumin, is injected

[10]The models and developments presented here have been extracted from E. B. Reeve and J. E. Roberts, "Kinetics of I^{131}-Albumin Distribution and Breakdown," *J. Gen. Physiol.* 43 (1959): 415.

into the vascular system, where it mixes with the vascular fluid. After a few minutes its behavior is assumed to follow that of the unlabeled albumin. At any time t thereafter, there is a certain amount of radioactive substance x in the plasma and a certain amount y in the extravascular fluids.

I^{131}-albumin, being radioactive, is continuously being broken down as a result of the liberation of radioactive breakdown products, and certain of the breakdown products are being excreted by the animal. All these actions are continuous in nature, and therefore to speak of the rates of change, we must use derivatives with respect to time. The compartment model associated with I^{131}-albumin is shown in Figure 3.5. The quantities x, y, z, and u in the figure represent fractions of the total amount originally injected. Since there is only one injection, no new protein is being synthesized in this model. The process described by this compartment model can be analyzed by studying the following system of differential equations:

$$\begin{aligned}\frac{dx}{dt} &= k_2y - (k_1 + k_3)x \\ \frac{dy}{dt} &= k_1x - (k_2 + k_4)y \\ \frac{dz}{dt} &= k_3x + k_4y - k_5z \\ \frac{du}{dt} &= k_5z.\end{aligned} \tag{16}$$

The first equation is arrived at by noticing in the model that x is changing due to gaining k_2y g/day and losing k_1x and k_3x g/day. The other equations are obtained by means of similar reasoning. This system also has initial and terminal conditions associated with it. Initially, all the radioactive substance is in the plasma; thus, $t = 0 \Rightarrow x = 1, y = z = u = 0$. After "infinite" time, all the substance has been excreted; thus, $t \to \infty \Rightarrow x = y = z = 0, u = 1$. The solution of this system is left to the exercises. We note, however, that (16), together with the initial conditions, determines a unique solution. This solution depends,

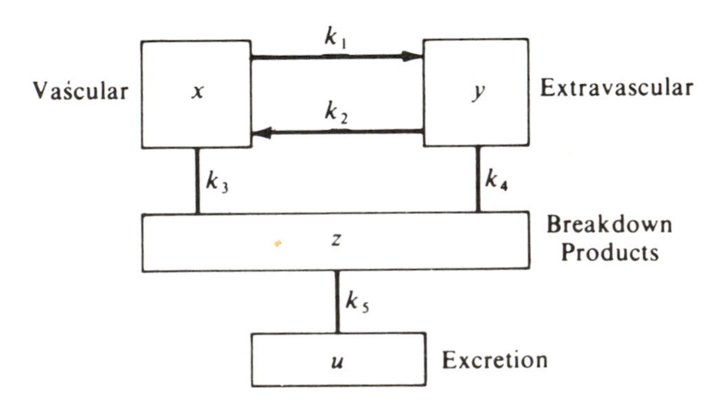

Figure 3.5

in general, on the rate constants $k_1, \ldots, k_5$. These constants are determined experimentally and by the terminal conditions.

Circuit Theory

■ In the electric circuit in Figure 3.6, assume that the currents are initially zero and that $R_1 = \frac{3}{2}$ ohms, $R_2 = 4$ ohms, $L_1 = L_2 = 1$ henry, and $V(t) = 2$ volts. Determine the currents $I_1(t)$ and $I_2(t)$ at any time t.

Solution From Eq. (18) of Section 3.1.1, the currents I_1 and I_2 satisfy the following linear system of differential equations:

$$\dot{I}_1 = -\frac{R_1}{L_1}I_1 + \frac{R_1}{L_1}I_2 + \frac{V(t)}{L_1} \tag{17}$$

$$\dot{I}_2 = \frac{R_1}{L_2}I_1 - \frac{R_1 + R_2}{L_2}I_2, \tag{18}$$

and the initial conditions

$$I_1(0) = 0, \qquad I_2(0) = 0. \tag{19}$$

Substituting the given values of R_1, R_2, L_1, L_2, and $V(t)$, we obtain

$$\dot{I}_1 = -\tfrac{3}{2}I_1 + \tfrac{3}{2}I_2 + 2 \tag{20}$$

$$\dot{I}_2 = \tfrac{3}{2}I_1 - \tfrac{11}{2}I_2. \tag{21}$$

Differentiating both sides of (20) and using (21) and then (20), we find that

$$\begin{aligned}
\ddot{I}_1 &= -\tfrac{3}{2}\dot{I}_1 + \tfrac{3}{2}\dot{I}_2 \\
&= -\tfrac{3}{2}\dot{I}_1 + \tfrac{3}{2}(\tfrac{3}{2}I_1 - \tfrac{11}{2}I_2) = -\tfrac{3}{2}\dot{I}_1 + \tfrac{9}{4}I_1 - \tfrac{33}{4}I_2 \\
&= -\tfrac{3}{2}\dot{I}_1 + \tfrac{9}{4}I_1 - \tfrac{11}{2}(\dot{I}_1 + \tfrac{3}{2}I_1 - 2) \\
\Rightarrow \quad & \ddot{I}_1 + 7\dot{I}_1 + 6I_1 = 11 \\
& \qquad \Rightarrow \quad I_1(t) = c_1e^{-t} + c_2e^{-6t} + \tfrac{11}{6}.
\end{aligned}$$

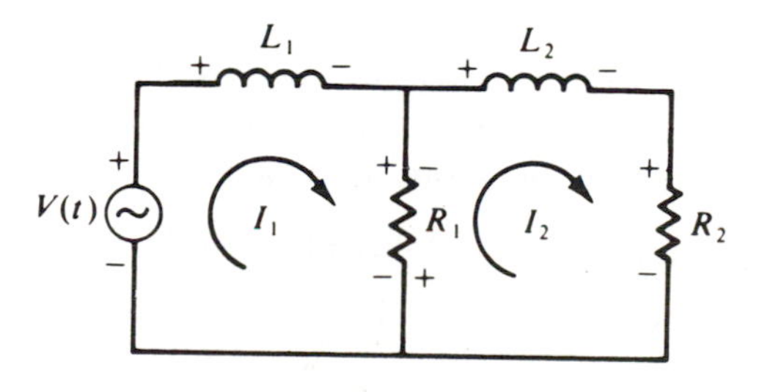

Figure 3.6

Substituting the value of I_1 into (20), we find that

$$I_2 = \tfrac{2}{3}(-c_1e^{-t} - 6c_2e^{-6t} + \tfrac{3}{2}c_1e^{-t} + \tfrac{3}{2}c_2e^{-6t} + \tfrac{33}{12} - 2)$$
$$= \tfrac{1}{3}c_1e^{-t} - 3c_2e^{-6t} + \tfrac{1}{2}.$$

Using the initial conditions (19), we have

$$c_1 + c_2 + \tfrac{11}{6} = 0$$
$$\tfrac{1}{3}c_1 - 3c_2 + \tfrac{1}{2} = 0$$
$$\Rightarrow \quad c_1 = -\tfrac{9}{5}, \qquad c_2 = -\tfrac{1}{30}.$$

Thus,

$$I_1(t) = -\tfrac{9}{5}e^{-t} - \tfrac{1}{30}e^{-6t} + \tfrac{11}{6}$$
$$I_2(t) = -\tfrac{3}{5}e^{-t} + \tfrac{1}{10}e^{-6t} + \tfrac{1}{2}.$$

EXERCISES

In Exercises 1 through 8, find the general solution of the system.

1. $\dot{x}_1 = 4x_1 - x_2$
$\dot{x}_2 = 2x_1 + x_2$

2. $\dot{x}_1 = 3x_1 - 2x_2$
$\dot{x}_2 = x_1$

3. $\dot{x}_1 = 4x_1 - x_2 + 3e^{2t}$
$\dot{x}_2 = 2x_1 + x_2 + 2t$

4. $\dot{x}_1 = 3x_1 - 2x_2 + 1$
$\dot{x}_2 = x_1 + t$

5. $\dot{x} = 5x - 6y + 1$
$\dot{y} = 6x - 7y + 1$

6. $t\dot{x} = 3x - 2y$
$t\dot{y} = x + y - t^2$

7. $\dot{x} = 3x - 2y + 2t^2$
$\dot{y} = 5x + y - 1$

8. $\dot{x}_1 = x_1 + 2x_2$
$\dot{x}_2 = 4x_1 - x_2$

9. Determine the currents in the electric network of Exercise 12 of Section 3.1 if $R_1 = 4$ ohms, $L_1 = 2$ henries, $L_2 = 4$ henries, $V_1 = (2 \sin t)$ volts, $V_2(t) = (2 \sin t)$ volts, and the currents are initially zero.

10. Assume that the system

$$\dot{N}_1 = 4N_1 - 6N_2$$
$$\dot{N}_2 = 8N_1 - 10N_2$$

represents two competing populations, with N_1 being the desirable population and N_2 a parasite. Compute the populations $N_1(t)$ and $N_2(t)$ at any time t. Show that both populations are headed for extinction.

In Exercises 11 through 18, find the solution of the IVP.

11. $\dot{x}_1 = 4x_1 - x_2$
$\dot{x}_2 = 2x_1 + x_2$
$x_1(0) = 1,\ x_2(0) = 3.$

12. $\dot{x}_1 = 3x_1 - 2x_2$
$\dot{x}_2 = x_1$
$x_1(0) = -1,\ x_2(0) = 0.$

13. $\dot{x}_1 = 4x_1 - x_2 + 3e^{2t}$
$\dot{x}_2 = 2x_1 + x_2 + 2t$
$x_1(0) = -\frac{5}{18}, x_2(0) = \frac{47}{9}.$

14. $\dot{x}_1 = 3x_1 - 2x_2 + 1$
$\dot{x}_2 = x_1 + t$
$x_1(0) = -\frac{1}{2}, x_2(0) = -\frac{1}{4}.$

15. $\dot{x} = 5x - 6y + 1$
$\dot{y} = 6x - 7y + 1$
$x(0) = 0, y(0) = 0.$

16. $t\dot{x} = 3x - 2y$
$t\dot{y} = x + y - t^2$
$x(1) = 1, y(1) = \frac{1}{2}.$

17. $\dot{x} = 3x - 2y + 2t^2$
$\dot{y} = 5x + y - 1$
$x(0) = \frac{534}{2197}, y(0) = \frac{567}{2197}.$

18. $\dot{x}_1 = x_1 + 2x_2$
$\dot{x}_2 = 4x_1 - x_2$
$x_1(0) = 1, x_2(0) = 1.$

In Exercises 19 through 28, find the general solution of the system.

19. $\dot{x} = -x, \dot{y} = -y$

20. $\dot{x} = -2x + y, \dot{y} = x - 2y$

21. $\dot{x} = 3x - 2y, \dot{y} = 2x - 2y$

22. $\dot{x} = -x, \dot{y} = -x - y$

23. $\dot{x} = -x + y, \dot{y} = -x - y$

24. $\dot{x} = y, \dot{y} = -x$

25. $\dot{x} = y, \dot{y} = -x + 2y$

26. $\dot{x} = 2x - y, \dot{y} = 9x + 2y$

27. $\dot{x} = 2x + y, \dot{y} = -3x + 6y$

28. $\dot{x} = y, \dot{y} = z, \dot{z} = x$

29. In the chemistry application of Section 1.7.1 we arrived at the homogeneous system [See Eqs. (10), Section 1.7.1]

$$\dot{c}_1 = -\frac{k}{V_1}c_1 + \frac{k}{V_1}c_2$$

$$\dot{c}_2 = \frac{k}{V_2}c_1 - \frac{k}{V_2}c_2.$$

Use the method of elimination to obtain the general solution of this system.

30. (a) Using the method of elimination on the first two of Eqs. (16) show that x satisfies the differential equation

$$\ddot{x} + (k_1 + k_2 + k_3 + k_4)\dot{x} + [(k_1 + k_3)(k_2 + k_4) - k_1k_2]x = 0.$$

Thus, if r_1 and r_2 represent the roots of the characteristic equation, then $x = c_1e^{r_1t} + c_2e^{r_2t}$. Find r_1 and r_2 algebraically.

(b) Use the terminal condition ($t \to \infty$) to infer that $r_1 < 0$ and $r_2 < 0$.

(c) Since $r_1 < 0$ and $r_2 < 0$ write $r_1 = -a$ and $r_2 = -b$, where $a > 0$ and $b > 0$. Using the differential equation $dx/dt = k_2y - (k_1 + k_3)x$, show that

$$k_2y - (k_1 + k_3)x = -c_1ae^{-at} - c_2be^{-bt}.$$

(d) Use the initial conditions to find a relation between k_1 and k_3. [*Hint:* Use the result of part (c).]

(e) Solve for y.

(f) Solve for z.

(g) Solve for u.

31. Learning Theory In an article with applications to learning theory, Grossberg[11] obtained the following IVP:

$$\dot{x} = a(b - x) - cfy$$
$$\dot{y} = d(x - y) - cfy - ay$$
$$x(0) = b, \qquad y(0) = \frac{db}{a + d},$$

where a, b, c, d, and f are positive constants. Show that for every fixed f the unknown function y decays monotonically to a positive minimum.

3.3 THE MATRIX METHOD

As we saw in Section 3.2, we have no difficulty solving systems with constant coefficients of *two* linear differential equations in *two* unknown functions. The method of elimination is a simple method to use for such systems. The method of this section, known as the *matrix method,* has the advantage that it can be easily extended to systems with constant coefficients of n linear differential equations in n unknown functions.

For the sake of simplicity, we present the main features of this method in the special case of linear homogeneous systems with constant coefficients of the form

$$\begin{aligned} \dot{x}_1 &= a_{11}x_1 + a_{12}x_2 \\ \dot{x}_2 &= a_{21}x_1 + a_{22}x_2. \end{aligned} \tag{1}$$

Illustrative examples will be given for systems of n equations in n unknown functions for $n = 2$, 3, and 4.

Mimicking the method of solving linear homogeneous differential equations with constant coefficients, which was discussed in Sections 2.4 and 2.5, let us look for a solution of system (1) in the form

$$\begin{bmatrix} x_1 \\ x_2 \end{bmatrix} = \begin{bmatrix} A_1e^{\lambda t} \\ A_2e^{\lambda t} \end{bmatrix}. \tag{2}$$

The constants A_1, A_2, and λ are to be determined by demanding that $x_1 = A_1e^{\lambda t}$ and $x_2 = A_2e^{\lambda t}$ satisfy system (1). Substituting (2) into (1), we find that A_1, A_2, and λ must satisfy the system of equations

$$A_1\lambda e^{\lambda t} = a_{11}A_1e^{\lambda t} + a_{12}A_2e^{\lambda t}$$
$$A_2\lambda e^{\lambda t} = a_{21}A_1e^{\lambda t} + a_{22}A_2e^{\lambda t}.$$

[11] S. Grossberg, *J. Theoret. Biol.* (1969): 325–64.

Dividing through by $e^{\lambda t}$ and rearranging terms, we obtain the equivalent system

$$\begin{aligned} (a_{11} - \lambda)A_1 + a_{12}A_2 &= 0 \\ a_{21}A_1 + (a_{22} - \lambda)A_2 &= 0. \end{aligned} \tag{3}$$

System (3) consists of only two equations, and there are three unknown quantities involved: λ, A_1, and A_2. Nevertheless, one can solve this system by taking the following approach. We view system (3) as a linear homogeneous system in the unknowns A_1 and A_2. Naturally, we seek a nontrivial solution of this system, since the solution $(A_1, A_2) = (0, 0)$ gives rise to the trivial solution $x_1(t) = 0$, $x_2(t) = 0$ of system (1). Thus, (A_1, A_2) should be different from (0, 0). We now recall (Appendix A) that a homogeneous system of algebraic equations has a nontrivial solution if and only if the determinant of its coefficients is equal to zero. In the case of system (3), this means that there is a solution $(A_1, A_2) = (0, 0)$ if and only if

$$\begin{vmatrix} a_{11} - \lambda & a_{12} \\ a_{21} & a_{22} - \lambda \end{vmatrix} = 0; \tag{4}$$

that is,

$$\lambda^2 - (a_{11} + a_{22})\lambda + a_{11}a_{22} - a_{12}a_{21} = 0. \tag{4'}$$

DEFINITION 1

Equation (4′) *is called the characteristic equation of system* (1), *and the roots of the characteristic equation are called the characteristic roots of the system.*

It will be useful to the reader to note at this point that the determinant in Eq. (4) is obtained from the determinant

$$\begin{vmatrix} a_{11} & a_{12} \\ a_{21} & a_{22} \end{vmatrix}$$

of the coefficients of system (1) by subtracting λ from the main diagonal. [In matrix analysis, Eq. (4) is called the *characteristic equation* of the matrix

$$\begin{vmatrix} a_{11} & a_{12} \\ a_{21} & a_{22} \end{vmatrix}, \tag{5}$$

and the roots of the characteristic equation are called the *characteristic roots* or *eigenvalues* of the matrix (5).]

Let λ_1 and λ_2 be the characteristic roots of system (1). If the λ in (3) is replaced by a characteristic root, the corresponding system (3) has a nontrivial solution (A_1, A_2). Substituting these values of A_1 and A_2 in (2), we obtain a nontrivial solution of system (1). If Eq. (4′) has two distinct roots (in other words, if $\lambda_1 \neq \lambda_2$), we can obtain two nontrivial solutions of system (1): one from $\lambda = \lambda_1$ and the other from $\lambda = \lambda_2$. Furthermore, it can be shown that these two solutions are linearly independent. On the other hand, if Eq. (4′) does not have two distinct roots (in other words, if $\lambda_1 = \lambda_2$), the above procedure gives, in general, one nontrivial solution (and its multiples) of system (1). As we are interested in finding the general solution of system (1), we still need a second linearly independent solution of system (1). The trick in this case is to look for a second linearly independent solution of (1) in the form

$$\begin{bmatrix} x_1 \\ x_2 \end{bmatrix} = \begin{bmatrix} (a_1 t + b_1)e^{\lambda_1 t} \\ (a_2 t + b_2)e^{\lambda_1 t} \end{bmatrix}, \tag{6}$$

where the coefficients a_1, b_1, a_2, and b_2 are to be determined in such a way that (6) is a solution of (1) which is linearly independent to the first solution

$$\begin{bmatrix} x_1 \\ x_2 \end{bmatrix} = \begin{bmatrix} A_1 e^{\lambda_1 t} \\ A_2 e^{\lambda_1 t} \end{bmatrix}.$$

This is always possible. See Remark 4.

EXAMPLE 1 Find the general solution of the homogeneous system

$$\begin{aligned} \dot{x}_1 &= 2x_1 + x_2 \\ \dot{x}_2 &= -3x_1 + 6x_2. \end{aligned} \tag{7}$$

Solution Let us look for a solution of system (7) of the form

$$\begin{bmatrix} x_1 \\ x_2 \end{bmatrix} = \begin{bmatrix} A_1 e^{\lambda t} \\ A_2 e^{\lambda t} \end{bmatrix}. \tag{8}$$

Then λ should be a root of the characteristic equation (note that $a_{11} = 2$, $a_{12} = 1$, $a_{21} = -3$, and $a_{22} = 6$)

$$\begin{vmatrix} 2-\lambda & 1 \\ -3 & 6-\lambda \end{vmatrix} = 0, \qquad \text{that is,} \qquad \lambda^2 - 8\lambda + 15 = 0.$$

The characteristic roots are $\lambda_1 = 3$ and $\lambda_2 = 5$. When $\lambda = \lambda_1 = 3$, the constants A_1 and A_2 of the solution (8) should satisfy the homogeneous system (3); that is,

$$\begin{aligned} -A_1 + A_2 &= 0 \\ -3A_1 + 3A_2 &= 0 \end{aligned} \quad \Rightarrow A_1 = A_2.$$

Choosing $A_1 = A_2 = 1$, we have the solution

$$\begin{bmatrix} e^{3t} \\ e^{3t} \end{bmatrix}. \tag{9}$$

[Clearly, any other nontrivial solution of this system (for example, $A_1 = A_2 = 2$) would just give us another solution with which (9) is linearly dependent.] When $\lambda = \lambda_2 = 5$, system (3) becomes

$$\begin{aligned} -3A_1 + A_2 &= 0 \\ -3A_1 + A_2 &= 0 \end{aligned} \quad \Rightarrow A_2 = 3A_1.$$

The choice $A_1 = 1$ yields $A_2 = 3$, and we obtain another solution of (7):

$$\begin{bmatrix} e^{5t} \\ 3e^{5t} \end{bmatrix}. \tag{10}$$

It is easily seen (by applying Theorem 2 of Section 3.1) that the two solutions (9) and (10) of system (7) are linearly independent, and therefore the general solution of (7) is

$$\begin{bmatrix} x_1 \\ x_2 \end{bmatrix} = c_1 \begin{bmatrix} e^{3t} \\ e^{3t} \end{bmatrix} + c_2 \begin{bmatrix} e^{5t} \\ 3e^{5t} \end{bmatrix};$$

that is,

$$x_1 = c_1 e^{3t} + c_2 e^{5t}$$

$$x_2 = c_1 e^{3t} + 3c_2 e^{5t}.$$

EXAMPLE 2 Find the general solution of the system

$$\dot{x}_1 = 2x_1 - x_2$$

$$\dot{x}_2 = 9x_1 + 2x_2. \tag{11}$$

Solution The characteristic equation of system (11) is

$$\begin{vmatrix} 2-\lambda & -1 \\ 9 & 2-\lambda \end{vmatrix} = 0, \qquad \text{that is,} \qquad \lambda^2 - 4\lambda + 13 = 0.$$

The characteristic roots are $\lambda_1 = 2 + 3i$ and $\lambda_2 = 2 - 3i$. When $\lambda_1 = 2 + 3i$, the constants A_1 and A_2 of the solution (8) satisfy the homogeneous system (3) with $a_{11} = 2$, $a_{12} = -1$, $a_{21} = 9$, $a_{22} = 2$, and $\lambda = 2 + 3i$. That is,

$$\begin{aligned} -3iA_1 - A_2 &= 0 \\ 9A_1 - 3iA_2 &= 0 \end{aligned} \quad \Rightarrow A_2 = -3iA_1.$$

Choosing $A_1 = 1$, we find that $A_2 = -3i$, and therefore

$$\begin{bmatrix} e^{(2+3i)t} \\ -3ie^{(2+3i)t} \end{bmatrix} \tag{12}$$

is a solution of (11). When $\lambda_2 = 2 - 3i$, the constants A_1 and A_2 of the solution (8) satisfy the homogeneous system (3) with $a_{11} = 2$, $a_{12} = -1$, $a_{21} = 9$, $a_{22} = 2$, and $\lambda = 2 - 3i$. That is,

$$\begin{aligned} 3iA_1 - A_2 &= 0 \\ 9A_1 + 3iA_2 &= 0 \end{aligned} \quad \Rightarrow A_2 = 3iA_1.$$

Choosing $A_1 = 1$, we find that $A_2 = 3i$; therefore

$$\begin{bmatrix} e^{(2-3i)t} \\ 3ie^{(2-3i)t} \end{bmatrix} \tag{13}$$

is another solution of (11). Applying Theorem 2 of Section 3.1, it is easily seen that the two solutions (12) and (13) of system (11) are linearly independent, and therefore the general solution of (11) is

$$\begin{bmatrix} x_1 \\ x_2 \end{bmatrix} = c_1 \begin{bmatrix} e^{(2+3i)t} \\ -3ie^{(2+3i)t} \end{bmatrix} + c_2 \begin{bmatrix} e^{(2-3i)t} \\ 3ie^{(2-3i)t} \end{bmatrix}, \tag{14}$$

where c_1 and c_2 are arbitrary constants.

REMARK 1 Using the Euler identity $e^{a+ib} = e^a(\cos b + i \sin b)$, the solution (14) can also be written as follows:

$$\begin{bmatrix} x_1 \\ x_2 \end{bmatrix} = c_1 \begin{bmatrix} e^{2t}\cos 3t + ie^{2t}\sin 3t \\ -3ie^{2t}\cos 3t + 3e^{2t}\sin 3t \end{bmatrix} + c_2 \begin{bmatrix} e^{2t}\cos 3t - ie^{2t}\sin 3t \\ 3ie^{2t}\cos 3t + 3e^{2t}\sin 3t \end{bmatrix}$$

$$= \begin{bmatrix} (c_1 + c_2)e^{2t}\cos 3t + i(c_1 - c_2)e^{2t}\sin 3t \\ 3(c_1 + c_2)e^{2t}\sin 3t - 3i(c_1 - c_2)e^{2t}\cos 3t \end{bmatrix}.$$

Setting $C_1 = c_1 + c_2$ and $C_2 = i(c_1 - c_2)$, we obtain

$$\begin{bmatrix} x_1 \\ x_2 \end{bmatrix} = \begin{bmatrix} C_1e^{2t}\cos 3t + C_2e^{2t}\sin 3t \\ 3C_1e^{2t}\sin 3t - 3C_2e^{2t}\cos 3t \end{bmatrix}.$$

where C_1 and C_2 are arbitrary constants.

REMARK 2 Comparing the solutions (12) and (13) of system (11), we observe that one is the complex conjugate of the other. This is always the case when the characteristic roots are complex conjugates and the coefficients of the system are real numbers.

EXAMPLE 3 Find the general solution of the system

$$\begin{aligned} \dot{x}_1 &= x_2 \\ \dot{x}_2 &= -x_1 + 2x_2. \end{aligned} \tag{15}$$

Solution The characteristic equation of (15) is

$$\begin{vmatrix} -\lambda & 1 \\ -1 & 2-\lambda \end{vmatrix} = 0, \qquad \text{that is,} \qquad \lambda^2 - 2\lambda + 1 = 0.$$

The characteristic roots are $\lambda_1 = \lambda_2 = 1$. Therefore, the constants A_1 and A_2 of the solution (8) satisfy the homogeneous system (3) with $a_{11} = 0$, $a_{12} = 1$, $a_{21} = -1$, $a_{22} = 2$, and $\lambda = 1$. That is,

$$\begin{aligned} -A_1 + A_2 &= 0 \\ -A_1 + A_2 &= 0 \end{aligned} \quad \Rightarrow A_1 = A_2.$$

Choosing $A_1 = A_2 = 1$, we obtain the solution

$$\begin{bmatrix} e^t \\ e^t \end{bmatrix}. \tag{16}$$

Since the roots of the characteristic equation are equal, we look for a second linearly independent solution of (15) of the form (6),

$$\begin{bmatrix} x_1 \\ x_2 \end{bmatrix} = \begin{bmatrix} (a_1t + b_1)e^t \\ (a_2t + b_2)e^t \end{bmatrix}, \tag{17}$$

where the constants a_1, b_1, a_2, and b_2 are to be determined in such a way that (17) is a solution of (15) that is linearly independent of (16). Substituting $x_1 = (a_1t + b_1)e^t$ and $x_2 = (a_2t + b_2)e^t$ into (15), we obtain

$$a_1e^t + (a_1t + b_1)e^t = (a_2t + b_2)e^t$$

$$a_2e^t + (a_2t + b_2)e^t = -(a_1t + b_1)e^t + 2(a_2t + b_2)e^t.$$

Dividing by e^t and equating coefficients of like powers of t, we have

$$\begin{aligned} a_1 + b_1 &= b_2, & a_1 &= a_2 \\ a_2 + b_2 &= -b_1 + 2b_2, & a_2 &= -a_1 + 2a_2 \end{aligned} \tag{18}$$

or [since the last two equations in (18) are equivalent to the first two]

$$a_1 + b_1 = b_2, \qquad a_1 = a_2. \tag{19}$$

Now any choice of a_1, b_1, a_2, and b_2 that satisfies (19) and gives a solution which is linearly independent to (16) is acceptable. For example, the choice $a_1 = a_2 = 1$, $b_1 = 0$, and $b_2 = 1$ yields the solution

$$\begin{bmatrix} te^t \\ (t + 1)e^t \end{bmatrix}, \tag{20}$$

which is linearly independent of (16). (Why?) Thus, the general solution of (15) is

$$\begin{bmatrix} x_1 \\ x_2 \end{bmatrix} = c_1 \begin{bmatrix} e^t \\ e^t \end{bmatrix} + c_2 \begin{bmatrix} te^t \\ (t + 1)e^t \end{bmatrix};$$

that is,

$$x_1 = c_1e^t + c_2te^t \qquad \text{and} \qquad x_2 = c_1e^t + c_2(t + 1)e^t,$$

where c_1 and c_2 are arbitrary constants.

REMARK 3 Notice that for a second linearly independent solution of (15), it is not enough to multiply (16) by t as in the case of linear homogeneous differential equations with multiple characteristic roots. Indeed,

$$t \begin{bmatrix} e^t \\ e^t \end{bmatrix} = \begin{bmatrix} te^t \\ te^t \end{bmatrix}$$

is not a solution of system (15). In fact, system (15) does not even have a nontrivial solution of the form

$$\begin{bmatrix} x_1 \\ x_2 \end{bmatrix} = \begin{bmatrix} a_1te^t \\ a_2te^t \end{bmatrix}.$$

Also, unlike the case of linear homogeneous differential equations with multiple characteristic roots, a system of the form (1) with a double characteristic root may have two linearly independent solutions of the form (2), as the example below indicates. This is also true for more general linear homogeneous systems, but we will not go into the details in this book. See Example 5 and Remark 4.

EXAMPLE 4 Find the general solution of the system

$$\begin{aligned} \dot{x}_1 &= x_1 \\ \dot{x}_2 &= x_2. \end{aligned} \tag{21}$$

Solution The characteristic equation of system (21) is

$$\begin{vmatrix} 1-\lambda & 0 \\ 0 & 1-\lambda \end{vmatrix} = 0, \qquad \text{that is,} \qquad \lambda^2 - 2\lambda + 1 = 0.$$

The characteristic roots are $\lambda_1 = \lambda_2 = 1$. Therefore, the constants A_1 and A_2 of the solution (8) satisfy the homogeneous system (3) with $a_{11} = 1$, $a_{12} = 0$, $a_{21} = 0$, $a_{22} = 1$, and $\lambda = 1$. That is,

$$\left.\begin{aligned} 0\cdot A_1 + 0\cdot A_2 &= 0 \\ 0\cdot A_1 + 0\cdot A_2 &= 0 \end{aligned}\right. \quad \Rightarrow A_1 \text{ and } A_2 \text{ are arbitrary constants.}$$

Choosing first $A_1 = 1$, $A_2 = 0$ and then $A_1 = 0$, $A_2 = 1$, we obtain the two linearly independent solutions

$$\begin{bmatrix} e^t \\ 0 \end{bmatrix} \quad \text{and} \quad \begin{bmatrix} 0 \\ e^t \end{bmatrix}.$$

Thus, the general solution of system (21) is

$$\begin{bmatrix} x_1 \\ x_2 \end{bmatrix} = c_1 \begin{bmatrix} e^t \\ 0 \end{bmatrix} + c_2 \begin{bmatrix} 0 \\ e^t \end{bmatrix};$$

that is,

$$x_1 = c_1 e^t \qquad \text{and} \qquad x_2 = c_2 e^t.$$

As we explained in the introductory paragraph of this section, the matrix method has the distinct advantage that it can be easily extended to homogeneous systems of n linear differential equations with constant coefficients in n unknowns. The method of elimination is highly recommended in the case $n = 2$. We shall now present two examples with $n = 3$ and $n = 4$, respectively, and show how to apply the matrix method to compute their general solution.

EXAMPLE 5 Find the general solution of the system

$$\begin{aligned} \dot{x}_1 &= 5x_1 + 2x_2 + 2x_3 \\ \dot{x}_2 &= 2x_1 + 2x_2 - 4x_3 \\ \dot{x}_3 &= 2x_1 - 4x_2 + 2x_3. \end{aligned} \tag{22}$$

Solution Let us look for a solution of system (22) of the form

$$\begin{bmatrix} x_1 \\ x_2 \\ x_3 \end{bmatrix} = \begin{bmatrix} A_1 e^{\lambda t} \\ A_2 e^{\lambda t} \\ A_3 e^{\lambda t} \end{bmatrix}. \tag{23}$$

Proceeding exactly as we did for system (1), we find that in order to obtain nontrivial solutions, λ should be a root of the characteristic equation

$$\begin{vmatrix} 5-\lambda & 2 & 2 \\ 2 & 2-\lambda & -4 \\ 2 & -4 & 2-\lambda \end{vmatrix} = 0, \qquad \text{that is,} \qquad -\lambda^3 + 9\lambda^2 - 108 = 0.$$

The roots of the characteristic equation are $\lambda_1 = -3$ and $\lambda_2 = \lambda_3 = 6$. When $\lambda = -3$, the constants A_1, A_2, and A_3 should satisfy the homogeneous system

$$\begin{aligned} 8A_1 + 2A_2 + 2A_3 &= 0 \\ 2A_1 + 5A_2 - 4A_3 &= 0 \\ 2A_1 - 4A_2 + 5A_3 &= 0 \end{aligned} \Rightarrow \begin{aligned} 4A_1 + A_2 + A_3 &= 0 \\ 2A_1 + 5A_2 - 4A_3 &= 0. \end{aligned}$$

Choosing $A_3 = 1$, we then obtain $A_2 = 1$ and $A_1 = -\frac{1}{2}$. Thus one solution of (22) is

$$\begin{bmatrix} -\frac{1}{2}e^{-3t} \\ e^{-3t} \\ e^{-3t} \end{bmatrix}. \tag{24}$$

When $\lambda = 6$, the constants A_1, A_2, and A_3 should satisfy the homogeneous system

$$\begin{aligned} -A_1 + 2A_2 + 2A_3 &= 0 \\ 2A_1 - 4A_2 - 4A_3 &= 0 \Rightarrow -A_1 + 2A_2 + 2A_3 = 0. \\ 2A_1 - 4A_2 - 4A_3 &= 0 \end{aligned} \tag{25}$$

Choosing $A_3 = 1$ and $A_2 = 1$, we obtain $A_1 = 4$, and

$$\begin{bmatrix} 4e^{6t} \\ e^{6t} \\ e^{6t} \end{bmatrix} \tag{26}$$

is a second solution of (22). On the other hand, the choice $A_1 = 2$ and $A_2 = 1$ yields $A_3 = 0$, and therefore

$$\begin{bmatrix} 2e^{6t} \\ e^{6t} \\ 0 \end{bmatrix} \tag{27}$$

is a third solution of system (22). Let us check the solutions (24), (26), and (27) for linear independence. We have

$$\begin{vmatrix} -\frac{1}{2}e^{-3t} & 4e^{6t} & 2e^{6t} \\ e^{-3t} & e^{6t} & e^{6t} \\ e^{-3t} & e^{6t} & 0 \end{vmatrix} = e^{9t} \begin{vmatrix} -\frac{1}{2} & 4 & 2 \\ 1 & 1 & 1 \\ 1 & 1 & 0 \end{vmatrix} = \frac{9}{2}e^{9t} \neq 0,$$

which proves that the solutions are, in fact, linearly independent. The general solution of system (22) is, therefore,

$$\begin{bmatrix} x_1 \\ x_2 \\ x_3 \end{bmatrix} = c_1 \begin{bmatrix} -\frac{1}{2}e^{-3t} \\ e^{-3t} \\ e^{-3t} \end{bmatrix} + c_2 \begin{bmatrix} 4e^{6t} \\ e^{6t} \\ e^{6t} \end{bmatrix} + c_3 \begin{bmatrix} 2e^{6t} \\ e^{6t} \\ 0 \end{bmatrix};$$

that is,

$$x_1 = -\tfrac{1}{2}c_1 e^{-3t} + 4c_2 e^{6t} + 2c_3 e^{6t}$$

$$x_2 = c_1 e^{-3t} + c_2 e^{6t} + c_3 e^{6t}$$

$$x_3 = c_1 e^{-3t} + c_2 e^{6t}.$$

REMARK 4 In Example 5 the characteristic root $\lambda = -3$ is a simple root, and therefore system (22) has exactly one solution of the form (23) with $\lambda = -3$. However, the characteristic root $\lambda = 6$ is a double root; therefore system (22) should have one or two linearly independent solutions of the form (23) with $\lambda = 6$. Actually, in Example 5, for $\lambda = 6$, we found two linearly independent solutions, (26) and (27). Otherwise, we should have looked for another linearly independent solution of the form

$$\begin{bmatrix} (a_1 t + b_1)e^{6t} \\ (a_2 t + b_2)e^{6t} \\ (a_3 t + b_3)e^{6t} \end{bmatrix}.$$

In general, if λ is a characteristic root of the linear homogeneous system

$$\begin{aligned} \dot{x}_1 &= a_{11}x_1 + a_{12}x_2 + \cdots + a_{1n}x_n \\ \dot{x}_2 &= a_{21}x_1 + a_{22}x_2 + \cdots + a_{2n}x_n \\ &\cdots\cdots\cdots\cdots\cdots\cdots \\ \dot{x}_n &= a_{n1}x_1 + a_{n2}x_2 + \cdots + a_{nn}x_n, \end{aligned} \tag{28}$$

with multiplicity k ($k \leq n$), then to this characteristic root there correspond k linearly independent solutions of system (28). These k solutions are found as follows. First, find all linearly independent solutions of (28) of the form

$$\begin{bmatrix} A_1 e^{\lambda t} \\ A_2 e^{\lambda t} \\ \cdot \\ \cdot \\ \cdot \\ A_n e^{\lambda t} \end{bmatrix}. \tag{29}$$

Then find all linearly independent solutions of (28) of the form

$$\begin{bmatrix} (A_1t + B_1)e^{\lambda t} \\ (A_2t + B_2)e^{\lambda t} \\ \cdot \\ \cdot \\ \cdot \\ (A_nt + B_n)e^{\lambda t} \end{bmatrix} \tag{30}$$

that are linearly independent to solutions (29). Next, find all linearly independent solutions of (28) of the form

$$\begin{bmatrix} (A_1t^2 + B_1t + C_1)e^{\lambda t} \\ (A_2t^2 + B_2t + C_2)e^{\lambda t} \\ \cdot \\ \cdot \\ \cdot \\ (A_nt^2 + B_nt + C_n)e^{\lambda t} \end{bmatrix} \tag{31}$$

that are linearly independent to solutions (29) and (30). Proceeding in this way, we will find exactly k linearly independent solutions of (28) corresponding to this characteristic root. If we apply this procedure to each characteristic root of (28), we will find n linearly independent solutions.[12]

EXAMPLE 6 Find the general solution of the system

$$\begin{aligned} \dot{x}_1 &= x_2 - x_3 + x_4 \\ \dot{x}_2 &= -x_2 + x_4 \\ \dot{x}_3 &= x_3 - x_4 \\ \dot{x}_4 &= 2x_4. \end{aligned} \tag{32}$$

Solution Let us look for a solution of system (32) of the form

$$\begin{bmatrix} x_1 \\ x_2 \\ x_3 \\ x_4 \end{bmatrix} = \begin{bmatrix} A_1e^{\lambda t} \\ A_2e^{\lambda t} \\ A_3e^{\lambda t} \\ A_4e^{\lambda t} \end{bmatrix}.$$

In order that we obtain nontrivial solutions, λ should be a root of the characteristic equation

$$\begin{vmatrix} -\lambda & 1 & -1 & 1 \\ 0 & -1-\lambda & 0 & 1 \\ 0 & 0 & 1-\lambda & -1 \\ 0 & 0 & 0 & 2-\lambda \end{vmatrix} = 0,$$

[12]For a detailed presentation, see R. H. Cole, *Theory of Ordinary Differential Equations* (New York: Appleton-Century-Crofts, 1968), chapter 4.

that is, $-\lambda(-1 - \lambda)(1 - \lambda)(2 - \lambda) = 0$. The roots of the characteristic equation are

$$\lambda_1 = 0, \qquad \lambda_2 = -1, \qquad \lambda_3 = 1, \qquad \lambda_4 = 2.$$

When $\lambda = 0$, the constants, A_1, A_2, A_3, and A_4 should satisfy the homogeneous system

$$\begin{aligned} A_2 - A_3 + A_4 &= 0 \\ -A_2 \qquad\quad + A_4 &= 0 \\ A_3 - A_4 &= 0 \\ 2A_4 &= 0 \end{aligned} \quad \Rightarrow A_4 = A_3 = A_2 = 0.$$

Choosing $A_1 = 1$, we have the solution

$$\begin{bmatrix} 1 \\ 0 \\ 0 \\ 0 \end{bmatrix}. \tag{33}$$

When $\lambda = -1$, the constants A_1, A_2, A_3, and A_4 should satisfy the homogeneous system

$$\begin{aligned} A_1 + A_2 - A_3 + A_4 &= 0 \\ A_4 &= 0 \\ 2A_3 - A_4 &= 0 \\ 3A_4 &= 0 \end{aligned} \quad \Rightarrow A_4 = A_3 = 0 \quad \text{and} \quad A_1 + A_2 = 0.$$

Choosing $A_1 = 1$, we have the solution

$$\begin{bmatrix} e^{-t} \\ -e^{-t} \\ 0 \\ 0 \end{bmatrix}. \tag{34}$$

When $\lambda = 1$, the constants A_1, A_2, A_3, and A_4 should satisfy the homogeneous system

$$\begin{aligned} -A_1 + A_2 - A_3 + A_4 &= 0 \\ -2A_2 \qquad\quad + A_4 &= 0 \\ -A_4 &= 0 \\ A_4 &= 0 \end{aligned} \quad \Rightarrow A_4 = A_2 = 0 \quad \text{and} \quad A_1 + A_3 = 0.$$

For this case, the choice $A_1 = 1$ yields the solution

$$\begin{bmatrix} e^t \\ 0 \\ -e^t \\ 0 \end{bmatrix}. \tag{35}$$

Finally, when $\lambda = 2$, the constants A_1, A_2, A_3, and A_4 should satisfy the homogeneous system

$$\begin{aligned} -2A_1 + A_2 - A_3 + A_4 &= 0 \\ -3A_2 \qquad + A_4 &= 0 \quad \Rightarrow -2A_1 + 7A_2 = 0, \qquad A_4 = 3A_2 = -A_3. \\ -A_3 - A_4 &= 0 \end{aligned}$$

Choosing $A_4 = 3$, we find that $A_2 = 1, A_1 = \frac{7}{2}$, and $A_3 = -3$. Thus, we have the solution

$$\begin{bmatrix} \frac{7}{2}e^{2t} \\ e^{2t} \\ -3e^{2t} \\ 3e^{2t} \end{bmatrix}. \tag{36}$$

The four solutions (33), (34), (35), and (36) are linearly independent. In fact,

$$\begin{vmatrix} 1 & e^{-t} & e^t & \frac{7}{2}e^{2t} \\ 0 & -e^{-t} & 0 & e^{2t} \\ 0 & 0 & -e^t & -3e^{2t} \\ 0 & 0 & 0 & 3e^{2t} \end{vmatrix} = 3e^{2t} \neq 0.$$

Hence, the general solution of (32) is

$$\begin{bmatrix} x_1 \\ x_2 \\ x_3 \\ x_4 \end{bmatrix} = c_1 \begin{bmatrix} 1 \\ 0 \\ 0 \\ 0 \end{bmatrix} + c_2 \begin{bmatrix} e^{-t} \\ -e^{-t} \\ 0 \\ 0 \end{bmatrix} + c_3 \begin{bmatrix} e^t \\ 0 \\ -e^t \\ 0 \end{bmatrix} + c_4 \begin{bmatrix} \frac{7}{2}e^{2t} \\ e^{2t} \\ -3e^{2t} \\ 3e^{2t} \end{bmatrix};$$

that is,

$$\begin{aligned} x_1 &= c_1 + c_2e^{-t} + c_3e^t + \tfrac{7}{2}c_4e^{2t} \\ x_2 &= \qquad -c_2e^{-t} \qquad\quad + c_4e^{2t} \\ x_3 &= \qquad\qquad\qquad - c_3e^t - 3c_4e^{2t} \\ x_4 &= \qquad\qquad\qquad\qquad\quad 3c_4e^{2t}. \end{aligned}$$

NONHOMOGENEOUS SYSTEMS—VARIATION OF PARAMETERS 3.3.1

Theorem 5 of Section 3.1 states that the general solution of a linear nonhomogeneous system is the sum of the general solution of the corresponding homogeneous system and a particular solution of the nonhomogeneous system. To find a particular solution of the nonhomogeneous system we can use a variation of parameters technique similar to the corresponding method for linear non-homogeneous differential equations. (See Section 2.12.) We illustrate the technique by means of an example.

Consider the nonhomogeneous linear system

$$\begin{aligned} \dot{x}_1 &= 2x_1 + x_2 + f(t) \\ \dot{x}_2 &= -3x_1 + 6x_2 + g(t). \end{aligned} \tag{37}$$

In Example 1 of Section 3.3 we found that the general solution of the homogeneous system corresponding to (37) is

$$\begin{bmatrix} x_{1h} \\ x_{2h} \end{bmatrix} = c_1 \begin{bmatrix} e^{3t} \\ e^{3t} \end{bmatrix} + c_2 \begin{bmatrix} e^{5t} \\ 3e^{5t} \end{bmatrix}, \tag{38}$$

where c_1 and c_2 are arbitrary constants. The main idea in the method of variation of parameters is to seek a particular solution of (37) of the form

$$\begin{bmatrix} x_{1p} \\ x_{2p} \end{bmatrix} = u_1(t) \begin{bmatrix} e^{3t} \\ e^{3t} \end{bmatrix} + u_2(t) \begin{bmatrix} e^{5t} \\ 3e^{5t} \end{bmatrix}, \tag{39}$$

where u_1 and u_2 are functions to be determined. That is,

$$\begin{aligned} x_{1p} &= u_1(t)e^{3t} + u_2(t)e^{5t} \\ x_{2p} &= u_1(t)e^{3t} + 3u_2(t)e^{5t}. \end{aligned}$$

Substituting into (37) and simplifying we obtain

$$\begin{aligned} \dot{u}_1(t)e^{3t} + \dot{u}_2(t)e^{5t} &= f(t) \\ \dot{u}_1(t)e^{3t} + 3\dot{u}_2(t)e^{5t} &= g(t). \end{aligned}$$

Using Cramer's rule, we find

$$\dot{u}_1(t) = \frac{\begin{vmatrix} f(t) & e^{5t} \\ g(t) & 3e^{5t} \end{vmatrix}}{\begin{vmatrix} e^{3t} & e^{5t} \\ e^{3t} & 3e^{5t} \end{vmatrix}} = \tfrac{1}{2}[3f(t) - g(t)]e^{-3t} \tag{40}$$

and

$$\dot{u}_2(t) = \frac{\begin{vmatrix} e^{3t} & f(t) \\ e^{3t} & g(t) \end{vmatrix}}{\begin{vmatrix} e^{3t} & e^{5t} \\ e^{3t} & 3e^{5t} \end{vmatrix}} = \tfrac{1}{2}[g(t) - f(t)]e^{-5t}. \tag{41}$$

From (40) and (41) we find $u_1(t)$ and $u_2(t)$ by integration, and, substituting the results into (39), we obtain a particular solution of (37). The general solution of (37) is given by

$$\begin{bmatrix} x_1 \\ x_2 \end{bmatrix} = \begin{bmatrix} x_{1h} \\ x_{2h} \end{bmatrix} + \begin{bmatrix} x_{1p} \\ x_{2p} \end{bmatrix}.$$

In the special case where $f(t) = e^{3t}$ and $g(t) = 2e^{3t}$, we find (omitting the constants of integration which are unnecessary)

$$\dot{u}_1(t) = \tfrac{1}{2} \Rightarrow u_1(t) = \tfrac{1}{2}t$$

and

$$\dot{u}_2(t) = \tfrac{1}{2}e^{-2t} \Rightarrow u_2(t) = -\tfrac{1}{4}e^{-2t}.$$

Thus, from (39), we have

$$\begin{bmatrix} x_{1p} \\ x_{2p} \end{bmatrix} = \tfrac{1}{2}t \begin{bmatrix} e^{3t} \\ e^{3t} \end{bmatrix} - \tfrac{1}{4}e^{-2t} \begin{bmatrix} e^{5t} \\ 3e^{5t} \end{bmatrix}.$$

Hence, the general solution of (37), for $f(t) = e^{3t}$ and $g(t) = 2e^{3t}$, is given by

$$\begin{bmatrix} x_1 \\ x_2 \end{bmatrix} = c_1 \begin{bmatrix} e^{3t} \\ e^{3t} \end{bmatrix} + c_2 \begin{bmatrix} e^{5t} \\ 3e^{5t} \end{bmatrix} + \tfrac{1}{2}t \begin{bmatrix} e^{3t} \\ e^{3t} \end{bmatrix} - \tfrac{1}{4}e^{-2t} \begin{bmatrix} e^{5t} \\ 3e^{5t} \end{bmatrix}$$

or

$$x_1(t) = c_1e^{3t} + c_2e^{5t} + \tfrac{1}{2}te^{3t} - \tfrac{1}{4}e^{3t}$$

$$x_2(t) = c_1e^{3t} + 3c_2e^{5t} + \tfrac{1}{2}te^{3t} - \tfrac{3}{4}e^{3t}.$$

APPLICATIONS 3.3.2

■ In the biokinetics application of Section 3.2.1, assume that the rate constants have the values

Biokinetics

$$k_1 = k_2 = k_3 = k_4 = 1 \quad \text{and} \quad k_5 = 2.$$

Then system (16) becomes

$$\begin{aligned} \dot{x} &= -2x + y \\ \dot{y} &= x - 2y \\ \dot{z} &= x + y - 2z \\ \dot{u} &= 2z. \end{aligned} \tag{42}$$

We also have the initial conditions

$$x(0) = 1, \qquad y(0) = z(0) = u(0) = 0 \tag{43}$$

and the terminal conditions

$$x = y = z = 0 \quad \text{and} \quad u = 1 \quad \text{as } t \to \infty. \tag{44}$$

We want to solve the problem (42)–(44).

Solution We use the matrix method. The characteristic equation of (42) is

$$\begin{vmatrix} -2-\lambda & 1 & 0 & 0 \\ 1 & -2-\lambda & 0 & 0 \\ 1 & 1 & -2-\lambda & 0 \\ 0 & 0 & 2 & -\lambda \end{vmatrix} = \lambda(\lambda + 1)(\lambda + 2)(\lambda + 3) = 0,$$

with characteristic roots

$$\lambda_1 = 0, \quad \lambda_2 = -1, \quad \lambda_3 = -2, \quad \lambda_4 = -3.$$

To each one of these four roots there corresponds a solution of (42) of the form

$$\begin{bmatrix} x \\ y \\ z \\ u \end{bmatrix} = \begin{bmatrix} A_1 e^{\lambda t} \\ A_2 e^{\lambda t} \\ A_3 e^{\lambda t} \\ A_4 e^{\lambda t} \end{bmatrix}, \tag{45}$$

where the constants A_1, A_2, A_3, and A_4 satisfy the system

$$\begin{aligned} (-2-\lambda)A_1 + A_2 &= 0 \\ A_1 + (-2-\lambda)A_2 &= 0 \\ A_1 + A_2 + (-2-\lambda)A_3 &= 0 \\ 2A_3 - \lambda A_4 &= 0. \end{aligned} \tag{46}$$

For $\lambda = 0$, we find $A_1 = A_2 = A_3 = 0$, $A_4 = 1$, and therefore, from (45),

$$\begin{bmatrix} 0 \\ 0 \\ 0 \\ 1 \end{bmatrix}$$

is one solution of (42). For $\lambda = -1, -2$, and -3, solving the system (46) and using (45), we find, respectively, the solutions

$$\begin{bmatrix} e^{-t} \\ e^{-t} \\ 2e^{-t} \\ -4e^{-t} \end{bmatrix}, \quad \begin{bmatrix} 0 \\ 0 \\ e^{-2t} \\ -e^{-2t} \end{bmatrix}, \quad \begin{bmatrix} e^{-3t} \\ -e^{-3t} \\ 0 \\ 0 \end{bmatrix}.$$

The four solutions are linearly independent, and therefore the general solution of (42) is

$$\begin{bmatrix} x \\ y \\ z \\ u \end{bmatrix} = c_1 \begin{bmatrix} 0 \\ 0 \\ 0 \\ 1 \end{bmatrix} + c_2 \begin{bmatrix} e^{-t} \\ e^{-t} \\ 2e^{-t} \\ -4e^{-t} \end{bmatrix} + c_3 \begin{bmatrix} 0 \\ 0 \\ e^{-2t} \\ -e^{-2t} \end{bmatrix} + c_4 \begin{bmatrix} e^{-3t} \\ -e^{-3t} \\ 0 \\ 0 \end{bmatrix};$$

that is,

$$\begin{aligned} x &= c_2 e^{-t} + c_4 e^{-3t} \\ y &= c_2 e^{-t} - c_4 e^{-3t} \\ z &= 2c_2 e^{-t} + c_3 e^{-2t} \\ u &= c_1 - 4c_2 e^{-t} - c_3 e^{-2t}. \end{aligned} \tag{47}$$

Using the initial conditions (43), we find that

$$\begin{aligned} c_2 + c_4 &= 1 \\ c_2 - c_4 &= 0 \\ 2c_2 + c_3 &= 0 \\ c_1 - 4c_2 - c_3 &= 0. \end{aligned} \tag{48}$$

Because of the negative exponents, the terminal conditions $x = y = z = 0$ as $t \to \infty$ are satisfied without any restriction on the constants c_1, c_2, c_3, and c_4. The terminal condition $u = 1$ as $t \to \infty$ implies [from the last equation in (47)] that

$$c_1 = 1. \tag{49}$$

From (48), we find that

$$c_1 = 1, \qquad c_2 = c_4 = \tfrac{1}{2}, \qquad c_3 = -1.$$

Thus, (49) is also satisfied and the only solution of problem (42)–(44) is

$$\begin{aligned} x &= \tfrac{1}{2}e^{-t} + \tfrac{1}{2}e^{-3t} \\ y &= \tfrac{1}{2}e^{-t} - \tfrac{1}{2}e^{-3t} \\ z &= e^{-t} - e^{-2t} \\ u &= 1 - 2e^{-t} + e^{-2t}. \end{aligned}$$

EXERCISES

In Exercises 1 through 6, find the general solution of the system by the matrix method.

1. $\dot{x}_1 = 4x_1 + 2x_2$
$\dot{x}_2 = -3x_1 - x_2$

2. $\dot{x}_1 = 4x_1 - 2x_2$
$\dot{x}_2 = x_1 + x_2$

3. $\dot{x}_1 = 3x_1 - 2x_2$
$\dot{x}_2 = 2x_1 - x_2$

4. $\dot{x}_1 = -4x_1 + x_2$
$\dot{x}_2 = -x_1 - 2x_2$

5. $\dot{x}_1 = 3x_1 - 2x_2$
$\dot{x}_2 = 17x_1 - 7x_2$

6. $\dot{x}_1 = 3x_1 - 5x_2$
$\dot{x}_2 = 4x_1 - 5x_2$

In Exercises 7 through 10, solve the initial value problem.

7. $\dot{x}_1 = 4x_1 - 2x_2$
$\dot{x}_2 = x_1 + x_2$
$x_1(0) = 1, x_2(0) = 0$

8. $\dot{x}_1 = 3x_1 - 2x_2$
$\dot{x}_2 = 2x_1 - x_2$
$x_1(0) = -1, x_2(0) = 1$

9. $\dot{x}_1 = 3x_1 - 2x_2$
$\dot{x}_2 = 17x_1 - 7x_2$
$x_1(0) = 0, x_2(0) = 3$

10. $\dot{x}_1 = 3x_1 - 5x_2$
$\dot{x}_2 = 4x_1 - 5x_2$
$x_1(\pi) = 1, x_2(\pi) = \frac{4}{5}$

In Exercises 11 through 16, find the general solution of the system by the matrix method.

11. $\dot{x}_1 = 2x_1$
$\dot{x}_2 = -3x_2$

12. $\dot{x}_1 = 4x_2$
$\dot{x}_2 = -x_1$

13. $\dot{x}_1 = -x_2$
$\dot{x}_2 = 3x_1 + 4x_2$

14. $\dot{x}_1 = -x_2$
$\dot{x}_2 = x_1 - 2x_2$

15. $\dot{x}_1 = 4x_1 + 3x_2$
$\dot{x}_2 = -x_1$

16. $\dot{x}_1 = -2x_1 + x_2$
$\dot{x}_2 = -x_1$

In Exercises 17 through 26, find the general solution of the system of differential equations.

17. $\dot{x}_1 = 3x_1 + 2x_2 + 2x_3$
$\dot{x}_2 = x_1 + 4x_2 + x_3$
$\dot{x}_3 = -2x_1 - 4x_2 - x_3$

18. $\dot{x}_1 = 3x_1 - x_2$
$\dot{x}_2 = -x_1 + 2x_2 - x_3$
$\dot{x}_3 = -x_2 + 3x_3$

19. $\dot{x}_1 = 2x_1 + x_2$
$\dot{x}_2 = -x_1 + x_3$
$\dot{x}_3 = x_1 + 3x_2 + x_3$

20. $\dot{x}_1 = 7x_1 + 4x_2 - 4x_3$
$\dot{x}_2 = 4x_1 - 8x_2 - x_3$
$\dot{x}_3 = -4x_1 - x_2 - 8x_3$

21. $\dot{x}_1 = x_1 + 2x_2 + x_3 - x_4$
$\dot{x}_2 = -x_2 + 2x_3 + 2x_4$
$\dot{x}_3 = 2x_2 + 2x_3 + 2x_4$
$\dot{x}_4 = -3x_2 - 6x_3 - 6x_4$

22. $\dot{x}_1 = 3x_1 + x_2$
$\dot{x}_2 = x_1 + 3x_2$
$\dot{x}_3 = 2x_3 + x_4 + x_5$
$\dot{x}_4 = x_3 + 2x_4 + x_5$
$\dot{x}_5 = x_3 + x_4 + 2x_5$

23. $\dot{x}_1 = -10x_1 + x_2 + 7x_3$
$\dot{x}_2 = -9x_1 + 4x_2 + 5x_3$
$\dot{x}_3 = -17x_1 + x_2 + 12x_3$

24. $\dot{x}_1 = x_2$
$\dot{x}_2 = x_3$
$\dot{x}_3 = x_1$

In Exercises 25 through 30, solve the IVP.

25. $\dot{x}_1 = 3x_1 + 2x_2 + 2x_3$
$\dot{x}_2 = x_1 + 4x_2 + x_3$
$\dot{x}_3 = -2x_1 - 4x_2 - x_3$
$x_1(0) = 1,\ x_2(0) = 0,$
$x_3(0) = -1$

26. $\dot{x}_1 = 3x_1 - x_2$
$\dot{x}_2 = -x_1 + 2x_2 - x_3$
$\dot{x}_3 = -x_2 + 3x_3$
$x_1(0) = 1,\ x_2(0) = 0,$
$x_3(0) = -1$

27. $\dot{x}_1 = 2x_1 + x_2$
$\dot{x}_2 = -x_1 + x_3$
$\dot{x}_3 = x_1 + 3x_2 + x_3$
$x_1(0) = 1,\ x_2(0) = 1,$
$x_3(0) = 3$

28. $\dot{x}_1 = 7x_1 + 4x_2 - 4x_3$
$\dot{x}_2 = 4x_1 - 8x_2 - x_3$
$\dot{x}_3 = -4x_1 - x_2 - 8x_3$
$x_1(0) = 3,\ x_2(0) = 5,$
$x_3(0) = -1$

29. $\dot{x}_1 = -10x_1 + x_2 + 7x_3$
$\dot{x}_2 = -9x_1 + 4x_2 + 5x_3$
$\dot{x}_3 = -17x_1 + x_2 + 12x_3$
$x_1(0) = 6,\ x_2(0) = 1,$
$x_3(0) = 10$

30. $\dot{x}_1 = x_2$
$\dot{x}_2 = x_3$
$\dot{x}_3 = x_1$
$x_1(0) = x_2(0) = x_3(0) = 1$

In Exercises 31 through 36, use the method of variation of parameters to find a particular solution. Then write down the general solution of the system.

31. $\dot{x}_1 = 2x_1 + x_2 + e^{5t}$
$\dot{x}_2 = -3x_1 + 6x_2 + e^{5t}$

32. $\dot{x}_1 = 4x_1 + 2x_2 + 4$
$\dot{x}_2 = -3x_1 - x_2 - 3$

33. $\dot{x}_1 = -x_2 + t$
$\dot{x}_2 = 3x_1 + 4x_2 - 2 - 4t$

34. $\dot{x}_1 = 4x_2 - 3\cos t$
$\dot{x}_2 = -x_1$

35. $\dot{x} = -4x + y$
$\dot{y} = -x - 2y + 9$

36. $\dot{x} = 4x + 3y - 2 - 4t$
$\dot{y} = -x + t$

37. In the chemistry application of Section 1.7.1, we arrived at the homogeneous system [see Eqs. (10), Section 1.7.1]

$$\dot{c}_1 = -\frac{k}{V_1}c_1 + \frac{k}{V_1}c_2$$

$$\dot{c}_2 = \frac{k}{V_2}c_1 - \frac{k}{V_2}c_2.$$

Use the matrix technique to obtain the general solution of this system.

38. In the biokinetics application of Section 3.2.1, assume that $k_1 = k_2 = k_3 = k_4 = 1$ and $k_5 = 5$. Then system (16) becomes

$$\dot{x} = -2x + y$$
$$\dot{y} = x - 2y$$
$$\dot{z} = x + y - 5z$$
$$\dot{u} = 5z.$$

Find the general solution of this system by the matrix method. Find the particular solution of the system above that satisfies the initial and terminal conditions mentioned in the biokinetics application,

$$x(0) = 1, \qquad y(0) = z(0) = u(0) = 0$$

and

$$x = y = z = 0 \quad \text{and} \quad u = 1 \quad \text{as} \quad t \to \infty.$$

REVIEW EXERCISES

In Exercises 1 through 10, find the general solution of the system.

1. $\dot{x} = y$
$\dot{y} = -2x + 3y$

2. $\dot{x} = -x + 4y$
$\dot{y} = 2x + y$

3. $\dot{x} = y + e^{3t}$
$\dot{y} = -2x + 3y$

4. $\dot{x} = -x + 4y + e^{3t}$
$\dot{y} = 2x + y$

5. $\dot{x}_1 = x_2 + t^2$
$\dot{x}_2 = x_3$
$\dot{x}_3 = x_4$
$\dot{x}_4 = x_1$

6. $\dot{x}_1 = 2x_1$
$\dot{x}_2 = x_1 + x_3 - x_4$
$\dot{x}_3 = x_1 - x_3$
$\dot{x}_4 = -x_1 + x_4$

7. $\dot{x} = 3x - 2y$
$\dot{y} = -2x$
$\dot{z} = x + 2z$

8. $\dot{x} = 3x + y$
$\dot{y} = x + 2y + z$
$\dot{z} = y + 3z$

9. $\dot{x}_1 = x_1 - x_2$
$\dot{x}_2 = -2x_2 - x_3$
$\dot{x}_3 = 3x_3$

10. $\dot{x}_1 = 2x_1 + x_2$
$\dot{x}_2 = -x_1 - x_3$
$\dot{x}_3 = -x_1 - 3x_2 + x_3$

In Exercises 11 through 15, solve the initial value problem.

11. $\dot{x} = y$
$\dot{y} = -2x + 3y$
$x(0) = 1$
$y(0) = 3$

12. $\dot{x} = y + e^{3t}$
$\dot{y} = -2x + 3y$
$x(0) = 0$
$y(0) = 0$

13. $\dot{x}_1 = x_2 + t^2$
$\dot{x}_2 = x_3$
$\dot{x}_3 = x_4$
$\dot{x}_4 = x_1$
$x_1(0) = 1, x_2(0) = 1, x_3(0) = 1, x_4(0) = -1$

14. $\dot{x} = 3x - 2y$
$\dot{y} = -2x$
$\dot{z} = x + 2z$
$x(0) = 2, y(0) = -1, z(0) = 0$

15. $\dot{x}_1 = x_1 - x_2$
$\dot{x}_2 = -2x_2 - x_3$
$\dot{x}_3 = 3x_3$
$x_1(0) = 5, x_2(0) = -3, x_3(0) = 1$

16. In the electric circuit in Figure 3.6, assume that the currents are initially zero and that $R_1 = 2$ ohms, $R_2 = 4$ ohms, $L_1 = 1$ henry, $L_2 = 2$ henries, and $V(t) = 4$ volts. Determine the currents $I_1(t)$ and $I_2(t)$ at any time t.

17. Assume that the system

$$\dot{N}_1 = 5N_1 - 2N_2$$
$$\dot{N}_2 = 6N_1 - 2N_2$$

represents two interracting populations. Are the populations headed for extinction? For explosion? Explain.

18. Two tanks, A and B, are connected as shown in Figure 3.7. Tank A contains 50 gallons of pure water, and tank B contains 50 gallons of brine in which 20 pounds of salt is dissolved. Liquid circulates through the tanks at the

rate of 3 gallons per minute, and the mixture in each tank is kept uniform by stirring.
(a) How much salt is in each tank after 20 minutes?
(b) How much salt is in each tank after a very long time?

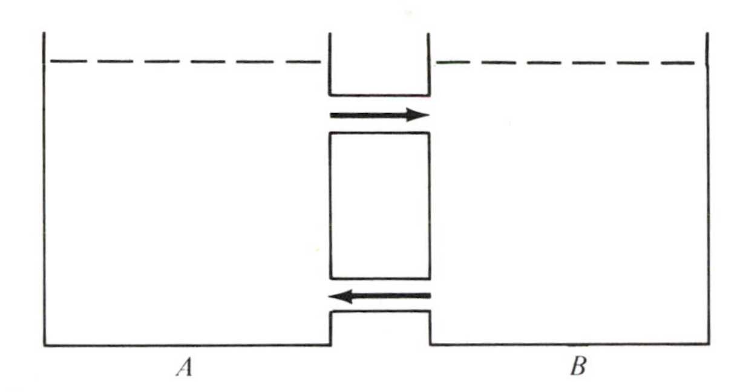

Figure 3.7

19. Solve[13] the system of differential equations

$$\frac{dx}{dt} = x + y - 3, \qquad \frac{dy}{dt} = -2x + 3y + 1,$$

subject to the condition $x = y = 0$ at $t = 0$.

20. In the network[14] shown in Figure 3.8, the switch is closed at $t = 0$ and the currents are initially zero. Find the current i_2 as a function of time.

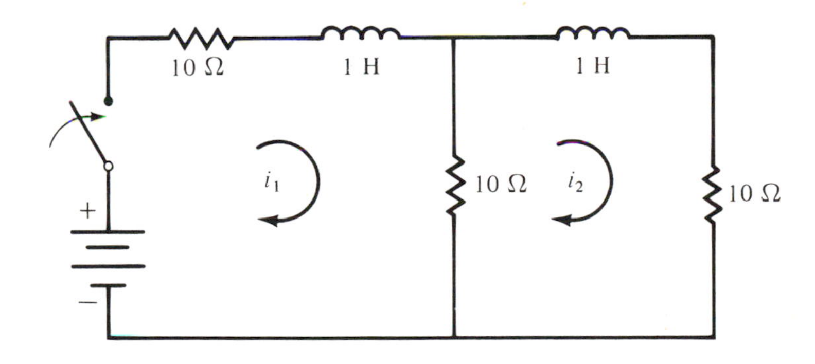

Figure 3.8

[13]This exercise is taken from the second William Lowell Putnam Mathematical Competition, *Amer. Math. Monthly* 46 (1939): 248.

[14]This is Example 14 in M. Van Valkenburg, *Network Analysis,* 2nd ed. (Englewood Cliffs, N.J.: Prentice-Hall, 1964), p. 179. © 1964. Reprinted by permission of Prentice-Hall, Inc.

CHAPTER 4

The Laplace Transform

4.1 INTRODUCTION

In this chapter we present another method for solving linear differential equations with constant coefficients and systems of such equations. It is called the method of Laplace transform. By this method an initial value problem is transformed into an algebraic equation or system of equations which we can solve by algebraic methods and a table of Laplace transforms. This method resembles in some ways the use of logarithms for solving exponential equations.

4.2 THE LAPLACE TRANSFORM AND ITS PROPERTIES

Assume that the function g is defined for $0 \leq t < \infty$ and is bounded and integrable in every finite interval $0 \leq t \leq b$. Then by definition

$$\int_0^\infty g(t)dt = \lim_{b\to\infty} \int_0^b g(t)dt.$$

We say that the *improper integral* on the left converges or diverges according to whether the limit on the right does or does not exist. For example, the improper integral $\int_0^\infty e^{-3t}dt$ converges but $\int_0^\infty e^{3t}dt$ diverges. In fact,

$$\begin{aligned}\int_0^\infty e^{-3t}dt &= \lim_{b\to\infty}\int_0^b e^{-3t}dt = \lim_{b\to\infty}\left(-\frac{1}{3}e^{-3t}\Big|_0^b\right)\\ &= \lim_{b\to\infty}\left(-\frac{1}{3}e^{-3b}+\frac{1}{3}\right) = \frac{1}{3},\end{aligned}$$

but

$$\begin{aligned}\int_0^\infty e^{3t}dt &= \lim_{b\to\infty}\int_0^b e^{3t}dt = \lim_{b\to\infty}\left(\frac{1}{3}e^{3t}\Big|_0^b\right)\\ &= \lim_{b\to\infty}\left(\frac{1}{3}e^{3b}-\frac{1}{3}\right) = \infty.\end{aligned}$$

DEFINITION 1

Assume that the function f is defined for $0 \leq t < \infty$. The Laplace transform of f, which we denote by F or $\mathscr{L}[f]$, is defined by the improper integral

$$F(s) = \mathscr{L}[f(t)] = \int_0^\infty e^{-st} f(t)\,dt, \tag{1}$$

provided that the integral in (1) exists for all s larger than or equal to some value s_0.

EXAMPLE 1 Compute the Laplace transform of $f(t) = 1$.

Solution

$$\begin{aligned}\int_0^\infty e^{-st} f(t)\,dt &= \int_0^\infty e^{-st}(1)\,dt = \int_0^\infty e^{-st}\,dt \\ &= \lim_{b\to\infty} \int_0^b e^{-st}\,dt = \lim_{b\to\infty} - \left[\frac{e^{-st}}{s}\Bigg|_0^b\right] \\ &= \lim_{b\to\infty} \left[\frac{1}{s} - \frac{e^{-bs}}{s}\right].\end{aligned}$$

If $s > 0$, the limit above exists and we obtain

$$\int_0^\infty e^{-st} f(t)\,dt = \frac{1}{s}, \qquad s > 0.$$

EXAMPLE 2 Compute the Laplace transform of $f(t) = e^{2t}$.

Solution

$$\begin{aligned}\int_0^\infty e^{-st} f(t)\,dt &= \int_0^\infty e^{-st} e^{2t}\,dt = \int_0^\infty e^{-(s-2)t}\,dt \\ &= \lim_{b\to\infty} \int_0^b e^{-(s-2)t}\,dt = \lim_{b\to\infty} - \left[\frac{e^{-(s-2)t}}{s-2}\Bigg|_0^b\right] \\ &= \lim_{b\to\infty} \left[\frac{1}{s-2} - \frac{e^{-(s-2)b}}{s-2}\right].\end{aligned}$$

This limit exists only when $s > 2$. Hence,

$$\int_0^\infty e^{-st} f(t)\,dt = \frac{1}{s-2}, \qquad s > 2.$$

In Examples 1 and 2 we observe that the Laplace transform is a function of s. We refer to f as the *inverse Laplace transform* of F. For the purposes of notation for the inverse Laplace transform, we introduce the alternative symbolism $\mathcal{L}[f(t)] = F(s)$. We also write $f(t) = \mathcal{L}^{-1}[F(s)]$. From Examples 1 and 2 we see that

$$\mathcal{L}^{-1}\left[\frac{1}{s}\right] = 1 \qquad \text{and} \qquad \mathcal{L}^{-1}\left[\frac{1}{s-2}\right] = e^{2t}.$$

We note that for a given f the corresponding F is uniquely determined (when it exists). Concerning inverse Laplace transforms, we will state Theorem 1, but first we give the following definitions.

DEFINITION 2

A function f is said to be piecewise continuous on an interval I if I can be subdivided into a finite number of subintervals, in each of which f is continuous and has finite left- and right-hand limits.

DEFINITION 3

A function f is of exponential order α as t tends to infinity if there exist numbers M, α, and T such that

$$|f(t)| \leq Me^{\alpha t} \qquad \textit{when} \qquad t \geq T.$$

Alternatively, f is of exponential order α if there exists an α such that $\lim_{t\to\infty}|f(t)|/e^{\alpha t} = L$, where $L = 0$ or L is a finite positive number.

We now quote without proof a theorem that guarantees the convergence of the integral (1).

THEOREM 1

If f is piecewise continuous on every finite interval in $[0,\infty)$, and if f is of exponential order α as t tends to infinity, the integral (1) converges for $s > \alpha$. Furthermore, if f and g are piecewise continuous functions whose Laplace transforms exist and satisfy $\mathcal{L}[f(t)] = \mathcal{L}[g(t)]$, then $f = g$ at their points of continuity. Thus, if $F(s)$ has a continuous inverse f, then f is unique.

The following theorems provide us with useful tools for manipulating Laplace transforms. The proof of the first theorem is a simple consequence of the definitions and will therefore be omitted (see Exercise 60). In what follows the symbols c_1, c_2, and k denote arbitrary constants.

THEOREM 2

$$\mathcal{L}[c_1 f(t) + c_2 g(t)] = c_1 F(s) + c_2 G(s).$$

THEOREM 3

$$\textit{If } F(s) = \mathcal{L}[f(t)], \textit{ then } F(s+k) = \mathcal{L}[e^{-kt}f(t)].$$

Proof

$$F(s+k) = \int_0^\infty e^{-(s+k)t} f(t)dt = \int_0^\infty e^{-st}e^{-kt}f(t)dt$$
$$= \int_0^\infty e^{-st}(e^{-kt}f(t))dt = \mathcal{L}[e^{-kt}f(t)].$$

EXAMPLE 3 Find (a) $\mathcal{L}[t^2]$. (b) $\mathcal{L}[\cosh kt]$.

Solution

(a)
$$\mathcal{L}[t^2] = \int_0^\infty t^2 e^{-st}dt = \lim_{t\to\infty}\left[\frac{t^2e^{-st}}{-s}\right] + \frac{2}{s}\int_0^\infty te^{-st}dt$$
$$= \lim_{t\to\infty}\left[\frac{t^2e^{-st}}{-s}\right] + \frac{2}{s}\lim_{t\to\infty}\left[\frac{te^{-st}}{s}\right] + \frac{2}{s^2}\int_0^\infty e^{-st}dt$$

If $s > 0$, then both of the above limits are zero as can be verified by use of l'Hospital's rule. Thus

$$\mathcal{L}[t^2] = \frac{2}{s^3},\ s > 0.$$

(b)
$$\mathcal{L}[\cosh kt] = \mathcal{L}\left[\frac{1}{2}e^{kt} + \frac{1}{2}e^{-kt}\right] = \frac{1}{2}\mathcal{L}[e^{kt}] + \frac{1}{2}\mathcal{L}[e^{-kt}]$$
$$= \frac{1}{2}\mathcal{L}[e^{kt}(1)] + \frac{1}{2}\mathcal{L}[e^{(-k)t}(1)] = \frac{1}{2}\frac{1}{s-k} + \frac{1}{2}\frac{1}{s-(-k)}$$
$$\mathcal{L}[\cosh kt] = \frac{1}{2}\frac{(s+k)+(s-k)}{(s-k)(s+k)} = \frac{1}{2}\frac{2s}{s^2-k^2} = \frac{s}{s^2-k^2}.$$

EXAMPLE 4 Find $\mathcal{L}[e^{-7t}t^2]$.

Solution Since $\mathcal{L}[t^2] = 2/s^3$, we have, by Theorem 3,

$$\mathcal{L}[e^{-7t}t^2] = \frac{2}{(s+7)^3}.$$

In the next section we illustrate how Laplace transforms are used to solve initial value problems. Quite often the Laplace transform of the solution is expressed as the product of two known functions of s. To find the solution of the IVP, one needs to find the inverse Laplace transform of this product. To do so, one frequently uses either partial fractions (see Appendix B) or Theorem 4. The property expressed in Theorem 4 is known as the *convolution property of Laplace transforms*. We state the theorem without proof.

THEOREM 4

$$G(s)F(s) = \mathcal{L}\left[\int_0^t g(t-\tau)f(\tau)d\tau\right]. \tag{2}$$

EXAMPLE 5 Find $\mathcal{L}[\int_0^t f(\tau)d\tau]$.

Solution In order to find $\mathcal{L}[\int_0^t f(\tau)d\tau]$, we apply the definition. (Alternatively we could apply Theorem 4 with $g(t) = 1$.

$$\mathcal{L}\left[\int_0^t f(\tau)d\tau\right] = \int_0^\infty \left[\int_0^t f(\tau)d\tau\right] e^{-st}dt$$

$$= \lim_{t\to\infty}\left[\frac{\left(\int_0^t f(\tau)d\tau\right)e^{-st}}{-s}\right] + \frac{1}{s}\int_0^\infty f(t)e^{-st}dt.$$

Since f is of exponential order, the integral of f is also of exponential order and the above limit is zero. Therefore

$$\mathcal{L}\left[\int_0^t f(\tau)d\tau\right] = \frac{1}{s}F(s).$$

The result of Example 5 enables us to use Laplace transform methods to solve certain types of integral equations. (See the population growth application in Section 4.6.)

We summarize our results in Table 4.1. Verification of some of the entries is left to the exercises.

Table 4.1

	$F(s)$	$f(t)$
(i)	$\frac{1}{s}, s>0$	1
(ii)	$\frac{1}{s^2}, s>0$	t
(iii)	$\frac{n!}{s^{n+1}}, s>0$	$t^n, n = 1, 2, 3, \ldots$
(iv)	$\frac{1}{s-k}, s>k$	e^{kt}
(v)	$\frac{n!}{(s-k)^{n+1}}, s>k$	$e^{kt}t^n, n = 1, 2, 3, \ldots$
(vi)	$\frac{k}{s^2+k^2}, s>0$	$\sin kt$
(vii)	$\frac{s}{s^2+k^2}, s>0$	$\cos kt$
(viii)	$\frac{m}{(s-k)^2+m^2}, s>k$	$e^{kt}\sin mt$
(ix)	$\frac{s-k}{(s-k)^2+m^2}, s>k$	$e^{kt}\cos mt$

(x)	$\dfrac{s}{s^2 - k^2}, s > k$	$\cosh kt$
(xi)	$\dfrac{k}{s^2 - k^2}, s > k$	$\sinh kt$
(xii)	$\dfrac{k_1 - k_2}{(s - k_1)(s - k_2)}, s > k_1, k_2$	$e^{k_1 t} - e^{k_2 t}$
(xiii)	$\dfrac{2ks}{(s^2 + k^2)^2}, s > 0$	$t \sin kt$
(xiv)	$\dfrac{s^2 - k^2}{(s^2 + k^2)^2}, s > 0$	$t \cos kt$
(xv)	$c_1F(s) + c_2G(s)$	$c_1f(t) + c_2g(t)$
(xvi)	$F(s + k)$	$e^{-kt}f(t)$
(xvii)	$F(ks)$	$\dfrac{1}{k}f\left(\dfrac{t}{k}\right), k > 0$
(xviii)	$F^{(n)}(s)$	$(-t)^n f(t)$
(xix)	$G(s)F(s)$	$\int_0^t g(t - \tau)f(\tau)d\tau = \int_0^t f(t - \tau)g(\tau)d\tau$
(xx)	$\dfrac{1}{s}F(s)$	$\int_0^t f(t)dt$
(xxi)	$\dfrac{1}{s^2}F(s)$	$\int_0^t \int_0^\tau f(u)du\, d\tau$
(xxii)	$s^nF(s) - s^{n-1}f(0) - s^{n-2}f'(0) - \cdots - f^{(n-1)}(0)$	$f^{(n)}(t)$

Our approach thus far has been to apply the definition of the Laplace transform to obtain the transform of certain specific functions or classes of functions. In much the same way as one uses a table of integrals or formulas for differentiation, one can also use a table of Laplace transforms. That is, we do not resort to the definition every time we wish to determine a particular Laplace transform, but rather we prefer to look up the transform in a table. Naturally, the reader realizes that not every function will appear in such a table. Therefore, if one is to be proficient at using the table, one must develop a capability for expressing the functions in the forms found in the tables. We take notice of the fact that all the specific functions that appear in Table 4.1, Laplace transforms, are continuous and of exponential order. Formula (xxii) of Table 4.1 is presented in Theorem 2 of Section 4.3.

EXAMPLE 6 Show that

$$\mathcal{L}[e^{k_1 t} - e^{k_2 t}] = \frac{k_1 - k_2}{(s - k_1)(s - k_2)}.$$

Solution Using formula (xv) of Table 4.1, we have

$$\mathcal{L}[e^{k_1 t} - e^{k_2 t}] = \mathcal{L}[e^{k_1 t}] - \mathcal{L}[e^{k_2 t}].$$

Next we apply (iv) to obtain

$$\mathcal{L}[e^{k_1t} - e^{k_2t}] = \frac{1}{s - k_1} - \frac{1}{s - k_2} = \frac{s - k_2 - (s - k_1)}{(s - k_1)(s - k_2)} = \frac{k_1 - k_2}{(s - k_1)(s - k_2)}.$$

EXAMPLE 7 Show that

$$\mathcal{L}[t \sin kt] = \frac{2ks}{(s^2 + k^2)^2}.$$

Solution From formula (vi) we see that

$$\mathcal{L}[\sin kt] = \frac{k}{s^2 + k^2}. \tag{3}$$

Apply formula (xviii) with $n = 1$ to Eq. (3):

$$\frac{d}{ds}\left(\frac{k}{s^2 + k^2}\right) = \mathcal{L}[-t \sin kt] = -\mathcal{L}[t \sin kt].$$

Thus,

$$-\frac{k(2s)}{(s^2 + k^2)^2} = -\mathcal{L}[t \sin kt]$$

$$\Rightarrow \frac{2ks}{(s^2 + k^2)^2} = \mathcal{L}[t \sin kt].$$

EXAMPLE 8 Find

$$\mathcal{L}^{-1}\left[\frac{5s^2 + 6s + 4}{(s+4)(s^2+4)}\right).$$

Solution Using partial-fraction decomposition (see Appendix B), we have

$$\frac{5s^2 + 6s + 4}{(s+4)(s^2+4)} = \frac{4}{s+4} + \frac{s-3}{s^2+4}.$$

Thus

$$\begin{aligned}
\mathcal{L}^{-1}\left[\frac{5s^2 + 6s + 4}{(s + 4)(s^2 + 4)}\right] &= \mathcal{L}^{-1}\left[\frac{4}{s + 4} + \frac{s-3}{s^2 + 4}\right] \\
&= 4\mathcal{L}^{-1}\left[\frac{1}{s + 4}\right] + \mathcal{L}^{-1}\left[\frac{s}{s^2 + 4}\right] - 3\mathcal{L}^{-1}\left[\frac{1}{s^2 + 4}\right] \\
&= 4\mathcal{L}^{-1}\left[\frac{1}{s + 4}\right] + \mathcal{L}^{-1}\left[\frac{s}{s^2 + 4}\right] - \frac{3}{2}\mathcal{L}^{-1}\left[\frac{2}{s^2 + 4}\right] \\
&= 4e^{-4t} + \cos 2t - \frac{3}{2}\sin 2t.
\end{aligned}$$

To obtain the last step we have used formulas iv, vi, and vii of Table 4.1.

We state the following theorem without proof.

THEOREM 5

If $f(t)$ is a periodic function with period T, then

$$\mathcal{L}[f(t)] = \frac{1}{1-e^{-sT}}\int_0^T e^{-sT}f(t)dt$$

EXERCISES

Using the definition, verify the Laplace transforms in Exercises 1 through 11. c, c_1, c_2, and k denote arbitrary constants.

1. $\mathcal{L}[t] = \dfrac{1}{s^2}$

2. $\mathcal{L}[t^2] = \dfrac{2}{s^3}$

3. $\mathcal{L}[t^3] = \dfrac{3!}{s^4}$

4. $\mathcal{L}[ct] = \dfrac{c}{s^2}$

5. $\mathcal{L}[c_1 + c_2t] = \dfrac{c_1}{s} + \dfrac{c_2}{s^2}$

6. $\mathcal{L}[\cos ct] = \dfrac{s}{s^2 + c^2}$

7. $\mathcal{L}[\sin ct] = \dfrac{c}{s^2 + c^2}$

8. $\mathcal{L}[\cosh kt] = \dfrac{s}{s^2 - k^2}, s > |k| \left(\text{recall that } \cosh kt = \dfrac{e^{kt} + e^{-kt}}{2}\right)$

9. $\mathcal{L}[\sinh kt] = \dfrac{k}{s^2 - k^2}, s > |k|$

10. $\mathcal{L}[e^{kt}] = \dfrac{1}{s - k}, s > k$

11. $\mathcal{L}[0] = 0$

Using Exercises 1 through 11 as formulas, find the inverse Laplace transforms in Exercises 12 through 18.

12. $\mathcal{L}^{-1}\left[\dfrac{5}{s^2}\right]$

13. $\mathcal{L}^{-1}\left[\dfrac{s}{s^2 + 16}\right]$

14. $\mathcal{L}^{-1}\left[\dfrac{1}{s - 5}\right]$

15. $\mathcal{L}^{-1}\left[\dfrac{5}{s^2 - 25}\right]$

16. $\mathcal{L}^{-1}\left[\dfrac{2}{s} - \dfrac{5}{s^2}\right]$

17. $\mathcal{L}^{-1}\left[\dfrac{2}{s^2 + 4}\right]$

18. $\mathcal{L}^{-1}\left[\dfrac{6}{s^4}\right]$

In Exercises 19 through 38, find the Laplace transform of each function.

19. $5 - 8t^3$

20. $t \cos 3t$

21. $\frac{1}{8} \cos \frac{3}{8} t$

22. $e^{(5/2)t} \cosh \frac{7}{2} t$

23. $e^{3t} \cos 2t - e^t \sinh 5t$

24. $-te^{5t} \cosh t$

25. $\cos t - \sin t$

26. $e^t \cos t$

27. $t^7 - t^4 + 5t^2$

28. $e^{2t}(\cos t + \sin t)$

29. $t \sinh t$

30. $t^2 \sin 4t$

31. $\dfrac{d}{dt}[te^{5t}]$

32. $\dfrac{d}{dt}[t^2e^{-t}]$

33. $\dfrac{d^2}{dt^2}[\cos t + te^t]$

34. $\dfrac{d^3}{dt^3}[t^3]$

35. $\displaystyle\int_0^t \cosh \tau \cos (t - \tau)\, d\tau$

36. $\displaystyle -t\int_0^t e^{-5(t-\tau)}\tau\, d\tau$

37. $\displaystyle\int_0^t e^\tau \cos 2\tau\, d\tau$

38. $\displaystyle\int_0^{t^{2/3}} \tau^{1/2}e^{\tau^{3/2}}\cos \tau^{3/2}\, d\tau$

In Exercises 39 through 58, find the inverse Laplace transform of each expression.

39. $\dfrac{2}{s^2 + k^2}$

40. $\dfrac{5}{s^2 - 5s + 4}$

41. $\dfrac{n!}{(s - k)^{n+1}}, n = 1, 2, 3, \ldots$

42. $\dfrac{5s^2 + 6s + 4}{(s + 4)(s^2 + 4)}$

43. $\dfrac{s}{s^2 - k^2}$

44. $\dfrac{3s + 2}{(s - 2)^2 + 4}$

45. $\dfrac{2}{(s^2 + 1)^2}$

46. $\dfrac{2s - 5}{s(s^2 - 6s + 34)}$

47. $\dfrac{s^2 + 3s + 36}{s(s^2 + 13s + 36)}$

48. $\dfrac{s^2 - 4s + 8}{s^2(s^2 - 4s + 4)}$

49. $\dfrac{s^2 + 4s + 36}{(s^2 - 4)^2}$

50. $\dfrac{2s^2 - s - 7}{(s + 1)^2(s^2 + 7s + 10)}$

51. $\dfrac{s^2 + 2s + 53}{(s + 2)(s^2 + 49)}$

52. $\dfrac{2}{(s - 1)(s^2 + 1)}$

53. $\dfrac{s^2 + 3s - 18}{s(s^2 - 6s + 9)}$

54. $\dfrac{s^2 + s + 4}{(s + 3)(s^2 - 10s + 25)}$

55. $\dfrac{s^2 - s + 1}{s^3(s + 1)}$

56. $\dfrac{s + 2}{s^2(s - 1)}$

57. $\dfrac{2s - 3}{(s + 1)^2 + 16}$

58. $\dfrac{5s + 7}{(s + 3)^2 - 16}$

59. Verify that $\displaystyle\int_0^t g(t - \tau)f(\tau)d\tau = \int_0^t f(t - \tau)g(\tau)d\tau$.

60. Prove Theorem 2.

61. If $F(s) = \mathscr{L}[f(t)]$, show that $F(ks) = \dfrac{1}{k}\mathscr{L}\left[f\left(\dfrac{t}{k}\right)\right]$, where $k > 0$ is a constant. [Hint: set $t = ku$.]

62. If $F(s) = \mathscr{L}[f(t)]$, show that $\dfrac{d}{ds}F(s) = \mathscr{L}[-tf(t)]$.

63. If n is a positive integer, show that $\dfrac{d^n}{ds^n}F(s) = \mathscr{L}[(-t)^n f(t)]$.

64. Show that $\mathscr{L}[t \cos kt] = \dfrac{s^2 - k^2}{(s^2 + k^2)^2}$.

65. Show that $\mathscr{L}\left[\int_0^t \int_0^\tau f(u)\, du\, d\tau\right] = \dfrac{1}{s^2}F(s)$.

66. Show that $\mathscr{L}[e^{kt} \sin mt] = \dfrac{m}{(s-k)^2 + m^2}$.

67. Show that $\mathscr{L}[e^{kt} \cos mt] = \dfrac{s-k}{(s-k)^2 + m^2}$.

68. Use Theorem 5 to compute $\mathscr{L}[\sin t]$.

69. Use Theorem 5 to compute $\mathscr{L}[\cos t]$.

70. Use Theorem 5 to compute $\mathscr{L}[\sin kt]$.

71. Use Theorem 5 to compute $\mathscr{L}[\cos kt]$.

4.3 THE LAPLACE TRANSFORM APPLIED TO DIFFERENTIAL EQUATIONS AND SYSTEMS

At the beginning of this chapter we emphasized that Laplace transforms provide us with a useful tool for solving certain types of differential equations. The theorems of Section 4.2 are helpful to us for the purpose of manipulating transforms. In order to apply Laplace transforms to differential equations, we need to know what the Laplace transform of a derivative (or second derivative, or higher) of a function is. The next two theorems provide us with this information. In both theorems we assume that all functions of t that appear satisfy the hypotheses of Theorem 1 of Section 4.2, so that their Laplace transform exists.

THEOREM 1

$$\mathscr{L}[f'(t)] = sF(s) - f(0).$$

Proof

$$\mathscr{L}[f'(t)] = \int_0^\infty e^{-st} f'(t)dt.$$

Integrate by parts setting $u = e^{-st}$, $dv = f'(t)dt$. Then $du = -se^{-st}\,dt$ and $v = f(t)$. Hence,

$$\mathcal{L}[f'(t)] = e^{-st}f(t)\Big|_0^\infty - \int_0^\infty f(t)(-se^{-st}dt)$$

$$= -f(0) + s\int_0^\infty e^{-st}f(t)dt$$

$$= sF(s) - f(0).$$

In the boundary term, $e^{-st}f(t)\,|_0^\infty$, we know that $\lim_{t\to\infty} e^{-st}f(t) = 0$, since f is of exponential order.

THEOREM 2

$$\mathcal{L}[f^{(n)}(t)] = s^nF(s) - s^{n-1}f(0) - s^{n-2}f'(0) - \cdots - f^{(n-1)}(0).$$

Proof We prove the result for $n = 2$ and leave the generalization to the reader.

$$\mathcal{L}[f''(t)] = \int_0^\infty e^{-st}f''(t)dt.$$

Integrate by parts, setting $u = e^{-st}$ and $dv = f''(t)dt$. Hence, $du = -se^{-st}dt$ and $v = f'(t)$. Therefore,

$$\mathcal{L}[f''(t)] = e^{-st}f'(t)\Big|_0^\infty - \int_0^\infty f'(t)(-se^{-st}dt)$$

$$= -f'(0) + s\int_0^\infty e^{-st}f'(t)dt$$

$$= -f'(0) + s\mathcal{L}[f'(t)] = -f'(0) + s(sF(s) - f(0))$$

$$= s^2F(s) - sf(0) - f'(0).$$

Note In the proof of Theorem 2 just given, it is assumed that $\lim_{t\to\infty} e^{-st}f'(t) = 0$. Similarly, in the general case it must be assumed that $\lim_{t\to\infty} e^{-st}f^{(k)}(t) = 0$, for $k = 1, 2, \ldots, n-1$. Thus, we assume that f and its derivatives are of exponential order as t tends to infinity.

Basically, the method of solving differential equations by Laplace transforms is outlined as follows: Given a differential equation, we "take the Laplace transform of it" (in other words, we apply the Laplace transformation to each term in the differential equation). This results in an equation (usually a linear algebraic equation) for the Laplace transform of the unknown function. We then solve this equation for this transform, and, in the final step, we find the inverse of this transform. For systems of differential equations the procedure is similar and leads to a system of algebraic equations.

Formula (xxii) of Table 4.1 indicates that in order to obtain explicit results, one must know $f(0), f'(0), \ldots, f^{(n-1)}(0)$. That is, we must have an initial value problem. It is a fact that the Laplace transform has a rather wide range of applicability; however, we will restrict our attention here to linear differential equations and systems with constant coefficients.

The method outlined above will be clarified via examples.

EXAMPLE 1 Solve the initial value problem

$$\ddot{y} - 3\dot{y} + 2y = 0, \tag{1}$$

$$y(0) = 1, \quad \dot{y}(0) = 0. \tag{2}$$

Solution Taking the Laplace transform of Eq. (1), we have

$$\mathcal{L}[\ddot{y} - 3\dot{y} + 2y] = \mathcal{L}[0]$$

$$\Rightarrow \mathcal{L}[\ddot{y}] - 3\mathcal{L}[\dot{y}] + 2\mathcal{L}[y] = 0$$

$$s^2Y(s) - sy(0) - \dot{y}(0) - 3(sY(s) - y(0)) + 2Y(s) = 0.$$

Note that we use $Y(s)$ to denote $\mathcal{L}[y(t)]$. Substituting the initial values (2), we have

$$s^2Y(s) - s(1) - 0 - 3(sY(s) - 1) + 2Y(s) = 0$$

$$s^2Y(s) - s - 3sY(s) + 3 + 2Y(s) = 0$$

$$(s^2 - 3s + 2)Y(s) = s - 3$$

$$\Rightarrow \quad Y(s) = \frac{s - 3}{s^2 - 3s + 2}$$

$$\Rightarrow \quad y(t) = \mathcal{L}^{-1}\left[\frac{s - 3}{s^2 - 3s + 2}\right].$$

Thus, we have solved the IVP if we can determine the indicated inverse Laplace transform. There are perhaps a couple of approaches to this end, but the simplest is to apply partial-fraction decomposition. We proceed as follows (see Appendix B):

$$\frac{s - 3}{s^2 - 3s + 2} = \frac{s - 3}{(s - 2)(s - 1)} = \frac{A}{s - 2} + \frac{B}{s - 1}.$$

Now,

$$A = \left.\frac{s - 3}{s - 1}\right|_{s=2} = -1 \quad \text{and} \quad B = \left.\frac{s - 3}{s - 2}\right|_{s=1} = 2.$$

Hence,

$$\mathcal{L}^{-1}\left[\frac{s - 3}{(s - 2)(s - 1)}\right] = \mathcal{L}^{-1}\left[\frac{-1}{s - 2} + \frac{2}{s - 1}\right] = -\mathcal{L}^{-1}\left[\frac{1}{s - 2}\right] + 2\mathcal{L}^{-1}\left[\frac{1}{s - 1}\right]$$

$$= -e^{2t} + 2e^{t}.$$

Consequently, the solution to the IVP (1)–(2) is

$$y(t) = 2e^t - e^{2t}.$$

The reader no doubt recognizes that the IVP (1)–(2) can be solved easily by the methods of Chapter 2. So why should we take a seemingly difficult approach to a simple problem? Keep in mind two facts. First, we are at this point trying to illustrate the method. Second, the Laplace transform provides a simpler method of solution in many cases.

EXAMPLE 2 Solve the IVP

$$\ddot{y} - 5\dot{y} + 6y = te^{2t} + e^{3t}$$
$$y(0) = 0, \quad \dot{y}(0) = 1.$$

Solution

$$\mathcal{L}[\ddot{y} - 5\dot{y} + 6y] = \mathcal{L}[te^{2t} + e^{3t}]$$

$$\Rightarrow \quad \mathcal{L}[\ddot{y}] - 5\mathcal{L}[\dot{y}] + 6\mathcal{L}[y] = \mathcal{L}[te^{2t}] + \mathcal{L}[e^{3t}]$$

$$\Rightarrow \quad s^2Y(s) - sy(0) - \dot{y}(0) - 5[sY(s) - y(0)] + 6Y(s) = \frac{1}{(s-2)^2} + \frac{1}{s-3}$$

$$(s^2 - 5s + 6)Y(s) - 1 = \frac{1}{(s-2)^2} + \frac{1}{s-3}$$

$$\Rightarrow \quad Y(s) = \frac{1}{(s^2 - 5s + 6)(s-2)^2} + \frac{1}{(s^2 - 5s + 6)(s-3)} + \frac{1}{s^2 - 5s + 6}$$

$$= \frac{1}{(s-2)^3(s-3)} + \frac{1}{(s-2)(s-3)^2} + \frac{1}{(s-2)(s-3)}$$

$$\Rightarrow \quad y(t) = \mathcal{L}^{-1}\left[\frac{1}{(s-2)^3(s-3)} + \frac{1}{(s-2)(s-3)^2} + \frac{1}{(s-2)(s-3)}\right]$$

$$= \mathcal{L}^{-1}\left[\frac{1}{(s-2)^3(s-3)}\right] + \mathcal{L}^{-1}\left[\frac{1}{(s-2)(s-3)^2}\right]$$

$$+ \mathcal{L}^{-1}\left[\frac{1}{(s-2)(s-3)}\right].$$

Now

$$\mathcal{L}^{-1}\left[\frac{1}{(s-2)(s-3)}\right] = \mathcal{L}^{-1}\left[\frac{3-2}{(s-3)(s-2)}\right] = e^{3t} - e^{2t}.$$

Also, using partial fraction decomposition (see Appendix B),

$$\frac{1}{(s-2)^3(s-3)} = \frac{A}{(s-2)^3} + \frac{B}{(s-2)^2} + \frac{C}{s-2} + \frac{D}{s-3}$$

and

$$\frac{1}{(s-2)(s-3)^2} = \frac{E}{s-2} + \frac{F}{s-3} + \frac{G}{(s-3)^2}.$$

We have,

$$A = \left.\frac{1}{s-3}\right|_{s=2} = -1, \qquad D = \left.\frac{1}{(s-2)^3}\right|_{s=3} = 1$$

$$E = \left.\frac{1}{(s-3)^2}\right|_{s=2} = 1, \qquad G = \left.\frac{1}{s-2}\right|_{s=3} = 1$$

and

$$C + D = 0 \Rightarrow C = -D = -1$$

$$-3A + 6B - 12C - 8D = 1 \Rightarrow B = \frac{1 + 12C + 8D + 3A}{6} = -1,$$

$$E + F = 0 \Rightarrow F = -E = -1.$$

Therefore,

$$\begin{aligned} y(t) &= \mathcal{L}^{-1}\left[\frac{-1}{(s-2)^3}\right] + \mathcal{L}^{-1}\left[\frac{-1}{(s-2)^2}\right] + \mathcal{L}^{-1}\left[\frac{-1}{s-2}\right] + \mathcal{L}^{-1}\left[\frac{1}{s-3}\right] \\ &\quad + \mathcal{L}^{-1}\left[\frac{-1}{s-3}\right] + \mathcal{L}^{-1}\left[\frac{1}{(s-3)^2}\right] + e^{3t} - e^{2t} \\ &= \frac{-e^{2t}t^2}{2!} + (-1)e^{2t}t + (-1)e^{2t} + e^{3t} + e^{2t} - e^{3t} + e^{3t}t + e^{3t} - e^{2t} \\ &= e^{2t}\left(-\frac{t^2}{2} - t - 1\right) + e^{3t}(t+1). \end{aligned}$$

The following example illustrates that the Laplace-transform method can still be applied, even though the initial value is given at a point different than $t = 0$.

EXAMPLE 3 Solve the IVP

$$\dot{y} + 3y = 0 \tag{3}$$

$$y(3) = 1. \tag{4}$$

Solution We note that the IVP can be attacked by first making the change of variables $\tau = t - 3$, thus obtaining an equivalent problem with the initial value given at $\tau = 0$. However, we can also proceed directly as follows.

$$\mathcal{L}[\dot{y} + 3y] = \mathcal{L}[0]$$
$$\Rightarrow \quad sY(s) - y(0) + 3Y(s) = 0$$
$$(s + 3)\,Y(s) = y(0)$$
$$Y(s) = \frac{y(0)}{s + 3}$$
$$\Rightarrow \quad y(t) = y(0)e^{-3t}$$
$$y(3) = 1 \Rightarrow 1 = y(0)e^{-3(3)} \Rightarrow y(0) = e^{9}$$
$$y(t) = e^{9}e^{-3t} = e^{9-3t}.$$

EXAMPLE 4 Solve the IVP

$$\dot{x}_1 = -3x_1 + 4x_2 \tag{5}$$
$$\dot{x}_2 = 2x_1 + 3x_2 \tag{6}$$
$$x_1(0) = -1, \quad x_2(0) = 3. \tag{7}$$

Solution Taking the Laplace transform of Eq. (5), we have

$$sX_1(s) - x_1(0) = -3X_1(s) + 4X_2(s). \tag{8}$$

Taking the Laplace transform of Eq. (6), we have

$$sX_2(s) - x_2(0) = -2X_1(s) + 3X_2(s). \tag{9}$$

Using the initial conditions (7), we can rewrite Eqs. (8) and (9) as follows:

$$\begin{aligned} (s + 3)X_1(s) - 4X_2(s) &= -1 \\ 2X_1(s) + (s - 3)X_2(s) &= 3. \end{aligned} \tag{10}$$

The system (10) is a linear system for $X_1(s)$, $X_2(s)$. Solving we have

$$X_1(s) = \frac{-s + 15}{s^2 - 1} = -\frac{s}{s^2 - 1} + \frac{15}{s^2 - 1}$$
$$X_2(s) = \frac{3s + 11}{s^2 - 1} = 3\frac{s}{s^2 - 1} + \frac{11}{s^2 - 1}.$$

Thus,

$$x_1(t) = -\cosh t + 15 \sinh t$$
$$x_2(t) = 3 \cosh t + 11 \sinh t,$$

or

$$x_1(t) = 7e^{t} - 8e^{-t},$$
$$x_2(t) = 7e^{t} - 4e^{-t}.$$

EXERCISES

In Exercises 1 through 36, solve the IVP by the method of Laplace transforms.

1. $\dot{y} + 2y = 0$
$y(0) = 1$

2. $\ddot{y} + 4y = 4\cos 2t$
$y(0) = 0,\ \dot{y}(0) = 6$

3. $\ddot{y} + 3\dot{y} - 4y = 0$
$y(0) = 0,\ \dot{y}(0) = -5$

4. $\dot{y} - 3y = 13\cos 2t$
$y(0) = -1$

5. $\ddot{y} + 2\dot{y} + y = 2te^{-t}$
$y(0) = 3,\ \dot{y}(0) = -3$

6. $\ddot{y} + 3\dot{y} + 2y = 0$
$y(0) = 1,\ \dot{y}(0) = 1$

7. $\ddot{y} - y = 6e^{t}$
$y(0) = 2,\ \dot{y}(0) = 3$

8. $\ddot{y} - 4y = -3e^{t}$
$y(0) = -1,\ \dot{y}(0) = 1$

9. $\ddot{y} + 10\dot{y} + 25y = 2e^{-5t}$
$y(0) = 0,\ \dot{y}(0) = -1$

10. $\dddot{y} - 27y = 0$
$y(0) = -1,\ \dot{y}(0) = 6,\ \ddot{y}(0) = 18$

11. $\ddot{y} - 9\dot{y} + 18y = 54$
$y(0) = 0,\ \dot{y}(0) = -3$

12. $\ddot{y} - 9y = 20\cos t$
$y(0) = 0,\ \dot{y}(0) = 18$

13. $\ddot{y} + 9y = e^{t}$
$y(0) = 0,\ \dot{y}(0) = 0$

14. $\ddot{y} - 3\dot{y} + 2y = 24\cosh t$
$y(0) = 6,\ \dot{y}(0) = -3$

15. $\ddot{y} + 10\dot{y} + 26y = 37e^{t}$
$y(0) = 1,\ \dot{y}(0) = 2$

16. $\dddot{y} - y = -1$
$y(0) = 1,\ \dot{y}(0) = -2,\ \ddot{y}(0) = 2$

17. $\dddot{y} + y = 1$
$y(0) = 1,\ \dot{y}(0) = 3,$
$\ddot{y}(0) = -3$

18. $\dot{y} - 3y = 2e^{t}$
$y(1) = e^{3} - e$

19. $\ddot{y} + 6\dot{y} + 9y = 27t$
$y(0) = 1,\ \dot{y}(0) = 0$

20. $\ddot{y} - \dot{y} - 6y = \cos t + 57\sin t$
$y(0) = 1,\ \dot{y}(0) = 7$

21. $\ddot{y} - 3\dot{y} - 4y = 25te^{-t}$
$y(0) = 0,\ \dot{y}(0) = 4$

22. $\ddot{y} + 13\dot{y} + 36y = 10 - 72t$
$y(0) = 2,\ \dot{y}(0) = -1$

23. $\ddot{y} + 2\dot{y} - 15y = 16te^{-t} - 15$
$y(0) = 1,\ \dot{y}(0) = -9$

24. $\ddot{y} - 10\dot{y} + 21y = 21t^{2} + t + 13$
$y(0) = 3,\ \dot{y}(0) = 11$

25. $\ddot{y} + 7\dot{y} + 10y = 3e^{-2t} - 6e^{-5t}$
$y(0) = 0,\ \dot{y}(0) = 0$

26. $4\ddot{y} - 3\dot{y} - y = 34\sin t$
$y(0) = 1,\ \dot{y}(0) = -2$

27. $\dddot{y} - y = 12\sinh t$
$y(0) = 6,\ \dot{y}(0) = -1,$
$\ddot{y}(0) = 7$

28. $\ddot{y} + 2\dot{y} - 3y = 3t^{3} - 9t^{2} - 5t + 1$
$y(0) = 0, \dot{y}(0) = 8$

29. $\ddot{y} + 4\dot{y} + 5y = 39e^{t}\sin t$
$y(0) = -1,\ \dot{y}(0) = -1$

30. $\ddot{y} + 2\dot{y} + 5y = 8e^{t} + 5t$
$y(0) = 3,\ \dot{y}(0) = 2$

31. $\ddot{y} - 4\dot{y} + 4y = 3te^{2t} - 4$
$y(0) = 0,\ \dot{y}(0) = 0$

32. $\dddot{y} + y = 18e^{2t}$
$y(0) = -1,\ \dot{y}(0) = 13,\ \ddot{y}(0) = -1$

33. $\dddot{y} + 8y = -12e^{-2t}$
$y(0) = -8,\ \dot{y}(0) = 24,$
$\ddot{y}(0) = -46$

34. $\ddot{y} - y = 2t^{2} + 2e^{-t}$
$y(0) = 0,\ \dot{y}(0) = 1$

35. $\ddot{y} + 7\dot{y} + 6y = 250e^t \cos t$
$y(0) = 2,\ \dot{y}(0) = -7$

36. $\ddot{y} + 4\dot{y} + 13y = 13t + 17 + 40 \sin t$
$y(0) = 30,\ \dot{y}(0) = 4.$

Solve each of the initial value problems 37 through 46, using the Laplace transform method.

37. $\dot{x}_1 = 4x_1 - x_2$
$\dot{x}_2 = 2x_1 + x_2$
$x_1(0) = 1,\ x_2(0) = 3$

38. $\dot{x}_1 = 3x_1 - 2x_2$
$\dot{x}_2 = x_1$
$x_1(0) = -1,\ x_2(0) = 0$

39. $\dot{x}_1 = 4x_1 - 2x_2$
$\dot{x}_2 = x_1 + x_2$
$x_1(0) = 1,\ x_2(0) = 0$

40. $\dot{x}_1 = 3x_1 - 2x_2$
$\dot{x}_2 = 2x_1 - x_2$
$x_1(0) = -1,\ x_2(0) = 1$

41. $\dot{x}_1 = 3x_1 - 2x_2$
$\dot{x}_2 = 17x_1 - 7x_2$
$x_1(0) = 0,\ x_2(0) = 3$

42. $\dot{x}_1 = 3x_1 - 5x_2$
$\dot{x}_2 = 4x_1 - 5x_2$
$x_1(0) = 1,\ x_2(0) = \frac{4}{5}$

43. $\dot{x}_1 = 4x_1 - x_2 + 3e^{2t}$
$\dot{x}_2 = 2x_1 + x_2 + 2t$
$x_1(0) = -\frac{5}{18},\ x_2(0) = \frac{47}{9}$

44. $\dot{x}_1 = 5x_1 - 6x_2 + 1$
$\dot{x}_2 = 6x_1 - 7x_2 + 1$
$x_1(0) = 0,\ x_2(0) = 0$

45. $\dot{x}_1 = 2x_1 + x_2$
$\dot{x}_2 = -x_1 + x_3$
$\dot{x}_3 = x_1 + 3x_2 + x_3$
$x_1(0) = 1,\ x_2(0) = 1$
$x_3(0) = 3$

46. $\dot{x}_1 = 3x_1 - x_2$
$\dot{x}_2 = -x_1 + 2x_2 - x_3$
$\dot{x}_3 = -x_2 + 3x_3$
$x_1(0) = 1,\ x_2(0) = 0$
$x_3(0) = -1$

4.4 THE UNIT STEP FUNCTION

The unit step function or the Heaviside function is defined as follows.

$$h_1(t-t_0) = \begin{cases} 0 & \text{if} \quad t < t_0 \\ 1 & \text{if} \quad t \geq t_0 \end{cases}. \tag{1}$$

We assume that $t_0 \geq 0$. The graph of $h_1(t-t_0)$ is given in Figure 4.1.

A generalization of the unit step function is the function $h_c(t-t_0)$, where

$$h_c(t-t_0) = \begin{cases} 0 & \text{if} \quad t < t_0 \\ c & \text{if} \quad t \geq t_0 \end{cases}.$$

It is easy to see that

$$h_c(t-t_0) = ch_1(t-t_0). \tag{2}$$

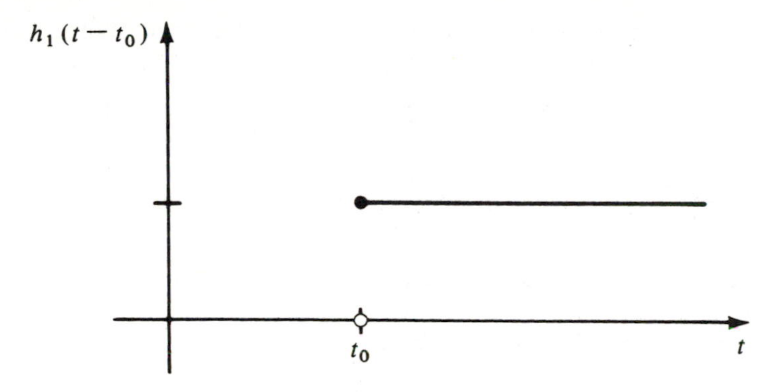

Figure 4.1

Applying the definition of the Laplace transform, we have

$$\begin{aligned}\mathscr{L}[h_1(t-t_0)] &= \int_0^\infty h_1(t-t_0)e^{-st}dt \\ &= \int_{t_0}^\infty e^{-st}dt = \left.\frac{-e^{-st}}{s}\right|_{t_0}^\infty \\ &= \frac{e^{-st_0}}{s}.\end{aligned} \tag{3}$$

For the special case $t_0 = 0$, we write $h_1(t) \equiv h_1(t-0)$, and from (3) we obtain

$$\mathscr{L}[h_1(t)] = \frac{1}{s}. \tag{4}$$

From (2) and (3) we obtain the Laplace transform of $h_c(t-t_0)$, namely,

$$\mathscr{L}[h_c(t-t_0)] = \frac{ce^{-st_0}}{s}.$$

One use of the Heaviside function is that many other functions can be expressed in terms of it. We illustrate with an example.

EXAMPLE 1 Express the square-wave function (Figure 4.2) in terms of Heaviside functions.

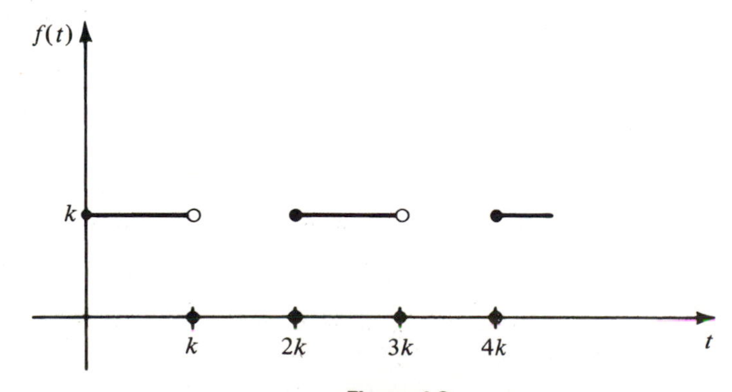

Figure 4.2

Solution Beginning at $t = 0$, and considering in succession the intervals $(0,k)$, $(k,2k)$, $(2k,3k)$, $(3k,4k)$, . . . , we find that

$$f(t) = h_k(t) - h_k(t-k) + h_k(t-2k) - h_k(t-3k) + h_k(t-4k) - \cdots$$
$$= \sum_{i=0}^{\infty} (-1)^i h_k(t-ik) = k \sum_{i=0}^{\infty} (-1)^i h_1(t-ik).$$

EXAMPLE 2 If $f(t)$ is the square-wave function of Example 1, find $\mathscr{L}[f(t)]$.

Solution From Example 1 we can write

$$\mathscr{L}[f(t)] = \mathscr{L}[k\sum_{i=0}^{\infty}(-1)^i h_1(t-ik)] = \int_0^{\infty} k\sum_{i=0}^{\infty}(-1)^i h_1(t-ik)e^{-st}dt.$$

If we assume that the integration and summation processes can be interchanged, we have

$$\mathscr{L}[f(t)] = k\sum_{i=0}^{\infty}(-1)^i \int_0^{\infty} h_1(t-ik)e^{-st}dt = k\sum_{i=0}^{\infty}(-1)^i \mathscr{L}[h_1(t-ik)].$$

Using (3) we obtain

$$\mathscr{L}[f(t)] = k\sum_{i=0}^{\infty}(-1)^i \frac{e^{-sik}}{s} = \frac{k}{s}\sum_{i=0}^{\infty}(-1)^i (e^{-sk})^i.$$

Since $sk > 0$, we know that $e^{-sk} < 1$; therefore, the infinite sum is a convergent geometric series with first term equal to 1 and ratio equal to $-e^{-sk}$. Consequently the sum is $1/[1-(-e^{-sk})] = 1/[1+e^{-sk}]$. Thus,

$$\mathscr{L}[f(t)] = \frac{k}{s(1+e^{-ks})}.$$

EXAMPLE 3 If $\mathscr{L}[f(t)] = F(s)$, show that

$$\mathscr{L}[f(t-t_0)h_1(t-t_0)] = e^{-st_0}F(s). \tag{5}$$

Solution

$$\mathscr{L}[f(t-t_0)h_1(t-t_0)] = \int_0^{\infty} f(t-t_0)h_1(t-t_0)e^{-st}dt$$
$$= \int_{t_0}^{\infty} f(t-t_0)e^{-st}dt.$$

In the latter integral make the substitution $z = t - t_0$ to obtain

$$\mathscr{L}[f(t-t_0)h_1(t-t_0)] = \int_0^{\infty} f(z)e^{-s(z+t_0)}dz$$
$$= e^{-st_0}\int_0^{\infty} f(z)e^{-sz}dz = e^{-st_0}F(s).$$

An important consequence of Eq. (5) is

$$\mathscr{L}^{-1}[e^{-st_0}F(s)] = f(t-t_0)h_1(t-t_0). \tag{6}$$

EXAMPLE 4 Solve the initial value problem

$$\begin{aligned} \ddot{y}-3\dot{y}+4y &= h_1(t-1) + h_1(t-2) \\ y(0) &= 0,\ \dot{y}(0) = 1. \end{aligned} \tag{7}$$

Solution Applying the Laplace transform to Eq. (7) we obtain

$$s^2Y - sy(0) - \dot{y}(0) - 3sY - 3y(0) + 4Y = \frac{e^{-s}}{s} + \frac{e^{-2s}}{s}$$

$$(s^2 - 3s + 4)Y = \frac{e^{-s}}{s} + \frac{e^{-2s}}{s} + 1$$

$$Y = \frac{e^{-s}}{s(s-4)(s+1)} + \frac{e^{-2s}}{s(s-4)(s-1)} + \frac{1}{(s-4)(s+1)}$$

$$Y = e^{-s}\left[\frac{-1/4}{s} + \frac{1/20}{s-4} + \frac{1/5}{s+1}\right] + e^{-2s}\left[\frac{-1/4}{s} + \frac{1/20}{s-4} + \frac{1/5}{s+1}\right] + \frac{1/5}{s-4} + \frac{-1/5}{s+1}.$$

Employing (6) we can write

$$\begin{aligned} y(t) = &-\frac{1}{4}h_1(t-1) + \frac{1}{20}e^{4(t-1)}h_1(t-1) + \frac{1}{5}e^{-(t-1)}h_1(t-1) \\ &-\frac{1}{4}h_1(t-2) + \frac{1}{20}e^{4(t-2)}h_1(t-2) + \frac{1}{5}e^{-(t-2)}h_1(t-2) \\ &+\frac{1}{5}e^{4t} - \frac{1}{5}e^{-t}. \end{aligned}$$

EXERCISES

In Exercises 1 through 14, solve the given initial value problem.

1. $\ddot{y} + 2\dot{y} + y = h_1(t)$
$y(0) = 0,\ \dot{y}(0) = -3$

2. $\dot{y} - 3y = 13\cos 2t + 3h_1(t)$
$y(0) = -1$

3. $\ddot{y} + 3\dot{y} - 4y = 20h_1(t-2)$
$y(0) = 0,\ \dot{y}(0) = 0$

4. $\ddot{y} - 9y = 20\cos t + 18h_1(t-1)$
$y(0) = 2,\ \dot{y}(0) = 0$

5. $\ddot{y} - 9\dot{y} + 18y = 54t - 9h_1(t)$
$y(0) = 3,\ \dot{y}(0) = 0$

6. $\ddot{y} + 10\dot{y} + 25y = 2e^{-5t} + 25h_1(t-5)$
$y(0) = 0,\ \dot{y}(0) = -1$

7. $\ddot{y} - y = 4e^{t} + 2h_1(t-3)$
$y(0) = 0,\ \dot{y}(0) = 0$

8. $\ddot{y} + 3\dot{y} + 2y = 8h_1(t) + 2h_1(t-1)$
$y(0) = 1,\ \dot{y}(0) = 0$

9. $\ddot{y} + 4y = 8h_1(t-1) - 4h_1(t-2)$
$y(0) = -1,\ \dot{y}(0) = 0$

10. $\ddot{y} - 3\dot{y} + 2y = 24\cosh t + h_2(t)$
$y(0) = 0,\ \dot{y}(0) = -18$

11. $\ddot{y} + 10\dot{y} + 26y = 52h_1(t) + 26h_1(t - 3)$
$y(0) = 1, \dot{y}(0) = -5$

12. $\ddot{y} + 6\dot{y} + 9y = 6h_3(t - 1) + 9h_1(t - 3)$
$y(0) = 0, \dot{y}(0) = 1$

13. $\ddot{y} - \dot{y} - 6y = 50 \cos t + 30h_1(t - 5)$
$y(0) = -7, \dot{y}(0) = 4$

14. $\ddot{y} - 3\dot{y} - 4y = 20h_1(t) + 40h_1(\mathrm{t} - 1) + 10h_2(t - 2)$
$y(0) = 0, \dot{y}(0) = 0$

15. Use Theorem 5 of Section 4.2 to compute $\mathcal{L}[f(t)]$ where $f(t)$ is the square-wave function of Example 1.

16. Use Theorem 5 of Section 4.2 to compute $\mathcal{L}[f(t)]$ where $f(t)$ is the "saw-tooth" function of Figure 4.3.

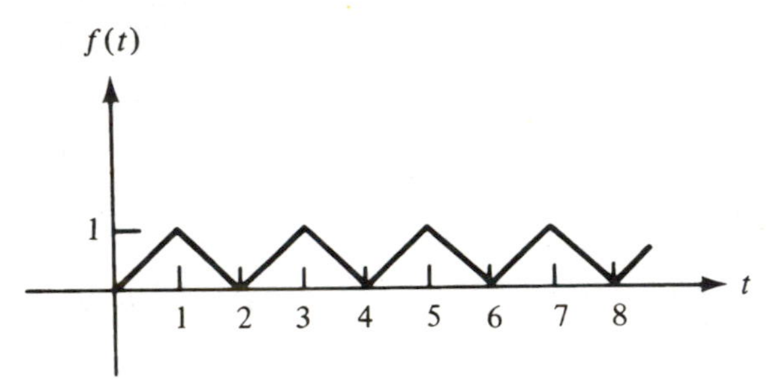

Figure 4.3

4.5 THE UNIT IMPULSE FUNCTION

The unit impulse function, or the Dirac delta function, happens not to be a function as the word is used in mathematics. For a first exposition, however, there are advantages to thinking of the unit impulse function as if it were a function. The idea behind the delta function is to provide an analytical model for certain types of physical phenomena that act for extremely short periods of time. For example, one might be interested in describing the force involved when a hammer strikes a nail. If the graph of this force were drawn, the ordinate would be zero up to the time the hammer struck the nail, say at $t = t_0$ ($t_0 \geq 0$); it then has some value for a very short period of time and is zero thereafter. Since the time interval of application of the force is small, it is reasonable to assume that during this period the force is a constant and that this constant is large in magnitude. It is conventional that the *impulse of the force* (that is, the integral of the force with respect to time) is unity. With this convention we can approximate the force graphically in Figure 4.4.

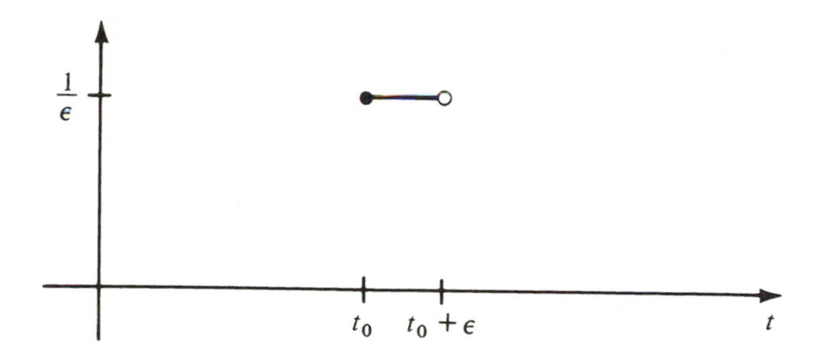

Figure 4.4

Utilizing the Heaviside function of Section 4.4, and denoting the force depicted in Figure 4.4 by $f_\epsilon(t - t_0)$, we can write

$$f_\epsilon(t - t_0) = \frac{1}{\epsilon}[h_1(t - t_0) - h_1(t - t_0 - \epsilon)]. \tag{1}$$

Note that

$$\int_{-\infty}^{\infty} f_\epsilon(t - t_0)dt = 1, \text{ for every } \epsilon > 0 \text{ and every } t_0 \geq 0. \tag{2}$$

The unit impulse function, or Dirac delta function $\delta(t - t_0)$ is defined by the conditions

$$\delta(t - t_0) = \lim_{\epsilon \to 0} f_\epsilon(t - t_0), \tag{3}$$

and

$$\int_{-\infty}^{\infty} \delta(t - t_0)dt = 1. \tag{4}$$

Thus we can think of $\delta(t - t_0)$ as representing a force of "infinite magnitude" that is applied instantaneously at $t = t_0$ and whose impulse is unity.

We use Eq. (3) to obtain the main properties of the unit impulse function.

EXAMPLE 1 Find the Laplace transform of $\delta(t - t_0)$.

Solution

$$\mathscr{L}[\delta(t - t_0)] = \int_0^{\infty} \delta(t - t_0)e^{-st}dt = \int_0^{\infty} \lim_{\epsilon \to 0} f_\epsilon(t - t_0)e^{-st}dt.$$

Assuming that the integration and limit processes can be interchanged, we have

$$\begin{aligned}\mathscr{L}[\delta(t-t_0)] &= \lim_{\epsilon\to 0}\int_0^\infty f_\epsilon(t-t_0)e^{-st}dt \\ &= \lim_{\epsilon\to 0}\int_0^\infty \frac{1}{\epsilon}[h_1(t-t_0)-h_1(t-t_0-\epsilon)]e^{-st}dt \\ &= \lim_{\epsilon\to 0}\frac{1}{\epsilon}\Big\{\mathscr{L}[h_1(t-t_0)]-\mathscr{L}[h_1(t-t_0-\epsilon)]\Big\} \\ &= \lim_{\epsilon\to 0}\frac{1}{\epsilon}\left[\frac{e^{-st_0}}{s}-\frac{e^{-s(t_0+\epsilon)}}{s}\right] = \frac{e^{-st_0}}{s}\lim_{\epsilon\to 0}\frac{1-e^{-s\epsilon}}{\epsilon}.\end{aligned}$$

Now $\lim_{\epsilon\to 0}\dfrac{1-e^{-s\epsilon}}{\epsilon}$ can be evaluated using l'Hospital's rule; thus

$$\lim_{\epsilon\to 0}\frac{1-e^{-s\epsilon}}{\epsilon} = \lim_{\epsilon\to 0}\frac{se^{-s\epsilon}}{1} = s.$$

Consequently,

$$\mathscr{L}[\delta(t-t_0)] = e^{-st_0}. \tag{5}$$

In the special case $t_0 = 0$ we write $\delta(t) \equiv \delta(t-0)$. From (5) we find that

$$\mathscr{L}[\delta(t)] = 1. \tag{6}$$

EXAMPLE 2 If g is any continuous function, show that

$$\int_{-\infty}^{\infty}\delta(t-t_0)g(t)dt = g(t_0). \tag{7}$$

Solution

$$\int_{-\infty}^{\infty}\delta(t-t_0)g(t)dt = \int_{-\infty}^{\infty}\lim_{\epsilon\to 0}f_\epsilon(t-t_0)g(t)dt.$$

Assuming that the integration and limit processes can be interchanged, we have

$$\begin{aligned}\int_{-\infty}^{\infty}\delta(t-t_0)g(t)dt &= \lim_{\epsilon\to 0}\int_{-\infty}^{\infty}f_\epsilon(t-t_0)g(t)dt = \lim_{\epsilon\to 0}\int_{t_0}^{t_0+\epsilon}\left(\frac{1}{\epsilon}\right)g(t)dt \\ &= \lim_{\epsilon\to 0}\frac{1}{\epsilon}\int_{t_0}^{t_0+\epsilon}g(t)dt.\end{aligned}$$

Since g is a continuous function, we can apply the mean value theorem of integral calculus to obtain

$$\int_{-\infty}^{\infty}\delta(t-t_0)g(t)dt = \lim_{\epsilon\to 0}\frac{1}{\epsilon}g(\alpha)[(t_0+\epsilon)-t_0],$$

where $t_0 \le \alpha \le t_0+\epsilon$. Now $\frac{1}{\epsilon}[(t_0+\epsilon)-t_0] = 1$, and since g is continuous, $\lim_{\epsilon \to 0} g(\alpha) = g(t_0)$; thus,

$$\int_{-\infty}^{\infty} \delta(t-t_0)g(t)dt = g(t_0).$$

EXAMPLE 3 Solve the initial value problem

$$\begin{aligned} \ddot{y} - 3\dot{y} + 2y &= \delta(t-1) \\ y(0) = 0,\ \dot{y}(0) &= 0. \end{aligned} \tag{8}$$

Solution Applying the Laplace transform to Eq. (8), we obtain

$$s^2Y - sy(0) - \dot{y}(0) - 3sY + 3y(0) + 2Y = e^{-s}$$

$$(s^2 - 3s + 2)Y = e^{-s}$$

$$Y = \frac{e^{-s}}{(s-1)(s-2)} = e^{-s}\left[\frac{-1}{s-1} + \frac{1}{s-2}\right].$$

Thus,

$$y(t) = -e^{(t-1)}h_1(t-1) + e^{2(t-1)}h_1(t-1).$$

EXERCISES

In Exercises 1 through 14, solve the given initial value problem.

1. $\ddot{y} - 3\dot{y} - 4y = 5\delta(t)$
$y(0) = 0,\ \dot{y}(0) = 0$

2. $\ddot{y} - \dot{y} - 6y = 52 \cos 2t + \delta(t)$
$y(0) = -5,\ \dot{y}(0) = -8$

3. $\ddot{y} + 6\dot{y} + 9y = \delta(t-2)$
$y(0) = 0,\ \dot{y}(0) = 1$

4. $\ddot{y} + 10\dot{y} + 26y = 145 \cos t + \delta(t-1)$
$y(0) = 1,\ \dot{y}(0) = -2$

5. $\ddot{y} - 3\dot{y} + 2y = 12 + \delta(t)$
$y(0) = 6,\ \dot{y}(0) = 0$

6. $\ddot{y} + 9y = 34e^{-5t} + 9\delta(t-5)$
$y(0) = 0,\ \dot{y}(0) = -1$

7. $\ddot{y} + 3\dot{y} + 2y = 6e^{t} + \delta(t-3)$
$y(0) = 1,\ \dot{y}(0) = 0$

8. $\ddot{y} - y = \delta(t) + \delta(t-1)$
$y(0) = 0,\ \dot{y}(0) = 3$

9. $\ddot{y} + 10\dot{y} + 25y = \delta(t-1) - 3\delta(t-2)$
$y(0) = 0,\ \dot{y}(0) = -1$

10. $\ddot{y} - 9\dot{y} + 18y = 280 \cosh t + \delta(t)$
$y(0) = 0,\ \dot{y}(0) = -1$

11. $\ddot{y} - 9y = 30\delta(t) + 6\delta(t-3)$
$y(0) = 8,\ \dot{y}(0) = 0$

12. $\ddot{y} + 3\dot{y} - 4y = 5\delta(t-1) + 10\delta(t-3)$
$y(0) = 0,\ \dot{y}(0) = 0$

13. $\ddot{y} - 2\dot{y} - 8y = 85 \cos t + 6\delta(t - 5)$
$y(0) = -9, \dot{y}(0) = 4$

14. $\ddot{y} + 2\dot{y} + y = \delta(t) + \delta(t - 1) + 3\delta(t - 2)$
$y(0) = 0, \dot{y}(0) = -3$

4.6 APPLICATIONS

In this section we illustrate a few practical applications of Laplace transforms. We do not restrict the treatment to differential equations exclusively. It is worth emphasizing that Laplace transforms have the nice feature that when they are applicable, they transform a given problem into a simpler problem.

Electric Circuits ■ In Section 2.11.1 we obtained, via Kirchhoff's voltage law, the equation

$$L\frac{di}{dt} + Ri + \frac{1}{C}\int_0^t i(\tau)d\tau = v(t). \tag{1}$$

Equation (1) is referred to as an *integrodifferential equation* for i. Here we have used lowercase i and v for current and voltage, respectively, to conform to the convention of labelling functions of t in lower case and their Laplace transform in upper case. Note that Eq. (1) is Eq. (12) of Section 2.11.1, in which we have used the relation (11) of that section. In Section 2.11.1 we differentiated Eq. (1) to obtain a differential equation for i. Naturally, this requires that $v(t)$ be a differentiable function. Using Laplace transforms, we can solve Eq. (1) directly.

EXAMPLE 1 Find $i(t)$ for the electric circuit with $L = 20$ henries, $R = 40$ ohms, $C = 0.05$ farad, and $v(t) = \sin t$.

Solution Applying the Laplace transform to Eq. (1), we obtain [recall that $i(0) = 0$]

$$LsI(s) + RI(s) + \frac{1}{Cs}I(s) = V(s)$$

$$\Rightarrow \quad I(s) = \frac{1}{Ls + R + 1/Cs}V(s)$$

$$= \frac{s/L}{s^2 + (R/L)s + 1/LC}V(s) \tag{2}$$

$$= G(s)V(s). \tag{3}$$

Substituting the given values for L, R, C, and $v(t)$ into Eq. (2), we obtain

$$I(s) = \frac{s/20}{s^2 + 2s + 1} \cdot \frac{1}{s^2 + 1} = \frac{s/20}{(s + 1)^2(s^2 + 1)}$$
$$= \frac{A}{s + 1} + \frac{B}{(s + 1)^2} + \frac{Cs + D}{s^2 + 1}.$$

Now (see Appendix B) $A = 0$, $B = -\frac{1}{40}$, $C = 0$, and $D = \frac{1}{40}$. Thus,

$$I(s) = \frac{-\frac{1}{40}}{(s + 1)^2} + \frac{\frac{1}{40}}{s^2 + 1}$$
$$i(t) = -\tfrac{1}{40}\mathcal{L}^{-1}\left[\frac{1}{(s + 1)^2}\right] + \tfrac{1}{40}\mathcal{L}^{-1}\left[\frac{1}{s^2 + 1}\right]$$
$$= -\tfrac{1}{40}te^{-t} + \tfrac{1}{40}\sin t.$$

REMARK 1 The phenomenon illustrated by the electric circuit can be viewed in the following way. A voltage (*input*) is applied to the circuit (*system*) and a current (*output*) is produced; that is, a system acts upon an input to produce an output. In an abstract way many processes (biological, chemical, physical) can be viewed in this fashion. Following this model, it would be convenient if one could obtain an equation relating output to input. Then it would be simply a matter of substituting given values for the input and determining the values of the output via this relation. Equation (1) is such a relation, but unfortunately the relationship is not simple enough to easily obtain output from input. On the other hand, Eq. (2) is better suited for these purposes. Of course, Eq. (2) relates the Laplace transform of the output to the Laplace transform of the input. Nevertheless, it has the advantage that given an input $v(t)$ to obtain the output $i(t)$ requires some algebraic manipulation, use of tables of Laplace transforms, and in some cases performing integrations. Equation (3) is frequently called the *input–output relation*. The function $G(s)$ is called the *transfer function* or *system function*. Relation (3) is viewed diagrammatically as in Figure 4.5.

Engineers and biologists take the attitude that knowledge of the transfer function provides complete information of the system. For the most part, engineers know the transfer function and wish to analyze the output. Biologists observe input and output and try to determine the transfer function from this information (for more discussion in this direction, see Exercises 14 through 17).

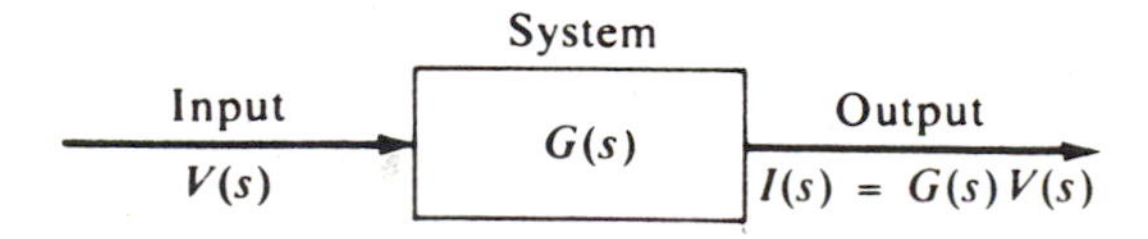

Figure 4.5

It is in this context [Eq. (3)] that the Laplace transform is utilized in present-day investigations.[1]

Mechanics ■ A particle is projected from a point on the surface of the earth with a given initial velocity. We wish to analyze the motion of this particle, taking into account the rotation of the earth. We choose the origin of a three-dimensional coordinate system to be at the point from which the particle was projected. The positive x axis is to point south, the positive y axis east, and the positive z axis points opposite to the direction of the acceleration due to gravity. The origin is taken to have latitude β and the angular velocity of the earth is denoted by ω. Neglecting the mass of the particle, it can be shown[2] that the equations of motion are as follows:

$$\begin{aligned} \ddot{x} &= 2\omega\dot{y}\sin\beta \\ \ddot{y} &= -2\omega(\dot{x}\sin\beta + \dot{z}\cos\beta) \\ \ddot{z} &= 2\omega\dot{y}\cos\beta - g, \end{aligned} \tag{4}$$

with initial values $x(0) = y(0) = z(0) = 0$ and $\dot{x}(0) = u$, $\dot{y}(0) = v$, $\dot{z}(0) = w$.

Set $\cos\beta = a$, $\sin\beta = b$, and apply the Laplace transform to system (4) to obtain the following system (after rearrangement) (see Exercise 1):

$$\begin{aligned} s^2X - 2\omega bsY &= u \\ 2\omega bsX + s^2Y + 2\omega asZ &= v \\ -2\omega asY + s^2Z &= w - \frac{g}{s}. \end{aligned} \tag{5}$$

System (5) can be solved for X, Y, and Z since the determinant of coefficients Δ has the value (see Exercise 2)

$$\Delta = s^4(s^2 + 4\omega^2). \tag{6}$$

Our restrictions on s guarantee that $\Delta \neq 0$. Thus (see Exercises 3, 4, and 6),

$$\begin{aligned} X(s) = {} & u\left(\frac{1}{s^2 + 4\omega^2} + \frac{4\omega^2a^2}{s^2(s^2 + 4\omega^2)}\right) + v\,\frac{2\omega b}{s(s^2 + 4\omega^2)} \\ & - w\,\frac{4\omega^2ab}{s^2(s^2 + 4\omega^2)} + g\,\frac{4\omega^2ab}{s^3(s^2 + 4\omega^2)} \end{aligned} \tag{7}$$

$$\begin{aligned} Y(s) = {} & -u\,\frac{2\omega b}{s(s^2 + 4\omega^2)} + v\,\frac{1}{s^2 + 4\omega^2} - w\,\frac{2\omega a}{s(s^2 + 4\omega^2)} \\ & + g\,\frac{2\omega a}{s^2(s^2 + 4\omega^2)} \end{aligned} \tag{8}$$

[1]For a more complete discussion, see R. Rosen, *Optimality Principles in Biology* (New York: Plenum Press, 1967).

[2]See B.M. Planck, *Einführung in die allgemeine Mechanik*, 4th ed. (Leipzig: S. Hirzel, 1928), p. 81.

$$Z(s) = -u\frac{4\omega^2ab}{s^2(s^2+4\omega^2)} + v\frac{2\omega a}{s(s^2+4\omega^2)} + w\left[\frac{1}{s^2+4\omega^2} + \frac{4\omega^2b^2}{s^2(s^2+4\omega^2)}\right] - g\left[\frac{1}{s(s^2+4\omega^2)} + \frac{4\omega^2b^2}{s^3(s^2+4\omega^2)}\right]. \tag{9}$$

Consequently (see Exercises 7, 8, and 9),

$$x(t) = \frac{u}{2\omega}(2a^2\omega t + b^2\sin 2\omega t) + \frac{v}{\omega}b\sin^2\omega t - \frac{w}{2\omega}ab(2\omega t - \sin 2\omega t) + \frac{g}{2\omega^2}ab(\omega^2t^2 - \sin^2\omega t) \tag{10}$$

$$y(t) = -\frac{u}{\omega}b\sin^2\omega t + \frac{v}{2\omega}\sin 2\omega t - \frac{w}{\omega}a\sin^2\omega t + \frac{g}{4\omega^2}a(2\omega t - \sin 2\omega t) \tag{11}$$

$$z(t) = -\frac{u}{2\omega}ab(2\omega t - \sin 2\omega t) + \frac{v}{\omega}a\sin^2\omega t + \frac{w}{2\omega}(2b^2\omega t + a^2\sin 2\omega t) - \frac{g}{2\omega^2}(b^2\omega^2t^2 + a^2\sin^2\omega t). \tag{12}$$

Note that if one considers the limit as ω tends to zero in Eqs. (10), (11), and (12), there results

$$x = ut, \qquad y = vt, \qquad z = wt - \tfrac{1}{2}gt^2. \tag{13}$$

Equations (13) would represent the motion of a particle relative to a still earth.[3]

Population Growth

■ In studying the population growth of some species, we assume that at time $t = 0$, there are a certain number of them, say $n(0)$, all of a certain age. For convenience, we refer to this age classification as zero age. At a future time t there are $n_1(t)$ of these specimens still in the population. Then $n_1(t)$ and $n(0)$ are connected, with the use of a *survival function* $f(t)$ by the relation

$$n_1(t) = n(0)f(t).$$

Furthermore, we consider that individuals of age zero placed in this population at some time $t_1 > 0$ also survive according to this rule. That is, if $m(t_1)$ of these individuals are placed into the population at time t_1, then at a future time $t > t_1$ there will be $m(t)$ of these individuals left in the population, where

$$m(t) = m(t_1)f(t - t_1).$$

[3]For a more complete discussion of this illustration and other applications of the Laplace transform, see Gustav Doetsch, *Guide to the Applications of the Laplace and Z-Transforms* (London: Van Nostrand, 1967).

Consider now that age-zero individuals are placed in the population at a rate $r(t)$, commonly called the *replacement rate*. In the time interval from τ to $\tau + \Delta\tau$, $r(\tau_1)\Delta\tau$, individuals are placed in the population. τ_1 must be such that $\tau \leq \tau_1 \leq \tau + \Delta\tau$. According to the survival law, $r(\tau_1)\Delta\tau\, f(t - \tau_1)$ of these individuals will still be present in the population at time t.

The consideration is as follows. At time $t = 0$ there are $n(0)$ zero-age individuals of a certain species given. As time goes on, zero-age individuals are placed into the population at the rate $r(t)$. We wish to determine the total population $n(t)$ at some later time t. Certainly one could approximate the answer for $n(t)$ by splitting the time interval $[0, t]$ into intervals of length $\Delta\tau$ and adding up the number of survivors for each interval. Naturally, the approximation is made more accurate by making the time interval $\Delta\tau$ smaller. In fact, we can say that the approximation tends to $n(t)$ as $\Delta\tau$ tends to zero. Thus,

$$n(t) = n(0)f(t) + \int_0^t r(\tau)\, f(t - \tau)d\tau. \tag{14}$$

Equation (14) is a model for studying the growth of populations governed by the considerations above.

EXAMPLE 2 Suppose that in 1976 there were 34,000 wild rabbits in Rhode Island, that the survival function for these rabbits is e^{-t}, and that wild rabbits are introduced into the population at a constant rate r_0. How many wild rabbits will be present at time t? What value should r_0 have if the population is to stay constant?

Solution

$$\begin{aligned}
n(\mathrm{t}) &= 34{,}000e^{-t} + \int_0^t r_0 e^{-(t-\tau)}d\tau \\
&= 34{,}000e^{-t} + r_0 e^{-t}\int_0^t e^{\tau}d\tau \\
&= 34{,}000e^{-t} + r_0 e^{-t}\left(e^{\tau}\Big|_0^t\right) \\
&= 34{,}000e^{-t} + r_0(1 - e^{-t}).
\end{aligned}$$

It is easy to see that if $r_0 = 34{,}000$, then $n(t) = 34{,}000$ for all t.

EXAMPLE 3 Suppose that in Example 2 it is desired to determine a rate function $r(t)$ such that the wild rabbit population is a linear function of time; that is,

$$n(t) = n(0) + ct.$$

Solution Here $n(0) = 34{,}000$ and c is a constant. Then $r(t)$ must satisfy

$$n(0) + ct = n(0)e^{-t} + \int_0^t r(\tau)e^{-(t-\tau)}d\tau. \tag{15}$$

Although Eq. (15) is not a differential equation, Laplace transforms can be employed to determine the unknown function $r(t)$. Taking the Laplace transform of Eq. (15), we obtain

$$\frac{n(0)}{s} + \frac{c}{s^2} = \left[\frac{n(0)}{s+1} + R(s)\frac{1}{s+1}\right].$$

Solving for $R(s)$, we have

$$R(s) = (s+1)\left[\frac{n(0)}{s} + \frac{c}{s^2} - \frac{n(0)}{s+1}\right].$$

Simplifying algebraically, we obtain

$$\begin{aligned} R(s) &= (s+1)\left[\frac{n(0)s(s+1) + c(s+1) - n(0)s^2}{s^2(s+1)}\right] \\ &= \frac{[n(0)+c]s + c}{s^2} \\ &= \frac{n(0)+c}{s} + \frac{c}{s^2}. \end{aligned}$$

Taking the inverse Laplace transform, we obtain the rate function

$$r(t) = n(0) + c + ct.$$

EXERCISES

1. Verify Eq. (5).
2. Verify Eq. (6). [*Hint:* $a^2 + b^2 = 1$.]
3. Verify Eq. (7).
4. Verify Eq. (8).
5. Using the method of Laplace transforms, find the replacement rate function for a population whose survival function is te^{-2t} and for which it is desired that the population numbers $n(0)t + 3t$ at any time $t > 0$.
6. Verify Eq. (9).
7. Verify Eq. (10).
8. Verify Eq. (11).
9. Verify Eq. (12). $\left[\textit{Hint: } \dfrac{4\omega^2}{s^3(s^2+4\omega^2)} = \dfrac{1}{s^3} - \dfrac{1}{s(s^2+4\omega^2)}.\right]$
10. Verify the asserted limits in (13). Recall l'Hospital's rule.
11. Show that Eq. (2) can be written in the form

$$I(s) = \frac{s/L}{(s+a)^2 + (\omega_0^2 - a^2)}V(s) \tag{16}$$

with a and ω_0 appropriately defined.

12. Suppose that $v(t) = v_0$ (a constant). Determine $i(t)$ from Eq. (16) in each of the following cases:

(a) $\omega_0^2 = a^2$ (b) $\omega_0^2 > a^2$ (c) $\omega_0^2 < a^2$

13. Suppose that $v(t) = v_0 \sin \omega t$ (v_0, ω constants; $\omega \neq \omega_0$). Determine $i(t)$ from Eq. (16) in each of the following cases:

(a) $\omega_0^2 = a^2$ (b) $\omega_0^2 > a^2$ (c) $\omega_0^2 > a^2$

Feedback Systems Feedback systems are comprised of two separate systems: a controlled system and a controlling system. A *controlled system* is one in which a specific activity is executed so as to maintain a certain fixed activity of the system. Such functions are performed by varying the input to the controlled system in amounts that depend on the discrepancy between the controlled system and its desired state at each time t. The amount of input to the controlled system is administered by another system whose function is to monitor the deviance of the controlled system and to supply the controlled system with inputs in such a way as to minimize the deviance. This other system is the *controlling system*. Such a feedback system can be represented diagrammatically as in Figure 4.6, where $Y_e(s) = Y_i(s) - Y_0(s)$ and $Y_c(s) = G_1(s)Y_e(s)$. From the transfer functions $G_1(s)$ and $G_2(s)$, a number of other transfer functions can be derived which characterize the behavior of the entire feedback system. Using the definition of transfer function, we can write

$$Y_0(s) = G_1(s)G_2(s)Y_e(s)$$

$$\Rightarrow \qquad \frac{Y_0(s)}{Y_e(s)} = G_1(s)G_2(s).$$

$G_1(s)G_2(s)$ is called the *open-loop* transfer function.

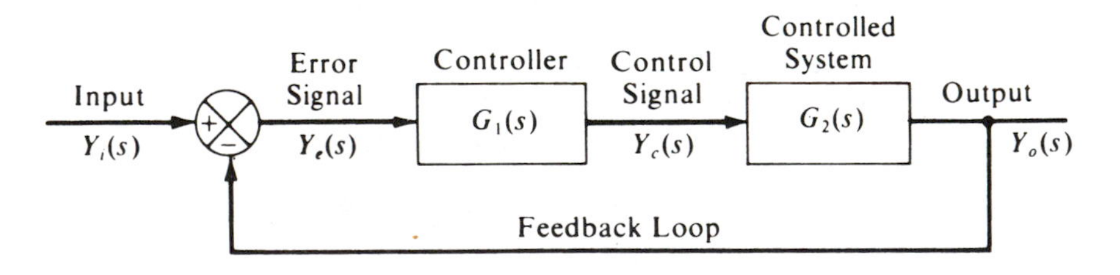

Figure 4.6

14. Show that

$$\frac{Y_0(s)}{Y_i(s)} = \frac{G_1(s)G_2(s)}{1 + G_1(s)G_2(s)}.$$

$$\left[\mathit{Hint:} \frac{a}{c} = \frac{a}{b} \cdot \frac{b}{c}. \right]$$

15. Show that

$$\frac{Y_e(s)}{Y_i(s)} = \frac{1}{1 + G_1(s)G_2(s)}.$$

Regulators For a "pure regulator" the input is a constant (thus it is generally chosen to be zero), and there is an additional signal which represents the effects of fluctuations in the environment of the system. The transform of this signal is denoted by $Y_f(s)$. Such feedback systems are represented diagrammatically as in Figure 4.7.

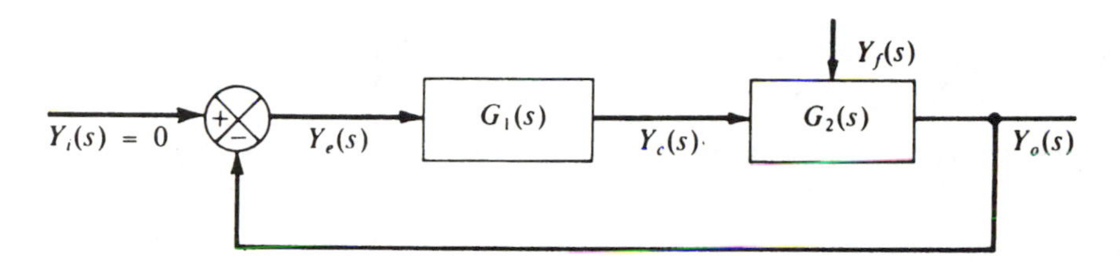

Figure 4.7

16. Show that

$$\frac{Y_0(s)}{Y_f(s)} = G_2(s)$$

is the open-loop transfer function.

[*Hint:* Follow an "initial signal" from input to output.]

17. Show that

$$\frac{Y_0(s)}{Y_f(s)} = \frac{G_2(s)}{1 + G_1(s)G_2(s)}$$

is the *closed-loop* transfer function.

REMARK 2 Many biological systems are modeled as feedback systems or as regulators. One such system is the eye.[4]

REVIEW EXERCISES

In Exercises 1 through 4, find the Laplace transform of the given function.

1. $4t^5 + 3t \sin t$ **2.** $h_1(t) \cos t$

[4]Many illustrations of this sort can be found in Rosen, *Optimality Principles*. The reader can see from the expressions developed in these exercises that researchers can attempt to determine the transfer function from the given input and the observed output. Efficient techniques for doing this are as yet not known.

3. $\delta(t) + h_1(t-2)$

4. $\delta(t-1) - h_1(t)e^{3t} + \cos t$

In Exercises 5 through 8, find the inverse Laplace transform of the given function.

5. $\dfrac{1}{s^2 - 10s + 9}$

6. $\dfrac{45s - 126}{s^3 - 5s^2 - 14s}$

7. $\dfrac{e^{-3(s-1)}}{s - 1}$

8. $7 + \dfrac{2}{s^2 + 4s + 13}$

In Exercises 9 through 16, solve the given initial value problem by the method of Laplace transforms.

9. $\ddot{y} - 7\dot{y} - 8y = 384t$
$y(0) = 0, \dot{y}(0) = 3$

10. $3\ddot{y} - 20\dot{y} - 7y = 925e^t \cos t$
$y(0) = 10, \dot{y}(0) = 2$

11. $5\ddot{y} + 17\dot{y} + 6y = 676 \sin 2t - 36t$
$y(0) = 0, \dot{y}(0) = -7$

12. $\ddot{y} - 8\dot{y} - 33y = 196e^{-3t}$
$y(0) = 0, \dot{y}(0) = 0$

13. $\dddot{y} + y = 6e^t$
$y(0) = 0, \dot{y}(0) = 0, \ddot{y}(0) = 0$

14. $\ddot{y} - 6\dot{y} + 9y = \delta(t) + 9$
$y(0) = 0, \dot{y}(0) = 1$

15. $\ddot{y} - 11\dot{y} + 24y = 120h_1(t-2)$
$y(0) = 5, \dot{y}(0) = 0$

16. $\ddot{y} + y = \delta(t-1)$
$y(0) = 0, \dot{y}(0) = 0$

17. For a "pure regulator" show that the open-loop output is given by $\displaystyle\int_0^t g_2(t-\tau)y_f(\tau)d\tau$.

18. Consider[5] the linear time invariant RLC-series circuit with input e_s and response i as shown in Figure 4.8. Calculate and sketch the impulse response ($q(0) = 0, \dot{q}(0) = 0$).

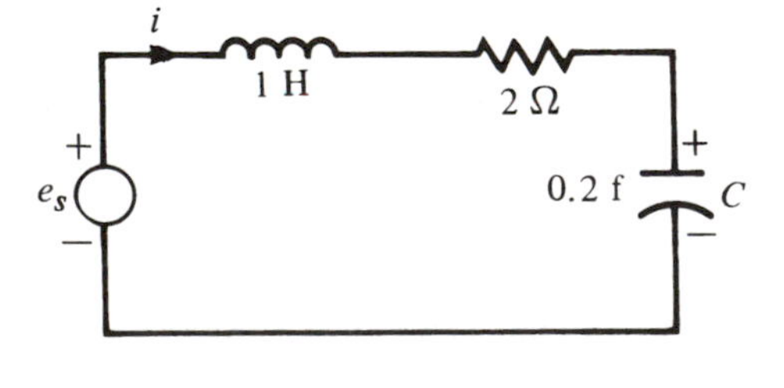

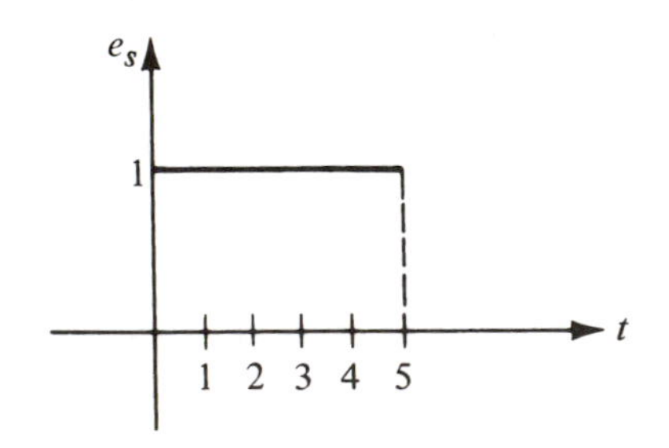

Figure 4.8

[5]This is Exercise 22a in Charles A. Desoer and Ernest S. Kuh, *Basic Circuit Theory* (New York: McGraw-Hill, 1969), p. 268. Reprinted by permission of McGraw-Hill Book Company.

19. A force $f(t) = Qte^{-bt}h_1(t)$ is applied[6] to a simple (harmonic) oscillator of mass m, (spring) stiffness K, and (damping) resistance R, initially at rest ($x(0) = \dot{x}(0) = 0$). Use the Laplace transform to calculate the displacement of the mass as a function of time for $t > 0$.

20. Show that

$$\mathscr{L}[f(t)] = \mathscr{L}\left[\sum_{n=0}^{\infty} \delta\left[t - \frac{\pi}{a}(2n+1)\right]\right] = \frac{1}{2\sinh\frac{\pi}{a}s},$$

where a is a constant. [*Hint:* assume that $\mathscr{L}\left[\sum_{n=0}^{\infty} \cdots\right] = \sum_{n=0}^{\infty} \mathscr{L}[\cdots].$]

In Exercises 21 through 30, solve the initial value problem by the method of the Laplace transformation.

21. $\dot{x}_1 = -x_1$
$\dot{x}_2 = -x_2$
$x_1(0) = 1,\ x_2(0) = 5$

22. $\dot{x}_1 = -2x_1 + x_2$
$\dot{x}_2 = x_1 - 2x_2$
$x_1(0) = 3,\ x_2(0) = 1$

23. $\dot{x}_1 = 3x_1 - 2x_2$
$\dot{x}_2 = 2x_1 - 2x_2$
$x_1(0) = -1,\ x_2(0) = 1$

24. $\dot{x}_1 = -x_1$
$\dot{x}_2 = -x_1 - x_2$
$x_1(0) = -4,\ x_2(0) = 6$

25. $\dot{x}_1 = -x_1 + x_2$
$\dot{x}_2 = -x_1 - x_2$
$x_1(0) = 7,\ x_2(0) = 3$

26. $\dot{x}_1 = x_2$
$\dot{x}_2 = -x_1$
$x_1(0) = 9,\ x_2(0) = -8$

27. $\dot{x}_1 = x_2$
$\dot{x}_2 = -x_1 + 2x_2$
$x_1(0) = 1,\ x_2(0) = -5$

28. $\dot{x}_1 = 2x_1 - x_2$
$\dot{x}_2 = 9x_1 + 2x_2$
$x_1(0) = 2,\ x_2(0) = -3$

29. $\dot{x}_1 = 2x_1 + x_2$
$\dot{x}_2 = -3x_1 + 6x_2$
$x_1(0) = 0,\ x_2(0) = -2$

30. $\dot{x}_1 = x_2$
$\dot{x}_2 = x_3$
$\dot{x}_3 = x_1$
$x_1(0) = -1,\ x_2(0) = 5,\ x_3(0) = -1$

[6] This is Exercise 11 in Philip M. Morse and K. Uno Ingard, *Theoretical Acoustics* (New York: McGraw-Hill, 1968), p. 62. Copyright © 1968 by McGraw-Hill Book Company. Reprinted by permission of McGraw-Hill Book Company.

APTER 5

Series Solutions of Second-Order Linear Equations

5.1 INTRODUCTION

In this chapter we present an effective method for solving many second-order linear differential equations with variable coefficients by means of infinite series. We shall refer to this as the method of *series solutions*. The method of series solutions can also be used to compute formal solutions and to approximate solutions of other linear and nonlinear differential equations with constant or variable coefficients. We concentrate here on series solutions of second-order linear differential equations with variable coefficients because of the importance of these equations in applications and because the method of series solutions has been well developed for such equations.

Second-order linear differential equations appear frequently in applied mathematics, especially in the process of solving some of the classical partial differential equations in mathematical physics. Following are some of the most important second-order linear differential equations with variable coefficients which occur in applications. It is common to refer to these equations by the name that appears to the left of the differential equation.

Airy's equation: $y'' - xy = 0$

Bessel's equation: $x^2y'' + xy' + (x^2 - p^2)y = 0$

Chebyshev's equation: $(1 - x^2)y'' - xy' + p^2y = 0$

Gauss's hypergeometric equation: $x(1 - x)y'' + [c - (a + b + 1)x]y' - aby = 0$

Hermite's equation: $y'' - 2xy' + 2py = 0$

Laguerre's equation: $xy'' + (1 - x)y' + py = 0$

Legendre's equation: $(1 - x^2)y'' - 2xy' + n(n + 1)y = 0$

With the exception of Airy's equation, each of these differential equations involves parameters whose description is associated with the problem that led to their formulation and the "separation constant" involved in the process of solving partial differential equations by separation of variables. All these equations can be solved by the method of series solutions. Before we explain how to obtain series solutions for the above as well as other differential equations with variable coefficients, we review some of the properties of power series that will be used in this chapter.

5.2 REVIEW OF POWER SERIES

A series of the form

$$a_0 + a_1(x - x_0) + a_2(x - x_0)^2 + \cdots + a_n(x - x_0)^n + \cdots$$

is called a *power series* in powers of $(x - x_0)$ and is denoted by

$$\sum_{n=0}^{\infty} a_n(x - x_0)^n. \tag{1}$$

The numbers $a_0, a_1, a_2, \ldots, a_n, \ldots$ are called the *coefficients* of the power series, and the point x_0 is called the *center* of the power series. We also say that (1) is a power series *about the point* x_0.

We say that a power series $\sum_{n=0}^{\infty} a_n(x - x_0)^n$ *converges* at a specific point x_1 if

$$\lim_{N\to\infty} \sum_{n=0}^{N} a_n(x_1 - x_0)^n$$

exists. In this case the value of the limit is called the *sum* of the series at the point x_1. If this limit does not exist, the series is said to *diverge* at the point x_1.

Given a power series (1), it is important to find all points x for which the series converges. To do this we compute the *radius of convergence* of the power series. This term is denoted by R and is given by the formula

$$R = \frac{1}{\lim_{n\to\infty} \sqrt[n]{|a_n|}} \tag{2}$$

or

$$R = \lim_{n\to\infty} \left| \frac{a_n}{a_{n+1}} \right|, \tag{3}$$

provided that the limit in (2) or (3) exists. (For series in a form other than (1), such as $\sum_{n=0}^{\infty} x^{2n+1}/2^n$, Eqs. (2) and (3) are not applicable in general. For such series the radius of convergence is determined by using the ratio test of calculus. See Exercises 3, 9, and 24.)

If $R = 0$, series (1) converges only at its center $x = x_0$. If $R = +\infty$, it converges for all x. Finally, if $0 < R < \infty$, the series converges in the interval $|x - x_0| < R$, that is, for

$$-R + x_0 < x < R + x_0, \tag{4}$$

and diverges for $|x - x_0| > R$. The interval (4), or the entire real line if $R = \infty$, is called the *interval of convergence* of series (1). It is within this interval that all operations that we will perform on series, in our attempt to find series solutions of differential equations, are legal. At the endpoints of the interval (4), that is, for $x = -R + x_0$ and $x = R + x_0$, the series may converge or diverge. To determine the behavior at these points, we set $x = -R + x_0$ in (1) and check the resulting series by one of the known tests, then repeat the procedure for $x = R + x_0$.

EXAMPLE 1 Determine the radius of convergence of each of the following power series:

(a) $\sum_{n=1}^{\infty} n^n x^n$ (b) $\sum_{n=0}^{\infty} (-1)^n (x-1)^n$ (c) $\sum_{n=0}^{\infty} \frac{x^n}{n!}$.

Solution (a) Here $a_n = n^n$ and from formula (2),

$$R = \frac{1}{\lim_{n\to\infty} \sqrt[n]{|a_n|}} = \frac{1}{\lim_{n\to\infty} n} = 0.$$

Hence series (a) converges only for $x = 0$ and diverges for any other x.

(b) Here $a_n = (-1)^n$ and from formula (2),

$$R = \frac{1}{\lim_{n\to\infty} \sqrt[n]{|(-1)^n|}} = \frac{1}{\lim_{n\to\infty} 1} = 1.$$

Thus, series (b) converges for all points x in the interval $|x - 1| < 1$; that is, $-1 < x - 1 < 1$ or $0 < x < 2$. The series diverges for $|x - 1| > 1$, that is, for $x < 0$ or $x > 2$. For $|x - 1| = 1$, that is, for $x = 0$ or $x = 2$, one can see directly that the series becomes

$$\sum_{n=0}^{\infty} (-1)^n(\pm 1)^n = \sum_{n=0}^{\infty} (\mp 1)^n,$$

and both of these series diverge.

(c) Here $a_n = 1/n!$. It is more convenient to use formula (3) in this case. Then

$$R = \lim_{n\to\infty} \left| \frac{1/n!}{1/(n+1)!} \right| = \lim_{n\to\infty} (n+1) = \infty.$$

Thus, series (c) converges for all x.

If R is the radius of convergence of a power series $\sum_{n=0}^{\infty} a_n(x - x_0)^n$, then for every x in the interval of convergence $|x - x_0| < R$, the sum of the series exists and defines a function

$$f(x) = \sum_{n=0}^{\infty} a_n(x - x_0)^n \qquad \text{for } |x - x_0| < R. \tag{5}$$

The function $f(x)$ defined by power series (5) is continuous and has derivatives of all orders. Furthermore, the derivatives $f'(x), f''(x), \ldots$ of the function $f(x)$ can be found by differentiating series (5) term by term. That is,

$$f'(x) = \sum_{n=1}^{\infty} na_n(x - x_0)^{n-1}$$

$$f''(x) = \sum_{n=2}^{\infty} n(n-1)a_n(x - x_0)^{n-2},$$

and so on. Finally, these series for $f'(x), f''(x), \ldots$ have the same radius of convergence R as the initial series (5).

In the process of finding power-series solutions of differential equations, in addition to taking derivatives of power series, we have to add, subtract, multiply, and equate two or more power series. These operations are performed in a fashion that resembles very much the corresponding operations with polynomials. The additional restriction for power series is to perform the operation within the interval of convergence of all series involved. For example,

(a) $\sum_{n=0}^{\infty} a_n(x - x_0)^n + \sum_{n=0}^{\infty} b_n(x - x_0)^n = \sum_{n=0}^{\infty} (a_n + b_n)(x - x_0)^n.$

(b) $\sum_{n=0}^{\infty} a_n(x - x_0)^n - \sum_{n=0}^{\infty} b_n(x - x_0)^n = \sum_{n=0}^{\infty} (a_n - b_n)(x - x_0)^n.$

(c) $a(x - x_0)^k \sum_{n=0}^{\infty} a_n(x - x_0)^n = \sum_{n=0}^{\infty} aa_n(x - x_0)^{n+k}.$

(d) If $\sum_{n=0}^{\infty} a_n(x - x_0)^n = \sum_{n=0}^{\infty} b_n(x - x_0)^n$

for all x in some interval $|x - x_0| < R$, then

$$a_n = b_n \qquad \text{for } n = 0, 1, 2, \ldots .$$

In particular, if a series is identically equal to zero, all the coefficients of the series must be zero.

The operations in (a), (b), and (d) were performed in one step because the general terms of the series involved were of the same power. However, in practice, we have to combine series whose general terms are not of the same power. In such cases we make an appropriate change in the index of summation of the series which does not change the sum of the series but makes the general terms of the same power. The basic idea behind the change of index is incorporated in the following identity:

$$\sum_{n=0}^{\infty} a_n(x - x_0)^n = \sum_{n=k}^{\infty} a_{n-k}(x - x_0)^{n-k}, \tag{6}$$

which holds for every integer k. The simplest way to prove (6) is to write out the two series term by term. In words, (6) says that we can decrease by k the n in the general term $a_n(x - x_0)^n$ provided that we increase by k the n under the summation symbol, and vice versa. For example,

$$\sum_{n=1}^{\infty} na_n(x - x_0)^{n-1} = \sum_{n=0}^{\infty} (n + 1)a_{n+1}(x - x_0)^n$$

and

$$\sum_{n=0}^{\infty} 3(n + 4)a_n x^{n+2} = \sum_{n=2}^{\infty} 3(n + 2)a_{n-2}x^n.$$

The reader should note that these changes in appearance of the summations are very much like making a simple substitution in a definite integral. Thus, in the first illustration the index n in the summation on the left is replaced by the new index j, where $j + 1 = n$. The resulting summation is that on the right with the index being called n again instead of j.

Finally, we define the concept of "analytic function," which will be widely used in this chapter.

A function f is called *analytic* at a point x_0 if it can be written as a power series

$$f(x) = \sum_{n=0}^{\infty} a_n(x - x_0)^n \tag{7}$$

with a positive radius of convergence.

Within its interval of convergence the power series (7) can be differentiated term by term. Evaluating $f(x)$, $f'(x)$, $f''(x)$, . . . at the point x_0 we obtain $f(x_0) = a_0$, $f'(x_0) = a_1$, $f''(x_0) = 2a_2$, . . . and in general

$$f^{(n)}(x_0) = n!a_n \qquad \text{for } n = 0, 1, 2, \ldots.$$

Hence, $a_n = f^{(n)}(x_0)/n!$ and the power series in (7) is the *Taylor-series expansion*

$$f(x) = \sum_{n=0}^{\infty} \frac{f^{(n)}(x_0)}{n!}(x - x_0)^n \tag{8}$$

of the function f at the point x_0. Thus, a function f is analytic at a point x_0 if its Taylor-series expansion (8) about x_0 exists and has a positive radius of convergence.

Let us give some examples of analytic functions. Every polynomial is an analytic function about any point x_0. Indeed, since the derivatives of order higher than n of a polynomial of degree n are equal to zero, the Taylor-series expansion of any polynomial function has only a finite number of nonzero terms, and so it converges everywhere. Also, a "rational function," that is, the quotient of two polynomials, is an analytic function at every point where the denominator is different from zero. For example, the function $3x^2 - 7x + 6$ is analytic everywhere, while the function

$$\frac{x^2 - 5x + 7}{x(x^2 - 9)}$$

is analytic about every point except $x = 0$, 3, and -3. Also, the functions e^x, $\sin x$, and $\cos x$ are analytic everywhere, as we can see from their Taylor-series expansions.

EXERCISES

In Exercises 1 through 6, determine the radius of convergence of the power series.

1. $\sum_{n=0}^{\infty} 3^n(x - 1)^n$ **2.** $\sum_{n=0}^{\infty} 3^n x^n$ **3.** $\sum_{n=0}^{\infty} \frac{(-1)^n}{(2n)!} x^{2n}$

4. $\sum_{n=0}^{\infty} \frac{x^n}{n!}$ **5.** $\sum_{n=0}^{\infty} \frac{(x + 1)^n}{n + 1}$ **6.** $\sum_{n=0}^{\infty} \frac{x^n}{5 \cdot 7 \cdots (2n + 3)}$

In Exercises 7 through 12, compute the first and second derivative of the series and find the radius of convergence of the resulting series.

7. $\sum_{n=0}^{\infty} 3^n(x-1)^n$ **8.** $\sum_{n=0}^{\infty} 3^n x^n$ **9.** $\sum_{n=0}^{\infty} \frac{(-1)^n}{(2n)!} x^{2n}$

10. $\sum_{n=0}^{\infty} \frac{x^n}{n!}$ **11.** $\sum_{n=0}^{\infty} \frac{(x+1)^n}{n+1}$ **12.** $\sum_{n=0}^{\infty} \frac{x^n}{5 \cdot 7 \cdots (2n+3)}$

13. If $y(x) = \sum_{n=0}^{\infty} a_n x^n$, show that

$$y'' - xy = 2a_2 + \sum_{n=3}^{\infty} [n(n-1)a_n - a_{n-3}]x^{n-2}.$$

14. If $y(x) = \sum_{n=0}^{\infty} a_n x^n$, show that

$$(1-x)y'' - y' + xy = (2a_2 - a_1) + \sum_{n=2}^{\infty} [(n+1)na_{n+1} - n^2 a_n + a_{n-2}]x^{n-1}.$$

15. If $y(x) = \sum_{n=0}^{\infty} a_n(x-1)^n$ is a solution of the differential equation $y' = 3y$, show that

$$a_{n+1} = \frac{3}{n+1} a_n, \qquad n = 0, 1, 2, \ldots$$

and find the solution.

In Exercises 16 through 23, show that the function is analytic at the indicated point x_0 by computing its Taylor series about the point x_0 and finding its interval of convergence.

16. $x^2 - 1,\ x_0 = -2$ **17.** $\cos x,\ x_0 = 0$ **18.** $e^x,\ x_0 = 1$

19. $\sin x,\ x_0 = \pi$ **20.** $\frac{1}{x+3},\ x_0 = 0$ **21.** $\frac{1}{x+3},\ x_0 = 1$

22. $\frac{1}{1-x^2},\ x_0 = 0$ **22.** $\frac{1}{x},\ x_0 = -3.$

24. Find the radius of convergence of the series $\sum_{n=0}^{\infty} \frac{x^{2n+1}}{2^n}$.

5.3 ORDINARY POINTS AND SINGULAR POINTS

Consider a second-order linear differential equation with variable coefficients of the form

$$a_2(x)y'' + a_1(x)y' + a_0(x)y = 0. \tag{1}$$

In the subsequent sections we will seek series solutions of the differential equation (1) in powers of $(x - x_0)$, where x_0 is a real number. As we will see, the form of the solutions will depend very much on the kind of point that x_0 is with respect to the differential equation. A point x_0 can be either an ordinary point or a singular point, according to the following definition.

DEFINITION 1

A point x_0 is called an ordinary point of the differential equation (1) *if the two functions*

$$\frac{a_1(x)}{a_2(x)} \quad \text{and} \quad \frac{a_0(x)}{a_2(x)} \tag{2}$$

are analytic at the point x_0. If at least one of the functions in (2) *is not analytic at the point x_0, then x_0 is called a singular point of the differential equation* (1).

In most differential equations of the form (1) that occur in applications, the coefficients $a_2(x)$, $a_1(x)$, and $a_0(x)$ are polynomials. After cancelling common factors, the rational functions $a_1(x)/a_2(x)$ and $a_0(x)/a_2(x)$ are analytic at every point except where the denominator vanishes. The points at which the denominator vanishes are singular points of the differential equation, and all other real numbers are ordinary points.

With reference to the differential equations mentioned in Section 5.1 see Table 5.1, which gives their ordinary and singular points in the finite real line.

In connection with the theory of series solutions it is important to classify the singular points of a differential equation into two categories according to the following definition.

DEFINITION 2

A point x_0 is called a regular singular point of the differential equation (1) *if it is a singular point [if at least one of the functions in* (2) *is not analytic at x_0] and the two functions*

$$(x - x_0)\frac{a_1(x)}{a_2(x)} \quad \text{and} \quad (x - x_0)^2\frac{a_0(x)}{a_2(x)} \tag{3}$$

are analytic at the point x_0. If at least one of the functions in (3) *is not analytic at the point x_0, then x_0 is called an irregular singular point of the differential equation* (1).

In Exercises 9 through 15 the student is asked to verify that all singular points of the differential equations of Table 5.1 are regular singular points.

EXAMPLE 1 Locate the ordinary points, regular singular points, and irregular singular points of the differential equation

$$(x^4 - x^2)y'' + (2x + 1)y' + x^2(x + 1)y = 0. \tag{4}$$

Differential equation	Ordinary points	Singular points
Airy	All points	None
Bessel	All points except $x_0 = 0$	0
Chebyshev	All points except $x_0 = \pm 1$	± 1
Gauss	All points except $x_0 = 0, 1$	0, 1
Hermite	All points	None
Laguerre	All points except $x_0 = 0$	0
Legendre	All points except $x_0 = \pm 1$	± 1

Table 5.1

Solution Here

$$a_2(x) = x^4 - x^2, \qquad a_1(x) = 2x + 1, \qquad a_0(x) = x^2(x + 1),$$

and so

$$\frac{a_1(x)}{a_2(x)} = \frac{2x + 1}{x^4 - x^2} = \frac{2x + 1}{x^2(x - 1)(x + 1)}, \frac{a_0(x)}{a_2(x)} = \frac{x^2(x + 1)}{x^4 - x^2} = \frac{1}{x - 1}. \tag{5}$$

It follows from (5) that every real number except 0, 1, and -1 is an ordinary point of the differential equation (4). To see which of the singular points 0, 1, and -1 is a regular singular point and which is an irregular singular point for the differential equation (4), we need to examine the two functions in (3).

For $x_0 = 0$, the two functions in (3) become

$$x\frac{2x + 1}{x^4 - x^2} = \frac{2x + 1}{x(x - 1)(x + 1)} \qquad \text{and} \qquad x^2\frac{x^2(x + 1)}{x^4 - x^2} = \frac{x^2}{x - 1}.$$

The first of these expressions is not analytic at $x = 0$, hence we conclude that the point $x_0 = 0$ is an irregular singular point for the differential equation (4). For $x_0 = 1$, the two functions in (3) become

$$(x - 1)\frac{2x + 1}{x^4 - x^2} = \frac{2x + 1}{x^2(x + 1)} \qquad \text{and} \qquad (x - 1)^2\frac{x^2(x + 1)}{x^4 - x^2} = x - 1.$$

Since both of these expressions are analytic at $x = 1$, we conclude that the point $x_0 = 1$ is a regular singular point for the differential equation (4). Finally, for $x_0 = -1$, the two functions in (3) become

$$(x + 1)\frac{2x + 1}{x^4 - x^2} = \frac{2x + 1}{x^2(x - 1)} \qquad \text{and} \qquad (x + 1)^2\frac{x^2(x + 1)}{x^4 - x^2} = \frac{(x + 1)^2}{x - 1},$$

and since both of them are analytic at $x = -1$ (their denominators do not vanish at $x = -1$), we conclude that the point $x_0 = -1$ is a regular singular point for the differential equation (4).

Our aim in the remaining sections of this chapter is to obtain series solutions about ordinary points and near regular singular points. The study of solutions near irregular singular points is difficult and beyond our scope.

EXERCISES

In Exercises 1 through 8, locate the ordinary points, regular singular points, and irregular singular points of the differential equation.

1. $xy'' - (2x + 1)y' + y = 0$

2. $y'' - 2(x - 1)y' + 2y = 0$

3. $(1 - x)y'' - y' + xy = 0$

4. $2x^2y'' + (x - x^2)y' - y = 0$

5. $(x - 1)^2y'' - (x^2 - x)y' + y = 0$

6. $x^2y'' - (x + 2)y = 0$

7. $x^3(1 - x^2)y'' + (2x - 3)y' + xy = 0$

8. $(x - 1)^4y'' - xy = 0$

In Exercises 9 through 15, verify that all singular points of the differential equation are regular singular points.

9. $y'' - xy = 0$ (Airy's equation)

10. $x^2y'' + xy' + (x^2 - p^2)y = 0$ (Bessel's equation)

11. $(1 - x^2)y'' - xy' + p^2y = 0$ (Chebyshev's equation)

12. $x(1 - x)y'' + [c - (a + b + 1)x]y' - aby = 0$ (Gauss's hypergeometric equation)

13. $y'' - 2xy' + 2py = 0$ (Hermite's equation)

14. $xy'' + (1 - x)y' + py = 0$ (Laguerre's equation)

15. $(1 - x^2)y'' - 2xy' + n(n + 1)y = 0$ (Legendre's equation)

In Exercises 16 through 21, answer true or false.

16. The point $x_0 = -1$ is a regular singular point for the differential equation $(1 - x^2)y'' - 2xy + 12y = 0$.

17. The point $x_0 = 0$ is an ordinary point for the differential equation $xy'' + (1 - x)y' + 2y = 0$.

18. The point $x_0 = 0$ is a singular point for the differential equation $(1 + x)y'' - 2y' + xy = 0$.

19. The point $x_0 = 0$ is an irregular singular point for the differential equation $x^3y'' - (x + 1)y = 0$.

20. The point $x_0 = 3$ is an ordinary point for the differential equation $(x + 3)y'' + xy' - y = 0$.

21. The point $x_0 = -3$ is a singular point for the differential equation $(x + 3)y'' + xy' - y = 0$.

5.4 POWER-SERIES SOLUTIONS ABOUT AN ORDINARY POINT

In this section we show how to solve any second-order linear differential equation with variable coefficients of the form

$$a_2(x)y'' + a_1(x)y' + a_0(x)y = 0 \tag{1}$$

in some interval about any *ordinary point* x_0. The point x_0 is usually dictated by the specific problem at hand, which requires that we find the solution of the differential equation (1) that satisfies two given initial conditions of the form

$$y(x_0) = y_0 \tag{2}$$

and

$$y'(x_0) = y_1. \tag{3}$$

Let us recall that if the coefficients $a_2(x)$, $a_1(x)$, and $a_0(x)$ are polynomials in x, then a point x_0 is an ordinary point of the differential equation (1) when $a_2(x_0) \neq 0$. In general, x_0 is an ordinary point of the differential equation (1) when the functions $a_1(x)/a_2(x)$ and $a_0(x)/a_2(x)$ have power-series expansions of the form

$$\frac{a_1(x)}{a_2(x)} = \sum_{n=0}^{\infty} A_n(x - x_0)^n \qquad \text{for } |x - x_0| < R_1 \tag{4}$$

and

$$\frac{a_0(x)}{a_2(x)} = \sum_{n=0}^{\infty} B_n(x - x_0)^n \qquad \text{for } |x - x_0| < R_2 \tag{5}$$

with positive radii of convergence R_1 and R_2. The functions (4) and (5) are in particular continuous in the interval $|x - x_0| < R$, where R is the smallest of R_1 and R_2, and therefore, by the existence theorem, Theorem 1 of Section 2.3, the IVP (1)–(3) has a unique solution throughout the interval $|x - x_0| < R$. Our task here is to compute (or approximate) this unique solution. The following theorem describes the form of any solution of the differential equation (1), in particular the form of the unique solution of the IVP (1)–(3).

THEOREM 1 (Solutions about an Ordinary Point)
If x_0 is an ordinary point of the differential equation (1), *then the general solution of the differential equation has a power-series expansion about x_0,*

$$y(x) = \sum_{n=0}^{\infty} a_n(x - x_0)^n, \tag{6}$$

with positive radius of convergence. More precisely, if R_1 and R_2 are the radii of convergence of the series (4) *and* (5), *then the radius of convergence of* (6) *is at least equal to the minimum of R_1 and R_2. The coefficients a_n for $n = 2, 3, \ldots$ of the series* (6) *can be obtained in terms of a_0 and a_1 by direct substitution of* (6) *into the differential equation* (1) *and by equating coefficients of the same power. Finally, if* (6) *is the solution of the IVP* (1)–(3), *then $a_0 = y_0$ and $a_1 = y_1$.*

EXAMPLE 1 Find the general solution of the differential equation

$$y'' - 2(x - 1)y' + 2y = 0 \tag{7}$$

about the ordinary point $x_0 = 1$.

Solution By Theorem 1 the general solution of Eq. (7) has a power-series expansion about $x_0 = 1$,

$$y(x) = \sum_{n=0}^{\infty} a_n(x-1)^n, \tag{8}$$

with a positive radius of convergence. To find a lower bound for the radius of convergence of the series (8), we need the radii of convergence R_1 and R_2 of the power-series expansions of the functions

$$\frac{a_1(x)}{a_2(x)} \quad \text{and} \quad \frac{a_0(x)}{a_2(x)}.$$

Here $a_2(x) = 1$, $a_1(x) = -2(x-1)$, and $a_0(x) = 2$. Hence,

$$\frac{a_1(x)}{a_2(x)} = -2(x-1) \quad \text{and} \quad \frac{a_0(x)}{a_2(x)} = 2,$$

and so $R_1 = R_2 = \infty$. Thus, the radius of convergence of the series (8) is also equal to ∞. That is, solution (8) will converge for all x. The coefficients of series (8) will be found by direct substitution of the series into the differential equation. Since (8) is the general solution of the second-order equation (7), it should contain two arbitrary constants. In fact, the coefficients a_0 and a_1 will remain unspecified, while every other coefficient $a_2, a_3, \ldots$ will be expressed in terms of a_0 and a_1. Differentiating series (8) term by term, we obtain

$$y' = \sum_{n=0}^{\infty} na_n(x-1)^{n-1} = \sum_{n=1}^{\infty} na_n(x-1)^{n-1}$$

and

$$y'' = \sum_{n=0}^{\infty} n(n-1)a_n(x-1)^{n-2} = \sum_{n=2}^{\infty} n(n-1)a_n(x-1)^{n-2}.$$

We are now ready to substitute y, y', and y'' into the differential equation (7). As we can see from Eq. (7), y' should be multiplied by $-2(x-1)$ and y by 2. For the sake of maintaining easy bookkeeping, we write the terms y'', $-2(x-1)y'$, and $2y$ of the differential equation in a column as follows:

$$y'' = \sum_{n=2}^{\infty} n(n-1)a_n(x-1)^{n-2}$$

$$-2(x-1)y' = -2(x-1)\sum_{n=1}^{\infty} na_n(x-1)^{n-1} = \sum_{n=1}^{\infty} -2na_n(x-1)^n$$

$$2y = 2\sum_{n=0}^{\infty} a_n(x-1)^n = \sum_{n=0}^{\infty} 2a_n(x-1)^n.$$

The sum of the terms on the left-hand sides is zero, because y is a solution of the differential equation (7). Hence, the sum of the three series on the right-hand side must be set equal to zero. Writing the expressions in columns is very handy in preparing the series in the summation process. It is easier to add the three series term by term if the general terms are of the same power in each series and that the lower index n under the summation symbol is the same in the three series. With this in mind, we rewrite the three series above in the following suitable and equivalent form:

$$y'' = \sum_{n=0}^{\infty} (n+2)(n+1)a_{n+2}(x-1)^n$$

$$= 2a_2 + \sum_{n=1}^{\infty} (n+2)(n+1)a_{n+2}(x-1)^n$$

$$-2(x-1)y' = \sum_{n=1}^{\infty} -2na_n(x-1)^n$$

$$2y = \sum_{n=0}^{\infty} 2a_n(x-1)^n = 2a_0 + \sum_{n=1}^{\infty} 2a_n(x-1)^n.$$

Adding the left-hand sides and the right-hand sides of these three equations, we obtain

$$0 = (2a_2 + 2a_0) + \sum_{n=1}^{\infty} [(n+2)(n+1)a_{n+2} - 2na_n + 2a_n](x-1)^n.$$

The right-hand side of this equation is a power series that is identically equal to zero. Hence, all its coefficients must be zero. That is,

$$2a_2 + 2a_0 = 0 \tag{9}$$

and

$$(n+2)(n+1)a_{n+2} - 2na_n + 2a_n = 0 \qquad \text{for} \qquad n = 1, 2, \ldots. \tag{10}$$

The condition (10) is called a *recurrence formula* because it allows a_{n+2} to be calculated once a_n is known. Using Eq. (9) and the recurrence formula (10), we can express the coefficients $a_2, a_3, \ldots$ of the power series in terms of the coefficients a_0 and a_1. In fact, from (9) we have

$$a_2 = -a_0, \tag{11}$$

and from (10) we obtain

$$a_{n+2} = \frac{2(n-1)}{(n+2)(n+1)} a_n \qquad \text{for} \qquad n = 1, 2, \ldots. \tag{12}$$

From Eq. (12) we find

$$a_3 = 0, \qquad a_4 = \frac{2}{4\cdot 3} a_2 = -\frac{2}{4\cdot 3} a_0 = -\frac{2^2}{4!} a_0$$

$$a_5 = 0, \qquad a_6 = \frac{2\cdot 3}{6\cdot 5} a_4 = -\frac{2^2\cdot 3}{6\cdot 5\cdot 4\cdot 3} a_0 = -\frac{2^3\cdot 3}{6!} a_0$$

$$a_7 = 0, \qquad a_8 = \frac{2\cdot 5}{8\cdot 7} a_6 = -\frac{2^3\cdot 5\cdot 3}{8\cdot 7\cdot 6\cdot 5\cdot 4\cdot 3} a_0 = -\frac{2^4\cdot 5\cdot 3\cdot 1}{8!} a_0$$

....

Thus,

$$a_{2n+1} = 0, \qquad n = 1, 2, \ldots$$

and

$$a_{2n} = -\frac{2^n \cdot 1 \cdot 3 \cdot 5 \cdots (2n-3)}{(2n)!} a_0, \qquad n = 2, 3, \ldots .$$

Hence, the general solution of the differential equation (7) is

$$\begin{aligned} y(x) &= a_0 + a_1(x-1) + a_2(x-1)^2 + a_4(x-1)^4 + a_6(x-1)^6 + \cdots \\ &= a_1(x-1) + a_0\left[1 - (x-1)^2 - \frac{2^2}{4!}(x-1)^4 - \frac{2^3 \cdot 3}{6!}(x-1)^6 - \cdots\right]. \end{aligned}$$

REMARK 1 As we expected, the general solution involves the two arbitrary constants a_1 and a_0. Therefore, the functions $x - 1$ and $1 - (x-1)^2 - (2^2/4!)(x-1)^4 - \cdots$ are two linearly independent solutions of Eq. (7).

EXAMPLE 2 Solve the IVP

$$(1-x)y'' - y' + xy = 0 \tag{13}$$

$$y(0) = 1 \tag{14}$$

$$y'(0) = 1. \tag{15}$$

Solution Since the initial conditions are given at the point 0, we are interested in a solution of the IVP (13)–(15) about the point $x_0 = 0$. The only singular point of the differential equation (13) is $x = 1$, and so the point $x_0 = 0$ is an ordinary point. Hence, the IVP (13)–(15) has a unique solution of the form:

$$y(x) = \sum_{n=0}^{\infty} a_n x^n. \tag{16}$$

If we want at this time to find a lower estimate of the radius of convergence of the power series (16), we must compute the radii of convergence of the power-series expansions of the functions $a_1(x)/a_2(x)$ and $a_0(x)/a_2(x)$. Observe that

$$\frac{a_1(x)}{a_2(x)} = -\frac{1}{1-x} = -\sum_{n=0}^{\infty} x^n, \qquad |x| < 1$$

and

$$\frac{a_0(x)}{a_2(x)} = \frac{x}{1-x} = x\sum_{n=0}^{\infty} x^n = \sum_{n=0}^{\infty} x^{n+1}, \qquad |x| < 1.$$

Thus, series (16) converges at least for $|x| < 1$.

By direct substitution of (16) into (13) and equating coefficients, the reader can verify that

$$2a_2 - a_1 = 0$$

and

$$a_{n+1} = \frac{n^2 a_n - a_{n-2}}{(n+1)n} \qquad n = 2, 3, \ldots .$$

From the initial condition (14) we obtain $a_0 = 1$, and from (15) we find that $a_1 = 1$.

$$\Rightarrow \qquad a_2 = \frac{1}{2!}, a_3 = \frac{1}{3!}, \ldots, a_n = \frac{1}{n!}, \ldots .$$

Thus, the solution of the IVP (13)–(15) is

$$y(x) = \sum_{n=0}^{\infty} a_n x^n = \sum_{n=0}^{\infty} \frac{1}{n!} x^n = e^x. \tag{17}$$

By Theorem 1 the radius of convergence of the power-series solution (17) is at least equal to 1. However, it can be larger. In fact, the radius of convergence of the solution (17) is equal to ∞.

REMARK 2 In Examples 1 and 2 we were able to compute all the coefficients a_n of the power-series solution. However, this is a luxury that is not always possible. Of course, we always have a recurrence formula that we can use to compute as many coefficients of the power-series solution as we please. In general, we compute enough coefficients a_n of the power-series solution to obtain a "good approximation" to the solution.

EXAMPLE 3 Compute the first five coefficients of the power-series solution $y(x) = \sum_{n=0}^{\infty} a_n x^n$ of the IVP

$$y'' - 2x^2 y' + 8y = 0 \tag{18}$$

$$y(0) = 0 \tag{19}$$

$$y'(0) = 1. \tag{20}$$

Solution We have

$$y'(x) = \sum_{n=0}^{\infty} n a_n x^{n-1} = \sum_{n=1}^{\infty} n a_n x^{n-1}$$

and

$$y''(x) = \sum_{n=0}^{\infty} n(n-1) a_n x^{n-2} = \sum_{n=2}^{\infty} n(n-1) a_n x^{n-2}.$$

Thus,

$$y'' = \sum_{n=2}^{\infty} n(n-1) a_n x^{n-2} = \sum_{n=0}^{\infty} (n+2)(n+1) a_{n+2} x^n \tag{21}$$

$$-2x^2 y' = \sum_{n=1}^{\infty} -2n a_n x^{n+1} = \sum_{n=2}^{\infty} -2(n-1) a_{n-1} x^n \tag{22}$$

$$8y = \sum_{n=0}^{\infty} 8a_n x^n \tag{23}$$

$$0 = (2a_2 + 8a_0) + (6a_3 + 8a_1)x + \sum_{n=2}^{\infty} [(n+2)(n+1)a_{n+2} - 2(n-1)a_{n-1} + 8a_n]x^n,$$

where the terms $(2a_2 + 8a_0)$ and $(6a_3 + 8a_1)x$ are the contributions of the series (21) and (23) for $n = 0$ and $n = 1$. Hence, $a_2 = -4a_0$, $a_3 = -\frac{4}{3}a_1$, and

$$(n+2)(n+1)a_{n+2} - 2(n-1)a_{n-1} + 8a_n = 0 \quad \text{for} \quad n = 2, 3, \ldots. \tag{24}$$

From the initial conditions we obtain $a_0 = 0$ and $a_1 = 1$. Then $a_2 = 0$ and $a_3 = -\frac{4}{3}$. Finally, from the recurrence formula (24) for $n = 2$, we find $12a_4 - 2a_1 + 8a_2 = 0$ and so $a_4 = \frac{1}{6}$. Thus,

$$\begin{aligned} y(x) &= a_0 + a_1x + a_2x^2 + a_3x^3 + a_4x^4 + \cdots \\ &= x - \tfrac{4}{3}x^3 + \tfrac{1}{6}x^4 + \cdots. \end{aligned}$$

In this example, using the recurrence formula, we can compute as many coefficients as time permits. However, there is no known closed-form expression for the general coefficients a_n for all n.

APPLICATIONS 5.4.1

The method of power-series solutions about an ordinary point provides an effective tool for obtaining the solutions of some differential equations that occur in applications.

Legendre's Equation

■ The differential equation

$$(1 - x^2)y'' - 2xy' + p(p+1)y = 0, \tag{25}$$

where p is a constant, is *Legendre's equation*. The solutions of Eq. (25) are very important in many branches of applied mathematics. For example, Legendre's equation appears in the study of the potential equation in spherical coordinates. In fact, the potential equation

$$\frac{\partial^2 V}{\partial x^2} + \frac{\partial^2 V}{\partial y^2} + \frac{\partial^2 V}{\partial z^2} = 0,$$

transformed to spherical polar coordinates

$$x = r \sin \theta \cos \phi, \qquad y = r \sin \theta \sin \phi, \qquad z = r \cos \theta,$$

becomes

$$\frac{\partial^2 V}{\partial r^2} + \frac{2}{r}\frac{\partial V}{\partial r} + \frac{1}{r^2}\frac{\partial^2 V}{\partial \theta^2} + \frac{\cot \theta}{r^2}\frac{\partial V}{\partial \theta} + \frac{1}{r^2 \sin^2 \theta}\frac{\partial^2 V}{\partial \phi^2} = 0.$$

If we are interested in a solution that is independent of ϕ of the form $V = r^p\Theta$, where Θ is a function of θ only, we find

$$\frac{d^2\Theta}{d\theta^2} + \cot \theta \frac{d\Theta}{d\theta} + p(p + 1)\Theta = 0.$$

Using the change of variables $x = \cos \theta$ and replacing Θ by y, we obtain Legendre's equation (25).

When p is a nonnegative integer, one of the solutions of Eq. (25) about the ordinary point $x_0 = 0$ is a polynomial. When suitably normalized (as we shall explain below), these polynomial solutions are called *Legendre's polynomials.* Legendre's polynomials are widely used in applications. They appear, for example, in quantum mechanics in the study of the hydrogen atom.

Now we proceed to obtain two linearly independent solutions of Legendre's equation about the point $x_0 = 0$. Here $a_2(x) = 1 - x^2$, $a_1(x) = -2x$, and $a_0(x) = p(p + 1)$. Since $a_2(0) = 1 \neq 0$, the point $x_0 = 0$ is an ordinary point for the differential equation (25). The form of any solution of the differential equation (25) about the point $x_0 = 0$ is

$$y(x) = \sum_{n=0}^{\infty} a_n x^n. \tag{26}$$

To find a lower bound for the radius of convergence of the solution (26), we need to compute the radii of convergence of the Taylor-series expansions about zero of the functions $a_1(x)/a_2(x)$ and $a_0(x)/a_2(x)$. We have

$$\begin{aligned}\frac{a_1(x)}{a_2(x)} &= -\frac{2x}{1 - x^2} = -2x(1 + x^2 + x^4 + \cdots) \\ &= \sum_{n=0}^{\infty} -2x^{2n+1}, \qquad x < 1\end{aligned}$$

and

$$\begin{aligned}\frac{a_0(x)}{a_2(x)} &= \frac{p(p + 1)}{1 - x^2} = p(p + 1)(1 + x^2 + x^4 + \cdots) \\ &= \sum_{n=0}^{\infty} p(p + 1)x^{2n}, \qquad x < 1.\end{aligned}$$

Hence, the radius of convergence of the solution (26) is at least equal to 1; that is, the series (26) converges at least for $|x| < 1$. From Eq. (26) we have

$$y'(x) = \sum_{n=1}^{\infty} na_n x^{n-1} \quad \text{and} \quad y''(x) = \sum_{n=2}^{\infty} n(n-1)a_n x^{n-2}$$

$$\Rightarrow \quad y'' = \sum_{n=2}^{\infty} n(n-1)a_n x^{n-2} = \sum_{n=0}^{\infty} (n+2)(n+1)a_{n+2}x^n$$

$$-x^2y'' = \sum_{n=2}^{\infty} -n(n-1)a_n x^n$$

$$-2xy' = \sum_{n=1}^{\infty} -2na_n x^n$$

$$p(p+1)y = \sum_{n=0}^{\infty} p(p+1)a_n x^n$$

$$0 = [2a_2 + p(p+1)a_0] + [6a_3 - 2a_1 + p(p+1)a_1]x + \sum_{n=2}^{\infty} [(n+2)(n+1)a_{n+2} - n(n-1)a_n - 2na_n + p(p+1)a_n]x^n$$

$$\Rightarrow \quad 2a_2 + p(p+1)a_0 = 0, \qquad 6a_3 - 2a_1 + p(p+1)a_1 = 0$$

and

$$(n+2)(n+1)a_{n+2} - n(n-1)a_n - 2na_n + p(p+1)a_n = 0, \qquad n = 2, 3, \ldots$$

or

$$a_2 = -\frac{p(p+1)}{2}a_0, \qquad a_3 = \frac{2 - p(p+1)}{6}a_1 = -\frac{(p-1)(p+2)}{3!}a_1$$

and

$$a_{n+2} = \frac{n(n-1) + 2n - p(p+1)}{(n+2)(n+1)}a_n = -\frac{(p-n)(p+n+1)}{(n+2)(n+1)}a_n, \qquad n = 2, 3, \ldots \tag{27}$$

$$\Rightarrow \quad a_4 = -\frac{(p-2)(p+3)}{4 \cdot 3}a_2 = \frac{p(p-2)(p+1)(p+3)}{4!}a_0,$$

and, in general,

$$a_{2n} = (-1)^n \frac{p(p-2)\cdots(p-2n+2)(p+1)(p+3)\cdots(p+2n-1)}{(2n)!}a_0, \qquad n = 1, 2, \ldots.$$

Also,

$$a_5 = -\frac{(p-3)(p+4)}{5 \cdot 4}a_3 = \frac{(p-1)(p-3)(p+2)(p+4)}{5!}a_1$$

and, in general,

$$a_{2n+1} = (-1)^n \frac{(p-1)(p-3)\cdots(p-2n+1)(p+2)(p+4)\cdots(p+2n)}{(2n+1)!} a_1,$$

$$n = 1, 2, \ldots . \qquad (28)$$

Hence, two linearly independent solutions of Legendre's equation about the point 0 are

$$y_1(x) = 1 + \sum_{n=0}^{\infty} (-1)^n \cdot \frac{p(p-2)\cdots(p-2n+2)(p+1)(p+3)\cdots(p+2n-1)}{(2n)!} x^{2n} \qquad (29)$$

and

$$y_2(x) = x + \sum_{n=1}^{\infty} (-1)^n \cdot \frac{(p-1)(p-3)\cdots(p-2n+1)(p+2)(p+4)\cdots(p+2n)}{(2n+1)!} x^{2n+1} \qquad (30)$$

and they converge for $x < 1$.

As we see from Eqs. (27) and (28), when p is equal to a nonnegative integer n, one of the solutions above is a polynomial of degree n. A multiple of this polynomial solution that takes the value 1 at $x = 1$ is called a *Legendre polynomial* and is denoted by $P_n(x)$. For example,

$$P_0(x) = 1, \qquad P_1(x) = x, \qquad P_2(x) = \tfrac{3}{2}x^2 - \tfrac{1}{2}, \qquad P_3(x) = \tfrac{5}{2}x^3 - \tfrac{3}{2}x$$

are Legendre polynomials.

Airy's Equation

■ The differential equation

$$y'' - xy = 0 \qquad (31)$$

is *Airy's equation.* The solutions of Airy's equation about the ordinary point $x_0 = 0$ are called *Airy functions* and have applications in the theory of diffraction. Airy functions were originally studied by Airy in connection with his calculations of light intensity in the neighborhood of a caustic surface.

From Theorem 1 it follows that any solution of the differential equation (31) about the point $x_0 = 0$ is of the form

$$y(x) = \sum_{n=0}^{\infty} a_n x^n \qquad (32)$$

and converges for all x. Substituting (32) into the differential equation (31) and equating coefficients, the reader can verify that $a_2 = 0$ and for $n = 1, 2, \ldots,$

$$a_{3n} = \frac{1 \cdot 4 \cdots (3n-2)}{(3n)!} a_0, \quad a_{3n+1} = \frac{2 \cdot 5 \cdots (3n-1)}{(3n+1)!} a_1, \quad a_{3n+2} = 0. \qquad (33)$$

Hence, the general solution of Airy's equation is

$$y(x) = \sum_{n=0}^{\infty} a_n x^n = \sum_{n=0}^{\infty} a_{3n} x^{3n} + \sum_{n=0}^{\infty} a_{3n+1} x^{3n+1} + \sum_{n=0}^{\infty} a_{3n+2} x^{3n+2}$$

$$= a_0 y_1(x) + a_1 y_2(x),$$

where

$$y_1(x) = 1 + \sum_{n=1}^{\infty} \frac{1 \cdot 4 \cdots (3n-2)}{(3n)!} x^{3n} \text{ and } y_2(x) = x + \sum_{n=1}^{\infty} \frac{2 \cdot 5 \cdots (3n-1)}{(3n+1)!} x^{3n+1}$$

are two linearly independent solutions of (31).

Chebyshev's Equation

■ The differential equation

$$(1 - x^2)y'' - xy' + p^2 y = 0, \tag{34}$$

where p is a constant, is *Chebyshev's equation* (the spelling Tschebyscheff is also used). As we will see, if the constant p is a nonnegative integer, Eq. (34) has a polynomial solution about the point $x_0 = 0$. When suitably normalized (when the leading coefficient is chosen as we will explain), these polynomial solutions are called *Chebyshev's polynomials.* The Chebyshev polynomials are very useful in numerical analysis. Chebyshev obtained the polynomials that bear his name in 1857 while seeking the polynomial of degree n and leading coefficient 1 that deviates least from zero on the interval $-1 \leq x \leq 1$.

From Theorem 1 it follows that any solution of the differential equation (34) about the point $x_0 = 0$ is of the form

$$y(x) = \sum_{n=0}^{\infty} a_n x^n \tag{35}$$

and converges for $x < 1$. By direct substitution of (35) into (34) and equating of coefficients, the reader can verify that for $n = 1, 2, \ldots,$

$$a_{2n} = (-1)^n \frac{p^2(p^2 - 2^2)(p^2 - 4^2) \cdots [p^2 - (2n-2)^2]}{(2n)!} a_0, \tag{36}$$

and

$$a_{2n+1} = (-1)^n \frac{(p^2 - 1^2)(p^2 - 3^2) \cdots [p^2 - (2n - (2n-1)^2]}{(2n+1)!} a_1. \tag{37}$$

Thus,

$$y(x) = \sum_{n=0}^{\infty} a_n x^n = \sum_{n=0}^{\infty} a_{2n} x^{2n} + \sum_{n=0}^{\infty} a_{2n+1} x^{2n+1}$$

$$= a_0 \left[1 + \sum_{n=1}^{\infty} (-1)^n \frac{p^2(p^2 - 2^2) \cdots [p^2 - (2n-2)^2]}{(2n)!} x^{2n} \right]$$

$$+ a_1 \left[x + \sum_{n=1}^{\infty} (-1)^n \frac{(p^2 - 1^2)(p^2 - 3^2) \cdots [p^2 - (2n-1)^2]}{(2n+1)!} x^{2n+1} \right].$$

Hence, two linearly independent solutions of Chebyshev's equation are

$$y_1(x) = 1 + \sum_{n=1}^{\infty} (-1)^n \frac{p^2(p^2 - 2^2) \cdots [p^2 - (2n-2)^2]}{(2n)!} x^{2n} \tag{38}$$

and

$$y_2(x) = x + \sum_{n=1}^{\infty} (-1)^n \frac{(p^2 - 1^2)(p^2 - 3^2) \cdots [p^2 - (2n-1)^2]}{(2n+1)!} x^{2n+1}. \tag{39}$$

It is clear from Eqs. (36) and (37) that if p is a nonnegative integer, one of the solutions is a polynomial of degree n. If we multiply this polynomial by 2^{n-1}, we obtain a polynomial solution called a *Chebyshev polynomial* and denoted by $T_n(x)$. For example, the polynomials 1, x, $2x^2 - 1$, and $4x^3 - 3x$ are the Chebyshev polynomials $T_0(x)$, $T_1(x)$, $T_2(x)$, and $T_3(x)$, respectively.

Hermite's Equation

■ The differential equation

$$y'' - 2xy' + 2py = 0, \tag{40}$$

where p is a constant, is *Hermite's equation.* As we shall see, if the constant p is a nonnegative integer, Eq. (40) has a polynomial solution about the point $x_0 = 0$. When suitably normalized, these polynomial solutions are called *Hermite's polynomials.* The Hermite polynomials are very important in quantum mechanics in the investigation of acceptable solutions of the Schrödinger equation for a harmonic oscillator. Hermite's polynomials are also useful in probability and statistics in obtaining the Gram–Charlier series expansions, that is, expansions in terms of Hermite's polynomials.

Clearly, the point $x_0 = 0$ is an ordinary point of the differential equation (40) and any solution is of the form

$$y(x) = \sum_{n=0}^{\infty} a_n x^n \tag{41}$$

and converges for all x. By direct substitution of (41) into (40) and equating coefficients, the reader can verify that for $n = 1, 2, \ldots,$

$$a_{2n} = (-1)^n \frac{2^n p(p-2) \cdots (p-2n+2)}{(2n)!} a_0 \tag{42}$$

and

$$a_{2n+1} = (-1)^n \frac{2^n (p-1)(p-3) \cdots (p-2n+1)}{(2n+1)!} a_1. \tag{43}$$

The general solution of the differential equation (40) is

$$y(x) = \sum_{n=0}^{\infty} a_n x^n = \sum_{n=0}^{\infty} a_{2n} x^{2n} + \sum_{n=0}^{\infty} a_{2n+1} x^{2n+1}$$

or

$$y(x) = a_0 \left[1 + \sum_{n=1}^{\infty} (-1)^n \frac{2^n p(p-2) \cdots (p-2n+2)}{(2n)!} x^{2n} \right]$$

$$+ a_1 \left[x + \sum_{n=1}^{\infty} (-1)^n \frac{2^n (p-1)(p-3) \cdots (p-2n+1)}{(2n+1)!} x^{2n+1} \right].$$

Hence, two linearly independent solutions of Hermite's differential equation are

$$y_1(x) = 1 + \sum_{n=1}^{\infty} (-1)^n \frac{2^n p(p-2) \cdots (p-2n+2)}{(2n)!} x^{2n} \tag{44}$$

and

$$y_2(x) = x + \sum_{n=1}^{\infty} (-1)^n \frac{2^n (p-1)(p-3) \cdots (p-2n+1)}{(2n+1)!} x^{2n+1}. \tag{45}$$

The series converge for all x.

As we see from Eq. (42), if p is zero or a positive even integer, say $p = 2k$, the coefficients a_{2n} vanish for $n \geq k + 1$, and so the solution (44) is a polynomial of degree $2k$. Similarly, from Eq. (43), we see that if p is a positive odd integer, say $p = 2k + 1$, the coefficients a_{2n+1} vanish for $n \geq k + 1$, and so the solution (45) is a polynomial of degree $2k + 1$.

Thus, when p is a nonnegative integer n, Hermite's differential equation has a polynomial solution of degree n. These polynomials, normalized so that the leading coefficient (the coefficient of x^n) is 2^n, are called the *Hermite polynomials* and are denoted by $H_n(x)$. For example,

$$H_0(x) = 1, \quad H_1(x) = 2x, \quad H_2(x) = 4x^2 - 2, \quad H_3(x) = 8x^3 - 12x, \text{ and so on.}$$

EXERCISES

In Exercises 1 through 10, solve the IVP by the method of power series about the initial point x_0.

1. $y'' - 2xy' + 4y = 0$, $y(0) = 1$, $y'(0) = 0$

2. $(1 - x^2)y'' - 2xy' + 6y = 0$, $y(0) = 1$, $y'(0) = 0$

3. $y'' - 2(x + 2)y' + 4y = 0$, $y(-2) = 1$, $y'(-2) = 0$

4. $(-x^2 + 4x - 3)y'' - 2(x - 2)y' + 6y = 0$, $y(2) = 1$, $y'(2) = 0$

5. $(1 - x^2)y'' - xy' + y = 0$, $y(0) = 0$, $y'(0) = 1$

6. $(1 - x^2)y'' - xy' + 4y = 0$, $y(0) = 1$, $y'(0) = 0$

7. $y'' - 2xy' + 2y = 0$, $y(0) = 0$, $y'(0) = 1$

8. $y'' - 2(x - 1)y' + 2y = 0$, $y(1) = 0$, $y'(1) = 1$

9. $(1 - x^2)y'' - 2xy' + 2y = 0$, $y(0) = 0$, $y'(0) = -1$

10. $(x^2 + 4x + 3)y'' + 2(x + 2)y' - 2y = 0$
$y(-2) = 0$
$y'(-2) = -1$

In Exercises 11 through 19, compute the first four coefficients of the power-series solution about the initial point.

11. $y'' - xy = 0$
$y(0) = 0$
$y'(0) = 1$

12. $y'' - xy = 0$
$y(0) = 1$
$y'(0) = 0$

13. $(x^2 + 2)y'' - 3y' + (x - 1)y = 0$
$y(1) = -20$
$y'(1) = -2$

14. $xy'' - 2(x + 1)y' + 2y = 0$
$y(3) = 2$
$y'(3) = 0$

15. $(x - 1)y'' - xy' + y = 0$
$y(0) = 0$
$y'(0) = 1$

16. $y'' - 2xy' + 4y = 0$
$y(0) = 0$
$y'(0) = 1$

17. $(1 - x^2)y'' - 2xy' + 6y = 0$
$y(0) = 0$
$y'(0) = 1$

18. $(1 - x^2)y'' - xy' + y = 0$
$y(0) = 1$
$y'(0) = 0$

19. $y'' - 2(x - 1)y' + 2y = 0$
$y(1) = 1$
$y'(1) = 0$

In Exercises 20 through 24, prove the statement without finding explicit solutions.

20. The power-series solution $\sum_{n=0}^{\infty} a_n x^n$ of the IVP

$$y'' + (x + 1)y' - y = 0$$
$$y(0) = 1$$
$$y'(0) = -1$$

converges for all x.

21. Any power-series solution of the differential equation

$$y'' + (x + 1)y' - y = 0$$

converges for all x.

22. The power-series solution $\sum_{n=0}^{\infty} a_n(x - 3)^n$ of the IVP

$$xy'' - 2y' + xy = 0$$
$$y(3) = 1$$
$$y'(3) = -2$$

converges for all x in the interval $0 < x < 6$.

23. The radius of convergence of the power-series solution of Exercise 18 is at least equal to 1.

24. The power-series solution of Exercise 15 converges for all x in the interval $-1 < x < 1$.

In Exercises 25 through 27, compute the Legendre polynomial corresponding to Legendre's equation.

25. $(1 - x^2)y'' - 2xy' + 2y = 0.$ **26.** $(1 - x^2)y'' - 2xy' + 6y = 0.$

27. $(1 - x^2)y'' - 2xy' + 12y = 0.$

28. Verify the recurrence formulas (33).

29. Verify the recurrence formulas (36) and (37).

30. Verify the recurrence formulas (42) and (43).

In Exercises 31 through 33, compute the Chebyshev polynomial corresponding to Chebyshev's equation.

31. $(1 - x^2)y'' - xy' + y = 0$ **32.** $(1 - x^2)y'' - xy' + 4y = 0$

33. $(1 - x^2)y'' - xy' + 9y = 0$

In Exercises 34 through 36, compute the Hermite polynomial corresponding to Hermite's equation.

34. $y'' - 2xy' + 2y = 0$ **35.** $y'' - 2xy' + 4y = 0$

36. $y'' - 2xy' + 6y = 0$

37. In an *RLC*-series circuit (see the electric circuit application in Section 2.11.1), assume that $L = 20$ henries, $R = (60 + 20t)$ ohms, $C = 0.05$ farad, and $V(t) = 0$. Compute a recurrence formula that can be used to evaluate approximately the current $I(t)$ in the circuit.

5.5 SERIES SOLUTIONS ABOUT A REGULAR SINGULAR POINT

In this section we show how to solve any second-order linear differential equation with variable coefficients of the form

$$a_2(x)y'' + a_1(x)y' + a_0(x)y = 0 \tag{1}$$

in a small deleted interval about a regular singular point x_0. A deleted interval about x_0 is a set of the form $0 < |x - x_0| < R$ for some positive number R. This set consists of the interval $|x - x_0| < R$, from which we delete its center x_0 (see Figure 5.1).

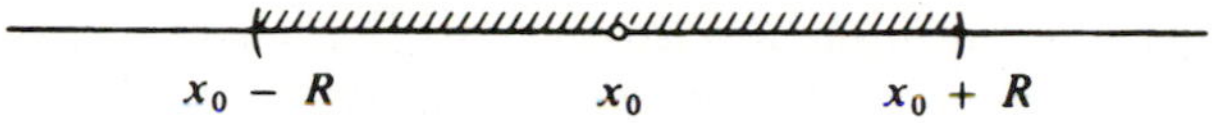

$x_0 - R$ x_0 $x_0 + R$

Figure 5.1

Let us recall that when the point x_0 is a regular singular point of the differential equation (1), then the functions

$$(x - x_0)\frac{a_1(x)}{a_2(x)} \quad \text{and} \quad (x - x_0)^2\frac{a_0(x)}{a_2(x)}$$

have power-series expansions of the form

$$(x - x_0)\frac{a_1(x)}{a_2(x)} = \sum_{n=0}^{\infty} A_n(x - x_0)^n \quad \text{for } |x - x_0| < R_1 \tag{2}$$

and

$$(x - x_0)^2\frac{a_0(x)}{a_2(x)} = \sum_{n=0}^{\infty} B_n(x - x_0)^n \quad \text{for } |x - x_0| < R_2, \tag{3}$$

with positive radii of convergence R_1 and R_2. Since the point x_0 is a singular point of the differential equation (1), its solutions, in general, are not defined at x_0. However, the differential equation (1) has two linearly independent solutions in the deleted interval $0 < |x - x_0| < R$, where R is the smallest of R_1 and R_2. Our problem in this section is to compute (or approximate) these two solutions near every regular singular point. Before we state a theorem that describes the form of the two linearly independent solutions of the differential equation (1) near a regular singular point, we need the following definition.

DEFINITION 1

Assume that x_0 is a regular singular point of the differential equation (1), *and assume that the expansions* (2) *and* (3) *hold. Then the quadratic equation*

$$\lambda^2 + (A_0 - 1)\lambda + B_0 = 0$$

is called the indicial equation of (1) *at x_0.*

THEOREM 1 (Solutions near a Regular Singular Point)

Assume that x_0 is a regular singular point of the differential equation (1) *and assume that the expansions* (2) *and* (3) *hold. Let λ_1 and λ_2 be the two roots of the indicial equation*

$$\lambda^2 + (A_0 - 1)\lambda + B_0 = 0, \tag{4}$$

indexed in such a way that $\lambda_1 \geq \lambda_2$ in case that both roots are real numbers. Then, one of the solutions of Eq. (1) *is of the form*

$$y_1(x) = |x - x_0|^{\lambda_1} \sum_{n=0}^{\infty} a_n(x - x_0)^n, \tag{5}$$

with $a_0 = 1$, *and is valid in the deleted interval* $0 < |x - x_0| < R$, *where* $R = \min(R_1, R_2)$. *A second linearly independent solution* $y_2(x)$ *of Eq.* (1) *in the deleted interval* $0 < |x - x_0| < R$ *is found as follows.*

CASE 1 *If* $\lambda_1 - \lambda_2 \neq$ *integer, then*

$$y_2(x) = |x - x_0|^{\lambda_2} \sum_{n=0}^{\infty} b_n(x - x_0)^n, \tag{6}$$

with $b_0 = 1$.

CASE 2 *If* $\lambda_1 = \lambda_2$, *then*

$$y_2(x) = y_1(x) \ln|x - x_0| + |x - x_0|^{\lambda_2} \sum_{n=0}^{\infty} b_n(x - x_0)^n, \tag{7}$$

with $b_0 = 1$.

CASE 3 *If* $\lambda_1 = \lambda_2 +$ (*positive integer*), *then*

$$y_2(x) = Cy_1(x) \ln|x - x_0| + |x - x_0|^{\lambda_2} \sum_{n=0}^{\infty} b_n(x - x_0)^n, \tag{8}$$

with $b_0 = 1$. *The constant C is sometimes equal to zero.* (*See Exercise 29.*)

As in the case of ordinary points, the coefficients of the series solutions above can be obtained by direct substitution of the solution into the differential equation and equating coefficients. First we compute the solution (5). Series of the form of Eq. (5) are called *Frobenius series,* and the method of finding such solutions of differential equations is customarily called the *method of Frobenius.* A second solution can be computed from (6), (7), or (8), as the case may be. A second solution can also be computed by using the method of reduction of order described in Section 2.8. The following three examples correspond to Cases 1, 2, and 3 of Theorem 1.

EXAMPLE 1 Compute the general solution of the differential equation

$$2x^2y'' + (x - x^2)y' - y = 0 \tag{9}$$

near the point $x_0 = 0$.

Solution Here $a_2(x) = 2x^2$, $a_1(x) = x - x^2$, and $a_0(x) = -1$. Since $a_2(0) = 0$, the point $x_0 = 0$ is a singular point of the differential equation (9). Since

$$(x - x_0)\frac{a_1(x)}{a_2(x)} = x\frac{x - x^2}{2x^2} = \frac{1}{2} - \frac{1}{2}x$$

and

$$(x - x_0)^2\frac{a_0(x)}{a_2(x)} = x^2\frac{-1}{2x^2} = -\frac{1}{2}$$

are analytic functions (with radius of convergence equal to ∞), the point $x_0 = 0$ is a regular singular point of the differential equation (9). Here $A_0 = \frac{1}{2}$ and $B_0 = -\frac{1}{2}$, and therefore the indicial equation of the differential equation (9) at the regular singular point 0 is $\lambda^2 + (\frac{1}{2} - 1)\lambda - \frac{1}{2} = 0$, that is,

$$2\lambda^2 - \lambda - 1 = 0.$$

The roots of the indicial equation are $-\frac{1}{2}$ and 1, and we must index them so that $\lambda_1 \geq \lambda_2$, that is,

$$\lambda_1 = 1 \qquad \text{and} \qquad \lambda_2 = -\tfrac{1}{2}.$$

By Theorem 1, one solution of the differential equation (9) is of the form

$$y_1(x) = x \sum_{n=0}^{\infty} a_n x^n, \tag{10}$$

with $a_0 = 1$. Since $\lambda_1 - \lambda_2 = \frac{3}{2}$, the two roots of the indicial equation do not differ by an integer, and so a second linearly independent solution $y_2(x)$ is of the form (Case 1)

$$y_2(x) = |x|^{-1/2} \sum_{n=0}^{\infty} b_n x^n, \tag{11}$$

with $b_0 = 1$. Since $R_1 = R_2 = \infty$, it follows that the power series in (10) converges for all x. However, the solution (11) is not defined at $x = 0$. It is defined (by Theorem 1) in the deleted interval $0 < |x| < \infty$, that is, for $x < 0$ or $x > 0$. We shall now compute the coefficients of the solutions (10) and (11) by direct substitution into the differential equation (9) and by equating coefficients of the same power of x.

We first compute the coefficients a_n of the solution (10). We have

$$y(x) = x \sum_{n=0}^{\infty} a_n x^n = \sum_{n=0}^{\infty} a_n x^{n+1}$$

$$y'(x) = \sum_{n=0}^{\infty} (n + 1)a_n x^n \qquad \text{and} \qquad y''(x) = \sum_{n=0}^{\infty} n(n + 1)a_n x^{n-1}$$

$$\Rightarrow 2x^2y'' = \sum_{n=0}^{\infty} 2n(n + 1)a_n x^{n+1}$$

$$xy' = \sum_{n=0}^{\infty} (n + 1)a_n x^{n+1}$$

$$-x^2y' = \sum_{n=0}^{\infty} -(n + 1)a_n x^{n+2} = \sum_{n=1}^{\infty} -na_{n-1}x^{n+1}$$

$$-y = \sum_{n=0}^{\infty} -a_n x^{n+1}$$

$$0 = (a_0 - a_0)x + \sum_{n=1}^{\infty} [2n(n + 1)a_n + (n + 1)a_n - na_{n-1} - a_n]x^{n+1}.$$

The term $(a_0 - a_0)x$ corresponds to the sum of terms for $n = 0$. Equating coefficients to zero, we obtain the recurrence formula

$$2n(n+1)a_n + (n+1)a_n - na_{n-1} - a_n = 0, \qquad n = 1, 2, \ldots$$

or

$$a_n = \frac{na_{n-1}}{2n(n+1) + (n+1) - 1} = \frac{1}{2n+3}a_{n-1}, \qquad n = 1, 2, \ldots .$$

We take $a_0 = 1$. Then,

$$\text{for } n = 1: \quad a_1 = \frac{1}{5}$$

$$\text{for } n = 2: \quad a_2 = \frac{1}{7}a_1 = \frac{1}{5 \cdot 7}$$

$$\text{for } n = 3: \quad a_3 = \frac{1}{9}a_2 = \frac{1}{5 \cdot 7 \cdot 9}$$

and, in general,

$$a_n = \frac{1}{5 \cdot 7 \cdots (2n+3)} \qquad \text{for } n = 1, 2, \ldots .$$

Thus, one of the solutions of the differential equation (9) near $x_0 = 0$ is

$$y_1(x) = x\left(1 + \frac{x}{5} + \frac{x^2}{5 \cdot 7} + \cdots + \frac{x^n}{5 \cdot 7 \cdots (2n+3)} + \cdots\right)$$

or

$$y_1(x) = x\left[1 + \sum_{n=1}^{\infty} \frac{x^n}{5 \cdot 7 \cdots (2n+3)}\right]. \tag{12}$$

Next, we compute the coefficients b_n of the solution (11). This solution is defined in the deleted interval $0 < |x|$, that is, for $x > 0$ or $x < 0$. We first assume that $x > 0$. Then

$$y_2(x) = x^{-1/2} \sum_{n=0}^{\infty} b_n x^n = \sum_{n=0}^{\infty} b_n x^{n-(1/2)},$$

$$y_2'(x) = \sum_{n=0}^{\infty} (n - \tfrac{1}{2}) b_n x^{n-(3/2)}$$

and

$$y_2''(x) = \sum_{n=0}^{\infty} (n - \tfrac{1}{2})(n - \tfrac{3}{2})b_n x^{n-(5/2)}.$$

$$\Rightarrow 2x^2y'' = \sum_{n=0}^{\infty} 2(n - \tfrac{1}{2})(n - \tfrac{3}{2})b_n x^{n-(1/2)}$$

$$xy' = \sum_{n=0}^{\infty} (n - \tfrac{1}{2})b_n x^{n-(1/2)}$$

$$-x^2y' = \sum_{n=0}^{\infty} -(n - \tfrac{1}{2})b_n x^{n+(1/2)} = \sum_{n=1}^{\infty} -(n - \tfrac{3}{2})b_{n-1} x^{n-(1/2)}$$

$$-y = \sum_{n=0}^{\infty} -b_n x^{n-(1/2)}$$

$$\begin{aligned} 0 = {} & (2 \cdot \tfrac{3}{4}b_0 - \tfrac{1}{2}b_0 - b_0)x^{-1/2} \\ & + \sum_{n=1}^{\infty} [2(n - \tfrac{1}{2})(n - \tfrac{3}{2})b_n + (n - \tfrac{1}{2})\, b_n \\ & - (n - \tfrac{3}{2})b_{n-1} - b_n]x^{n-(1/2)}. \end{aligned}$$

As before, the first term on the right-hand side is identically zero. Equating each coefficient of the series to zero, we obtain the recurrence formula

$$2(n - \tfrac{1}{2})(n - \tfrac{3}{2})b_n + (n - \tfrac{1}{2})b_n - (n - \tfrac{3}{2})b_{n-1} - b_n = 0, \qquad n = 1, 2, \ldots$$

or

$$b_n = \frac{(n - \tfrac{3}{2})b_{n-1}}{2(n - \tfrac{1}{2})(n - \tfrac{3}{2}) + (n - \tfrac{1}{2}) - 1} = \frac{1}{2n} b_{n-1}, \qquad n = 1, 2, \ldots.$$

We take $b_0 = 1$. Then

$$\text{for } n = 1: \quad b_1 = \frac{1}{2}$$

$$\text{for } n = 2: \quad b_2 = \frac{1}{2 \cdot 2} b_1 = \frac{1}{2^2} \cdot \frac{1}{2}$$

$$\text{for } n = 3: \quad b_3 = \frac{1}{2 \cdot 3} b_2 = \frac{1}{2 \cdot 3} \frac{1}{2^2 \cdot 2} = \frac{1}{2^3 \cdot 3!}$$

and, in general,

$$b_n = \frac{1}{2^n \cdot n!} \qquad n = 1, 2, \ldots.$$

Thus,

$$y_2(x) = x^{-1/2}\left(1 + \frac{x}{2} + \frac{x^2}{2^2 \cdot 2!} + \cdots + \frac{x^n}{2^n \cdot n!} + \cdots\right)$$

$$= x^{-1/2}\sum_{n=0}^{\infty}\frac{x^n}{2^n n!} = x^{-1/2}\sum_{n=0}^{\infty}\frac{(x/2)^n}{n!}.$$

or

$$y_2(x) = x^{-1/2}e^{x/2}. \tag{13}$$

Now we must compute the solution (11) when $x < 0$. To do this, we use the transformation $x = -t$ in the differential equation (9). By the chain rule and using dots for derivatives with respect to t, we obtain

$$y' = \frac{dy}{dx} = \frac{dy}{dt}\cdot\frac{dt}{dx} = -\dot{y}$$

and

$$y'' = \frac{d}{dt}(-\dot{y})\frac{dt}{dx} = \ddot{y}.$$

Thus, Eq. (9) becomes

$$2t^2\ddot{y} - (-t - t^2)\,\dot{y} - y = 0. \tag{9$'$}$$

Since $x < 0$ in Eq. (9), we have $t > 0$ in Eq. (9′). The indicial equation of Eq. (9′) is the same as for Eq. (9). Its roots are again

$$\lambda_1 = 1 \qquad \text{and} \qquad \lambda_2 = -\tfrac{1}{2}.$$

We only have to find the solution corresponding to $\lambda_2 = -\frac{1}{2}$ because for $\lambda_1 = 1$ we have already found a solution. As before, we look for a solution in the form

$$y_2(t) = |\,t\,|^{-1/2}\sum_{n=0}^{\infty} b_n t^n.$$

Because $t > 0$ here, we have

$$y_2(t) = t^{-1/2}\sum_{n=0}^{\infty} b_n t^n.$$

Computing the coefficients by direct substitution of $y_2(t)$ into the differential equation (9′), we obtain (the calculations are omitted)

$$y_2(t) = t^{-1/2}e^{-t/2}, \qquad t > 0.$$

But $t = -x$, and so

$$y_2(x) = (-x)^{-1/2}e^{x/2}, \qquad x < 0. \tag{13$'$}$$

Combining (13) and (13′), we see that for $x > 0$ or $x < 0$, we have

$$y_2(x) = |\,x\,|^{-1/2}e^{x/2}. \tag{14}$$

Thus, the general solution of Eq. (9) is

$$y(x) = c_1 x \left[1 + \sum_{n=1}^{\infty} \frac{x^n}{5 \cdot 7 \cdots (2n+3)}\right] + c_2 |x|^{-1/2} e^{x/2},$$

where c_1 and c_2 are arbitrary constants.

It will be very convenient to use the following remark in the sequel.

REMARK 1 It can be shown that if

$$y(x) = x^{\lambda} \sum_{n=0}^{\infty} a_n x^n \tag{15}$$

is a solution of Eq. (1) for $x > 0$, then

$$y(x) = (-x)^{\lambda} \sum_{n=0}^{\infty} a_n x^n \tag{16}$$

is also a solution for $x < 0$. This was observed in the previous example in Eqs. (13) and (13′). Combining (15) and (16), we obtain for $x > 0$ or $x < 0$,

$$y(x) = |x|^{\lambda} \sum_{n=0}^{\infty} a_n x^n.$$

EXAMPLE 2 Compute the general solution of the differential equation

$$(x-1)^2 y'' - (x^2 - x)y' + y = 0 \tag{17}$$

near the point $x_0 = 1$.

Solution Since computations about the point zero are simpler, we set $t = x - 1$ and find the general solution of the resulting equation near 0. Since $t = x - 1$, we obtain $x = t + 1$, $y' = \dot{y}$, and $y'' = \ddot{y}$. Thus, Eq. (17) becomes

$$t^2 \ddot{y} - (t^2 + t)\dot{y} + y = 0. \tag{18}$$

The point $t_0 = 0$ is a singular point of Eq. (18), and because

$$t\frac{a_1(t)}{a_2(t)} = t\frac{-(t^2+t)}{t^2} = -1 - t$$

and

$$t^2 \frac{a_0(t)}{a_2(t)} = t^2 \frac{1}{t^2} = 1,$$

it follows that $t_0 = 0$ is a regular singular point. Here $A_0 = -1$ and $B_0 = 1$. Therefore, the indicial equation is

$$\lambda^2 - 2\lambda + 1 = 0 \qquad \text{and} \qquad \lambda_1 = \lambda_2 = 1.$$

By Theorem 1, one solution of the differential equation (17) is of the form

$$y_1(t) = t\sum_{n=0}^{\infty} a_n t^n, \tag{19}$$

with $a_0 = 1$. Since $\lambda_1 - \lambda_2 = 0$, the two roots of the indicial equation are equal, and so a second linearly independent solution $y_2(t)$ is of the form (Case 2)

$$y_2(t) = y_1(t)\ln|t| + |t|\sum_{n=0}^{\infty} b_n t^n. \tag{20}$$

We first compute the coefficients a_n of the solution (19). By direct substitution of (19) into (18) and equating coefficients, the reader can verify that

$$a_n = \frac{1}{n!}a_0, \qquad n = 1, 2, \ldots .$$

Thus, taking $a_0 = 1$, we find that

$$y_1(t) = t\left(1 + \frac{t}{1!} + \frac{t^2}{2!} + \cdots + \frac{t^n}{n!} + \cdots\right) = t\sum_{n=0}^{\infty}\frac{t^n}{n!} \tag{21}$$

or

$$y_1(t) = te^t. \tag{22}$$

Next, we compute the coefficients b_n of the solution (20). In Eq. (20) we substitute $y_1(t)$ from Eq. (21), and for $t > 0$, we obtain

$$y_2(t) = t\left(\sum_{n=0}^{\infty}\frac{t^n}{n!}\right)\ln t + t\sum_{n=0}^{\infty} b_n t^n = \left(\sum_{n=0}^{\infty}\frac{t^{n+1}}{n!}\right)\ln t + \sum_{n=0}^{\infty} b_n t^{n+1}$$

$$\dot{y}_2(t) = \left(\sum_{n=0}^{\infty}\frac{n+1}{n!}t^n\right)\ln t + \sum_{n=0}^{\infty}\frac{t^n}{n!} + \sum_{n=0}^{\infty}(n+1)b_n t^n$$

$$\ddot{y}_2(t) = \left(\sum_{n=0}^{\infty}\frac{n(n+1)}{n!}t^{n-1}\right)\ln t + \sum_{n=0}^{\infty}\frac{n+1}{n!}t^{n-1} + \sum_{n=0}^{\infty}\frac{n}{n!}t^{n-1}$$

$$+ \sum_{n=0}^{\infty}(n+1)nb_n t^{n-1}$$

$$\Rightarrow \quad t^2\ddot{y}_2 = \left(\sum_{n=0}^{\infty}\frac{n(n+1)}{n!}t^{n+1}\right)\ln t + \sum_{n=0}^{\infty}\frac{n+1}{n!}t^{n+1} + \sum_{n=0}^{\infty}\frac{n}{n!}t^{n+1}$$

$$+ \sum_{n=0}^{\infty}(n+1)nb_n t^{n+1}$$

$$-t^2\dot{y}_2 = \left(\sum_{n=0}^{\infty} -\frac{n+1}{n!}t^{n+2}\right)\ln t + \sum_{n=0}^{\infty} -\frac{1}{n!}t^{n+1}$$

$$+ \sum_{n=0}^{\infty} -(n+1)b_n t^{n+2}$$

$$-t\dot{y}_2 = \left(\sum_{n=0}^{\infty} -\frac{n+1}{n!}t^{n+1}\right)\ln t + \sum_{n=0}^{\infty} -\frac{1}{n!}t^{n+1}$$

$$+ \sum_{n=0}^{\infty} -(n+1)b_n t^{n+1}$$

$$y_2 = \left(\sum_{n=0}^{\infty} \frac{1}{n!}t^{n+1}\right)\ln t + \sum_{n=0}^{\infty} b_n t^{n+1}.$$

The sum of the coefficients of $\ln t$ is easily seen to be equal to zero. (This is the case in all examples when we are in Case 2.) Also, as always, the sum of all terms corresponding to $n = 0$ is zero. Thus, equating coefficients to zero, we obtain

$$\frac{n+1}{n!} + \frac{n}{n!} - \frac{1}{(n-1)!} - \frac{1}{n!} + (n+1)nb_n - nb_{n-1} - (n+1)b_n + b_n = 0,$$

$$n = 1, 2, \ldots .$$

Simplifying, we obtain

$$\frac{n}{n!} + n^2 b_n - nb_{n-1} = 0$$

or

$$b_n = \frac{1}{n}b_{n-1} - \frac{1}{n \cdot n!}, \qquad n = 1, 2, \ldots .$$

Thus, with $b_0 = 1$, we obtain

$$b_1 = \frac{1}{1} - \frac{1}{1 \cdot 1!} = 0, \qquad b_2 = -\frac{1}{2 \cdot 2!} = -\frac{1}{4},$$

$$b_3 = \frac{1}{3}b_2 - \frac{1}{3 \cdot 3!} = -\frac{5}{36}, \qquad b_4 = \frac{1}{4}b_3 - \frac{1}{4 \cdot 4!} = -\frac{13}{288},$$

and so on. Hence,

$$y_2(t) = \left(t\sum_{n=0}^{\infty} \frac{t^n}{n!}\right)\ln t + t\left(1 - \frac{t^2}{4} - \frac{5t^3}{36} - \frac{13t^4}{288} + \cdots\right)$$

$$= te^t \ln t + t - \frac{t^3}{4} - \frac{5t^4}{36} - \frac{13t^5}{288} - \cdots . \tag{23}$$

Equations (22) and (23) gives two linearly independent solutions of Eq. (18) for $t > 0$. Setting $t = x - 1$ in these equations, we get two linearly independent solutions of Eq. (17) near $x_0 = 1$. The general solution of Eq. (17) is (using Remark 1)

$$y(x) = c_1(x-1)e^{x-1} + c_2\left[(x-1)e^{x-1}\ln|x-1| + (x-1) - \frac{(x-1)^3}{4} - \cdots\right].$$

EXAMPLE 3 Compute two linearly independent solutions of the differential equation

$$x^2y'' - (x + 2)y = 0 \tag{24}$$

near the point $x_0 = 0$.

Solution Here $a_2(x) = x^2$, $a_1(x) = 0$, and $a_0(x) = -(x + 2)$. The point $x_0 = 0$ is a singular point of the differential equation (24), because $a_2(0) = 0$. Since

$$(x - x_0)\frac{a_1(x)}{a_2(x)} = 0$$

and

$$(x - x_0)^2\frac{a_0(x)}{a_2(x)} = x^2\frac{-(x + 2)}{x^2} = -2 - x,$$

the point $x_0 = 0$ is a regular singular point. The indicial equation is (here $A_0 = 0$ and $B_0 = -2$) $\lambda^2 - \lambda - 2 = 0$. Its roots are

$$\lambda_1 = 2 \quad \text{and} \quad \lambda_2 = -1.$$

By Theorem 1, one solution of the differential equation (24) is of the form

$$y_1(x) = x^2 \sum_{n=0}^{\infty} a_n x^n, \tag{25}$$

with $a_0 = 1$. Since $\lambda_1 - \lambda_2 = +3$, the two roots of the indicial equation differ by a positive integer, and so a second linearly independent solution $y_2(x)$ is of the form (Case 3)

$$y_2(x) = Cy_1(x)\ln|x| + x^{-1}\sum_{n=0}^{\infty} b_n x^n, \tag{26}$$

with $b_0 = 1$ and C possibly equal to zero.

We first compute the coefficients a_n of the solution (25). We have

$$y_1(x) = x^2\sum_{n=0}^{\infty} a_n x^n = \sum_{n=0}^{\infty} a_n x^{n+2}$$

$$y_1'(x) = \sum_{n=0}^{\infty}(n + 2)a_n x^{n+1} \quad \text{and} \quad y_1''(x) = \sum_{n=0}^{\infty}(n + 2)(n + 1)a_n x^n$$

$$\Rightarrow x^2y_1'' = \sum_{n=0}^{\infty}(n + 2)(n + 1)a_n x^{n+2}$$

$$-xy_1 = \sum_{n=0}^{\infty} -a_n x^{n+3} = \sum_{n=1}^{\infty} -a_{n-1}x^{n+2}$$

$$-2y_1 = \sum_{n=0}^{\infty} -2a_n x^{n+2}$$

$$0 = (2a_0 - 2a_0)x^2 + \sum_{n=1}^{\infty}[(n + 2)(n + 1)a_n - a_{n-1} - 2a_n]x^{n+2}.$$

Equating coefficients to zero, we obtain the recurrence formula

$$(n+2)(n+1)a_n - a_{n-1} - 2a_n = 0, \qquad n = 1, 2, \ldots$$

or

$$a_n = \frac{a_{n-1}}{(n+2)(n+1) - 2} = \frac{1}{n(n+3)} a_{n-1}, \qquad n = 1, 2, \ldots.$$

We take $a_0 = 1$. Then $a_1 = \frac{1}{4}, a_2 = \frac{1}{40}, \ldots$. Therefore,

$$y_1(x) = x^2 + \frac{x^3}{4} + \frac{x^4}{40} + \cdots.$$

Next, we compute the coefficients b_n of the solution (26). We do it for $x > 0$. Substituting y_1 in (26), we obtain

$$y_2(x) = C\left(x^2 + \frac{x^3}{4} + \frac{x^4}{40} + \cdots\right)\ln x + \sum_{n=0}^{\infty} b_n x^{n-1}$$

$$y_2'(x) = C\left(2x + \frac{3}{4}x^2 + \frac{1}{10}x^3 + \cdots\right)\ln x$$
$$+ C\left(x + \frac{1}{4}x^2 + \frac{1}{40}x^3 + \cdots\right) + \sum_{n=0}^{\infty} (n-1)b_n x^{n-2}$$

$$y_2''(x) = C\left(2 + \frac{3}{2}x + \frac{3}{10}x^2 + \cdots\right)\ln x + C\left(2 + \frac{3}{4}x + \frac{1}{10}x^2 + \cdots\right)$$
$$+ C\left(1 + \frac{1}{2}x + \frac{3}{40}x^2 + \cdots\right) + \sum_{n=0}^{\infty} (n-1)(n-2)b_n x^{n-3}$$

$$\Rightarrow \quad x^2y_2'' = C\left(2x^2 + \frac{3}{2}x^3 + \cdots\right)\ln x + C\left(3x^2 + \frac{5}{4}x^3 + \cdots\right)$$
$$+ \left(2b_0\frac{1}{x} + 2b_3x^2 + \cdots\right)$$

$$-xy_2 = C\left(-x^3 - \frac{1}{4}x^4 - \cdots\right)\ln x + (-b_0 - b_1x - b_2x^2 - \cdots)$$

$$-2y_2 = C\left(-2x^2 - \frac{1}{2}x^3 - \cdots\right)\ln x$$
$$+ \left(-2b_0\frac{1}{x} - 2b_1 - 2b_2x - 2b_3x^2 - \cdots\right)$$

$$0 = C\left(3x^2 + \frac{5}{4}x^3 + \cdots\right)$$
$$+ [(-b_0 - 2b_1) + (-b_1 - 2b_2)x - b_2x^2 - \cdots]$$

or

$$(-b_0 - 2b_1) + (-b_1 - 2b_2)x + (3C - b_2)x^2 + \cdots = 0.$$

Taking $b_0 = 1$ and equating coefficients to zero, we obtain

$$b_1 = -\tfrac{1}{2}, \qquad b_2 = \tfrac{1}{4}, \qquad C = \tfrac{1}{12}, \cdots.$$

Therefore,

$$y_2(x) = \frac{1}{12}\left(x^2 + \frac{x^3}{4} + \frac{x^4}{40} + \cdots\right)\ln|x| + \left(\frac{1}{x} - \frac{1}{2} + \frac{1}{4}x + \cdots\right).$$

APPLICATIONS 5.5.1

The method of Frobenius is a powerful technique for obtaining solutions of certain differential equations which occur in applications.

Bessel's Equation

■ The differential equation

$$x^2y'' + xy' + (x^2 - p^2)y = 0, \tag{27}$$

where p is a constant, is *Bessel's equation* of *order* p. The solutions of Bessel's equation, called *Bessel functions,* are very important in applied mathematics and especially in mathematical physics. Before we study the solutions of Eq. (27) we show briefly how Bessel's equation appears in a specific application. The important method of separation of variables in solving partial differential equations is also mentioned in the process.

Temperature Distribution in a Cylinder

■ If we know the temperature distribution in a cylinder at time $t = 0$, then it is proved in physics that the temperature $u = u(r, \theta, t)$ at the point (r, θ) at any time t satisfies the following partial differential equation (in polar coordinates):

$$u_{rr} + \frac{1}{r}u_r + \frac{1}{r^2}u_{\theta\theta} = \frac{1}{k}u_t. \tag{28}$$

Here we assume that the temperature is independent of the height of the cylinder. The quantity k in Eq. (28) is a constant that depends on the thermal conductivity and, in general, the materials used in making the cylinder. We now employ the method of *separation of variables* to solve the partial differential equation (28). For a discussion of this method see Section 11.4. According to this method, we assume that Eq. (28) has a solution $u(r, \theta, t)$ which is a product of a function of r, a function of θ, and a function of t. That is,

$$u(r, \theta, t) = R(r)\Theta(\theta)T(t), \tag{29}$$

where the functions R, Θ, and T are to be determined. As we will see below, the function R will satisfy Bessel's equation (27). Substituting (29) in (28) and suppressing the arguments r, θ, and t, we obtain

$$R''\Theta T + \frac{1}{r}R'\Theta T + \frac{1}{r^2}R\Theta''T = \frac{1}{k}R\Theta T'. \tag{30}$$

The crucial step in the method of separation of variables is to be able to rewrite Eq. (30) in such a way that the variables separate. In fact, we can write Eq. (30) as follows:

$$r^2\frac{R''}{R} + r\frac{R'}{R} - \frac{r^2}{k}\frac{T'}{T} = -\frac{\Theta''}{\Theta}. \tag{31}$$

The left side of Eq. (31) is a function independent of θ, while the right side is a function of θ alone. The only way that this can happen is if both sides of Eq. (31) are equal to a constant, say p^2. Then we obtain the equations

$$\Theta'' + p^2\Theta = 0 \tag{32}$$

and

$$r^2\frac{R''}{R} + r\frac{R'}{R} - \frac{r^2}{k}\frac{T'}{T} = p^2$$

or

$$\frac{R''}{R} + \frac{1}{r}\frac{R'}{R} - \frac{P^2}{r^2} = \frac{1}{k}\frac{T'}{T} \tag{33}$$

Now the left side of Eq. (33) is a function of r alone, while the right side is a function of t. Hence, both sides of Eq. (33) are equal to a constant, say $-\lambda^2$. Thus,

$$T' + \lambda^2 kT = 0 \tag{34}$$

and

$$r^2R'' + rR' + (\lambda^2r^2 - p^2)R = 0. \tag{35}$$

The functions $\Theta(\theta)$ and $T(t)$ of the solution (29) are easily found by solving the simple differential equations (32) and (34). To find the function $R(r)$ of the solution (29), we must also solve Eq. (35). If we make the transformation $x = \lambda r$ and set $y(x) = R(r)$, we obtain

$$R' = \frac{dR}{dx}\cdot\frac{dx}{dr} = \lambda y' = \frac{x}{r}y' \qquad \text{and} \qquad R'' = \frac{d}{dx}(\lambda y')\frac{dx}{dr} = \lambda^2 y'' = \frac{x^2}{r^2}y''.$$

Substituting R' and R'' in Eq. (35), we find Bessel's equation (27).

We now return to find solutions of Eq. (27) near the point $x_0 = 0$. Although p could be a complex number, we assume for simplicity that $p \geq 0$. We have

$$a_2(x) = x^2, \qquad a_1(x) = x, \qquad a_0(x) = x^2 - p^2.$$

Since $a_2(0) = 0$, the point $x_0 = 0$ is a singular point. But

$$x\frac{a_1(x)}{a_2(x)} = 1 \qquad \text{and} \qquad x^2\frac{a_0(x)}{a_2(x)} = -p^2 + x^2, \tag{36}$$

and so $x_0 = 0$ is a regular singular point with indicial equation $\lambda^2 - p^2 = 0$. The two roots of the indicial equation are

$$\lambda_1 = p \qquad \text{and} \qquad \lambda_2 = -p.$$

Hence, one solution of Bessel's equation is of the form

$$y_1(x) = |x|^p \sum_{n=0}^{\infty} a_n x^n \tag{37}$$

From Eqs. (36) and Theorem 1 the solutions of Eq. (27) near 0 are valid for $|x| > 0$, that is, for $x > 0$ or $x < 0$. The form of a second linearly independent solution $y_2(x)$ of Eq. (27) depends on the value of $\lambda_1 - \lambda_2 = 2p$ in accordance with Cases 1, 2, and 3 of Theorem 1. More precisely, $y_2(x)$ is given as follows.

CASE 1 If $2p \neq$ integer, then

$$y_2(x) = |x|^{-p} \sum_{n=0}^{\infty} b_n x^n \tag{38}$$

In fact, this solution can be obtained from (37), replacing p by $-p$.

CASE 2 If $p = 0$, then

$$y_2(x) = y_1(x) \ln |x| + \sum_{n=0}^{\infty} b_n x^n. \tag{39}$$

CASE 3 If $2p =$ positive integer, then

$$y_2(x) = Cy_1(x) \ln |x| + |x|^{-p} \sum_{n=0}^{\infty} b_n x^n. \tag{40}$$

Let us compute the coefficients a_n of the solution (37). We have (for $x > 0$)

$$y_1(x) = \sum_{n=0}^{\infty} a_n x^{n+p}, \quad y_1'(x) = \sum_{n=0}^{\infty} (n + p)a_n x^{n+p-1},$$

and

$$y_1''(x) = \sum_{n=0}^{\infty} (n + p)(n + p - 1)a_n x^{n+p-2}$$

$$\Rightarrow x^2 y_1'' = \sum_{n=0}^{\infty} (n + p)(n + p - 1)a_n x^{n+p}$$

$$xy_1' = \sum_{n=0}^{\infty} (n + p)a_n x^{n+p}$$

$$x^2 y_1 = \sum_{n=0}^{\infty} a_n x^{n+p+2} = \sum_{n=2}^{\infty} a_{n-2} x^{n+p}$$

$$-p^2 y_1 = \sum_{n=0}^{\infty} - p^2 a_n x^{n+p}$$

$$0 = [p(p-1)a_0 + pa_0 - p^2a_0]x^p + [(1+p)pa_1 + (1+p)a_1 - p^2a_1]x^{1+p}$$
$$+ \sum_{n=2}^{\infty} [(n+p)(n+p-1)a_n + (n+p)a_n + a_{n-2} - p^2a_n)]x^{n+p}$$

$$\Rightarrow (1+2p)a_1 = 0 \quad \text{and}$$
$$(n+p)(n+p-1)a_n + (n+p)a_n + a_{n-2} - p^2a_n = 0, \qquad n = 2, 3, \ldots$$

$$\Rightarrow a_1 = 0 \quad \text{and} \quad a_n = -\frac{1}{n(n+2p)}a_{n-2}, \qquad n = 2, 3, \ldots$$

$$\Rightarrow a_1 = a_3 = a_5 = \cdots = 0 \quad \text{and}$$
$$a_{2n} = (-1)^n \frac{1}{2^{2n}n!(p+1)(p+2)\cdots(p+n)}a_0, \qquad n = 1, 2, \ldots.$$

Hence, the solution (37) is given by

$$y_1(x) = a_0 |x|^p \left[1 + \sum_{n=1}^{\infty} \frac{(-1)^n}{2^{2n}n!(p+1)(p+2)\cdots(p+n)}x^{2n}\right], \tag{41}$$

and the series converges for all x. When $2p \neq$ integer, a second linearly independent solution of Eq. (27) can be found if we replace p by $-p$ in Eq. (41). Thus,

$$y_2(x) = a_0 |x|^{-p} \left[1 + \sum_{n=1}^{\infty} \frac{(-1)^n}{2^{2n}n!(-p+1)(-p+2)\cdots(-p+n)}x^{2n}\right], \tag{42}$$

and the series converges for $x > 0$ or $x < 0$.

In the theory of Bessel functions the constant a_0 in the solution (41) is taken equal to

$$a_0 = \frac{1}{2^p\Gamma(p+1)}, \tag{43}$$

where Γ is the *gamma function*, that is the function defined by the integral

$$\Gamma(p) = \int_0^{\infty} x^{p-1}e^{-x}dx. \tag{44}$$

With this choice of constant a_0, the solution (41) is called a *Bessel function of the first kind of order p* and is denoted by $J_p(x)$. If we use the identity

$$\Gamma(n+p+1) = (n+p)(n+p-1)\cdots(p+2)(p+1)\Gamma(p+1), \tag{45}$$

we obtain the formula

$$J_p(x) = \sum_{n=0}^{\infty} \frac{(-1)^n}{n!\Gamma(n+p+1)}\left(\frac{x}{2}\right)^{2n+p}. \tag{46}$$

The solution (42) with $a_0 = 1/2^{-p}\Gamma(-p+1)$ is also a Bessel function of the first kind but of order $-p$ and is denoted by $J_{-p}(x)$. Clearly,

$$J_{-p}(x) = \sum_{n=0}^{\infty} \frac{(-1)^n}{n!\Gamma(n-p+1)}\left(\frac{x}{2}\right)^{2n-p}. \tag{47}$$

Thus, $J_p(x)$ is always a solution of Bessel's equation (27). Furthermore, when $2p$ is not an integer, the functions $J_p(x)$ and $J_{-p}(x)$ are linearly independent solutions of Bessel's equation.

Setting $p = 0$ in Eq. (46) and observing that $\Gamma(n+1) = n!$, we obtain

$$J_0(x) = \sum_{n=0}^{\infty} \frac{(-1)^n}{(n!)^2}\left(\frac{x}{2}\right)^{2n}. \tag{48}$$

The function $J_0(x)$ is a solution of Bessel's equation (27) for $p = 0$.

When $p = 0$ or when $2p$ is a positive integer, a second linearly independent solution of Bessel's equation is of the form (39) or (40), respectively, and can be obtained by direct substitution of the appropriate series into the differential equation. Such solutions of Eq. (27) with an appropriate choice of the constant b_0 are known as *Bessel functions of the second kind.*

Laguerre's Equation

■ The differential equation

$$xy'' + (1-x)y' + py = 0, \tag{49}$$

where p is a constant, is *Laguerre's equation.* As we will see below, if the constant p is a nonnegative integer, one of the solutions of the differential equation (49), near the point $x_0 = 0$, is a polynomial. When suitably normalized, these polynomial solutions are called *Laguerre polynomials.* The Laguerre polynomials are useful in the quantum mechanics of the hydrogen atom.

The point $x_0 = 0$ is a regular singular point of the differential equation (49) with indicial equation $\lambda^2 = 0$. Since $\lambda_1 = \lambda_2 = 0$, the differential equation has a solution of the form

$$y_1(x) = \sum_{n=0}^{\infty} a_n x^n \qquad \text{for } -\infty < x < \infty.$$

Choosing $a_0 = 1$, the reader can verify that

$$a_n = (-1)^n \frac{p(p-1)\cdots(p-n+1)}{(n!)^2} \qquad \text{for } n = 1, 2, \ldots. \tag{50}$$

Thus, one solution of Laguerre's equation is

$$y_1(x) = 1 + \sum_{n=1}^{\infty} (-1)^n \frac{p(p-1)\cdots(p-n+1)}{(n!)^2} x^n \tag{51}$$

and converges for all x.

From Eq. (50) we see that if p is a nonnegative integer k, the coefficients a_n vanish for $n \geq k+1$. In this case the solution (51) is a polynomial of degree k. This polynomial multiplied by $k!$ is denoted by $L_k(x)$ and is called a *Laguerre polynomial.* For example, 1, $1-x$, $2-4x+x^2$, and $6-18x+9x^2-x^3$ are the Laguerre polynomials $L_0(x)$, $L_1(x)$, $L_2(x)$, and $L_3(x)$, respectively.

Gauss's Hypergeometric Equation

■ The differential equation

$$x(1-x)y'' + [c-(a+b+1)x]y' - aby = 0, \tag{52}$$

where a, b, and c are constants, is *Gauss's hypergeometric equation* or simply the hypergeometric equation. Equation (52) is of great theoretical and practical importance because many second-order linear differential equations are reducible to it and because many important special functions are closely related to its solutions.

Let us assume that c is not equal to an integer. The indicial equation of (52) near the regular singular point $x_0 = 0$ is

$$\lambda^2 + (c-1)\lambda = 0.$$

The two roots of the indicial equation are 0 and $1-c$. Since c is not an integer, the two roots do not differ by an integer. Thus, by Theorem 1, Eq. (52) has two linearly independent solutions of the form

$$y_1(x) = \sum_{n=0}^{\infty} a_n x^n \quad \text{and} \quad y_2(x) = |x|^{1-c} \sum_{n=0}^{\infty} b_n x^n.$$

The reader can verify that

$$a_n = \frac{a(a+1)\cdots(a+n-1)b(b+1)\cdots(b+n-1)}{n!c(c+1)\cdots(c+n-1)}a_0,$$

$$n = 1, 2, 3, \ldots. \tag{53}$$

Hence, choosing $a_0 = 1$, we obtain the solution

$$y_1(x) = 1 + \frac{ab}{1\cdot c}x + \frac{a(a+1)b(b+1)}{1\cdot 2c(c+1)}x^2 + \cdots, \quad |x|<1. \tag{54}$$

The series solution (54) is called the *hypergeometric series* and is denoted by $F(a, b, c; x)$. Thus,

$$F(a, b, c; x) = 1 + \sum_{n=1}^{\infty} \frac{a(a+1)\cdots(a+n-1)b(b+1)\cdots(b+n-1)}{c(c+1)\cdots(c+n-1)}\frac{x^n}{n!} \tag{55}$$

and converges in the interval $|x|<1$. It is important to note that many functions can be obtained from the hypergeometric functions for various values of the constants a, b, and c. For example,

$$F(a, b, c; 0) = 1$$

$$\lim_{a\to\infty} F\left(a, b, b; \frac{x}{a}\right) = e^x$$

$$F(-a, b, b; -x) = (1+x)^a$$

$F(n+1, -n, 1; x)$ is a polynomial of degree n for n a nonnegative integer

$$xF(1, 1, 2; -x) = \ln(1+x).$$

EXERCISES

In Exercises 1 through 11, find the form of two linearly independent solutions near the point $x_0 = 0$.

1. $x^2y'' + xy' + (x^2 - \frac{1}{9})y = 0$

2. $x^2y'' + xy' + (x^2 - 4)y = 0$

3. $x^2y'' + xy' + x^2y = 0$

4. $x(1 - x)y'' + (\frac{1}{2} - 3x)y' - y = 0$

5. $xy'' + (1 - x)y' + y = 0$

6. $x^2y'' + 4x(1 - x)y' + 2y = 0$

7. $x^2y'' + x(1 - x)y' - \frac{1}{16}y = 0$

8. $x^2y'' + 3x(1 - x)y' + y = 0$

9. $xy'' + (1 - x)y' + 2y = 0$

10. $x^2y'' + xy' + (x^2 - 1)y = 0$

11. $xy'' + 3y' + 4x^3y = 0$

In Exercises 12 through 17, compute a nontrivial solution, free of logarithms, near the point $x_0 = 0$.

12. $x^2y'' + xy' + (x^2 - \frac{1}{9})y = 0$

13. $x^2y'' + xy' + (x^2 - 4)y = 0$

14. $x^2y'' + xy' + x^2y = 0$

15. $x(1 - x)y'' + (\frac{1}{2} - 3x)y' - y = 0$

16. $xy'' + (1 - x)y' + y = 0$

17. $xy'' + (1 - x)y' + 2y = 0$

In Exercises 18 and 19, compute two linearly independent solutions near the point $x_0 = 0$.

18. $x^2y'' + xy' + (x^2 - \frac{1}{9})y = 0$

19. $xy'' + (1 - x)y' + y = 0$

In Exercises 20 through 28, answer true or false.

20. The indicial equation of the differential equation

$$x^2y'' + xy' + (x^2 - p^2)y = 0$$

at $x_0 = 0$ is

$$\lambda^2 - p^2 = 0.$$

21. The indicial equation of the differential equation

$$xy'' + (1 - x)y' + 3y = 0$$

at $x_0 = 0$ is

$$\lambda^2 = 0.$$

22. The differential equation

$$xy'' + (1 - x)y' + 3y = 0$$

has two linearly independent solutions near $x_0 = 0$ of the form

$$y_1(x) = \sum_{n=0}^{\infty} a_n x^n \quad \text{and} \quad y_2(x) = \sum_{n=0}^{\infty} b_n x^n.$$

23. One of the solutions of the differential equation

$$x^2y'' + xy' + (x^2 - \tfrac{1}{16})y = 0$$

near the point $x_0 = 0$ involves a logarithm.

24. The indicial equation of the differential equation

$$x^2y'' - (x^2 + x)y' + y = 0$$

at $x_0 = 0$ is

$$\lambda^2 - 2\lambda + 1 = 0.$$

25. The indicial equation of the differential equation

$$x^2y'' - (x^3 + x^2 + x)y' + (4x + 1)y = 0$$

at $x_0 = 0$ is

$$\lambda^2 - 2\lambda + 1 = 0.$$

26. The differential equation

$$x^2y'' + xy' + (x^2 - \tfrac{1}{9})y = 0$$

has two linearly independent solutions near $x = 1$ of the form

$$\sum_{n=0}^{\infty} a_n(x - 1)^n.$$

27. The differential equation

$$x(1 - x)y'' + (\tfrac{2}{3} - 3x)y' - y = 0$$

has two linearly independent solutions near $x_0 = 0$ of the form

$$y_1(x) = \sum_{n=0}^{\infty} a_n x^n, \qquad y_2(x) = y_1(x)\ln|x| + \sum_{n=0}^{\infty} b_n x^n.$$

28. The differential equation

$$(x - 1)y'' + (2 - x)y' + y = 0$$

has two linearly independent solutions near $x = 1$ of the form

$$y_1(x) = \sum_{n=0}^{\infty} a_n(x - 1)^n, \qquad y_2(x) = y_1(x)\ln|x - 1| + \sum_{n=0}^{\infty} b_n(x - 1)^n.$$

29. Show that the point $x_0 = 0$ is a regular singular point for the differential equation

$$xy'' - y' + 4x^3y = 0.$$

In this case, the roots of the indicial equation differ by an integer. Find the two linearly independent solutions and in the process verify that $C = 0$ in formula (8).

In Exercises 30 through 33, obtain a recurrence formula for the coefficients of the Frobenius-series solution [Eq. (5)] near the indicated point x_0.

30. $(x + 1)^2y'' - (x + 3)y = 0$, near $x_0 = -1$

31. $(x - 1)^2y'' - (x + 1)y = 0$, near $x_0 = 1$

32. $x^2y'' - (x^2 + x)y' + y = 0$, near $x_0 = 0$

33. $2(x + 3)^2y'' - (x^2 + 5x + 6)y' - y = 0$, near $x_0 = -3$

34. Verify the recurrence formula (50).

35. Verify the recurrence formula (53).

In Exercises 36 through 38, compute the Bessel functions of the first kind corresponding to Bessel's equation.

36. $x^2y'' + xy' + (x^2 - \frac{1}{9})y = 0$ **37.** $x^2y'' + xy' + x^2y = 0$

38. $x^2y'' + xy' + (x^2 - \frac{1}{4})y = 0$

In Exercises 39 through 41, compute the Laguerre polynomial corresponding to Laguerre's equation.

39. $xy'' + (1 - x)y' + y = 0$ **40.** $xy'' + (1 - x)y' + 2y = 0$

41. $xy'' - (x - 1)y' + 3y = 0$

In Exercises 42 and 43, compute the hypergeometric series corresponding to Gauss's equation.

42. $x(1 - x)y'' + (\frac{1}{2} - 3x)y' - y = 0$

43. $x(1 - x)y'' + (\frac{3}{4} - 4x)y' - 2y = 0$

44. Show that any differential equation of the form

$$(x - A)(x - B)y'' + (Cx + D)y' + Ey = 0,$$

where A, B, C, and D are constants and $A \neq B$ can be transformed into Gauss's equation (52) by means of the transformation $x = A + (B - A)t$.

In Exercises 45 and 46, reduce the given differential equation to Gauss's hypergeometric equation by means of the transformation described in Exercise 44.

45. $(x - 1)(x + 2)y'' + (x + \frac{1}{2})y' + 2y = 0$

46. $(x^2 - \frac{1}{4})y'' + 2y' - 6y = 0$

REVIEW EXERCISES

In Exercises 1 through 4, solve the IVP by the method of power series about the indicated initial point.

1. $y'' - xy = 0$, $y(0) = -1$, $y'(0) = 0$

2. $y'' - 2xy' + 6y = 0$, $y(0) = 0$, $y'(0) = -12$

3. $(1 - x^2)y'' - xy' + 9y = 0$
$y(0) = 0$
$y'(0) = -3$

4. $(1 - x^2)y'' - 2xy' + 12y = 0$
$y(0) = 0$
$y'(0) = -3$

In Exercises 5 through 8, compute the first four coefficients of the power-series solution about the initial point.

5. $x^2y'' + xy' + x^2y = 0$
$y(1) = 1$
$y'(1) = 0$

6. $(1 - x^2)y'' + xy' + y = 0$
$y(6) = 4$
$y'(6) = -1$

7. $(1 - x^2)y'' - xy' + y = 0$
$y(-6) = 4$
$y'(-6) = 1$

8. $xy'' + (1 - x)y' + 2y = 0$
$y(2) = 0$
$y'(2) = 0$

In Exercises 9 through 14, find the form of two linearly independent solutions about or near the indicated point x_0. What can you say about the radius of convergence of the solutions without solving the equation?

9. $y'' - xy = 0; x_0 = 0$

10. $x^2y'' + xy' + (x^2 - \frac{1}{4})y = 0; x_0 = 0$

11. $x^2y'' + xy' + (x^2 - \frac{1}{4})y = 0; x_0 = 1$

12. $(1 - x^2)y'' - xy' + 16y = 0; x_0 = 0$

13. $(1 - x^2)y'' - xy' + 16y = 0; x_0 = -1$

14. $x(3 - x)y'' + (1 - 3x)y' - y = 0; x_0 = 8$

In Exercises 15 through 18, compute the Frobenius-series solution near the point $x_0 = 0$.

15. $x^2y'' + xy' + (x^2 - 9)y = 0$

16. $x^2y'' + xy' + (x^2 - \frac{1}{4})y = 0$

17. $xy' + (1 - x)y' + 3y = 0$

18. $x(1 - x)y'' + (\frac{1}{3} - 2x)y' + 2y = 0$

19. Derive the Taylor series for the function $\cos x$ by solving the IVP

$$y'' + y = 0, \qquad y(0) = 1, \qquad y'(0) = 0$$

by the method of power series.

20. Derive the Taylor series for the function e^{-x} by solving the IVP

$$y'' - y = 0, \qquad y(0) = 1, \qquad y'(0) = -1$$

by the method of power series.

21. Show that the solutions of the Euler differential equation

$$x^2y'' + A_0xy' + B_0y = 0$$

as found in Section 2.7 are as described by Theorem 1, Section 5.5. The above can be used as a motivation for series solutions about a regular singular point.[1]

[1]See also *Amer. Math. Monthly* 67(1960): 278–79.

CHAPTER 6

Boundary Value Problems

6.1 INTRODUCTION AND SOLUTION OF BOUNDARY VALUE PROBLEMS

In this chapter we present a brief introduction to boundary value problems for linear second-order differential equations. There is a large body of theoretical material that has been developed for boundary value problems. Our presentation will bring out some essential features of the basic theory by means of simple yet instructive examples. Consider the second-order linear differential equation

$$a_2(x)y'' + a_1(x)y' + a_0(x)y = f(x), \tag{1}$$

where the coefficients $a_2(x)$, $a_1(x)$, $a_0(x)$ and the function $f(x)$ are continuous in some interval $a \le x \le b$ with $a_2(x) \ne 0$ in this interval. Until now we were mainly concerned with finding a solution $y(x)$ of the differential equation (1) which at some point $x = x_0$ in the interval $a \le x \le b$ satisfied two given initial conditions

$$y(x_0) = y_0 \quad \text{and} \quad y'(x_0) = y_1. \tag{2}$$

The differential equation (1), together with the initial conditions (2), constitutes an initial value problem (IVP). In most IVPs the independent variable x of the differential equation usually represents time, x_0 being the initial time and y_0 and y_1 the initial conditions. However, when the independent variable x represents a space variable (as in the diffusion applications in Section 2.1), we usually want to find a solution $y(x)$ of the differential equation (1) that satisfies a condition at each end point of the interval $a \le x \le b$. For example,

$$y(a) = A \quad \text{and} \quad y(b) = B, \tag{3}$$

where A and B are two constants. The conditions (3) which are given at the end points (or boundary points) of the interval $a \le x \le b$ are called *boundary conditions.* The differential equation (1), together with the boundary conditions (3), constitutes a *boundary value problem* (*BVP*). The form of the boundary conditions at the end points a and b may vary widely, as we shall see in the examples and exercises below.

The procedure for solving a BVP is similar to the procedure used in an IVP. First, we find the general solution of the differential equation and then we use the boundary conditions to determine the arbitrary constants in the general solution. However, the striking difference between an IVP and a BVP is that while the IVP (1) and (2) has exactly one solution (as we know from Theorem 1 of Section 2.3), a BVP may have one, none, or infinitely many solutions. The following examples illustrate this point.

EXAMPLE 1 Solve the BVP

$$y'' + y = x \qquad \text{for} \qquad 0 \le x \le \frac{\pi}{2} \tag{4}$$

$$y(0) = 2, \qquad y\left(\frac{\pi}{2}\right) = 1. \tag{5}$$

Solution First, we compute the general solution of the differential equation (4). The homogeneous solution is $y_h = c_1 \cos x + c_2 \sin x$. Using the method of undetermined coefficients, we find that a particular solution of the differential equation (4) is of the form $y_p = Ax + B$. Substituting y_p into the differential equation (4), we find $Ax + B = x$. Thus, $A = 1$ and $B = 0$ and a particular solution is $y_p = x$. Hence, the general solution of the differential equation (4) is

$$y(x) = c_1 \cos x + c_2 \sin x + x. \tag{6}$$

Using the boundary conditions (5), we find that

$$y(0) = 2 \Rightarrow c_1 = 2$$

and

$$y\left(\frac{\pi}{2}\right) = 1 \Rightarrow c_2 + \frac{\pi}{2} = 1 \Rightarrow c_2 = 1 - \frac{\pi}{2}.$$

Thus, the BVP (4)–(5) has the *unique* solution

$$y(x) = 2 \cos x + \left(1 - \frac{\pi}{2}\right) \sin x + x. \tag{7}$$

EXAMPLE 2 Solve the BVP

$$y'' + y = x \qquad \text{for} \qquad 0 \le x \le \pi \tag{8}$$

$$y(0) = 2, \qquad y(\pi) = 1. \tag{9}$$

Solution The general solution of the differential equation (8) was computed in Example 1 and is

$$y(x) = c_1 \cos x + c_2 \sin x + x.$$

Next, we examine for what values of the constants c_1 and c_2 the boundary conditions (9) are satisfied. We have

$$y(0) = 2 \Rightarrow c_1 = 2$$

and

$$y(\pi) = 1 \Rightarrow -c_1 + \pi = 1 \Rightarrow c_1 = \pi - 1.$$

Clearly, these conditions are inconsistent, since they imply that the constant c_1 has the values 2 and $\pi - 1$ simultaneously. Thus, we conclude in this case that the BVP (8)–(9) has *no* solution.

EXAMPLE 3 Solve the BVP

$$y'' + y = x \quad \text{for} \quad 0 \le x \le \pi \tag{10}$$

$$y(0) = 2, \quad y(\pi) = \pi - 2. \tag{11}$$

Solution The general solution of the differential equation (10) is

$$y(x) = c_1 \cos x + c_2 \sin x + x.$$

Now,

$$y(0) = 2 \Rightarrow c_1 = 2$$

and

$$y(\pi) = \pi - 2 \Rightarrow -c_1 + \pi = \pi - 2 \Rightarrow c_1 = 2.$$

Thus, $c_1 = 2$ while c_2 remains arbitrary. In this case, the BVP (10)–(11) has *infinitely many* solutions:

$$y(x) = 2 \cos x + c_2 \sin x + x. \tag{12}$$

The boundary conditions at the end points a and b need not always be of the type considered in the previous examples. They can contain combinations of y and its derivatives at the points a and b. We give a simple example.

EXAMPLE 4 Solve the BVP

$$y'' + 4y = 0 \quad \text{for} \quad 0 \le x \le \pi \tag{13}$$

$$y(0) - 2y'(0) = -2. \tag{14}$$

$$y(\pi) + 3y'(\pi) = 3. \tag{15}$$

Solution The general solution of the differential equation (13) is

$$y(x) = c_1 \cos 2x + c_2 \sin 2x.$$

We have $y'(x) = -2c_1 \sin 2x + 2c_2 \cos 2x$, and using the boundary conditions, we find that

$$y(0) - 2y'(0) = -2 \Rightarrow c_1 - 4c_2 = -2$$

and

$$y(\pi) + 3y'(\pi) = 3 \Rightarrow c_1 + 6c_2 = 3.$$

Solving this system we obtain $c_1 = 0$ and $c_2 = \frac{1}{2}$. Therefore, the BVP (13)–(15) has the unique solution

$$y(x) = \tfrac{1}{2} \sin 2x.$$

Of course, it is very important to know under what conditions a BVP has a unique solution, no solution, or infinitely many solutions. This problem, in general, is very difficult and beyond our scope. We only develop here a simple but interesting result about the BVP consisting of the differential equation (1) and the simple boundary conditions (3).

Let $y_1(x)$ and $y_2(x)$ be two linearly independent solutions of the homogeneous differential equation corresponding to the differential equation (1), and let $y_p(x)$ be a particular solution of the differential equation (1). Then the general solution of the differential equation (1) is given by

$$y(x) = c_1y_1(x) + c_2y_2(x) + y_p(x). \tag{16}$$

In view of the boundary conditions (3), the constants c_1 and c_2 of the general solution should satisfy the following linear system of algebraic equations:

$$\left.\begin{aligned} c_1y_1(a) + c_2y_2(a) + y_p(a) &= A \\ c_1y_1(b) + c_2y_2(b) + y_p(b) &= B \end{aligned}\right\}$$

or

$$\left.\begin{aligned} c_1y_1(a) + c_2y_2(a) &= A - y_p(a) \\ c_1y_1(b) + c_2y_2(b) &= B - y_p(b) \end{aligned}\right\}. \tag{17}$$

Now it is clear that the BVP (1) and (3) has as many solutions as the system (17). We recall (Appendix A) that a linear system of the form

$$\left.\begin{aligned} a_{11}x + a_{12}\,y &= b_1 \\ a_{21}x + a_{22}\,y &= b_2 \end{aligned}\right\} \tag{18}$$

has exactly one solution if the determinant of the coefficients satisfies

$$\begin{vmatrix} a_{11} & a_{12} \\ a_{21} & a_{22} \end{vmatrix} \neq 0. \tag{19}$$

On the other hand, if the determinant in (19) is equal to zero, the system (18) has no solution when

$$\begin{vmatrix} a_{11} & b_1 \\ a_{21} & b_2 \end{vmatrix} \neq 0 \tag{20}$$

and has infinitely many solutions when the determinant in (20) is equal to zero.

Summarizing the facts above, we have the following.

THEOREM 1

Let $y_1(x)$ and $y_2(x)$ be two linearly independent solutions of the homogeneous differential equation corresponding to the differential equation (1) *and let $y_p(x)$ be a particular solution of the differential equation* (1). *Then, the following statements are valid:*

(a) If

$$\begin{vmatrix} y_1(a) & y_2(a) \\ y_1(b) & y_2(b) \end{vmatrix} \neq 0, \tag{21}$$

the BVP (1) *and* (3) *has one and only one solution in the interval* $a \le x \le b$.

(*b*) *If the determinant in* (21) *is equal to zero, the BVP* (1) *and* (3) *has no solution or infinitely many solutions in the interval* $a \le x \le b$, *depending on whether the determinant*

$$\begin{vmatrix} y_1(a) & A - y_p(a) \\ y_1(b) & B - y_p(b) \end{vmatrix} \tag{22}$$

is $\neq 0$ *or equal to zero, respectively.*

Let us apply Theorem 1 to the BVPs in Examples 1, 2, and 3, where $y_1(x) = \cos x$, $y_2(x) = \sin x$, and $y_p(x) = x$. The BVP (4)–(5) has a unique solution because

$$\begin{vmatrix} y_1(a) & y_2(a) \\ y_1(b) & y_2(b) \end{vmatrix} = \begin{vmatrix} \cos 0 & \sin 0 \\ \cos\frac{\pi}{2} & \sin\frac{\pi}{2} \end{vmatrix} = \begin{vmatrix} 1 & 0 \\ 0 & 1 \end{vmatrix} = 1 \neq 0.$$

The BVP (8)–(9) has no solution, because

$$\begin{vmatrix} y_1(a) & y_2(b) \\ y_1(b) & y_2(b) \end{vmatrix} = \begin{vmatrix} \cos 0 & \sin 0 \\ \cos\pi & \sin\pi \end{vmatrix} = \begin{vmatrix} 1 & 0 \\ -1 & 0 \end{vmatrix} = 0$$

and

$$\begin{vmatrix} y_1(a) & A - y_p(a) \\ y_1(b) & B - y_p(b) \end{vmatrix} = \begin{vmatrix} \cos 0 & 2 - 0 \\ \cos\pi & 1 - \pi \end{vmatrix} = \begin{vmatrix} 1 & 2 \\ -1 & 1 - \pi \end{vmatrix} \neq 0.$$

Finally, the BVP (10)–(11) has infinitely many solutions, because

$$\begin{vmatrix} y_1(a) & y_2(a) \\ y_1(b) & y_2(b) \end{vmatrix} = \begin{vmatrix} \cos 0 & \sin 0 \\ \cos\pi & \sin\pi \end{vmatrix} = \begin{vmatrix} 1 & 0 \\ -1 & 0 \end{vmatrix} = 0$$

and

$$\begin{vmatrix} y_1(a) & A - y_p(a) \\ y_1(b) & B - y_p(b) \end{vmatrix} = \begin{vmatrix} \cos 0 & 2 - 0 \\ \cos\pi & \pi - 2 - \pi \end{vmatrix} = \begin{vmatrix} 1 & 2 \\ -1 & -2 \end{vmatrix} = 0.$$

APPLICATIONS 6.1.1

Diffusion

■ Suppose that a gas diffuses into a liquid in a long, narrow pipe (Figure 6.1). Suppose that this process takes place for such a long period of time that the concentration $y(x)$ of gas in the pipe depends only on the distance x from some

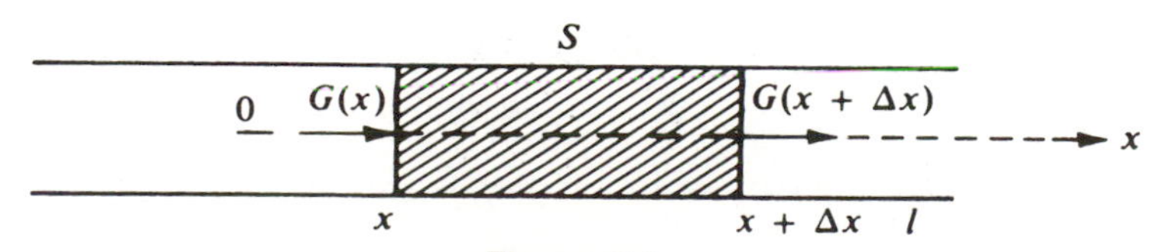

Figure 6.1

initial point 0 (and is independent of time). Then, as we proved in Section 2.1.1, $y(x)$ satisfies the BVP

$$y'' - \frac{k}{D}y = 0 \tag{23}$$

$$y(0) = A, \qquad y(l) = 0. \tag{24}$$

In Exercise 16 the reader is asked to solve the BVP (23)–(24).

Deflection of a Taut String

■ A taut string, lying on the xy plane, is stretched, under a tension T, between two fixed points $x = 0$ and $x = L$ on the x axis. A transverse (perpendicular to the x axis) load $f(x)$ is applied to the string and produces a deflection $u(x)$. Then it can be shown that the deflection $u(x)$ satisfies the differential equation

$$-Tu'' = f(x). \tag{25}$$

Since the ends of the string are kept fixed at the points $x = 0$ and $x = L$ of the x axis, $u(x)$ must also satisfy the boundary conditions

$$u(0) = 0, \qquad u(L) = 0. \tag{26}$$

EXAMPLE 5 Solve the BVP (25)–(26) when $f(x) \equiv f_0$ is constant.

Solution The general solution of the differential equation (25) is

$$u(x) = c_1 + c_2 x - \frac{f_0}{2T}x^2.$$

But

$$u(0) = 0 \Rightarrow c_1 = 0$$

and

$$u(L) = 0 \Rightarrow c_1 + c_2 L - \frac{f_0}{2T}L^2 = 0.$$

Hence, $c_1 = 0$, $c_2 = f_0 L/2T$, and the BVP (25)–(26) has the unique solution

$$u(x) = \frac{f_0 L}{2T}x - \frac{f_0}{2T}x^2 = \frac{f_0}{2T}x(L - x).$$

EXERCISES

In Exercises 1 through 6, answer true or false.

1. $y(x) = \cos 3x$ is a solution of the BVP

$$y'' + 9y = 0$$

$$y(0) = 1, \qquad y(\pi) = -1.$$

2. $y(x) = \cos 3x - \sin 3x$ is a solution of the BVP

$$y'' + 9y = 0$$

$$y(0) = 1, \qquad y(\pi) = -1.$$

3. The BVP

$$y'' + 9y = 0$$
$$y(0) = 1, \qquad y(\pi) = -1$$

has a unique solution.

4. The BVP

$$y'' + 9y = 0$$
$$y(0) = 1, \qquad y(\pi) = B$$

has no solution if $B \neq -1$.

5. $y(x) = \cos 3x - \sin 3x$ is a solution of the BVP

$$y'' + 9y = 0$$
$$y(0) = 1, \qquad y\left(\frac{\pi}{2}\right) = -1.$$

6. The BVP

$$y'' + 9y = 0$$
$$y(0) = 1, \qquad y\left(\frac{\pi}{2}\right) = -1$$

has a unique solution.

In Exercises 7 through 11, solve the BVP.

7. $y'' + 9y = 0, 0 \le x \le \pi$
$y(0) = 1, y(\pi) = -1$

8. $y'' + 9y = 0, 0 \le x \le \pi$
$y(0) = 1, y(\pi) = 2$

9. $y'' + 9y = 0, 0 \le x \le \dfrac{\pi}{2}$
$y(0) = 1, y\left(\dfrac{\pi}{2}\right) = -1$

10. $y'' - 3y' + 2y = e^x, 0 \le x \le 1$
$y'(0) = 0, y(1) = 0$

11. $y'' - 3y' + 2y = 0, 0 \le x \le 1$
$y(0) = 0, y'(1) = 0$

12. Show that the BVP

$$y'' + 16y = 0, \qquad 0 \le x \le \pi$$
$$4y(0) - y'(0) = A, \qquad 4y(\pi) - y'(\pi) = B$$

has infinitely many solutions if $A = B$, and no solution if $A \neq B$.

13. Solve the BVP

$$y'' - y = 2e^x$$
$$y(0) - 2y'(0) = -2, \qquad 3y(1) - y'(1) = e.$$

14. Solve the BVP

$$x^2y'' - 3xy' + 3y = 0, \qquad 1 \le x \le 2$$

$$y(1) - y'(1) = 2, \qquad y(2) - 2y'(2) = 4.$$

15. Show that the BVP

$$x^2y'' - 3xy' + 3y = \ln x, \qquad 1 \le x \le 2$$

$$y(1) = A, \qquad y(2) = B$$

has a unique solution for all values of A and B.

16. Solve the BVP consisting of the differential equation (23) and the boundary conditions (24).

17. A (one-dimensional and steady-state) diffusion process in a medium bounded by the planes $x = 0$ and $x = l$ which are maintained at constant concentrations c_1 and c_2, respectively, is governed by the BVP

$$y''(x) = 0$$

$$y(0) = c_1, \qquad y(l) = c_2$$

(provided that the diffusion coefficient D is constant). Solve this BVP and show that the concentration $y(x)$ changes linearly from c_1 to c_2 through the medium.

18. Solve the BVP (25)–(26) when

$$f(x) = x, \qquad T = \tfrac{1}{2}, \qquad L = 1.$$

6.2 EIGENVALUES AND EIGENFUNCTIONS

From the point of view of applications, especially problems in mathematical physics, there is a very important class of BVPs in which the differential equation is of the form

$$(p(x)y')' + (q(x) + \lambda)y = 0, \tag{1}$$

where $p(x)$ (> 0) and $q(x)$ are continuous in some interval $a \le x \le b$, λ is a real parameter, and the two boundary conditions at the end points a and b are of the form

$$\begin{aligned} \alpha_1 y(a) + \alpha_2 y'(a) &= 0 \\ \beta_1 y(b) + \beta_2 y'(b) &= 0. \end{aligned} \tag{2}$$

The BVP (1)–(2) is always satisfied by the trivial solution $y(x) \equiv 0$. However, the important issue in this type of BVP is to determine values of λ for which the BVP (1)–(2) has a nontrivial solution. These values of λ are called the *eigenvalues* of the BVP (1)–(2). The corresponding nontrivial solutions are called *eigenfunctions*. The BVP (1)–(2) is also called an *eigenvalue problem* (*EVP*).

In this section we shall compute the eigenvalues and eigenfunctions of some simple but representative EVPs and state some of their properties.

EXAMPLE 1 Compute the eigenvalues and eigenfunctions of the EVP

$$y'' + \lambda y = 0 \qquad \text{for} \qquad 0 \le x \le \pi \tag{3}$$

$$y(0) = 0, \qquad y(\pi) = 0. \tag{4}$$

Solution Our problem here is to find all the values of the real parameter λ for which the EVP (3)–(4) has a nontrivial solution and to find the corresponding nontrivial solutions. The characteristic roots of the differential equation (3) are $\pm\sqrt{-\lambda}$. Hence, the form of the solutions of the differential equation (3) depends on whether λ is negative, zero, or positive. So in solving the eigenvalue problem (3)–(4) we have to examine three cases.

CASE 1 $\lambda < 0$ Set $\lambda = -k^2$, where $k > 0$. Then the differential equation (3) becomes $y'' - k^2y = 0$, and its general solution is

$$y(x) = c_1e^{-kx} + c_2e^{kx}. \tag{5}$$

Next, we use the boundary conditions (4) to determine the constants c_1 and c_2. Remember that we want to find solutions that are not identically equal to zero. We have

$$y(0) = 0 \Rightarrow c_1 + c_2 = 0 \tag{6}$$

and

$$y(\pi) = 0 \Rightarrow c_1e^{-k\pi} + c_2e^{k\pi} = 0. \tag{7}$$

Solving the system of simultaneous equations for c_1 and c_2, we find $c_1 = c_2 = 0$. In fact, from Eq. (6) we obtain $c_2 = -c_1$, and inserting this value into Eq. (7), we find that $c_1(e^{-k\pi} - e^{k\pi}) = 0$. Since $k > 0$, $e^{-k\pi} \neq e^{k\pi}$, and so $c_1 = 0$. Then $c_2 = 0$ and the solution (5) is identically zero. This implies that the EVP (3)–(4) does not have negative eigenvalues.

CASE 2 $\lambda = 0$ In this case the differential equation (3) becomes $y'' = 0$, and its general solution is

$$y(x) = c_1 + c_2x. \tag{8}$$

Using the boundary conditions (4), we see that

$$y(0) = 0 \Rightarrow c_1 = 0$$

and

$$y(\pi) = 0 \Rightarrow c_1 + c_2\pi = 0 \Rightarrow c_2\pi = 0 \Rightarrow c_2 = 0.$$

Thus, the solution (8) is identically zero, which implies that $\lambda = 0$ is not an eigenvalue.

CASE 3 $\lambda > 0$ Set $\lambda = k^2$, where $k > 0$. Then the differential equation (3) becomes $y'' + k^2y = 0$, and its general solution is

$$y(x) = c_1 \cos kx + c_2 \sin kx. \tag{9}$$

We now employ the boundary conditions

$$y(0) = 0 \Rightarrow c_1 = 0$$

and

$$y(\pi) = 0 \Rightarrow c_1 \cos k\pi + c_2 \sin k\pi = 0 \Rightarrow c_2 \sin k\pi = 0.$$

Thus, $c_1 = 0$ and $c_2 \sin k\pi = 0$. Recall once more that we try to compute nontrivial solutions of the differential equation (3). In this case, c_1 must be zero; hence, a nontrivial solution is possible only if c_2 can be chosen different from zero. Indeed, $c_2 \sin k\pi = 0$ if $\sin k\pi = 0$ or $k\pi = n\pi$ for $n = 1, 2, \ldots$; that is, $k = n$ for $n = 1, 2, \ldots$. Now $\lambda = k^2$, and therefore the eigenvalues are given by

$$\lambda = n^2 \qquad \text{for} \qquad n = 1, 2, \ldots. \tag{10}$$

Equation (10) gives all the eigenvalues of the EVP (3)–(4). To find the eigenfunctions corresponding to these eigenvalues, we use (9) with $c_1 = 0$, $c_2 \neq 0$, and $k = n$. Then $y = c_2 \sin nx$. Thus, up to a nonzero constant multiple (that is, c_2), the eigenfunction of the EVP (3)–(4) that corresponds to the eigenvalue $\lambda = n^2$ is given by

$$y = \sin nx \qquad \text{for} \quad n = 1, 2, \ldots. \tag{11}$$

Hence, the eigenvalues and eigenfunctions of the EVP (3)–(4) are given by

$$\lambda_n = n^2 \qquad \text{for} \quad n = 1, 2, \ldots \tag{12}$$

and

$$y_n = \sin nx \qquad \text{for} \quad n = 1, 2, \ldots, \tag{13}$$

where the subscript n is attached to indicate the correspondence between eigenvalues and eigenfunctions.

EXAMPLE 2 Determine the eigenvalues and eigenfunctions of the EVP

$$y'' + (-4 + \lambda)y = 0 \qquad \text{for} \quad 0 \le x \le 1 \tag{14}$$

$$y'(0) = 0, \qquad y'(1) = 0. \tag{15}$$

Solution The characteristic roots of the differential equation (14) are $\pm\sqrt{4 - \lambda}$. Hence, the form of the solutions of the differential equation (14) depends on whether the quantity $4 - \lambda$ is negative, zero, or positive.

CASE 1 $4 - \lambda < 0$ Set $4 - \lambda = -k^2$, where $k > 0$. Then the differential equation (14) becomes $y'' + k^2y = 0$, and its general solution is

$$y(x) = c_1 \cos kx + c_2 \sin kx.$$

We have $y'(x) = -c_1k \sin kx + c_2k \cos kx$ and, using the boundary conditions (15), we find that

$$y'(0) = 0 \Rightarrow c_2k = 0 \Rightarrow c_2 = 0$$

and

$$y'(1) = 0 \Rightarrow -c_1k \sin k = 0 \Rightarrow \sin k = 0 \Rightarrow k = n\pi, \qquad \text{for} \quad n = 1, 2, \ldots .$$

Here $\lambda = 4 + k^2$, and therefore

$$\lambda_n = 4 + n^2\pi^2 \qquad \text{for} \quad n = 1, 2, \ldots \tag{16}$$

and

$$y_n = \cos n\pi x \qquad \text{for} \quad n = 1, 2, \ldots \tag{17}$$

are eigenvalues and eigenfunctions of the EVP (14)–(15).

CASE 2 $4 - \lambda = 0$ Then the differential equation (14) becomes $y'' = 0$, and its general solution is

$$y(x) = c_1 + c_2x.$$

Using the boundary conditions (15), we find that

$$y'(0) = 0 \Rightarrow c_2 = 0$$

and

$$y'(1) = 0 \Rightarrow c_2 = 0.$$

Hence, in this case we obtained the nontrivial solution $y(x) = c_1$ or up to a constant multiple $y(x) = 1$. The value $\lambda = 4$ is therefore an eigenvalue of the EVP (14)–(15) with corresponding eigenfunction $y(x) = 1$.

CASE 3 $4 - \lambda > 0$ Set $4 - \lambda = k^2$, where $k > 0$. Then the differential equation (14) becomes $y'' - k^2y = 0$, and its general solution is

$$y(x) = c_1e^{-kx} + c_2e^{kx}.$$

Using the boundary conditions (15), we find that

$$y'(0) = 0 \Rightarrow -c_1k + c_2k = 0 \Rightarrow -c_1 + c_2 = 0$$

and

$$y'(1) = 0 \Rightarrow -c_1ke^{-k} + c_2ke^{k} = 0 \Rightarrow -c_1e^{-k} + c_2e^{k} = 0.$$

Thus, $c_1 = c_2 = 0$, and there is no eigenvalue in this case.

If we observe that Eqs. (16) and (17) for $n = 0$ give us $\lambda_0 = 4$ and $y_0 = 1$, that is, the eigenvalue and eigenfunction found in Case 2, we conclude that the eigenvalues and eigenfunctions of the EVP (14)–(15) are given by the equations

$$\lambda_n = 4 + n^2\pi^2 \quad \text{for} \quad n = 0, 1, 2, \ldots \tag{18}$$

and

$$y_n = \cos n\pi x \quad \text{for} \quad n = 0, 1, 2, \ldots . \tag{19}$$

Finally, we state a theorem that summarizes some important properties of the eigenvalues and eigenfunctions of the EVP (1)–(2). To this end we first define the concepts of inner product and orthogonal functions.

DEFINITION 1

Let f and g be two continuous functions defined in the interval $a \le x \le b$. The inner product of the functions f and g is denoted by (f, g) and is defined by

$$(f, g) = \int_a^b f(x)g(x)dx.$$

When $(f, g) = 0$, we say that the functions f and g are orthogonal.

THEOREM 1

The EVP (1)–(2) possesses an infinite sequence of eigenvalues

$$\lambda_1, \lambda_2, \ldots, \lambda_n, \ldots \rightarrow +\infty \tag{20}$$

and a corresponding sequence of eigenfunctions

$$y_1, y_2, \ldots, y_n, \ldots$$

with the property that in the interval $a \le x \le b$,

$$(y_n, y_m) = \begin{cases} 0, & \text{if} \quad n = m \\ 1, & \text{if} \quad n = m. \end{cases} \tag{21}$$

Furthermore, any function $f(x)$ that is twice continuously differentiable and satisfies the boundary conditions (2) has the following eigenfunction expansion:

$$f(x) = \sum_{n=1}^{\infty} (f, y_n)y_n, \tag{22}$$

where series (22) converges uniformly in the interval $a \le x \le b$.

In Example 1 it is clear that property (20) is satisfied. Let us verify (21). In the EVP (3)–(4) we found that [see Eq. (13)]

$$y_n = \sin nx \quad \text{for} \quad n = 1, 2, \ldots . \tag{13}$$

Computing the inner product (y_n, y_m) we obtain that

$$(y_n, y_m) = \int_0^\pi \sin nx \sin mx\, dx = \begin{cases} 0, & \text{if} \quad n \neq m \\ \dfrac{\pi}{2}, & \text{if} \quad n = m. \end{cases}$$

In order that Eq. (21) be satisfied, we have to multiply each eigenfunction in (13) by $(2/\pi)^{1/2}$, which is acceptable since the eigenfunctions are defined up to a nonzero constant multiple. Thus, for Theorem 1 we have to replace the eigenfunction (13) of the EVP (3)–(4) by

$$y_n = \left(\frac{2}{\pi}\right)^{1/2} \sin nx \qquad \text{for} \quad n = 1, 2, \ldots. \tag{13'}$$

With respect to these eigenfunctions, property (22) indicates that if $f(x)$ is a function that is twice continuously differentiable (in other words, f'' exists and is continuous) and satisfies the boundary conditions (4), that is, $f(0) = f(\pi) = 0$, then

$$f(x) = \frac{2}{\pi} \sum_{n=1}^{\infty} (f(x), \sin nx) \sin nx = \frac{2}{\pi} \sum_{n=1}^{\infty} \left(\int_a^b f(x) \sin nx\, dx \right) \sin nx, \tag{23}$$

and this series converges uniformly in the interval $0 \leq x \leq \pi$. The series (23) is known as the *Fourier sine series*[1] of the function $f(x)$ in the interval $0 \leq x \leq \pi$.

APPLICATION 6.2.1

Eigenvalue problems occur frequently in problems of mathematical physics. In particular, they appear in the process of solving, by the method of separation of variables, initial–boundary value problems which involve one of the following partial differential equations:

$$u_t = \alpha^2 u_{xx} \qquad \text{(heat equation)}, \tag{24}$$

$$u_{tt} - c^2 u_{xx} = 0 \qquad \text{(wave equation)}, \tag{25}$$

and

$$u_{xx} + u_{yy} = 0 \qquad \text{(Laplace's equation)}. \tag{26}$$

Let us demonstrate this in the case of Eq. (24).

The Heat Equation

■ Equation (24) appears in the study of heat conduction and many other diffusion processes (for a derivation of this equation and more discussion of its solution and some applications of it, see Section 11.7). For example, consider an insulated uniform bar of length l. Assume that the bar is oriented so that the x axis coincides with the axis of the bar. Let $u(x, t)$ denote the temperature in

[1]See R. V. Churchill, *Fourier Series and Boundary Value Problems*, 2nd ed. (New York: McGraw-Hill, 1963).

the bar at the point x at time t. Then u satisfies Eq. (24) for $0 < x < l$. The constant α^2 is called the *thermal diffusivity* and depends only on the material of which the bar is made. To determine the flow of heat $u(x, t)$ in the bar, we need to know some initial and boundary conditions. For example, assume that the initial temperature $f(x)$ in the bar is a known function of x, that is,

$$u(x, 0) = f(x), \qquad 0 \leq x \leq l, \tag{27}$$

and that the ends of the bar are kept at zero temperature, that is,

$$u(0, t) = u(l, t) = 0. \tag{28}$$

Using the method of separation of variables (see Section 5.5.1 or Section 11.4), we look for a solution $u(x, t)$ of the initial–boundary value problem (24), (27), and (28) of the form

$$u(x, t) = X(x)T(t). \tag{29}$$

We have

$$u_t = XT' \quad \text{and} \quad u_{xx} = X''T,$$

and Eq. (24) becomes

$$XT' = \alpha^2 X''T$$

or

$$\frac{X''}{X} = \frac{1}{\alpha^2}\frac{T'}{T}. \tag{30}$$

The left-hand side of Eq. (30) is a function of x alone, and its right-hand side is a function of t alone. The only way that this can happen is if both sides of Eq. (30) are equal to some constant λ. That is,

$$\frac{X''}{X} = \lambda$$

and

$$\frac{1}{\alpha^2}\frac{T'}{T} = \lambda$$

or

$$X'' - \lambda X = 0 \tag{31}$$

and

$$T' - \alpha^2 \lambda T = 0. \tag{32}$$

Using the boundary conditions (28), we find

$$0 = u(0,t) = X(0)T(t) \qquad \text{and} \qquad 0 = u(l,t) = X(l)T(t),$$

which imply that

$$X(0) = X(l) = 0. \tag{33}$$

Thus, *the function $X(x)$ of the solution (29) satisfies the EVP consisting of the differential equation (31) and the boundary conditions (33).* We leave it to the student to show that the eigenvalues and eigenfunctions of the EVP (31), (33) are given by

$$\lambda_n = -\frac{n^2\pi^2}{l^2}, \qquad n = 1, 2, \ldots \tag{34}$$

and

$$X_n(x) = \sin\frac{n\pi x}{l}, \qquad n = 1, 2, \ldots. \tag{35}$$

For the sake of completeness, we now indicate what remains to be done to find the solution $u(x, t)$ to the initial boundary value problem (24), (27), and (28). Since λ has one of the values given by Eq. (34), the solution of Eq. (32), up to a constant multiple, is given by

$$T_n(t) = \exp(-\alpha^2 n^2\pi^2 l^{-2} t), \qquad n = 1, 2, \ldots. \tag{36}$$

In view of the equations (29), (35), and (36), each of the functions

$$u_n(x, t) = \exp(-\alpha^2 n^2\pi^2 l^{-2} t)\sin\frac{n\pi x}{l}, \qquad n = 1, 2, \ldots \tag{37}$$

is a solution of the heat equation (24) and satisfies the boundary conditions (28). Furthermore, it can be shown that any linear combination

$$\sum_{n=1}^{\infty} c_n u_n(x, t) \tag{38}$$

of the functions in (37) is a solution of (24) and (28) provided that the series (38) converges and can be differentiated term by term a sufficient number of times with respect to t and x. All that we have to do now is to choose the constants c_n, $n = 1, 2, \ldots$, in such a way that the series (38) satisfies the initial condition (27). That is,

$$f(x) = \sum_{n=1}^{\infty} c_n u_n(x, 0) = \sum_{n=1}^{\infty} c_n \sin\frac{n\pi x}{l}, \qquad 0 \le x \le l, \tag{39}$$

which means that the constants c_n must be the coefficients of the Fourier sine series of the function $f(x)$ in the interval $0 \le x \le l$. The procedure just outlined is carried out in detail in Section 11.7.

EXERCISES

In Exercises 1 through 8, compute the eigenvalues and eigenfunctions of the EVP.

1. $y'' + \lambda y = 0$ for $0 \le x \le \frac{1}{2}$
$y(0) = 0$, $y(\frac{1}{2}) = 0$

2. $y'' + \lambda y = 0$ for $0 \le x \le 1$
$y'(0) = 0$, $y(1) = 0$

3. $y'' + \lambda y = 0$ for $-\frac{1}{2} \le x \le \frac{1}{2}$
$y(-\frac{1}{2}) = 0$, $y(\frac{1}{2}) = 0$

4. $y'' + \lambda y = 0$ for $0 \le x \le \pi$
$y(0) = y'(0)$, $y(\pi) = y'(\pi)$

5. $y'' + 2y + \lambda y = 0$ for $0 \le x \le 2$
$y(0) = 0,\ y(2) = 0$

6. $x^2y'' + 3xy' + \lambda y = 0$ for $1 \le x \le e^2$
$y(1) = 0,\ y(e^2) = 0$

[*Hint:* Set $x = e^t$.]

7. $y'' + (2 + \lambda)y = 0$ for $0 \le x \le 1$
$y(0) = 0,\ y(1) = 0$

8. $y'' + (-2 + \lambda)y = 0$ for $0 \le x \le 1$
$y(0) = 0,\ y(1) = 0$

9. Show that the eigenfunctions of the EVP (14)–(15) are pairwise-orthogonal in the interval $0 \le x \le 1$; that is, $(y_n, y_m) = 0$ for $n \neq m$.

10. Show that the eigenvalues and eigenfunctions of the EVP (31), (33) are given by (34) and (35), respectively.

11. The Wave Equation Equation (25) appears in the study of phenomena involving the propagation of waves (for example, water waves, acoustic waves, electromagnetic waves). Consider the initial boundary value problem

$$u_{tt} - c^2u_{xx} = 0, \qquad 0 < x < l, \qquad t > 0$$
$$u(x, 0) = f(x), \qquad u_t(x, 0) = g(x), \qquad 0 \le x \le l$$
$$u(0, t) = u(l, t) = 0, \qquad t \ge 0,$$

where c is a constant sometimes called the *wave speed*. Assume that this problem has a solution of the form

$$u(x, t) = X(x)T(t).$$

Show that the function X satisfies the EVP

$$X'' - \lambda X = 0$$

$$X(0) = X(l) = 0$$

and that the function T satisfies the differential equation

$$T'' - \lambda c^2 T = 0.$$

Indicate a formal procedure for obtaining the solution of this problem.

12. Laplace's Equation Equation (26) appears in the study of steady-state heat flows, in potential theory, and so on. Consider the BVP

$$u_{xx} + u_{yy} = 0, \qquad 0 < x < A, \qquad 0 < y < B$$
$$u(x, 0) = u(x, B) = 0, \qquad 0 < x < A$$
$$u(0, y) = 0, \qquad u(A, y) = f(y), \qquad 0 \le y \le B.$$

Assume that this problem has a solution of the form

$$u(x, y) = X(x)Y(y).$$

Show that the function Y satisfies the EVP

$$Y'' + \lambda Y = 0$$
$$Y(0) = Y(B) = 0,$$

and the function X satisfies the differential equation

$$X'' - \lambda X = 0$$

and the initial condition

$$X(0) = 0.$$

Indicate a formal procedure for obtaining the solution of this problem.

REVIEW EXERCISES

In Exercises 1 through 4, solve the BVP.

1. $y'' + 16y = 32x, \quad 0 \le x \le \pi$
$y(0) = 0, y(\pi) = 1$

2. $y'' + 16y = 32x, \quad 0 \le x \le \pi$
$y(0) = 0, y(\pi) = 2\pi$

3. $y'' + 16y = 32x, \quad 0 \le x \le \frac{\pi}{8}$
$y(0) = 0, y\left(\frac{\pi}{8}\right) = 0$

4. $y'' + 16y = 32x, \quad 0 \le x \le \pi$
$y(0) - y'(0) = 1, y'(\pi) = 0$

5. Under what conditions on A and B does the BVP

$$y'' + 16y = 32x, \quad 0 \le x \le \pi$$
$$y(0) = A, \quad y(\pi) = B$$

have exactly one solution, no solution, infinitely many solutions?

6. Show that the BVP

$$y'' + 16y = 32x, \quad 0 \le x \le \pi$$
$$y(0) = A, \quad y'(\pi) = B$$

has exactly one solution, no matter what the values of A and B are. Find the solution.

In Exercises 7 through 9, compute the eigenvalues and eigenfunctions of the EVP.

7. $y'' - \lambda y = 0, \quad 0 \le x \le \pi$
$y(0) = 0, y(\pi) = 0$

8. $y'' + \lambda y = 0, \quad 0 \le x \le \pi$
$y(0) = 0, y'(\pi) = 0$

9. $y'' + \lambda y = 0, \quad 0 \le x \le \pi$
$y'(0) = 0, y'(\pi) = 0$

CHAPTER 7

Numerical Solutions of Differential Equations

7.1 INTRODUCTION

Up to now our treatment of differential equations has centered on the determination of solutions of the differential equation. More specifically, our methods have provided us with formulas for the solution. For example, the differential equation $y'' + y = 0$ has $y = c_1 \sin x + c_2 \cos x$ as a formula for its solutions, the differential equation $y'' - y = 0$ has $y = c_1 e^x + c_2 e^{-x}$ as a formula for its solutions, and the differential equation $y'' - xy = 0$ has

$$y = c_1\left[1 + \sum_{n=1}^{\infty} \frac{1 \cdot 4 \cdot 7 \cdots (3n-2)}{(3n)!} x^{3n}\right] + c_2\left[x + \sum_{n=1}^{\infty} \frac{2 \cdot 5 \cdot 8 \cdots (3n-1)}{(3n+1)!} x^{3n+1}\right]$$

as a formula for its solutions. It is natural to adopt the attitude that these formulas supply all the information about the solution that we need. For instance, if we want the value of y corresponding to a specific value of x in the cases above, we simply substitute the given value of x into the appropriate formula and compute the corresponding value of y. Unfortunately, matters are not always that simple. Even in those cases where the solution is represented by such relatively simple functions as e^x, $\sin x$, or $\cos x$, the accuracy desired in the numerical value of y may cause some computational problems. Furthermore, the formula for y may itself impose computational difficulties. This is the case in the description of the solution of $y'' - xy = 0$ above. Of course, with the present-day proliferation of pocket calculators and the widespread availability of digital computers, such computational difficulties can be easily overcome if the accuracy desired is within the capability of the device used.

Our description of determining the numerical value of the solution of a differential equation, although it does point out one aspect of the problem, is *not* what is meant by numerical solutions of differential equations. *Rather, this terminology applies to methods associated with the determination of approximate numerical values of the solution when it is generally impossible to obtain a formula for the solution.* The reader perhaps recalls from calculus that many functions do not have antiderivatives, and hence it is impossible to represent the indefinite integral of these functions by a formula in terms of elementary functions. In this same context there are many differential equations for which it is impossible to obtain a formula for the solution. In such cases we "solve" the differential equation by applying certain *numerical methods*.

There are at present a large number of numerical methods for solving differential equations. Since our treatment is admittedly introductory and brief, we touch upon only a few of the elementary methods.

Any numerical method involves approximations of one sort or another. Consequently, questions arise concerning the accuracy of the results. In other words, how much error is involved in the approximation? Although we present a few definitions and results concerning the error involved, any reasonable attempt at answering such questions here would take us far beyond our goal of presenting a brief introduction to numerical methods.[1]

In our subsequent discussion, we assume that the differential equation in question is of the form $y' = f(x, y)$. We note that all differential equations can be written in this form or as a system of differential equations of this form. Hence, from one point of view there is no loss of generality in making this assumption. Furthermore, to avoid cumbersome treatment of integration constants, we concentrate on initial value problems. Thus we investigate the IVP

$$y' = f(x,y) \tag{1}$$

$$y(x_0) = y_0. \tag{2}$$

To obtain a feeling for the goal of numerical methods, let us imagine that the curve in Figure 7.1 represents the solution to the IVP (1)–(2). Suppose that we want to know the coordinates of the point (x_1,y_1). In the case that y is given explicitly by a formula involving x, it is simply a matter of knowing x_1.

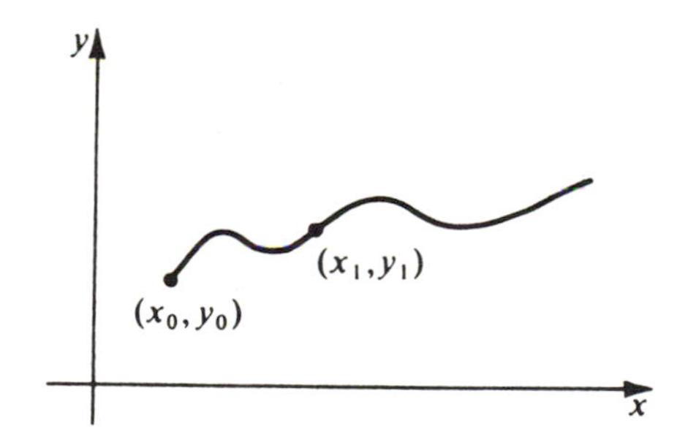

Figure 7.1

If y is not known in terms of x and the solution curve of Figure 7.1 exists only in theory (from the existence-uniqueness theorem), is it still possible to determine y_1 knowing x_1? The answer to this question is the basic goal of any numerical method.

[1]For some answers to questions about the accuracy of the methods, the interested reader is referred to E. Isaacson and H. B. Keller, *The Analysis of Numerical Methods* (New York: John Wiley & Sons, 1966).

We will say that we have solved a differential equation numerically on the interval $[x_0,b]$ when we have obtained a set of points $\{(x_0,y_0), (x_1,y_1), \ldots, (x_n,y_n)\}$ with $x_n = b$ and y_i an approximation of the solution when x has the value x_i.

7.2 EULER METHOD

Referring to Figure 7.2, we can consider that in transferring from the initial point (x_0, y_0) to the new point (x_1, y_1) we have induced changes in the coordinates. These changes are denoted by Δx and Δy, where $\Delta x = x_1 - x_0$ and $\Delta y = y_1 - y_0$. Equivalently, we can write $x_1 = x_0 + \Delta x$ and $y_1 = y_0 + \Delta y$. We can then approximate y_1 by approximating Δy. Naturally, one considers x_1 and hence Δx to be known exactly. The *Euler method* is to approximate Δy by the differential dy. Now $dy = (dy/dx)dx = f(x,y)dx \approx f(x,y)\Delta x$. Of course, if this approximation is to be reasonably accurate, then Δx should be quite small (see Figure 7.2). Given the IVP (1)–(2) of Section 7.1 and a specific value of x, say x_1, then y_1 is given by the Euler method to be $y_0 + f(x_0,y_0)\Delta x = y_0 + f(x_0, y_0)(x_1 - x_0)$. Thus, one can calculate y_1 directly from the information given. One can then use (x_1,y_1) to approximate y_2 for some given x_2 $(> x_1)$. In this case, $y_2 = y_1 + f(x_1,y_1)\Delta x = y_1 + f(x_1,y_1)(x_2 - x_1)$. This process can then be repeated to obtain additional points. Usually, this process is carried out in a specified interval with a prescribed spacing between the x coordinates. The spacing is normally taken to be uniform (in which case the spacing is often called the "mesh width" or simply "mesh") and is denoted by h. If it is desired to obtain n points in addition to the initial point, then h is given by the formula $(b - x_0)/n$. The sequence of points $\{(x_i,y_i)\}_{i=0}^{n}$, when plotted, hopefully lie very close to the solution curve.

The Euler method can be summarized by the following formulas of the coordinates of the points (x_i,y_i):

$$x_i = x_{i-1} + h, \qquad i = 1, 2, 3, \ldots, n \tag{1}$$

$$y_i = y_{i-1} + f(x_{i-1},y_{i-1})h, \qquad i = 1, 2, 3, \ldots, n. \tag{2}$$

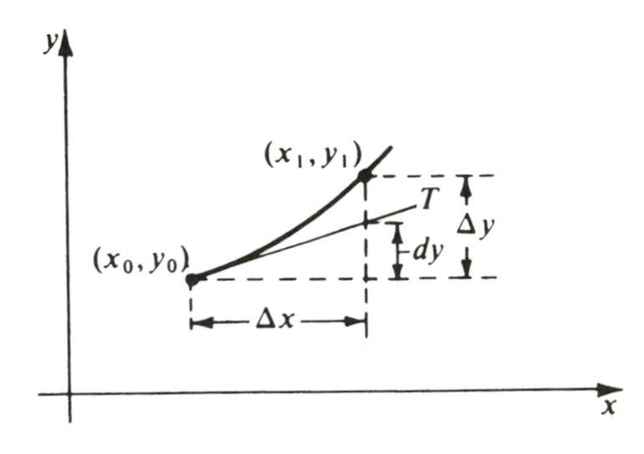

Figure 7.2

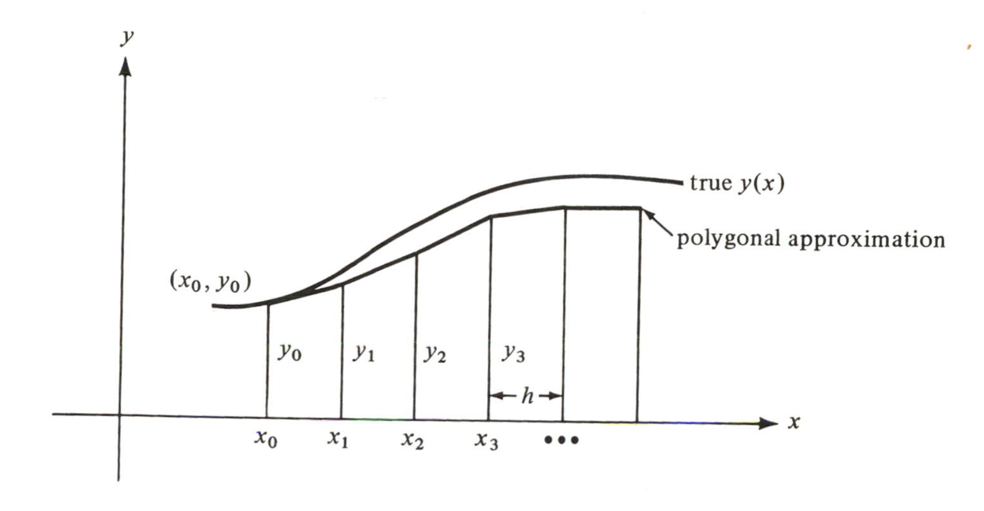

Figure 7.3

The "solution" to the IVP is then the sequence of points $\{(x_i, y_i)\}_{i=0}^n$. This solution is usually displayed in tabular form, although it could be displayed geometrically by plotting each of the points. (See Figure 7.3.)

EXAMPLE 1 Use the Euler method to solve numerically the IVP

$$y' = x^2 + y^2 \tag{3}$$

$$y(0) = 0 \tag{4}$$

on the interval $0 \leq x \leq 1$.

Solution We were not told how many points to obtain; therefore, the choice is at our disposal. The more points we ask for, the smaller the value of h will be and so the more accurate our results will be. The price to be paid, of course, is that the smaller h is, the more calculations we require. We choose $n = 5$. Thus, $h = (1 - 0)/5 = 0.2$. The results are displayed in Table 7.1. Even though we are only interested in x_i and y_i, we present the intermediate calculations to assist the reader in following the details. Where necessary we rounded numbers off to six decimal places. (There are problems of accuracy associated with round-off also, but again we skirt the issue since it would take us beyond our treatment.)

Thus, the collection of points $\{(0,0), (0.2,0), (0.4,0.008), (0.6,0.040013), (0.8,0.112333), (1,0.242857)\}$ is our numerical solution (by the Euler method with $n = 5$) of the IVP (3)–(4) in the interval $0 \leq x \leq 1$. Here and in subsequent applications, it is to be considered that if one were to plot the points of our numerical solution and connect these points with straight lines, the resulting polygonal curve would be an approximation to the graph of the solution on the interval $0 \leq x \leq 1$.

Table 7.1 Numerical Solution of the IVP (3)–(4) by the Euler Method

i	x_{i-1}	y_{i-1}	$f(x_{i-1},y_{i-1})$	$f(x_{i-1},y_{i-1})h$	x_i	y_i
1	0	0	$0^2 + 0^2 = 0$	$0(0.2) = 0$	0.2	0
2	0.2	0	$(0.2)^2 + 0^2 = 0.04$	$(0.04)(0.2) = 0.008$	0.4	0.008
3	0.4	0.008	$(0.4)^2 + (0.008)^2 = 0.160064$	$(0.160064)(0.2) = 0.032013$	0.6	0.040013
4	0.6	0.040013	$(0.6)^2 + (0.040013)^2 = 0.361601$	$(0.361601)(0.2) = 0.072320$	0.8	0.112333
5	0.8	0.112333	$(0.8)^2 + (0.112333)^2 = 0.652619$	$(0.652619)(0.2) = 0.130524$	1.0	0.242857

The calculations presented in Table 7.1 can be carried out with the aid of a computer or programmable calculator.[2] A flowchart of the steps involved is shown in Figure 7.4. A program for Example 1 would use the initial values $x = 0$, $y = 0$, $b = 1$, and $n = 5$. Once a program has been written it is a simple matter to produce a solution with any other set of initial values. In particular, if we increase the value of n, we decrease the value of $h = \Delta x$ and generally improve the accuracy of the method.

The error resulting from the use of a numerical method, such as the Euler method, to approximate a true solution is called *truncation* or *discretization error*. Generally, increasing n (decreasing the step size) reduces the truncation error. However, another type of error, called *round-off error*, tends to become more significant as h gets smaller. This is due to the limitation of all computers that can carry only a specified number of digits. This implies that the final digit in any number represented in the computer may be in error due to round-off. As n increases, the accumulated round-off error can more than offset the gains made in improving the accuracy of the method. Numerical analysts have spent and continue to spend much time searching for the proper balance between reducing the truncation error (increasing n) and the round-off error (decreasing n).

One fairly straightforward procedure to suggest a possible best solution for a given computer is to write a program that repeats the sequence of steps illustrated in the flowchart several times, each time doubling the size of n, until the computed results [that is, (x_n, y_n)] no longer differ in the number of decimal places required. While this is no guarantee of the accuracy of the results, many programmers find this practical rule very useful for a first guess.

Of course, one way to obtain a better approximation of the solution is to use another, more accurate method with fewer steps. We shall do this in the remaining sections.

With regard to the error in the Euler approximation, we state the following theorem without proof.

THEOREM 1.

Suppose that the functions $f(x, y)$, $\dfrac{\partial f}{\partial x}$ *and* $\dfrac{\partial f}{\partial y}$ *are defined and continuous on the rectangle* $R = \{(x, y);\ x_0 \le x \le x_0 + a,\ y_0 - b \le y \le y_0 + b\}$. *Let* $M = \max\limits_{(x,y)\in R} |f(x, y)|$ *and* $\gamma = \min\left(a, \dfrac{b}{M}\right)$. *Let K and L represent two positive numbers such that*

$$\max_{(x,y)\in R} \left|\frac{\partial f}{\partial y}\right| \le K, \qquad \max_{(x,y)\in R} \left|\frac{\partial f}{\partial x} + f\frac{\partial f}{\partial y}\right| \le L.$$

[2]The authors wish to thank Thomas M. Green for the development of the flowcharts and the computer and calculator exercises.

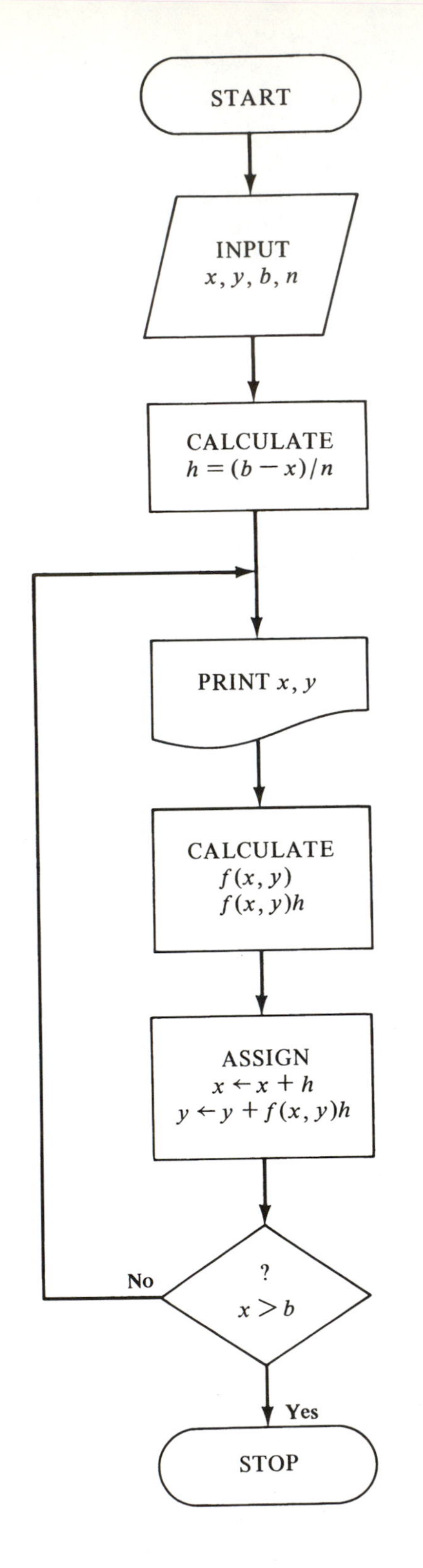

1. Input initial values:
   $x \leftarrow x_0$
   $y \leftarrow y_0$

2. $h = \Delta x$

3. Print $\{(x_i, y_i)\}_i^n = 0$

4. $y' = f(x, y)$
   $dy = f(x, y)h$

5. Increment the values of x and y:
   $x_i = x_{i-1} + h$
   $y_i = y_{i-1} + (dy)_{i-1}$
   $i = 1, 2, 3, \ldots, n$

6. Test value of x.

Figure 7.4 Annotated Flowchart for the Euler Method.

If E_i represents the error made in the ith step by approximating the actual solution $y(x_i)$ by the Euler approximation $y_i = hf(x_{i-1}, y_{i-1})$, then

$$E_i \leq \alpha h, \qquad i = 1, 2, \ldots, n$$

where $\alpha = L(e^{\lambda k} - 1)/2k$.

EXERCISES

In all problems, maintain four-decimal-place accuracy. In Exercises 1 through 6, use the Euler method with $h = 0.2$ to solve each of the IVPs on $0 \leq x \leq 1$.

1. $y' = x - 2y$, $y(0) = 3$
2. $y' = e^{-y} - y + 1$, $y(0) = 1$
3. $y' = 3x^2 - 2y$, $y(0) = 0$
4. $y' = \sin y + x$, $y(0) = 0$
5. $y' = y^2 - 2x^2 + 1$, $y(0) = 0$
6. $y' = 2xy^{-1}$, $y(0) = 1$
7. Write a computer or calculator program to repeat Exercise 1 using (a) $h = 0.1$ and (b) $h = 0.01$ and compare with the results of Exercise 1.
8. Write a computer or calculator program to repeat Exercise 3 using (a) $h = 0.1$ and (b) $h = 0.01$ and compare with the results of Exercise 3.
9. Write a computer or calculator program to repeat Exercise 5 using (a) $h = 0.1$ and (b) $h = 0.01$ and compare with the results of Exercise 5.
10. Solve Exercise 6 exactly and compare the actual set of points (x_i, y_i), $i = 1, 2, 3, 4, 5$ with the approximate set obtained in Exercise 6.
11. Given the IVP $y' = 3xy^2$, $y(\frac{1}{10}) = 0.2$, use the Euler method with $h = 0.25$ to approximate $y(0.85)$. Compare your result with the actual value.
12. Solve the IVP $y' = x + y, y(0) = 0$, on the interval $0 \leq x \leq 1$. Use the Euler method with $h = 0.2$. Compare with the exact results.
13. Repeat Exercise 12 with $h = 0.1$.
14. Write a computer or calculator program to repeat Exercise 12 with $h = 0.5, 0.25, 0.125, 0.0625, \ldots$, allowing h to decrease until there is no change in the first three decimal places of the approximations for $y(1)$. Adjust the program so that only (x_n, y_n) is printed and not the other points on the interval for $i < n$.

7.3 TAYLOR-SERIES METHOD

In Section 2.9 and in Chapter 5 we demonstrated how the Taylor series can be used to generate series representations of solutions of a differential equation. In some cases it is possible to generate the complete series, and in other cases it is not possible (or perhaps not feasible) to obtain the complete series. The

Taylor series can be used in another way to obtain an approximation to the value of the solution of an IVP at specific values of x. Given the IVP

$$y' = f(x,y), \qquad x_0 \leq x \leq b \tag{1}$$

$$y(x_0) = y_0, \tag{2}$$

we again wish to obtain a sequence of points $\{(x_i, y_i)\}_{i=0}^{n}$ that lie close to the solution curve on the interval $[x_0, b]$. We recall that the Taylor-series representation of the solution y about x_0 is

$$y(x) = \sum_{n=0}^{\infty} \frac{y^{(n)}(x_0)}{n!}(x - x_0)^n.$$

Setting $x = x_1 = x_0 + h$, we have

$$\begin{aligned} y(x_1) = y(x_0 + h) &= \sum_{n=0}^{\infty} \frac{y^{(n)}(x_0)}{n!} h^n \\ &= y(x_0) + y'(x_0)h + \frac{y''(x_0)}{2!}h^2 + \frac{y'''(x_0)}{3!}h^3 + \cdots \\ &= y_0 + f(x_0, y_0)h + \frac{y''(x_0)}{2!}h^2 + \frac{y'''(x_0)}{3!}h^3 + \cdots \end{aligned}$$

The Euler method can be thought of as an approximation of the Taylor series by retaining only the first two terms of this series. In general, if the Taylor series is approximated by

$$y(x_0) + y'(x_0)h + \cdots + \frac{y^{(l)}(x_0)}{l!}h^l,$$

we say that the approximation is a *Taylor approximation of order l.* In this language the Euler method is a Taylor approximation of order 1.

If we assume that all partial derivatives of $f(x,y)$ of any order exist at (x_0, y_0), then the differential equation (1) and the initial condition (2) can be used to determine each $y^{(k)}(x_0)$, $k = 2, 3, \ldots$, and consequently y_1. For example,

$$y'(x_0) = y'(x)\Big|_{\substack{x=x_0\\ y=y_0}} = f(x_0, y_0)$$

$$y''(x_0) = \frac{d}{dx}(y'(x))\Big|_{\substack{x=x_0\\ y=y_0}} = \frac{d}{dx}(f(x, y))\Big|_{\substack{x=x_0\\ y=y_0}}$$

$$= \left[\frac{\partial f}{\partial x} + \frac{\partial f}{\partial y}\frac{dy}{dx}\right]\Big|_{\substack{x=x_0\\ y=y_0}} = f_x(x_0, y_0) + f_y(x_0, y_0) \cdot f(x_0, y_0).$$

Subsequent derivatives can be obtained in like fashion. Knowing (x_1,y_1) and x_2, we can then calculate y_2 in a similar manner. Thus,

$$y_2 = y(x_2) = y(x_1 + h) = \sum_{n=0}^{\infty} \frac{y^{(n)}(x_1)}{n!} h^n,$$

where $y(x_1) = y_1$, $y'(x_1) = f(x_1,y_1)$, $y''(x_1) = f_x(x_1,y_1) + f_y(x_1,y_1) \cdot f(x_1,y_1)$, and so on. Additional points are obtained in the same way.

In any application of the Taylor method, one must specify the order l of the approximation. The larger l is, the more accurate we anticipate the results to be. Once again the computations involved increase with increasing l. We illustrate the method by repeating Example 1 of Section 7.2.

EXAMPLE 1 Use the Taylor approximation of order 2 to obtain a numerical solution of the IVP

$$y' = x^2 + y^2 \tag{3}$$

$$y(0) = 0 \tag{4}$$

on $0 \le x \le 1$.

Solution We choose $h = 0.2$. From the differential equation (3), we obtain

$$\begin{aligned} y' &= x^2 + y^2 \\ y'' &= 2x + 2yy' = 2x + 2y(x^2 + y^2) \\ &= 2x + 2x^2y + 2y^3. \end{aligned}$$

In general, we have

$$\begin{aligned} y_i = y(x_i) = y(x_{i-1} + h) &\approx y(x_{i-1}) + y'(x_{i-1})h + \frac{y''(x_{i-1})}{2}h^2 \\ &\approx y_{i-1} + f(x_{i-1},y_{i-1})h + \{f_x(x_{i-1},y_{i-1}) \\ &\quad + f_y(x_{i-1},y_{i-1}) \cdot f(x_{i-1},y_{i-1})\}\frac{h^2}{2} \\ &\approx y_{i-1} + (x_{i-1}^2 + y_{i-1}^2)h \\ &\quad + (2x_{i-1} + 2x_{i-1}^2y_{i-1} + 2y_{i-1}^3)\frac{h^2}{2}. \end{aligned}$$

We present the results in Table 7.2, incorporating the intermediate calculations.

Thus, our numerical solution of the IVP (3)–(4), by the Taylor approximation of order 2 with $h = 0.2$, on the interval $0 \le x \le 1$, is the set

$$\{(0,0), (0.2,0), (0.4,0.016), (0.6, 0.064154), (0.8, 0.161911), (1, 0.331469)\}.$$

For the Taylor approximation of order l it can be shown that the error is proportional to h^l. Thus we anticipate that the Taylor approximation of order

Table 7.2 Numerical Solution of the IVP (3)–(4) by the Taylor Approximation of Order 2

i	x_i	y_i	$[x_i^2 + y_i^2]h$	$[2x_i + 2x_i^2y_i + 2y_i^3]\frac{h^2}{2}$	x_{i+1}	y_{i+1}
0	0	0	$[0^2 + 0^2]\ (0.2) = 0$	$[2(0) + 2(0)^2(0) + 2(0)^3]\frac{(0.2)^2}{2} = 0$	0.2	0
1	0.2	0	$[(0.2)^2 + 0^2](0.2) = 0.008$	$[2(0.2) + 2(0.2)^2(0) + 2(0)^3]\frac{(0.2)^2}{2} = 0.008$	0.4	0.016
2	0.4	0.016	$[(0.4)^2 + (0.016)^2](0.2) = 0.032051$	$[2(0.4) + 2(0.4)^2(0.016) + 2(0.016)^3]\frac{(0.2)^2}{2} = 0.016103$	0.6	0.064154
3	0.6	0.064154	$[(0.6)^2 + (0.064154)^2](0.2) = 0.072823$	$[2(0.6) + 2(0.6)^2(0.064154) + 2(0.064154)^3]\frac{(0.2)^2}{2} = 0.024934$	0.8	0.161911
4	0.8	0.161911	$[(0.8)^2 + (0.161911)^2](0.2) = 0.133243$	$[2(0.8) + 2(0.8)^2(0.161911) + 2(0.161911)^3]\frac{(0.2)^3}{2} = 0.036315$	1.0	0.331469

2 is more accurate than the Taylor approximation of order 1. Of course a higher-order Taylor approximation would be even more accurate. Naturally, the higher the order, the more laborious the calculations. Special forms of "higher-order methods" are presented in the next section. The forms presented are those commonly called the Runge–Kutta methods.

The reader will be asked to compare the various methods of this chapter in Review Exercises. Some exercises involve IVPs that can be solved exactly, and therefore the numerical results of each method can be compared to the actual results as well as comparisons with the other methods.

We have only to make minor modifications in our flowchart for the Euler method to produce a flowchart for the Taylor-series method. Namely, we change steps 4 and 5 as in Figure 7.5.

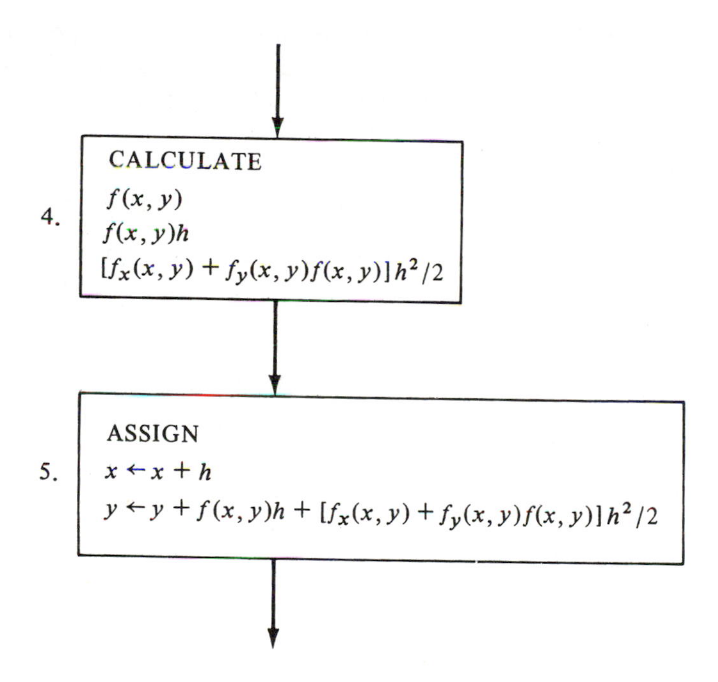

Figure 7.5. Changes in Flowchart Presented in Section 7.2 to Produce Flowchart for Taylor-Series Approximation of Order 2.

EXERCISES

In all problems, maintain four-decimal-place accuracy. In Exercises 1 through 8, use the Taylor approximation of order 2 with $h = 0.2$ to solve the IVPs on $0 \le x \le 1$.

1. $y' = x^2 + y$, $y(0) = 0$

2. $y' = x^2 + e^y$, $y(0) = 0$

3. $y' = x - y$, $y(0) = 1$

4. $y' = \sin y + x$, $y(0) = 0$

5. $y' = xy$, $y(0) = 1$

6. $y' = 3x + 2y$, $y(0) = -1$

7. $y' = x^2 y$
$y(0) = 2$

8. $y' = e^y \sin y$
$y(0) = 0.1$

9. Use the Taylor approximation of order 2 with $h = 0.2$ to solve the IVP

$$y' = x + y$$
$$y(0) = 0$$

on $0 \leq x \leq 1$. Compare with the exact results. Also compare with the Euler method (Exercise 12 in Section 7.2).

10. Use the Taylor approximation of order 3 with $h = 0.2$ to solve Exercise 9. Compare with the results of Exercise 9 and with the exact results. [*Hint:* Calculate y'' and y''' directly from the differential equation.]

11. Repeat Exercise 9 with $h = 0.1$ and compare with the corresponding results of Exercise 9 and with the exact values. For instance, $y(0.2)$ of this exercise is to be compared with $y(0.2)$ of Exercise 9 and with the exact value of $y(0.2)$.

12. Repeat Exercise 10 with $h = 0.1$ and compare with the corresponding results of Exercise 10 and with the exact results (see Exercise 11).

In Exercises 13 through 15 write a computer or calculator program for the Taylor approximation of order 2.

13. Repeat Exercise 1 with (a) $h = 0.1$ and (b) $h = 0.01$.

14. Repeat Exercise 3 with (a) $h = 0.1$ and (b) $h = 0.01$.

15. Repeat Exercise 6 with (a) $h = 0.1$ and (b) $h = 0.01$.

7.4 RUNGE–KUTTA METHODS

In the description of the Taylor approximation of order 2, we wrote

$$y(x_0 + h) \approx y(x_0) + f(x_0, y_0)h + [f_x(x_0, y_0) + f_y(x_0, y_0) \cdot f(x_0, y_0)]\frac{h^2}{2}.$$

The Taylor approximation of order 3 would have the additional term

$$\{f_{xx}(x_0, y_0) + 2f_{xy}(x_0, y_0) \cdot f(x_0, y_0) + f_{yy}(x_0, y_0)[f(x_0, y_0)]^2 + f_x(x_0, y_0)f_y(x_0, y_0) + [f_y(x_0, y_0)]^2 f(x_0, y_0)\}\frac{h^3}{6}.$$

Based on our knowledge of the Taylor series (or convergent series in general) from calculus, we know that the accuracy of our approximation improves with the number of terms retained. On the other hand, the description of the higher-order terms gets more and more complicated and the associated calculations more profuse. The basic idea of the *Runge–Kutta methods*[3] is to preserve the order of a Taylor approximation (in the sense of the error involved) while eliminating the necessity of calculating the various partial derivatives of f that are involved. The alternative proposed by these methods involves evaluating the function f at certain judicious points rather than evaluating the specific partial derivatives. We present here the Runge–Kutta formulas of order 3 and order 4.[4]

The Runge–Kutta approximation of order 3 is defined by the following collection of formulas:

$$k_1 = hf(x_i, y_i) \tag{1}$$

$$k_2 = hf(x_i + \tfrac{1}{2}h, y_i + \tfrac{1}{2}k_1) \tag{2}$$

$$k_3 = hf(x_i + h, y_i + 2k_2 - k_1) \tag{3}$$

$$y_{i+1} = y_i + \tfrac{1}{6}(k_1 + 4k_2 + k_3). \tag{4}$$

The scheme then is to obtain y_1 from formulas (1)–(4) by setting $i = 0$. With this value of y_1 and with x_1 we obtain y_2 from formulas (1)–(4) by setting $i = 1$. This process is repeated until we have the desired number of points.

EXAMPLE 1 Use the Runge–Kutta approximation of order 3 with $h = 0.2$ to solve the IVP

$$y' = x^2 + y^2 \tag{5}$$

$$y(0) = 0 \tag{6}$$

on $0 \le x \le 1$.

Solution As before, it is convenient to display our solution (see Table 7.3). Once again we show the intermediate calculations, in the hope that they help the reader to follow the development.

Our numerical solution using the Runge–Kutta approximation of order 3 with $h = 0.2$ of the IVP (5)–(6) on $0 \le x \le 1$ is given by the set

$$\{(0, 0), (0.2, 0.002667), (0.4, 0.021371), (0.6, 0.072509), (0.8, 0.174273), (1, 0.350721)\}.$$

[3]Runge first developed the approximation of order 3, and Kutta generalized and refined the approximations of orders 3 and 4.

[4]For a development of these formulas and an analysis of their accuracy, the reader is referred to K. S. Kunz, *Numerical Analysis* (New York: McGraw-Hill, 1957).

Table 7.3 Numerical Solution of the IVP (5)–(6) Using the Runge–Kutta Method of Order 3

i	x_i	y_i	k_1	k_2	k_3	$\frac{1}{6}(k_1 + 4k_2 + k_3)$	x_{i+1}	y_{i+1}
0	0	0	$(0.2)[0^2 + 0^2] = 0$	$(0.2)[(.1)^2 + 0^2] = 0.002$	$(0.2)[(.2)^2 + (.004)^2] = 0.008003$	$\frac{1}{6}[0 + 4(.002) + .008003] = 0.002667$	0.2	0.002667
1	0.2	0.002667	$0.2[(.2)^2 + (.002667)^2] = 0.008001$	$0.2[(.3)^2 + (.006668)^2] = 0.018009$	$0.2[(.4)^2 + (.030684)^2] = 0.032188$	$\frac{1}{6}[.008001 + 4(.018009) + .032188] = 0.018704$	0.4	0.021371
2	0.4	0.021371	$0.2[(.4)^2 + (.021371)^2] = 0.032091$	$0.2[(.5)^2 + (.037417)^2] = 0.050280$	$0.2[(.6)^2 + (.089840)^2] = 0.073614$	$\frac{1}{6}[.032091 + 4(.050280) + .073614] = 0.051138$	0.6	0.072509
3	0.6	0.072509	$0.2[(.6)^2 + (.072509)^2] = 0.073052$	$0.2[(.7)^2 + (.109035)^2] = 0.100378$	$0.2[(.8)^2 + (.200213)^2] = 0.136017$	$\frac{1}{6}[.073052 + 4(.100378) + .136017] = 0.101764$	0.8	0.174273
4	0.8	0.174273	$0.2[(.8)^2 + (.174273)^2] = 0.134074$	$0.2[(.9)^2 + (.241388)^2] = 0.173646$	$0.2[(1)^2 + (.387491)^2] = 0.230030$	$\frac{1}{6}[.134074 + 4(.173646) + .230030] = 0.176448$	1.0	0.350721

The Runge–Kutta approximation of order 4 is defined by the following formulas:

$$l_1 = hf(x_i, y_i) \tag{7}$$

$$l_2 = hf(x_i + \tfrac{1}{2}h, y_i + \tfrac{1}{2}l_1) \tag{8}$$

$$l_3 = hf(x_i + \tfrac{1}{2}h, y_i + \tfrac{1}{2}l_2) \tag{9}$$

$$l_4 = hf(x_i + h, y_i + l_3) \tag{10}$$

$$y_{i+1} = y_i + \tfrac{1}{6}(l_1 + 2l_2 + 2l_3 + l_4). \tag{11}$$

We employ formulas (7)–(11) in exactly the same manner as the Runge–Kutta formulas of order 3. We illustrate with an example.

EXAMPLE 2 Solve the IVP

$$y' = x^2 + y^2 \tag{12}$$

$$y(0) = 0 \tag{13}$$

and $0 \leq x \leq 1$. Use the Runge–Kutta approximation of order 4 with $h = 0.2$.

Solution As in the previous example, we demonstrate our results in a table (see Table 7.4).

Our solution of the IVP (12)–(13) on $0 \leq x \leq 1$ using the Runge–Kutta approximation of order 4 is the set

$$\{(0, 0), (0.2, 0.002667), (0.4, 0.021360), (0.6, 0.072451), (0.8, 0.174090), (1, 0.350257)\}.$$

Steps 4 and 5 of our flowchart given in Section 7.2 would be changed to those in Figure 7.6 in order to produce a flowchart for the Runge–Kutta method of order 4.

The Runge–Kutta method is highly accurate (small truncation error) even when n is relatively small (small round-off error). For these reasons it is a widely used method.

It can be shown that the error involved in the Runge–Kutta approximation of order $k(k = 3, 4)$ is proportional to h^k. Thus the Runge–Kutta method is more accurate than the Euler method and although it has the same order of accuracy as the Taylor approximation of order k, it has the feature that the calculations depend on knowledge of the function f only and not on derivatives of f as in the case of the Taylor method.

EXERCISES

In all problems, maintain four-decimal-place accuracy. In Exercises 1 through 8, use the Runge–Kutta approximation of order 3 with $h = 0.2$ to solve the IVPs on $0 \leq x \leq 1$.

1. $y' = 3x + 2y$, $y(0) = 1$

2. $y' = y^2 + 2y - x$, $y(0) = 0$

3. $y' = x - 4y$, $y(0) = 0$

4. $y' = 2xy - x^2$, $y(0) = 0$

5. $y' = y^2 - 2x$, $y(0) = 0$

6. $y' = 2yx^2 + 4$, $y(0) = 1$

7. $y' = 3y + 2x^2 - x$, $y(0) = 0$

8. $y' = y + y^2$, $y(0) = -1$

9. Use the Runge–Kutta approximation of order 4 with $h = 0.2$ to solve the IVP of Exercise 1 on $0 \le x \le 1$.

10. Use the Runge–Kutta approximation of order 4 with $h = 0.2$ to solve the IVP of Exercise 3 on $0 \le x \le 1$.

11. Use the Runge–Kutta approximation of order 4 with $h = 0.2$ to solve the IVP of Exercise 5 on $0 \le x \le 1$.

Table 7.4 Numerical Solution of the IVP (12)–(13) Using the Runge–Kutta Method of Order 4

i	x_i	y_i	l_1	l_2	l_3
0	0	0	$0.2[0^2 + 0^2] = 0$	$0.2[(.1)^2 + 0^2] = 0.002$	$0.2[(.1)^2 + (.001)^2] = 0.002$
1	0.2	0.002667	$0.2[(.2)^2 + (.002667)^2] = 0.008001$	$0.2[(.3)^2 + (.006668)^2] = 0.018009$	$\frac{1}{5}[(.3)^2 + (.011672)^2] = 0.018027$
2	0.4	0.021360	$\frac{1}{5}[(.4)^2 + (.021360)^2] = 0.032091$	$\frac{1}{5}[(.5)^2 + (.037406)^2] = 0.050280$	$\frac{1}{5}[(.5)^2 + (.046500)^2] = 0.050432$
3	0.6	0.072451	$\frac{1}{5}[(.6)^2 + (.072451)^2] = 0.073050$	$\frac{1}{5}[(.7)^2 + (.108976)^2] = 0.100375$	$\frac{1}{5}[(.7)^2 + (.122639)^2] = 0.101008$
4	0.8	0.174090	$\frac{1}{5}[(.8)^2 + (.174090)^2] = 0.134061$	$\frac{1}{5}[(.9)^2 + (.241121)^2] = 0.173628$	$\frac{1}{5}[(.9)^2 + (.260904)^2] = 0.175614$

l_4	$\frac{1}{6}(l_1 + 2l_2 + 2l_3 + l_4)$	x_{i+1}	x_{i+1}
$0.2[(.2)^2 + (.002)^2] = 0.008001$	$\frac{1}{6}[(0 + 2(.002) + 2(.002) + 0.008001] = 0.002667$	0.2	0.002667
$\frac{1}{5}[(.4)^2 + (.020694)^2] = 0.032086$	$\frac{1}{6}[(.008001 + 2(.018009) + 2(.018027) + 0.032086] = 0.0018693$	0.4	0.021360
$\frac{1}{5}[(.6)^2 + (.071792)^2] = 0.073031$	$\frac{1}{6}[(0.032091 + 2(.050280) + 2(.050432) + 0.073031] = 0.051091$	0.6	0.072451
$\frac{1}{5}[(.8)^2 + (.173459)^2] = 0.134018$	$\frac{1}{6}[.073050 + 2(.100375) + 2(.101008) + 0.134018] = 0.101639$	0.8	0.174090
$\frac{1}{5}[(1)^2 + (.349704)^2] = 0.224459$	$\frac{1}{6}[.134061 + 2(.173629) + 2(.175614) + 0.224459] = 0.176167$	1.0	0.350257

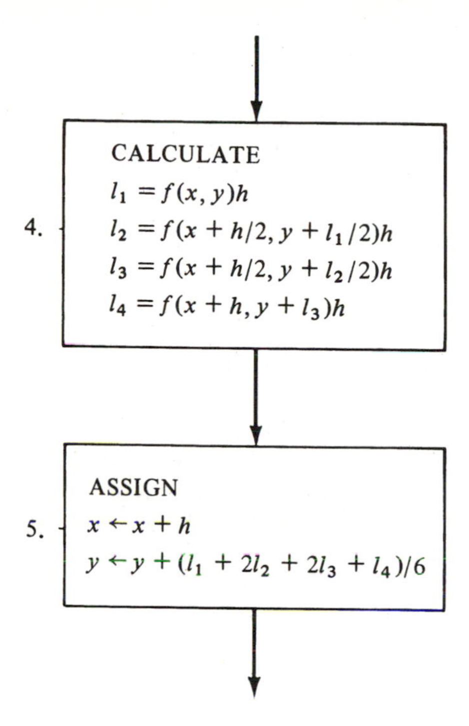

Figure 7.6 Changes in Flowchart Presented in Section 7.2 to Produce Flowchart for the Runge–Kutta Approximation of Order 4.

12. Use the Runge–Kutta approximation of order 4 with $h = 0.2$ to solve the IVP of Exercise 4 on $0 \leq x \leq 1$.

13. Use the Runge–Kutta approximation of order 3 with $h = 0.2$ to solve the IVP

$$y' = x + y$$
$$y(0) = 0$$

on $0 \leq x \leq 1$. Compare with the exact results.

14. Use the Runge–Kutta approximation of order 4 with $h = 0.2$ to solve the IVP

$$y' = x + y$$
$$y(0) = 0$$

on $0 \leq x \leq 1$. Compare with the exact results and compare with the results of Exercise 13. Also compare with the Euler method (Exercise 12 in Section 7.2).

In Exercises 15 through 18 write a computer or calculator program for the Runge–Kutta approximation.

15. Repeat Exercise 13 with (a) $h = 0.1$ and (b) $h = 0.01$.

16. Repeat Exercise 14 with (a) $h = 0.1$ and (b) $h = 0.01$.

17. Repeat Exercise 1 with (a) $h = 0.1$ and (b) $h = 0.01$.

18. Repeat Exercise 3 with (a) $h = 0.1$ and (b) $h = 0.01$.

7.5 SYSTEMS OF FIRST-ORDER DIFFERENTIAL EQUATIONS

The methods presented in Sections 7.2 through 7.4 can also be applied to initial value problems for a system of first-order differential equations. We restrict our discussion to systems with two unknown functions, but the methods apply to systems with any number of unknown functions.

Thus, we consider the initial value problem

$$x' = f_1(t, x, y) \tag{1}$$

$$y' = f_2(t, x, y)$$

$$x(t_0) = x_0, \qquad y(t_0) = y_0. \tag{2}$$

In this system x and y are the unknown functions and t is the independent variable. By a numerical solution of the initial value problem (1)–(2) on the interval $t_0 \le t \le b$, we mean a set of points $\{(t_i, x_i, y_i)\}_{i=0}^{n}$ where $t_n = b$, and x_i, y_i are approximations to the solution of the system (1) when $t = t_i$. We assume that Δt_i is the same for all i; thus, $\Delta t_i = h = (b - t_0)/n$.

For the initial value problem (1)–(2), the Euler formulas are as follows:

$$x_i = x_{i-1} + f_1(t_{i-1}, x_{i-1}, y_{i-1})h, \qquad i = 1, 2, \ldots, n \tag{3}$$

$$y_i = y_{i-1} + f_2(t_{i-1}, x_{i-1}, y_{i-1})h, \qquad i = 1, 2, \ldots, n. \tag{4}$$

EXAMPLE 1 Assume $1 \le t \le 2$. Use the Euler method with $h = 0.2$ to solve the initial value problem

$$x' = t + x^2 - y^2 \tag{5}$$

$$y' = t^2 - 2x^2 + 3y^2 \tag{6}$$

$$x(1) = 0, \qquad y(1) = 0. \tag{7}$$

Solution We exhibit the details of the solution in tabular form (Table 7.5). Thus, the numerical solution is the set

$$\{(1, 0, 0), (1.2, 0.2, 0.2), (1.4, 0.44, 0.496), (1.6, 0.7095, 0.9582),$$
$$(1.8, 0.9466, 1.8197), (2.0, 0.8236, 4.0959)\}.$$

Table 7.5 Numerical Solution of the IVP (5)–(7) Using the Euler Method

t_i	x_i	y_i	$f_1(t_i, x_i, y_i)$	$f_2(t_i, x_i, y_i)$	$x_i + f_1(t_i, x_i, y_i)h$	$y_i + f_2(t_i, x_i, y_i)h$
1	0	0	$1 + 0^2 - 0^2 = 1$	$1^2 - 2(0)^2 + 3(0)^2 = 1$	$0 + 1(0.2) = 0.2$	$0 + 1(0.2) = 0.2$
1.2	0.2	0.2	$1.2 + (0.2)^2 - (0.2)^2 = 1.2$	$(1.2)^2 - 2(0.2)^2 + 3(0.2)^2 = 1.48$	$0.2 + (1.2)(0.2) = 0.44$	$0.2 + (1.48)(.2) = 0.496$
1.4	0.44	0.496	$1.4 + (0.44)^2 - (.496)^2 = 1.3476$	$(1.4)^2 - 2(0.44)^2 + 3(.496)^2 = 2.3108$	$0.44 + (1.3476)(0.2) = 0.7095$	$0.496 + (2.3108)(.2) = 0.9582$
1.6	0.7095	0.9582	$1.6 + (.7095)^2 - (.9582)^2 = 1.8533$	$(1.6)^2 - 2(.7095)^2 + 3(.9582)^2 = 4.3074$	$0.7095 + (1.8533)(0.2) = 0.9466$	$0.9582 + (4.3074)(.2) = 1.8197$
1.8	0.9466	1.8197	$1.8 + (.9466)^2 - (1.8197)^2 = -0.6151$	$(1.8)^2 - 2(.9466)^2 + 3(1.8197)^2 = 11.3814$	$0.9466 + (-0.6151)(0.2) = 0.8236$	$1.8197 + (11.3814)(.2) = 4.0959$
2	0.8236	4.0959				

The Taylor formulas corresponding to systems are rather complicated in general. For example, suppose we desire $d^2y/dt^2\,|_{t=t_0}$, then

$$\left.\frac{d^2y}{dt^2}\right|_{t=t_0} = \left.\frac{d}{dt}[y'(t)]\right|_{\substack{t=t_0\\x=x_0\\y=y_0}} = \left.\frac{d}{dt}[f_2(t,x,y)]\right|_{\substack{t=t_0\\x=x_0\\y=y_0}}$$

$$= \left.\left[\frac{\partial f_2}{\partial t} + \frac{\partial f_2}{\partial x}\frac{dx}{dt} + \frac{\partial f_2}{\partial y}\frac{dy}{dt}\right]\right|_{\substack{t=t_0\\x=x_0\\y=y_0}}$$

$$= \frac{\partial f_2}{\partial t}(t_0, x_0, y_0) + \frac{\partial f_2}{\partial x}(t_0, x_0, y_0)\cdot f_1(t_0, x_0, y_0)$$

$$+ \frac{\partial f_2}{\partial y}(t_0, x_0, y_0)\cdot f_2(t_0, x_0, y_0).$$

Naturally, the higher the order of the Taylor approximation, the more complex the calculations become. We give here the formulas for a Taylor expansion of order 2.

$$x_{i+1} = x_i + f_1(t_i, x_i, y_i)h + \left[\frac{\partial f_1}{\partial t}(t_i, x_i, y_i) + \frac{\partial f_1}{\partial x}(t_i, x_i, y_i)\right.$$

$$\left.\cdot f_1(t_i, x_i, y_i) + \frac{\partial f_1}{\partial y}(t_i, x_i, y_i)\cdot f_2(t_i, x_i, y_i)\right]\frac{h^2}{2}$$

$$y_{i+1} = y_i + f_2(t_i, x_i, y_i)h + \left[\frac{\partial f_2}{\partial t}(t_i, x_i, y_i) + \frac{\partial f_2}{\partial x}(t_i, x_i, y_i)\right.$$

$$\left.\cdot f_1(t_i, x_i, y_i) + \frac{\partial f_2}{\partial y}(t_i, x_i, y_i)\cdot f_2(t_i, x_i, y_i)\right]\frac{h^2}{2}.$$

For the Runge–Kutta method, the generalization for systems is more straightforward. For example, the Runge–Kutta approximation of order 3 would be as follows.

$$\alpha_1 = hf_1(t_i, x_i, y_i)$$

$$\beta_1 = hf_2(t_i, x_i, y_i)$$

$$\alpha_2 = hf_1(t_i + \tfrac{1}{2}h, x_i + \tfrac{1}{2}\alpha_1, y_i + \tfrac{1}{2}\beta_1)$$

$$\beta_2 = hf_2(t_i + \tfrac{1}{2}h, x_i + \tfrac{1}{2}\alpha_1, y_i + \tfrac{1}{2}\beta_1)$$

$$\alpha_3 = hf_1(t_i + h, x_i + 2\alpha_2 - \alpha_1, y_i + 2\beta_2 - \beta_1)$$

$$\beta_3 = hf_2(t_i + h, x_i + 2\alpha_2 - \alpha_1, y_i + 2\beta_2 - \beta_1)$$

$$x_{i+1} = x_i + \tfrac{1}{6}(\alpha_1 + 4\alpha_2 + \alpha_3)$$

$$y_{i+1} = y_i + \tfrac{1}{6}(\beta_1 + 4\beta_2 + \beta_3).$$

The Runge–Kutta approximations of order 4 follow in a similar fashion and are left as an exercise.

EXERCISES

Maintain four-decimal-place accuracy. In each exercise it is assumed that $0 \le t \le 1$.

1. Use the Euler method with $h = 0.2$ to obtain a numerical solution to the initial value problem

$$x' = 1 + t + y$$
$$y' = y + x^2$$
$$x(0) = 0, \qquad y(0) = 0.$$

2. Use the Taylor approximation of order 2 with $h = 0.2$ to obtain a numerical solution of the initial value problem of Exercise 1.

3. Use the Runge–Kutta approximation of order 3 with $h = 0.2$ to solve the initial value problem of Exercise 1.

4. Write the Runge–Kutta formulas of order 4 for the initial value problem (1)–(2).

For each of the initial value problems in Exercises 5 through 10, use the Euler method with $h = 0.2$ to obtain a numerical solution on $0 \le t \le 1$.

5. $x' = t - y + x$
   $y' = x + 2y$
   $x(0) = 1,\ y(0) = 0$

6. $x' = t^2 - 3x^2 + y$
   $y' = 2t + x - y$
   $x(0) = 0,\ y(0) = 1$

7. $x' = x + y^2$
   $y' = -t + x - y$
   $x(0) = 1,\ y(0) = 1$

8. $x' = 5t + 6xy$
   $y' = -t - y + 3x$
   $x(0) = 1,\ y(0) = 0$

9. $x' = t^3 - y^2$
   $y' = t + x - xy$
   $x(0) = 0,\ y(0) = 1$

10. $x' = y$
    $y' = -2x + 3y$
    $x(0) = 2,\ y(0) = 3$

11. Exercise 10 can be solved explicitly (see Example 1, Section 3.2). Use this explicit solution to evaluate $x(1)$ and $y(1)$. Compare with the results of Exercise 10.

12. Solve Exercise 10 by the Taylor approximation of order 2 with $h = 0.2$, and compare the results with the actual values (see Exercise 11).

13. Solve Exercise 10 by the Runge–Kutta approximation of order 3 with $h = 0.2$ and compare the results with the actual values (see Exercise 11).

14. Write a computer or calculator program for the Runge–Kutta approximation of order 3 with $h = 0.1$ to solve Exercise 10.

7.6 APPLICATIONS

Flood Waves ■ In his study of the behavior of flood waves in rivers, Stoker,[5] derives the following differential equation:

$$(v_0 + c_0)^2 \frac{dc_1}{d\xi} - 3c_1^2 + c_1 gS\left(\frac{1}{v_0} - \frac{2}{3c_0}\frac{gB}{gB + 2c_0^2}\right) = 0. \tag{1}$$

In this equation g represents the acceleration due to gravity ($g = 32.16$ ft/sec^2), B represents the width of the river (any cross section of the river perpendicular to the direction of the river flow is assumed to be a rectangle of constant width B but variable depth y), S represents the slope of the river bed, v_0 is the initial velocity of the wave, c_0 is related to the initial depth y_0 by $c_0 = (gy_0)^{1/2}$ (note that c_0 has units of velocity), c_1 is an unknown function that is related to the depth y of the river by the approximation $c_1(\xi) \approx (\sqrt{gy} - c_0)/\tau$. Furthermore, $\xi = x$, $\tau = (v_0 + c_0)t - x$, and t is time, x is a space variable whose axis is parallel to the direction of flow of the river.

Equation (1) is a Bernoulli equation and hence could be solved by the method associated with that type of differential equation (see Exercise 19). The results are rather cumbersome, however, so numerical methods might be more appropriate as a means of solution (see Exercises 1 through 4).

An Outstanding Long Jump ■ A novel and entertaining application of differential equations is presented in an article by Brearley.[6]

During the 1968 Olympic Games in Mexico City, R. Beamon of the United States set a world record of 29 feet $2\frac{1}{2}$ inches in the long jump. This improved the existing world mark by an astounding 2 feet $7\frac{5}{8}$ inches. Many critics attributed the length of the jump to the thinness of the air at Mexico City. In his article Brearley attempts to disclaim these critics. We present here Brearley's model only and recommend that the reader consult the article for Brearley's complete argument.

After takeoff, the center of mass G of the long jumper describes a path that one can study as a projectile moving through a resisting medium. The equation of motion in vector form is

$$M\frac{d\mathbf{V}}{dt} = M\mathbf{g} + \mathbf{D}, \tag{2}$$

where M is the mass of the long jumper, $\mathbf{V}$ is the velocity vector of G at any instant, $\mathbf{g}$ is the acceleration vector due to gravity, and $\mathbf{D}$ is the air drag on the long jumper.

The direction of $\mathbf{D}$ is opposite to that of $\mathbf{V}$, and the magnitude of $\mathbf{D}$ is given experimentally as $k\rho V^2$, where ρ is air density, $V = |\mathbf{V}|$, and k is constant for fixed body posture. Hence, $\mathbf{D} = -\rho k V^2 \mathbf{U}$, where $\mathbf{U}$ is a unit vector in the direction of $\mathbf{V}$.

[5] J. J. Stoker, *Water Waves* (New York: Interscience, 1957).

[6] M. N. Brearley, "The Long Jump Miracle of Mexico City," *Math. Magazine* 45 (1972).

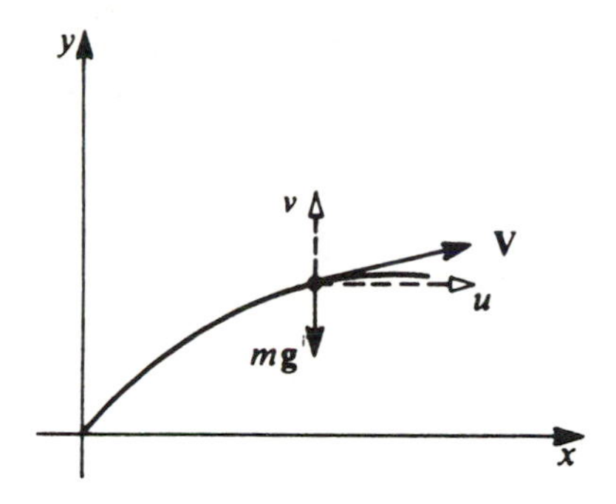

Figure 7.7

The path of G is considered relative to a two-dimensional coordinate system (x, y) with the origin located at the position of G at $t = 0$ (takeoff). At time t after takeoff, u and v represent the velocity components of G parallel to the x and y axes, respectively (see Figure 7.7).

Equation (2) can be written as the following system

$$M\frac{du}{dt} = -\rho k u(u^2 + v^2)^{1/2} \tag{3}$$

$$M\frac{dv}{dt} = -Mg - \rho k v(u^2 + v^2)^{1/2}. \tag{4}$$

We also have the initial conditions $u(0) = u_0$, $v(0) = v_0$. Brearley gives the values

$$u_0 = 9.45 \text{ m/sec} \tag{5}$$

$$v_0 = 4.15 \text{ m/sec} \tag{6}$$

$$k = 0.182 \text{ m}^2 \tag{7}$$

$$\rho = 0.984 \text{ kg/m}^3 \quad \text{(at Mexico City)} \tag{8}$$

$$M = 80 \text{ kg} \quad \text{(approximate mass of R. Beamon).} \tag{9}$$

Equations (3) and (4) are difficult to solve in their present form and hence can be studied numerically. Equations (3) and (4), treated as a system, can more conveniently be solved numerically as indicated in Section 7.5. However, we shall modify the problem by making an assumption that Brearley did not make. (See Exercises 5 through 8.)

EXERCISES

For the differential equation (1) applied to the Pawcatuck River in Rhode Island, take the following values: $B = 500$ ft, $S = 0.4$ ft/mile, $y_0 = 15$ ft, $v_0 = 2.19$

mph, and $c_0 = 15.6$ mph. With this information and $c_1(0) = 22.3$, solve Eq. (1) numerically on the interval $0 \le \xi \le 1$ using $h = 0.2$ by:

1. the Euler method;

2. the Taylor approximation of order 2;

3. the Runge–Kutta method of order 3;

4. the Runge–Kutta method of order 4.

In Eq. (3) we assume that $v = \frac{1}{2}(u - 1.15)$, so we obtain

$$M\frac{du}{dt} = -\rho k u[u^2 + \tfrac{1}{4}(u - 1.15)^2]^{1/2}, \tag{10}$$

which is a differential equation involving only the unknown function u. Using the values given in (5), (7), (8), and (9), solve the differential equation (10) numerically on the interval $0 \le t \le 1$ with $h = 0.2$ by:

5. the Euler method;

6. the Taylor approximation of order 2;

7. the Runge–Kutta method of order 3;

8. the Runge–Kutta method of order 4.

Heat Loss Consider a mass M of water of specific heat unity that is mixed so well that its temperature T is the same throughout. T_1 represents the temperature of the medium surrounding the mass. The rate of loss of heat from the mass is given by the differential equation

$$M\frac{dT}{dt} = -A(T^4 - T_1^4) \tag{11}$$

Assuming that $M = 100$, $A = 3 \times 10^{-4}$, $T_1 = 10$, and $T_0 = T(0) = 30$, solve Eq. (11) numerically on $0 \le t \le 1$ with $h = 0.2$ by:

9. the Euler method;

10. the Taylor approximation of order 2;

11. the Runge–Kutta method of order 3;

12. the Runge–Kutta method of order 4.

13. The differential equation (11) can be solved by separation of variables and partial fractions. Solve Eq. (11) by these techniques. Observe that the expression involved in the solution is not convenient for obtaining numerical values for T.

14. **Gas Ionization** A gas is to be ionized in such a way that the number of electrons per unit volume equals the number of positive ions per unit volume. Positive ions and electrons recombine to form neutral molecules at a rate equal to kn^2, where the constant k is called the *constant of recombination.* We assume that the gas is initially ($t = 0$) un-ionized, and for $t > 0$, A (a constant) ions per unit volume are produced. This leads to the IVP

$$\frac{dn}{dt} = A - kn^2 \qquad (12)$$

$$n(0) = 0. \qquad (13)$$

If we take $A = 100{,}000$ and $k = 5 \times 10^{-6}$, solve the IVP (12)–(13) numerically on $0 \leq t \leq 1$ with $h = 0.2$ by Euler's method.

15. Solve the IVP (12)–(13) exactly by separation of variables. Using the values of A and k given in Exercise 14, compare the numerical results of Exercise 14 with the actual results.

16. Physics In the study of the simple pendulum (see Chapter 8), the following differential equation occurs:

$$\frac{d\theta}{dt} = -\sqrt{\frac{g}{l}}(2\cos\theta - 2\cos\alpha)^{1/2}, \qquad (14)$$

where θ is the angle that the pendulum makes with the vertical, g is the acceleration due to gravity, l is the length of the pendulum, and $\theta(0) = \theta_0 = \alpha$. If we assume that $\sqrt{g/l} = 2$ and $\alpha = \pi/6$, solve Eq. (14) numerically on $0 \leq t \leq 1$ with $h = 0.2$ using the Taylor approximation of order 2.

17. Astronomy If one takes into account the general theory of relativity when studying planetary orbits, the following differential equation occurs[7]:

$$\frac{d^2u}{d\theta^2} + u = \frac{GM}{h^2}(1 + \epsilon u^2). \qquad (15)$$

In this differential equation r, θ are polar coordinates, the equation of the orbit is assumed in the form $r = r(\theta)$, $u = r^{-1}$, M is the mass of the sun, G is a universal constant, h is twice the area swept out by the radius vector in unit time, and ϵ is a small parameter. If we set $v = du/d\theta$, we can write $d^2u/d\theta^2 = v(dv/du)$ and the differential equation (15) takes the form

$$\frac{dv}{du} = \frac{(GM/h^2)(1 + \epsilon u^2) - u}{v}. \qquad (16)$$

For simplicity we take $GM/h^2 = 1$, $\epsilon = 0.01$ and $v(1) = 1$. Solve Eq. (16) numerically on $1 \leq u \leq 2$ with $h = 0.2$ by the Runge–Kutta method of order 3.

[7]C. C. Lin and L.A. Segel, *Mathematics Applied to Deterministic Problems in the Natural Sciences* (New York: Macmillan, 1974).

18. The following differential equation occurs in nonlinear oscillation theory in the consideration of limit cycles[8]:

$$\frac{d\nu}{d\xi} = \alpha^2 \frac{(\nu - \nu^2/3) - \xi}{\nu}, \tag{17}$$

where ν and ξ are space variables and α is a parameter. Choose $\alpha = 0.1$, $\nu(0) = 1$, and solve Eq. (17) on $0 \leq \xi \leq 1$ with $h = 0.2$ using the Runge–Kutta method of order 4.

19. Equation (1) can be written in the form $(dc_1/d\xi) + ac_1 = bc_1^2$, where the constants a and b are given by

$$a = \frac{gS}{(\nu_0 + c_0)^2}\left(\frac{1}{\nu_0} - \frac{2}{3c_0}\frac{gB}{gB + 2c_0^2}\right)$$

$$b = \frac{3}{(\nu_0 + c_0)^2}.$$

Solve the differential equation $(dc_1/d\xi) + ac_1 = bc_1^2$ by the method associated with Bernoulli's differential equation leaving the answer in terms of a and b. (See Exercise 27, Section 1.4.)

REVIEW EXERCISES

Exercises 1 through 5 relate to the following initial value problem:

$$\frac{dy}{dx} = x + y + y^2 \tag{A}$$

$$y(1) = 0. \tag{B}$$

1. Solve the initial value problem (A)–(B) on the interval $1 \leq x \leq 2$ by the Euler method with $h = 0.2$.

2. Solve the initial value problem (A)–(B) on the interval $1 \leq x \leq 2$ by the Taylor approximation of order 2 with $h = 0.2$.

3. Solve the initial value problem (A)–(B) on the interval $1 \leq x \leq 2$ by the Taylor approximation of order 3 with $h = 0.2$.

4. Solve the initial value problem (A)–(B) on the interval $1 \leq x \leq 2$ by the Runge–Kutta approximation of order 3 with $h = 0.2$.

5. Solve the initial value problem (A)–(B) on the interval $1 \leq x \leq 2$ by the Runge–Kutta approximation of order 4 with $h = 0.2$.

Exercises 6 through 10 relate to the following initial value problem:

$$\dot{x} = -2xy, \qquad \dot{y} = x^2 + y^2 \tag{C}$$

$$x(1) = 1, \qquad y(1) = -1. \tag{D}$$

[8]J. J. Stoker, *Nonlinear Vibrations in Mechanical and Electrical Systems* (New York: Interscience, 1950).

6. Solve the initial value problem (C)–(D) on the interval $1 \leq t \leq 2$ by the Euler method with $h = 0.2$.

7. Solve the initial value problem (C)–(D) on the interval $1 \leq t \leq 2$ by the Runge–Kutta approximation of order 3 with $h = 0.2$.

8. Solve the initial value problem (C)–(D) on the interval $1 \leq t \leq 2$ by the Taylor approximation of order 2 with $h = 0.2$.

9. Solve the initial value problem (C)–(D) on the interval $1 \leq t \leq 2$ by the Runge–Kutta approximation of order 4 with $h = 0.2$.

10. If $(x(t),y(t))$ is a solution of the initial value problem (C)–(D), show that

$$y = -\sqrt{\frac{4-x^3}{3x}}.$$

11. Use the results of either Exercise 6, 7, 8, or 9 to determine whether or not the relationship in Exercise 10 is valid for all $t \geq 1$.

If we account[9] for the nonlinear dissipative effects in the neck of a Helmholtz cavity resonator, an approximate equation for free oscillations of such a resonator is of the form

$$m\frac{d^2x}{dt^2} + R\frac{dx}{dt}\left[1 + \beta\left(\frac{dx}{dt}\right)^2\right] + Kx = 0. \tag{E}$$

12. Setting $v = \dfrac{dx}{dt}$ and $v\dfrac{dv}{dx} = \dfrac{d^2x}{dt^2}$ in Eq. (E), show that this equation can be rewritten in the form

$$\frac{dv}{dx} = -\frac{\left(\frac{K}{R}\right)x + v(1 + \beta v^2)}{\left(\frac{m}{R}\right)v}.$$

Taking $\dfrac{m}{R} = 900/(K/R)$, $K/R = 100$, and $\beta = 2 \cdot 10^{-5}$, solve this equation on the interval $0 \leq x \leq 1$ by the Runge–Kutta approximation of order 3 with $h = 0.2$ if it is given that $v = 700$ cm/sec when $x = 0$.

Exercises 13 through 15 relate to the following initial value problem (see Example 2, Section 1.5):

$$\frac{dy}{dx} = \frac{(4-y)[(x\cos x)\ln(2y-8) + 1]}{x\sin x}, \tag{F}$$

$$y(1) = \frac{9}{2}. \tag{G}$$

[9]This is Exercise 3 in Philip M. Morse and K. Uno Ingard, *Theoretical Acoustics* (New York: McGraw-Hill, 1968), p. 883. Reprinted by permission of McGraw-Hill Book Company.

13. Solve the initial value problem (F)–(G) on the interval $1 \leq x \leq 2$ by the Euler method with $h = 0.2$ and compare the approximate results with the exact results.

14. Solve the initial value problem (F)–(G) on the interval $1 \leq x \leq 2$ by the Taylor approximation of order 2 with $h = 0.2$ and compare the approximate results with the exact results.

15. Solve the initial value problem (F)–(G) on the interval $1 \leq x \leq 2$ by the Runge–Kutta approximation of order 3 with $h = 0.2$ and compare the approximate results with the exact results.

CHAPTER 8

Nonlinear Differential Equations and Systems

8.1 INTRODUCTION

Nonlinear differential equations and systems of nonlinear differential equations occur frequently in applications. However, only a few types of nonlinear differential equations (for example, separable, homogeneous, exact) can be solved explicitly. The same is true for nonlinear systems. Fortunately, the situation is not as hopeless as it seems. We are often primarily interested in certain properties of the solutions and not the solutions themselves. For example, are the solutions bounded for all time? Are they periodic? Does the limit of the solution exist as $t \to +\infty$? A very effective technique in studying nonlinear differential equations and systems is to "linearize" them, that is, to approximate them by linear differential equations. Also, the coefficients and initial conditions of differential equations are often determined approximately. It is therefore very important to know what effect "small" changes in the coefficients of differential equations and their initial conditions have on their exact solutions. It is often possible to answer some of the questions above by utilizing the form of the differential equation and properties of its coefficients without the luxury of knowing the exact solution.

The questions raised above belong to that branch of differential equations known as *qualitative theory*. Our aim in this chapter is to present some elementary qualitative results for nonlinear differential equations and systems.

8.2 EXISTENCE AND UNIQUENESS THEOREMS

Here we state an existence and uniqueness theorem for systems of two differential equations of the first order with two unknown functions of the form

$$\begin{aligned} \dot{x}_1 &= f_1(t, x_1, x_2) \\ \dot{x}_2 &= f_2(t, x_1, x_2), \end{aligned} \tag{1}$$

subject to the initial conditions

$$x_1(t_0) = x_1^0, \qquad x_2(t_0) = x_2^0. \tag{2}$$

The same theorem applies to second-order differential equations of the form

$$\ddot{y} = f(t, y, \dot{y}), \tag{3}$$

subject to the initial conditions

$$y(t_0) = y^0, \qquad \dot{y}(t_0) = y^1. \tag{4}$$

In fact, if we set

$$y = x_1 \quad \text{and} \quad \dot{y} = x_2, \tag{5}$$

the differential equation (3) is equivalent to the system

$$\begin{aligned} \dot{x}_1 &= x_2 \\ \dot{x}_2 &= f(t, x_1, x_2), \end{aligned} \tag{6}$$

and the initial conditions (4) are equivalent to

$$x_1(t_0) = y^0, \qquad x_2(t_0) = y^1. \tag{7}$$

Before we state the existence and uniqueness theorem we explain what is meant by a *Lipschitz condition.*

DEFINITION 1

Let $f(t, x_1, x_2)$ be a function defined in a region $\mathcal{R}$ of the three-dimensional Euclidean space R^3. We say that $f(t, x_1, x_2)$ satisfies a Lipschitz condition (with respect to x_1 and x_2) in $\mathcal{R}$ if there exist constants L_1 and L_2 such that

$$|f(t, x_1, x_2) - f(t, y_1, y_2)| \le L_1 |x_1 - y_1| + L_2 |x_2 - y_2|$$

for all points (t, x_1, x_2), (t, y_1, y_2) in $\mathcal{R}$. The constants L_1 and L_2 are called Lipschitz constants. When the partial derivatives $\partial f/\partial x_1$ and $\partial f/\partial x_2$ exist and are continuous, it can be shown that f satisfies a Lipschitz condition.

THEOREM 1 (Existence and Uniqueness)

Let each of the functions $f_1(t, x_1, x_2)$ and $f_2(t, x_1, x_2)$ be continuous and satisfy a Lipschitz condition (with respect to x_1 and x_2) in the region

$$\mathcal{R} = \left\{ (t, x_1, x_2)\colon \begin{array}{r} |t - t_0| \le A \\ |x_1 - x_1^0| \le B \\ |x_2 - x_2^0| \le C \end{array} \right\}. \tag{8}$$

Then the initial value problem (1)–(2) has a unique solution defined in some interval $a < t < b$ about the point t_0.

COROLLARY 1

Let the function $f(t, x_1, x_2)$ be continuous and satisfy a Lipschitz condition (with respect to x_1 and x_2) in the region $\mathcal{R}$ defined by Eq. (8) with $x_1^0 = y^0$ and $x_2^0 = y^1$. Then the IVP (3)–(4) has a unique solution defined in some interval $a < t < b$ about the point t_0.

Theorem 1 and its corollary can be easily extended to systems of n differential equations with n unknown functions and to differential equations of order n, respectively.

With the hypotheses of Theorem 1 it can be shown that the solutions of system (1) depend continuously on the initial conditions (2). Therefore, "small" changes in the initial conditions (2) will result in small changes in the solutions of system (1). This result is very important in physical applications, because we can never measure the initial conditions exactly.

EXERCISES

1. Show that the function $f(t, x_1, x_2) = x_1^{1/2} + x_2$ does not satisfy a Lipschitz condition (with respect to x_1 and x_2) in the region

$$-1 \leq t \leq 1, \qquad 0 \leq x_1 \leq 1, \qquad 0 \leq x_2 \leq 1.$$

How about in the region

$$-1 \leq t \leq 1, \qquad 3 \leq x_1 \leq 4, \qquad 0 \leq x_2 \leq 1?$$

2. For what points (t_0, y^0, y^1) does Corollary 1 imply that the IVP

$$\ddot{y} = \dot{y} + y^{1/2}$$
$$y(t_0) = y^0$$
$$\dot{y}(t_0) = y^1$$

has a unique solution?

3. Suppose that we make an error of magnitude 10^{-3} in measuring the initial conditions A and B in the IVP

$$\ddot{y} + y = 0$$
$$y(0) = A$$
$$\dot{y}(0) = B.$$

What is the largest error we make in evaluating the solution $y(t)$?

8.3 SOLUTIONS AND TRAJECTORIES OF AUTONOMOUS SYSTEMS

A system of differential equations of the form

$$\dot{x} = f(x, y), \qquad \dot{y} = g(x, y), \tag{1}$$

where the functions f and g are not time-dependent is called an *autonomous system*. Second-order differential equations of the form

$$\ddot{x} = F(x, \dot{x}) \tag{2}$$

can be also written as an autonomous system by setting

$$\dot{x} = y, \qquad \dot{y} = F(x, y).$$

For the sake of existence and uniqueness of solutions, we assume that each of the functions f and g is continuous and satisfies a Lipschitz condition (with respect to x and y) in some region $\mathcal{R}$ of the xy plane. Then by Theorem 1 of Section 8.2, if (x_0, y_0) is any point in $\mathcal{R}$ and if t_0 is any real number, there exists a unique solution

$$x = x(t), \qquad y = y(t) \tag{3}$$

of system (1), defined in some interval $a < t < b$ which contains the point t_0, and satisfying the initial conditions

$$x(t_0) = x_0, \qquad y(t_0) = y_0. \tag{4}$$

Clearly, the solution (3) defines a curve in the three-dimensional space t, x, y. If we regard t as a parameter, then as t varies in the interval $a < t < b$, the points $(x(t), y(t))$ trace out a curve in the xy plane called the *trajectory* or *orbit* of the solution (3). In the study of physical systems, the pair (x, y) is called a *phase* of the system and, therefore, the xy plane is usually called the *phase plane*. Let us emphasize again that a solution of system (1) is a curve in the three-dimensional space t, x, y, while a trajectory is a curve in the phase plane that is described parametrically by a solution of system (1).

For example, $x = \cos t$, $y = \sin t$, $-\infty < t < \infty$ is a solution of the system

$$\dot{x} = -y, \qquad \dot{y} = x. \tag{5}$$

This solution is a helix in the three-dimensional space t, x, y. And, as t varies in the interval $-\infty < t < \infty$, the points $(\cos t, \sin t)$ trace out the unit circle $x^2 + y^2 = 1$ an infinite number of times. Thus, the trajectory of this solution is the circle $x^2 + y^2 = 1$ in the xy plane. Another solution of system (5) is $x = \sin t$, $y = -\cos t$. The trajectory of this solution is again the unit circle $x^2 + y^2 = 1$. As this example shows, different solutions of a system may have the same trajectory.

With the assumptions on f and g made above, it can be shown that system (1) has exactly one trajectory through any point $(x_0, y_0) \in \mathcal{R}$. If we know a solution (3) of system (1), we can find the corresponding trajectory by eliminating, if possible, the parameter t from the two equations in (3). This was actually done in the example above. However, we can get a lot of information about the trajectories of a system even when we cannot find the solutions explicitly. In fact, in some cases we can find the trajectories explicitly, while an explicit solution of the system is impossible. To see this, assume that in some region D, $f(x, y) \neq 0$. Then, using the two equations in (1) and the chain rule, we obtain the first-order differential equation

$$\frac{dy}{dx} = \frac{dy}{dt} \cdot \frac{dt}{dx} = \frac{g(x, y)}{f(x, y)}. \tag{6}$$

The solutions of Eq. (6) give the trajectories of system (1) through the points of D. If we can solve Eq. (6), then we obtain explicitly the trajectories. Even if we are unable to find explicit solutions, we can use Eq. (6) to compute the slope of any trajectory through a point of D. Similarly, if $g(x, y) \neq 0$ in some region S, we can utilize the first-order differential equation

$$\frac{dx}{dy} = \frac{f(x, y)}{g(x, y)} \tag{7}$$

to compute the trajectories through the points of S.

The points (x_0, y_0) in the xy plane (the phase plane), where both functions f and g are zero, are of special interest. For such points neither Eq. (6) nor Eq. (7) is valid. Such points are called *critical points* (or *equilibrium points*) of system (1). Clearly, if (x_0, y_0) is a critical point of system (1), then $x(t) = x_0$, $y(t) = y_0$ is a solution for all t. The trajectory corresponding to this solution is the single point (x_0, y_0) in the phase plane.

As we mentioned above, Eq. (2) can be easily put in the form of system (1) by setting $\dot{x} = y$. Then

$$\dot{x} = y, \qquad \dot{y} = F(x, y). \tag{8}$$

If (x_0, y_0) is a critical point of (8), it follows that $y_0 = 0$, and hence

$$x(t) = x_0, \qquad y(t) = 0 \tag{9}$$

is a solution of (8) for all t. Thus, if x represents the position and y the velocity of a particle moving according to the differential equation (2), it follows that $x = x_0$ is a position of equilibrium of the motion. With such an interpretation, we see that critical points are identified with positions of equilibrium.

EXAMPLE 1 Find the critical points and the trajectories of the following system of differential equations:

$$\dot{x} = -y, \qquad \dot{y} = x. \tag{10}$$

In particular, draw the trajectory corresponding to the solution of (10) with initial conditions

$$x(0) = 1, \qquad y(0) = \sqrt{3}. \tag{11}$$

Solution The critical points of (10) are determined by the two equations $-y = 0$, $x = 0$. Thus, $(0, 0)$ is the only critical point of (10). Using Eq. (6), we see that the trajectories of (10) for $y \neq 0$ are the solution curves of the differential equation

$$\frac{dy}{dx} = -\frac{x}{y}, \qquad y \neq 0.$$

The general solution of this separable differential equation is

$$x^2 + y^2 = c^2. \tag{12}$$

For $y = 0$ we see from (10) that $x = 0$. Thus, $(0, 0)$ is also a trajectory that is contained in Eq. (12) for $c = 0$. Hence, Eq. (12), which is a one-parameter family of circles centered at the origin, gives all the trajectories of system (10)

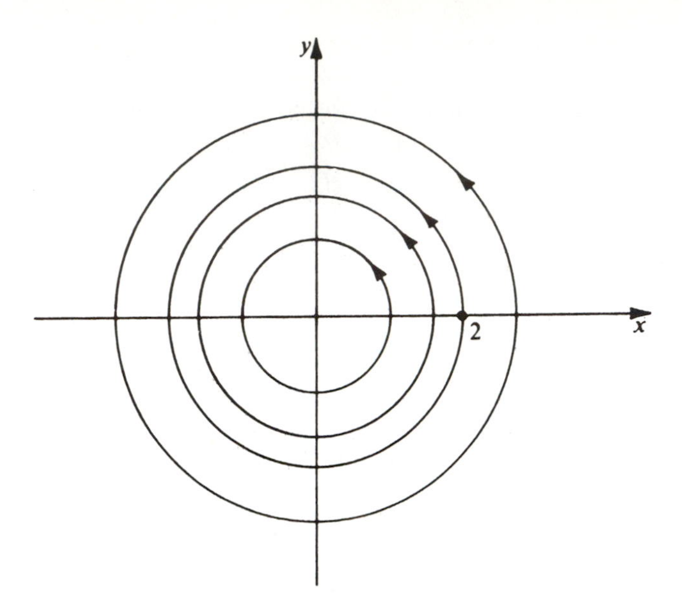

Figure 8.1

(see Figure 8.1). The arrows on the trajectories indicate the direction of increasing t and can be found from system (10). For example, when $y > 0$, the first equation in (10) implies that $\dot{x} < 0$. This means that $x(t)$ decreases (and so it moves in the counterclockwise direction).

From Eq. (12) we see that the trajectory of the solution of the IVP (10)–(11) is the circle $x^2 + y^2 = 4$, centered at the origin with radius equal to 2.

Section 8.5 contains many worked-out examples in which we find the shape of all trajectories of linear systems.

EXERCISES

In Exercises 1 through 4, find the critical points and the trajectories of the system. Indicate the direction of increasing t.

1. $\dot{x} = x,\ \dot{y} = y$ **2.** $\dot{x} = -x,\ \dot{y} = 2y$

3. $\dot{x} = y,\ \dot{y} = x$ **4.** $\dot{x} = y,\ \dot{y} = -x$

In Exercises 5 through 10, compute the trajectory of the solution of the IVP. Indicate the direction of increasing t.

5. $\dot{x} = x,\ \dot{y} = 2y,\quad x(0) = 1,\quad y(0) = -1$

6. $\dot{x} = x,\ \dot{y} = -2y,\quad x(0) = 1,\quad y(0) = 1$

7. $\dot{x} = x,\ \dot{y} = x + y,\quad x(0) = 1,\quad y(0) = 0$

8. $\dot{x} = -x,\ \dot{y} = -x - y,\quad x(0) = 1,\quad y(0) = 0$

9. $\dot{x} = -x + y,\ \dot{y} = -x - y, \quad x(0) = -1,\ y(0) = 1$

10. $\dot{x} = -y,\ \dot{y} = x, \quad x(0) = 1, \quad y(0) = -1$

In Exercises 11 through 14, find the critical points and the equations of the trajectories of the solutions of the system.

11. $\dot{x} = -x^2 + y^2,\ \dot{y} = 2xy$ **12.** $\dot{x} = y,\ \dot{y} = -\sin x$

13. $\dot{x} = y,\ \dot{y} = -4 \sin x$ **14.** $\dot{x} = x - xy,\ \dot{y} = -y + xy$

In Exercises 15 through 18, transform the differential equation into an equivalent system (by setting $\dot{x} = y$) and compute the equation of the trajectories.

15. $\ddot{x} + x = 0$ **16.** $\ddot{x} + \sin x = 0$ **17.** $\ddot{x} - x = 0$ **18.** $\ddot{x} - x + x^3 = 0$

19. (a) Show that the point (0, 0) is the only critical point of the system

$$\dot{x} = ax + by, \qquad \dot{y} = cx + dy$$

if and only if $ad - bc \neq 0$.

(b) Show that the system has a line of critical points if $ad - bc = 0$.

8.4 STABILITY OF CRITICAL POINTS OF AUTONOMOUS SYSTEMS

Consider again the autonomous system

$$\dot{x} = f(x, y), \qquad \dot{y} = g(x, y). \tag{1}$$

We recall that a point (x_0, y_0) is a critical point (or an equilibrium point) of system (1) if $f(x_0, y_0) = 0$ and $g(x_0, y_0) = 0$. Since the derivative of a constant is zero, it follows that if the point (x_0, y_0) is a critical point of (1), then the pair of constant functions

$$x(t) \equiv x_0, \qquad y(t) \equiv y_0 \tag{2}$$

is a solution of (1) for all t.

In many situations it is important to know whether every solution of (1) that starts sufficiently close to the solution (2) at $t = 0$ will remain close to (2) for all future time $t > 0$. If this is true, the solution (2), or the critical point (x_0, y_0), is called stable. More precisely, we give the following definitions.

DEFINITION 1

The critical point (x_0, y_0) (or the constant solution (2)) of system (1) is called stable if for every positive number ϵ there corresponds a positive number δ such that, every solution $(x(t), y(t))$ of (1) which at $t = 0$ satisfies

$$[x(0) - x_0]^2 + [y(0) - y_0]^2 < \delta \tag{3}$$

exists and satisfies

$$[x(t) - x_0]^2 + [y(t) - y_0]^2 < \epsilon \tag{4}$$

for all $t \geq 0$.

DEFINITION 2

A critical point (x_0, y_0) [or the constant solution (2)] is called asymptotically stable if it is stable and if in addition there exists a positive number δ_0 such that, every solution $(x(t), y(t))$ of (1) which at $t = 0$ satisfies

$$[x(0) - x_0]^2 + [y(0) - y_0]^2 < \delta_0 \tag{5}$$

exists for all $t \geq 0$ and satisfies

$$\lim_{t\to\infty} x(t) = x_0, \qquad \lim_{t\to\infty} y(t) = y_0. \tag{6}$$

DEFINITION 3

A critical point that is not stable is called unstable.

Roughly speaking, stability means that a small change in the initial conditions causes only a small effect on the solution, asymptotic stability means that the effect of a small change tends to die out, while instability means that a small change in the initial conditions has a large effect on the solution.

The concepts of stable, asymptotically stable, and unstable critical point are illustrated in Figure 8.2(a), (b), and (c), respectively.

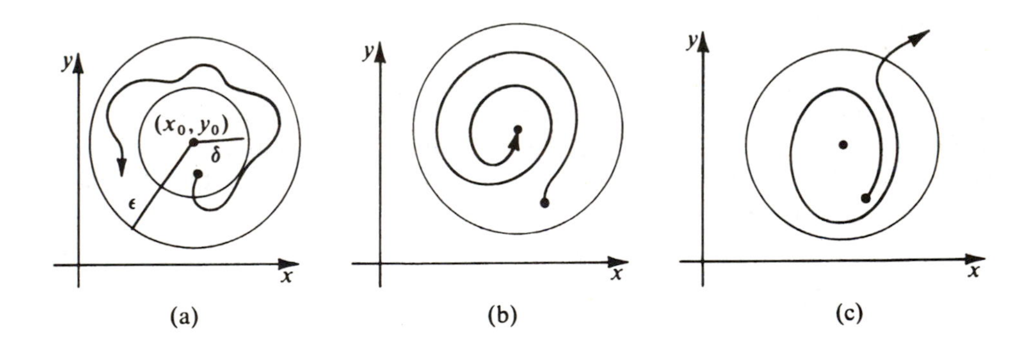

Figure 8.2

EXAMPLE 1 Show that the critical point (0, 0) of the system

$$\dot{x} = -y, \qquad \dot{y} = x \tag{7}$$

is stable. Is it asymptotically stable?

Solution We shall apply Definition 1. Let $\epsilon > 0$ be given. Choose $\delta = \epsilon$. Every solution of (7) is of the form

$$x(t) = c_1 \cos t + c_2 \sin t$$

$$y(t) = c_1 \sin t - c_2 \cos t,$$

where c_1 and c_2 are arbitrary constants. Here $x_0 = y_0 = 0$, $x(0) = c_1$, and $y(0) = -c_2$. We must show that (3) implies (4). In fact, $c_1^2 + c_2^2 < \delta$ implies that

$$(c_1 \cos t + c_2 \sin t)^2 + (c_1 \sin t - c_2 \cos t)^2 = c_1^2 + c_2^2 < \delta = \epsilon, \tag{8}$$

and the proof that the critical point (0, 0) of system (7) is stable is complete. From Eq. (8) we see that the trajectories of the system (7) are circles centered at the critical point (0, 0). Hence, they do not approach the critical point (0, 0) as $t \to \infty$. That is, Eq. (6) does not hold. Therefore, (0, 0) is not an asymptotically stable critical point of (7).

EXAMPLE 2 Show that the critical point (0, 0) of the system

$$\dot{x} = -x, \qquad \dot{y} = -y \tag{9}$$

is asymptotically stable.

Solution First, we must prove that (0, 0) is stable. Let $\epsilon > 0$ be given. Choose $\delta = \epsilon$. The general solution of (9) is

$$x(t) = c_1 e^{-t}, \qquad y(t) = c_2 e^{-t},$$

where c_1 and c_2 are arbitrary constants. Here $x_0 = y_0 = 0$, $x(0) = c_1$, and $y(0) = c_2$. Again we must show that (3) implies (4). In fact, $c_1^2 + c_2^2 < \delta$ implies that

$$(c_1 e^{-t})^2 + (c_2 e^{-t})^2 = (c_1^2 + c_2^2)e^{-2t} \le c_1^2 + c_2^2 < \delta = \epsilon.$$

Thus, (0, 0) is stable. Since (for any c_1 and c_2)

$$\lim_{t\to\infty} x(t) = \lim_{t\to\infty} c_1 e^{-t} = 0, \qquad \lim_{t\to\infty} y(t) = \lim_{t\to\infty} c_2 e^{-t} = 0,$$

it follows that the critical point (0, 0) is asymptotically stable.

EXAMPLE 3 Show that the critical point (0, 0) of the system

$$\dot{x} = -3x + 4y, \qquad \dot{y} = -2x + 3y \tag{10}$$

is unstable.

Solution If the critical point (0, 0) were stable, then given $\epsilon > 0$ there should exist a $\delta > 0$ such that (3) implies (4). Consider the solution

$$x(t) = \frac{\sqrt{\delta}}{2} e^t, \qquad y(t) = \frac{\sqrt{\delta}}{2} e^t$$

of (10). Here $x_0 = y_0 = 0$, $x(0) = y(0) = \sqrt{\delta}/2$, and therefore Eq. (3) is satisfied. However, Eq. (4) becomes

$$\frac{\delta}{2} e^{2t} < \epsilon,$$

which cannot be true for all $t \geq 0$. Hence, (0, 0) is an unstable critical point of the system (10).

When the autonomous system (1) is linear with constant coefficients, that is, when

$$\dot{x} = ax + by, \qquad \dot{y} = cx + dy, \tag{11}$$

where a, b, c, and d are constants, we can obtain the solution explicitly. Therefore, it is not surprising that the stability character of the critical point (0, 0) of system (11) is easy to study. We shall assume that $ad - bc \neq 0$. Then the point (0, 0) is the only critical point of (11). [See Exercise 19(a) in Section 8.3.] As we explained in Section 3.3, the solutions of system (11) are of the form

$$x = Ae^{\lambda t}, \qquad y = Be^{\lambda t},$$

where λ is a root of the characteristic equation

$$\lambda^2 - (a + d)\lambda + ad - bc = 0. \tag{12}$$

The stability character of the critical point (0, 0) of system (11) depends almost entirely on the roots of Eq. (12). In fact, we have the following theorem.

THEOREM 1

(a) The critical point (0, 0) *of system* (11) *is stable if, and only if, both roots of Eq.* (12) *are real and negative or have nonpositive real parts.*

(b) The critical point (0, 0) *of system* (11) *is asymptotically stable if, and only if, both roots of Eq.* (12) *are real and negative or have negative real parts.*

(c) The critical point (0, 0) *of system* (11) *is unstable if one (or both) of the roots of Eq.* (12) *is real and positive or if at least one of the roots has a positive real part.*

The three examples considered were all linear. In Example 1, the characteristic equation is (here $a = 0, b = -1, c = 1$, and $d = 0$) $\lambda^2 + 1 = 0$. Its roots are $\pm i$. Since they have zero real parts, using Theorem 1(a), it follows that the critical point (0, 0) of system (7) is stable.

In Example 2, the characteristic equation is (here $a = -1, b = 0, c = 0$, and $d = -1$) $\lambda^2 + 2\lambda + 1 = 0$. Its roots are $\lambda_1 = \lambda_2 = -1$ (a double root). Since they are real and negative, using Theorem 1(b) we see that the critical point (0, 0) of system (9) is asymptotically stable.

In Example 3, the characteristic equation is (here $a = -3, b = 4, c = -2$, and $d = 3$) $\lambda^2 - 1 = 0$. Its roots are $\lambda_1 = 1$ and $\lambda_2 = -1$. Since one of the roots is real and positive, using Theorem 1(c) it follows that the critical point (0, 0) of system (10) is unstable.

Consider again the autonomous system (1) and assume that (x_0, y_0) is a critical point of the system. Since a transformation of the form $X = x - x_0$, $Y = y - y_0$ transforms the autonomous system (1) into an equivalent system with (0, 0) as a critical point, we assume, without loss of generality, that (0, 0) is a critical point of system (1). An effective technique in studying the autonomous system (1) near the critical point (0, 0) is to approximate it by a linear system of the form (11). In many cases one can prove that if the approximation is "good", the linear system (11) has solutions which themselves are "good" approximations to the solutions of system (1). The result that we are about to state is of this nature.

Assume that system (1) is of the form

$$\begin{aligned} \dot{x} &= ax + by + F(x, y) \\ \dot{y} &= cx + dy + G(x, y), \end{aligned} \tag{13}$$

with $ad - bc \neq 0$ and $F(0, 0) = G(0, 0) = 0$. [Hence (0, 0) is a critical point of (13).] Assume, furthermore, that near the origin (0, 0), the functions F and G are continuous, have continuous first partial derivatives, and that

$$\lim_{\substack{x\to 0\\ y\to 0}} \frac{F(x, y)}{\sqrt{x^2 + y^2}} = \lim_{\substack{x\to 0\\ y\to 0}} \frac{G(x, y)}{\sqrt{x^2 + y^2}} = 0. \tag{14}$$

[Roughly speaking, the condition (14) means that the linear system (11) is a good approximation of system (13).] Then the following result holds.

THEOREM 2

(a) *The critical point* (0, 0) *of the nonlinear system* (13) *is asymptotically stable if the critical point* (0, 0) *of the "linearized" system* (11) *is asymptotically stable.*

(b) *The critical point* (0, 0) *of the nonlinear system* (13) *is unstable if the critical point* (0, 0) *of system* (11) *is unstable.*

This theorem offers no conclusion about system (13) when (0, 0) is just a stable point of system (11).

EXAMPLE 4 Show that the critical point (0, 0) of the nonlinear system

$$\begin{aligned} \dot{x} &= -x + y + (x^2 + y^2) \\ \dot{y} &= -2y - (x^2 + y^2)^{3/2} \end{aligned} \tag{15}$$

is asymptotically stable.

Solution Here $a = -1$, $b = 1$, $c = 0$, $d = -2$, and $ad - bc = 2 \neq 0$. $F(x, y) = (x^2 + y^2)$, $G(x, y) = -(x^2 + y^2)^{3/2}$, and $F(0, 0) = G(0, 0) = 0$. Clearly, the conditions (14) are satisfied. The linearized system is

$$\dot{x} = -x + y, \qquad \dot{y} = -2y. \tag{16}$$

The characteristic equation of system (16) is $\lambda^2 + 3\lambda + 2 = 0$. Its roots are $\lambda_1 = -1$ and $\lambda_2 = -2$. Since they are both negative, the point (0, 0) is an asymptotically stable critical point of (16). Hence, by Theorem 2(a), the point (0, 0) is also an asymptotically stable critical point of the nonlinear system (15).

EXAMPLE 5 Show that the critical point (0, 0) of the nonlinear system

$$\begin{aligned} \dot{x} &= -3x + 4y + x^2 - y^2 \\ \dot{y} &= -2x + 3y - xy \end{aligned} \tag{17}$$

is unstable.

Solution Here $a = -3$, $b = 4$, $c = -2$, and $d = 3$ with $ad - bc = -1 \neq 0$. $F(x, y) = x^2 - y^2$, $G(x, y) = -xy$, with $F(0, 0) = G(0, 0) = 0$. We express x and y in polar coordinates: $x = r\cos\theta$, $y = r\sin\theta$. Then ($x \to 0$ and $y \to 0$ is equivalent to $r \to 0$)

$$\frac{F(x, y)}{\sqrt{x^2 + y^2}} = \frac{r^2(\cos^2\theta - \sin^2\theta)}{r} = r\cos 2\theta \to 0 \qquad \text{as } r \to 0$$

and

$$\frac{G(x, y)}{\sqrt{x^2 + y^2}} = -\frac{r^2\cos\theta\sin\theta}{r} = -r\cos\theta\sin\theta \to 0 \qquad \text{as } r \to 0.$$

Hence, the conditions (14) are satisfied. The linearized system is

$$\dot{x} = -3x + 4y, \qquad \dot{y} = -2x + 3y. \tag{18}$$

The characteristic equation of system (18) is $\lambda^2 - 1 = 0$. Its roots are $\lambda_1 = 1$ and $\lambda_2 = -1$. Since one of them is positive, the point (0, 0) is an unstable critical point of (18). By Theorem 2(b), the point (0, 0) is also an unstable critical point of the nonlinear system (17).

EXERCISES

By using the definition, determine whether the critical point (0, 0) of each of the systems in Exercises 1 through 3 is stable, asymptotically stable, or unstable.

1. $\dot{x} = y$, $\dot{y} = -x$ **2.** $\dot{x} = -y$, $\dot{y} = -x$ **3.** $\dot{x} = -x + y$, $\dot{y} = -2y$.

In Exercises 4 through 10, determine whether the critical point (0, 0) of the system is stable, asymptotically stable, or unstable.

4. $\dot{x} = -x + y$, $\dot{y} = -x - y$

5. $\dot{x} = x - y$, $\dot{y} = x + 3y$

6. $\dot{x} = x - y$, $\dot{y} = 5x - y$

7. $\dot{x} = -3x + 4y$, $\dot{y} = -2x + 3y$

8. $\dot{x} = 5x - 6y$, $\dot{y} = 6x - 7y$

9. $\dot{x} = -3x + 5y$, $\dot{y} = -x + y$

10. $\dot{x} = 3x - 2y$, $\dot{y} = 4x - y$

11. Assume that the system

$$\dot{x} = 4x - 6y, \qquad \dot{y} = 8x - 10y$$

represents two competing populations, where x is the desirable population and y is a parasite. Show that the critical point (0, 0) of this system is asymptotically stable and therefore both populations are headed for extinction.

12. Show that the critical point (1, 1) of the system

$$\dot{x} = 5x - 6y + 1, \qquad \dot{y} = 6x - 7y + 1,$$

which appears in Turing's theory of morphogenesis is asymptotically stable. [*Hint:* Transform the critical point to (0, 0) by means of $X = x - 1$, $Y = y - 1$.]

In Exercises 13 through 17, show that the point (0, 0) is a critical point for the given system and determine, if possible, whether it is asymptotically stable or unstable.

13. $\dot{x} = 5x - 6y + xy$, $\dot{y} = 6x - 7y - xy$

14. $\dot{x} = 3x - 2y + (x^2 + y^2)^2$, $\dot{y} = 4x - y + (x^2 - y^2)^5$

15. $\dot{x} = y + x^2 - xy$, $\dot{y} = -2x + 3y + y^2$

16. $\dot{x} = x - xy$, $\dot{y} = -y + xy$

17. $\dot{x} = -x - x^2 + y^2$, $\dot{y} = -y + 2xy$

Transform the differential equation $\ddot{x} + p\dot{x} + qx = 0$ into an equivalent system by setting $\dot{x} = y$, and investigate the stability character of the resulting system at the critical point (0, 0) in each of the following cases.

18. $p^2 > 4q$, $q > 0$, and $p < 0$

19. $p^2 > 4q$, $q > 0$, and $p > 0$

20. $p^2 > 4q$ and $q < 0$

21. $p^2 = 4q$ and $p > 0$

22. $p^2 = 4q$ and $p < 0$

23. $p^2 < 4q$ and $p > 0$

24. $p^2 < 4q$ and $p < 0$

25. $p = 0$ and $q > 0$

8.5 PHASE PORTRAITS OF AUTONOMOUS SYSTEMS

The picture of all trajectories of a system is called the *phase portrait* of the system. In this section we first present, by means of examples, all possible phase portraits of the linear autonomous system

$$\dot{x} = ax + by, \qquad \dot{y} = cx + dy \tag{1}$$

near the critical point $x = 0$, $y = 0$. We shall assume that (0, 0) is the only critical point of system (1), and this is equivalent to assuming that $ad - bc \neq 0$. (See Exercise 19, Section 8.3.) At the end of this section we state a theorem about the phase portraits of some nonlinear systems.

As we explained in Section 3.3, the solutions of system (1) are of the form

$$x = Ae^{\lambda t}, \qquad y = Be^{\lambda t},$$

where λ is an eigenvalue of the matrix

$$\begin{pmatrix} a & b \\ c & d \end{pmatrix};$$

that is, λ is a root of the characteristic equation

$$\lambda^2 - (a + d)\lambda + ad - bc = 0. \tag{2}$$

The phase portrait of system (1) depends almost entirely on the roots λ_1 and λ_2 of Eq. (2). [The assumption $ad - bc \neq 0$ implies that $\lambda = 0$ is not a root of Eq. (2).] There are five different cases that must be studied separately.

CASE 1 *Real and distinct roots of the same sign,* that is,

$$\lambda_2 < \lambda_1 < 0 \qquad \text{or} \qquad \lambda_1 > \lambda_2 > 0.$$

CASE 2 *Real roots with opposite sign,* that is,

$$\lambda_1 < 0 < \lambda_2.$$

CASE 3 *Equal roots,* that is,

$$\lambda_1 = \lambda_2 = \lambda \qquad \text{with} \qquad \lambda < 0 \qquad \text{or} \qquad \lambda > 0.$$

In this case we obtain two distinct portraits, depending on whether or not the matrix

$$\begin{pmatrix} a - \lambda & b \\ c & d - \lambda \end{pmatrix}$$

is equal to zero.

CASE 4 *Complex conjugate roots but not pure imaginary,* that is,

$$\lambda_1 = k + il, \qquad \lambda_2 = k - il \qquad \text{with} \qquad k < 0 \qquad \text{or} \qquad k > 0.$$

CASE 5 *Pure imaginary roots,* that is,

$$\lambda_1 = il, \qquad \lambda_2 = -il \qquad \text{with} \qquad l \neq 0.$$

It is sufficient to illustrate these cases by means of specific examples. This is true because any two systems falling into one and the same of these cases (and corresponding subcase) can be transformed into each other by means of a linear change of variables.

EXAMPLE 1 $(\lambda_2 < \lambda_1 < 0)$ Draw the phase portrait of the linear system

$$\dot{x} = -2x + y, \qquad \dot{y} = x - 2y. \tag{3}$$

Solution Here the characteristic roots are $\lambda_1 = -1$ and $\lambda_2 = -3$. The general solution of system (3) is

$$x = c_1e^{-t} + c_2e^{-3t}, \qquad y = c_1e^{-t} - c_2e^{-3t}, \tag{4}$$

where c_1 and c_2 are arbitrary constants. We want to find the trajectories of all the solutions given by (4) for all different values of the constants c_1 and c_2. When $c_1 = c_2 = 0$, we have the solution $x = 0$, $y = 0$, whose trajectory is the origin (0, 0) in Figure 8.3. When $c_1 \neq 0$ and $c_2 = 0$, we find the solutions

$$x = c_1e^{-t}, \qquad y = c_1e^{-t}, \tag{5}$$

and when $c_1 = 0$ and $c_2 \neq 0$, we have the solutions

$$x = c_2e^{-3t}, \qquad y = -c_2e^{-3t}. \tag{6}$$

For $c_1 > 0$, all the solutions (5) have the same trajectory, $y = x > 0$. Similarly, for $c_1 < 0$, all the solutions (5) have the common trajectory, $y = x < 0$. From (6) with $c_2 > 0$ and $c_2 < 0$, we also find the trajectories $y = -x < 0$ and $y = -x > 0$, respectively. These four trajectories are half-lines shown in Figure 8.3. The arrows on the half-lines indicate the direction of motion on the trajectories as t increases. To obtain the other trajectories explicitly, we must eliminate t from the two equations in (4) and investigate the resulting curves for all nonzero values of the constants c_1 and c_2. When this approach is complicated (as in this example), we can still obtain a good picture of the phase portrait of the system as follows. First, from (4), it is clear that every trajectory of system (3) approaches the origin (0, 0) as t approaches infinity. Furthermore, for $c_1 \neq 0$ and $c_2 \neq 0$, we have

$$\frac{y}{x} = \frac{c_1e^{-t} - c_2e^{-3t}}{c_1e^{-t} + c_2e^{-3t}} = \frac{c_1 - c_2e^{-2t}}{c_1 + c_2e^{-2t}} \to 1 \qquad \text{as } t \to \infty.$$

Thus, all these trajectories approach the origin tangent to the line $y = x$. In fact, all the trajectories of system (3), except the pair $y = -x > 0$ and $y = -x < 0$, approach the origin tangent to the line $y = x$. Figure 8.3 shows a few

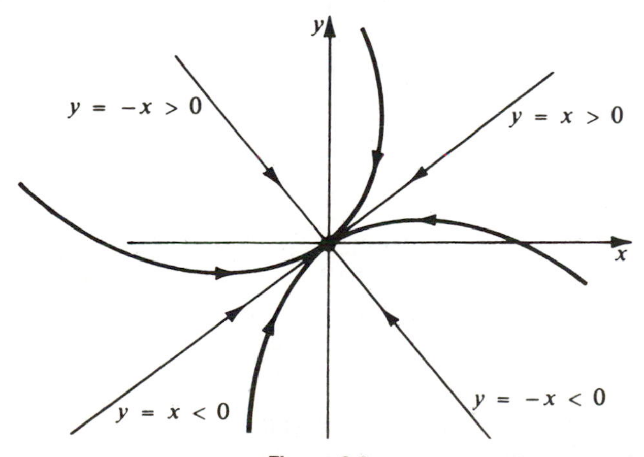

Figure 8.3

trajectories of the phase portrait of system (3). The arrows on the trajectories indicate the direction of increasing t. This type of critical point is called a *stable node.*

When $\lambda_1 > \lambda_2 > 0$, as in Exercise 1, the phase portrait is exactly the same, except that the direction of the arrows on the trajectories is reversed and the trajectories approach the origin as $t \to -\infty$. This type of critical point is called an *unstable node.* In the case of an unstable node, all the solutions, except the solution $x = 0$, $y = 0$, approach infinity as $t \to \infty$.

EXAMPLE 2 ($\lambda_1 < 0 < \lambda_2$) Draw the phase portrait of the linear system

$$\dot{x} = 3x - 2y, \qquad \dot{y} = 2x - 2y. \tag{7}$$

Solution Here the characteristic roots are $\lambda_1 = -1$ and $\lambda_2 = 2$. The general solution of (7) is

$$x = c_1e^{-t} + c_2e^{2t}, \qquad y = 2c_1e^{-t} + \tfrac{1}{2}c_2e^{2t}, \tag{8}$$

where c_1 and c_2 are arbitrary constants. When $c_1 \neq 0$ and $c_2 = 0$, we obtain the solutions

$$x = c_1e^{-t}, \qquad y = 2c_1e^{-t}, \tag{9}$$

and when $c_1 = 0$ and $c_2 \neq 0$, we have the solutions

$$x = c_2e^{2t}, \qquad y = \tfrac{1}{2}c_2e^{2t}. \tag{10}$$

For $c_1 > 0$, the trajectory of the solutions (9) is the half-line $y = 2x > 0$, while for $c_1 < 0$, the trajectory is the half-line $y = 2x < 0$ (see Figure 8.4). The arrows on these trajectories point toward the origin because both x and y approach zero as $t \to \infty$. This observation follows from (9).

For $c_2 > 0$, the trajectory of the solutions (10) is the half-line $y = \frac{1}{2}x > 0$, while for $c_2 < 0$, the trajectory is the half-line $y = \frac{1}{2}x < 0$ (see Figure 8.4). The arrows on these trajectories point away from the origin, since in this case both x and y tend to $\pm\infty$ as $t \to \infty$, as can be seen from (10).

For $c_1 \neq 0$ and $c_2 \neq 0$, we have

$$\frac{y}{x} = \frac{2c_1e^{-t} + \frac{1}{2}c_2e^{2t}}{c_1e^{-t} + c_2e^{2t}} = \frac{2c_1e^{-3t} + \frac{1}{2}c_2}{c_1e^{-3t} + c_2} \to \frac{1}{2} \qquad \text{as } t \to \infty.$$

Hence, all trajectories are asymptotic to the line $y = \frac{1}{2}x$ as t tends to ∞.

On the other hand,

$$\frac{y}{x} = \frac{2c_1e^{-t} + \frac{1}{2}c_2e^{2t}}{c_1e^{-t} + c_2e^{2t}} = \frac{2c_1 + \frac{1}{2}c_2e^{3t}}{c_1 + c_2e^{3t}} \to 2 \qquad \text{as } t \to -\infty,$$

which implies that all trajectories are asymptotic to the line $y = 2x$ as t tends to $-\infty$. Figure 8.4 shows a few trajectories of the phase portrait of system (7).

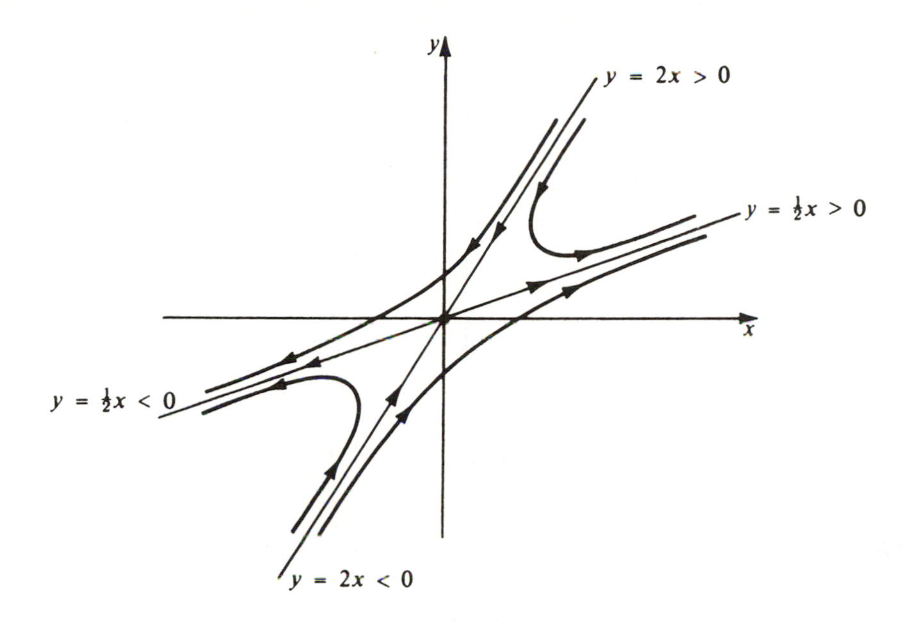

Figure 8.4

This type of critical point is called a *saddle point.* Since one of the roots of the characteristic equation is positive, it follows that a saddle point is an unstable point of the system. As we see from Figure 8.4, only two trajectories approach the origin; the rest tend away from it.

EXAMPLE 3A ($\lambda_1 = \lambda_2 = \lambda < 0$ and the matrix $\begin{pmatrix} a-\lambda & b \\ c & d-\lambda \end{pmatrix}$ is zero)

Draw the phase portrait of the linear system

$$\dot{x} = -2x, \qquad \dot{y} = -2y. \tag{11}$$

Solution Here the characteristic roots are $\lambda_1 = \lambda_2 = -2 < 0$, and the general solution of the system is

$$x = c_1e^{-2t}, \qquad y = c_2e^{-2t}, \tag{12}$$

where c_1 and c_2 are arbitrary constants. The trajectories of the solutions (12) are the half-lines $c_1y = c_2x$ with slope c_2/c_1. (Thus, all slopes are possible.) The phase portrait of system (11) has the form described in Figure 8.5. This type of critical point is called a *stable node* when $\lambda < 0$ and an unstable node when $\lambda > 0$. The phase portrait of an unstable node is as in Figure 8.5 except that the direction of the arrows on the trajectories is reversed.

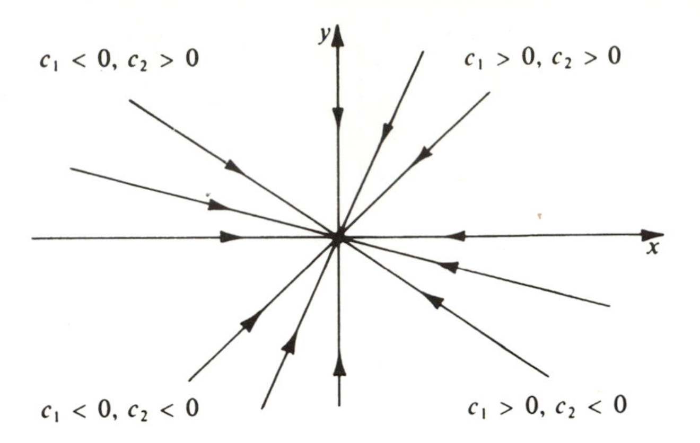

Figure 8.5

EXAMPLE 3B ($\lambda_1 = \lambda_2 = \lambda < 0$ and the matrix $\begin{pmatrix} a-\lambda & b \\ c & d-\lambda \end{pmatrix}$ is not zero)

Draw the phase portrait of the system

$$\dot{x} = -x, \qquad \dot{y} = -x - y. \tag{13}$$

Solution Here the characteristic roots are $\lambda_1 = \lambda_2 = -1 < 0$, and the general solution of the system is

$$x = c_1 e^{-t}, \qquad y = -c_1 te^{-t} + c_2 e^{-t}, \tag{14}$$

where c_1 and c_2 are arbitrary constants. From (14) we see that all trajectories approach the origin as $t \to \infty$. When $c_1 = 0$, $c_2 \neq 0$, we find the solutions

$$x = 0, \qquad y = c_2 e^{-t}. \tag{15}$$

Clearly, the positive and negative y axes are the trajectories of the solutions (15) (see Figure 8.6). For $c_1 \neq 0$, we can eliminate t between the two equations in (14), obtaining

$$y = \frac{c_2}{c_1} x + x \ln \left| \frac{x}{c_1} \right| . \tag{16}$$

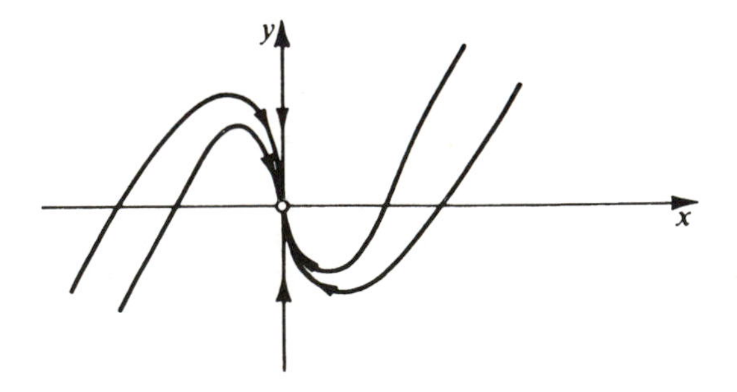

Figure 8.6

Equation (16) gives explicitly the trajectories of the solutions (14) for $c_1 = 0$. All these trajectories approach the origin tangent to the y axis (see Figure 8.6). This type of critical point is called a *stable node* when $\lambda < 0$ and an *unstable node* when $\lambda > 0$. The phase portrait of the unstable node is as in Figure 8.6 except that the arrows on the trajectories are reversed.

EXAMPLE 4 ($\lambda_1 = k + il$, $\lambda_2 = k - il$ with $k < 0$) Draw the phase portrait of the linear system

$$\begin{aligned} \dot{x} &= -x + y \\ \dot{y} &= -x - y. \end{aligned} \tag{17}$$

Solution Here the characteristic roots are $\lambda_1 = -1 + i$ and $\lambda_2 = -1 - i$. The general solution of the system is

$$\begin{aligned} x &= c_1 e^{-t} \cos t + c_2 e^{-t} \sin t \\ y &= -c_1 e^{-t} \sin t + c_2 e^{-t} \cos t. \end{aligned}$$

It seems complicated to find the trajectories through this form of the solution. For this reason let us introduce polar coordinates in (17) by setting

$$x = r \cos \phi, \qquad y = r \sin \phi.$$

Then

$$\dot{x} = \dot{r} \cos \phi - (r \sin \phi)\dot{\phi} \qquad \text{and} \qquad \dot{y} = \dot{r} \sin \phi + (r \cos \phi)\dot{\phi}.$$

Substituting these values of x, y, $\dot{x}$, and $\dot{y}$ into (17) and solving for $\dot{r}$ and $\dot{\phi}$, we obtain the equivalent system

$$\dot{r} = -r, \qquad \phi = -1. \tag{18}$$

The general solution of (18) is

$$r = Ae^{-t}, \qquad \phi = -t + B. \tag{19}$$

The trajectories are therefore logarithmic spirals approaching the origin as t tends to infinity. The phase portrait of system (17) has the form described in Figure 8.7. This type of critical point is called a *stable focus* when $k < 0$ and an *unstable focus* when $k > 0$ (see Figure 8.8). As is expected, in the unstable focus the arrows have opposite direction.

EXAMPLE 5 ($\lambda_1 = il$, $\lambda_2 = -il$ with $l \neq 0$) Draw the phase portrait of the linear system

$$\dot{x} = y, \qquad \dot{y} = -x. \tag{20}$$

Solution Here the characteristic roots are $\lambda_1 = i$ and $\lambda_2 = -i$, and the general solution of the system is

$$x = c_1 \cos t + c_2 \sin t, \qquad y = -c_1 \sin t + c_2 \cos t, \tag{21}$$

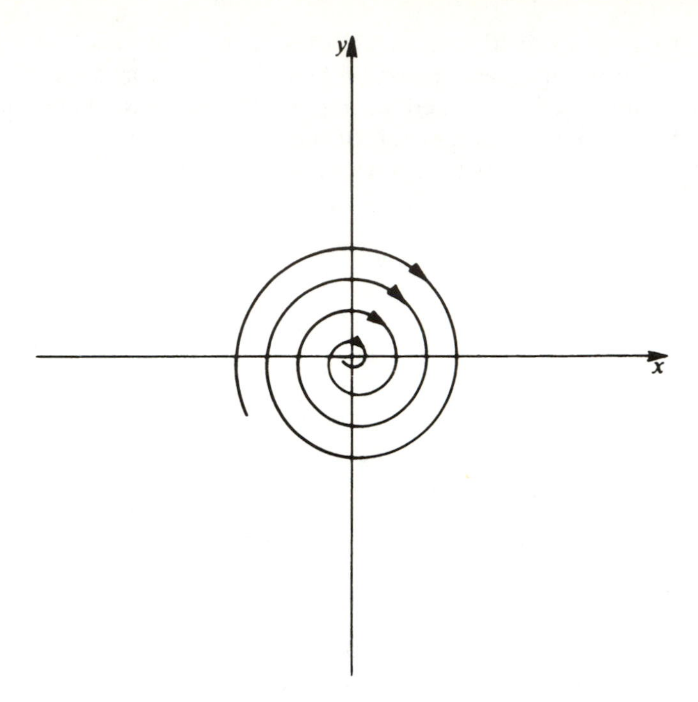

Figure 8.7

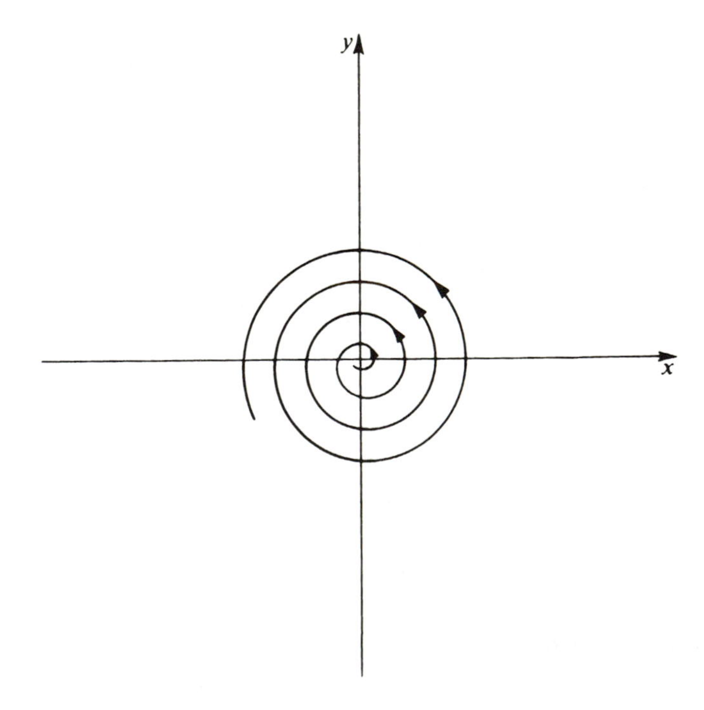

Figure 8.8

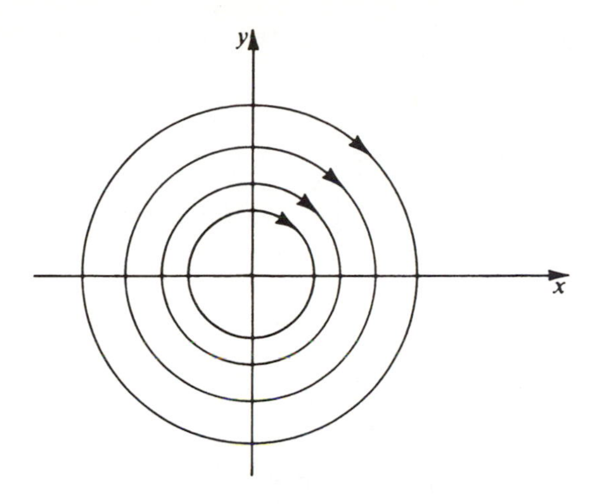

Figure 8.9

where c_1 and c_2 are arbitrary constants. From (21) we see that $x^2 + y^2 = c_1^2 + c_2^2$. Therefore, the trajectories of the system are circles centered at the origin with radius $\sqrt{c_1^2 + c_2^2}$. The phase portrait of system (20) has the form described in Figure 8.9. This type of critical point is called a *center*. The direction of the arrows in Figure 8.9 is found from system (20). From the first equation in (20) we see that $\dot{x} > 0$ when $y > 0$; that is, x increased when y is positive. This implies that the motion along the trajectories is in the clockwise direction.

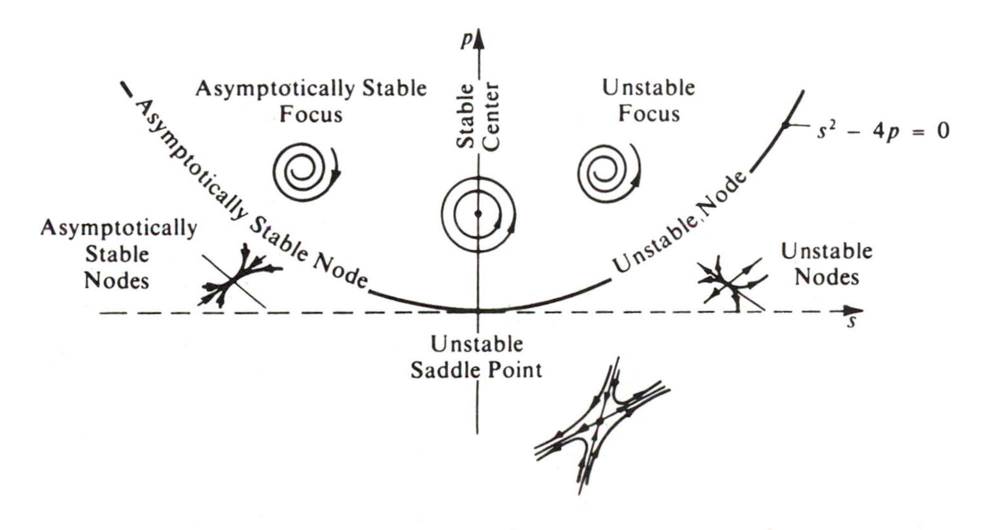

Figure 8.10 Summary of Portraits

Summary In Figure 8.10 we summarize the nature and stability of the critical point (0, 0) of the system

$$\dot{x} = ax + by, \qquad \dot{y} = cx + dy,$$

when $ad - bc \neq 0$. If λ_1 and λ_2 are the characteristic roots, then $\lambda_1 + \lambda_2 = a + d$ and $\lambda_1\lambda_2 = ad - bc$. We set $s = \lambda_1 + \lambda_2$, $p = \lambda_1\lambda_2$, and $D = s^2 - 4p$. In the sp plane we have graphed the parabola $s^2 - 4p = 0$. Since we have assumed that $ad - bc \neq 0$, the entire s axis is excluded from the diagram.

Finally, we state a theorem about the phase portrait, near a critical point (0, 0), of a nonlinear system of the form

$$\begin{aligned} \dot{x} &= ax + by + F(x, y) \\ \dot{y} &= cx + dy + G(x, y), \end{aligned} \tag{22}$$

where a, b, c, d, and the functions F and G satisfy the conditions mentioned before Theorem 2 in Section 8.4. Let λ_1 and λ_2 be the two roots of the characteristic equation (2) of the linearized system (1). Under these hypotheses the following result holds.

THEOREM 1

The critical point (0, 0) *of the nonlinear system* (22) *is*

(a) a node if λ_1 and λ_2 are real, distinct, and of the same sign;
(b) a saddle if λ_1 and λ_2 are real and of opposite sign;
(c) a focus if λ_1 and λ_2 are complex conjugates but not pure imaginary;
(d) a focus or center if λ_1 and λ_2 are pure imaginary.

In Example 5 of Section 8.4 we verified that the functions $F(x, y) = x^2 - y^2$ and $G(x, y) = -xy$ satisfy the hypotheses of Theorem 1. Then from Theorem 1 and the Examples 1, 2, and 4, we conclude that the critical point (0, 0) of the systems

(a) $\dot{x} = -2x + y + x^2 - y^2$, $\dot{y} = x - 2y - xy$,
(b) $\dot{x} = 3x - 2y + x^2 - y^2$, $\dot{y} = 2x - 2y - xy$, and
(c) $\dot{x} = -x + y + x^2 - y^2$, $\dot{y} = -x - y - xy$

is a node, a saddle, and a focus point, respectively.

EXERCISES

In Exercises 1 through 10, determine whether the critical point (0, 0) is a node, saddle point, focus, or center, and draw the phase portrait.

1. $\dot{x} = x$, $\dot{y} = 3y$

2. $\dot{x} = -x$, $\dot{y} = -3y$

3. $\dot{x} = x$, $\dot{y} = -y$

4. $\dot{x} = -x$, $\dot{y} = y$

5. $\dot{x} = 2x$, $\dot{y} = 2y$

6. $\dot{x} = -x$, $\dot{y} = -y$

7. $\dot{x} = -x, \dot{y} = x - y$

8. $\dot{x} = x, \dot{y} = -x + y$

9. $\dot{x} = -2x + y, \dot{y} = -x - 2y$

10. $\dot{x} = -y, \dot{y} = x.$

In Exercises 11 through 14, determine whether the critical point (0, 0) is a node, saddle point, focus, or center, and graph a few trajectories.

11. $\dot{x} = -x + y, \dot{y} = -x - y$

12. $\dot{x} = -x + y, \dot{y} = -x - 3y$

13. $\dot{x} = 4x + 6y, \dot{y} = -7x - 9y$

14. $\dot{x} = 2x - y, \dot{y} = -x + 2y$

In Exercises 15 through 19, determine whether the critical point (0, 0) is a node, a saddle, or a focus.

15. $\dot{x} = -2x + y - x^2 + 2y^2, \dot{y} = 3x + 2y + x^2y^2$

16. $\dot{x} = -x + x^2, \dot{y} = -3y + xy$

17. $\dot{x} = -x + xy, \dot{y} = y + (x^2 + y^2)^2$

18. $\dot{x} = 2x + y^2, \dot{y} = 3y - x^2$

19. $\dot{x} = x - xy, \dot{y} = -y + xy$

Transform the differential equation $\ddot{x} + p\dot{x} + qx = 0$ into an equivalent system by setting $\dot{x} = y$ and investigate the type of the phase portrait of the resulting system near the critical point (0, 0) in each of the following cases:

20. $p^2 - 4q > 0, q > 0$, and $p < 0$

21. $p^2 - 4q > 0, q > 0$, and $p > 0$

22. $p^2 - 4q > 0$ and $q < 0$

23. $p^2 = 4q$ and $p > 0$

24. $p^2 = 4q$ and $p < 0$

25. $p^2 - 4q < 0$ and $p > 0$

26. $p = 0$ and $q > 0$

In Exercises 27 through 34, determine the nature and stability of the critical point (0, 0) by using Figure 8.10.

27. $\dot{x} = x + y, \dot{y} = x - 3y$

28. $\dot{x} = 3x - 2y, \dot{y} = 2x + 3y$

29. $\dot{x} = -2x - 3y, \dot{y} = -3x + 2y$

30. $\dot{x} = -3x + 2y, \dot{y} = -2x - 3y$

31. $\dot{x} = x, \dot{y} = y$

32. $\dot{x} = -x, \dot{y} = -y$

33. $\dot{x} = -x, \dot{y} = -x - y$

34. $\dot{x} = -3x + 2y, \dot{y} = -2x + 2y$

8.6 APPLICATIONS

In this section we present a few applications of nonlinear differential equations and systems. In comparison to mathematical models described by linear equations and systems, nonlinear equations and systems describe situations that are closer to reality in many cases.

Biology
Interacting Populations

■ Consider two species and assume that one of them, called prey, has an abundant supply of food while the other species, called predator, feeds exclusively on the first. The mathematical study of such an ecosystem was initiated independently by Lotka and Volterra in the mid 1920s. Let us denote by $x(t)$ the number of prey and by $y(t)$ the number of predators at time t.

If the two species were in isolation from each other, they would vary at a rate proportional to their number present,

$$\dot{x} = ax \quad \text{and} \quad \dot{y} = -cy. \tag{1}$$

In Eq. (1), $a > 0$ because the prey population has an abundant supply of food and therefore increases, while $-c < 0$ because the predator population has no food and hence decreases.

However, we have assumed that the two species interact in such a way that the predator population eats the prey. It is reasonable to assume that the number of fatal encounters per unit time is proportional to x and y, and therefore to xy. Then the prey will decrease while the predators will increase at rates proportional to xy. That is, the two interacting populations satisfy the nonlinear system

$$\dot{x} = ax - bxy, \qquad \dot{y} = -cy + dxy, \tag{2}$$

where b and d are positive constants.

Although system (2) is nonlinear and there is no known way to solve it explicitly, it is nevertheless possible, using the qualitative theory of such systems, to obtain many properties of its solutions, and in turn to make useful predictions about the behavior of the two species.

Solving the system of simultaneous equations $ax - bxy = 0$ and $-cy + dxy = 0$, we find the critical points $(0, 0)$ and $(c/d, a/b)$ of system (2). Thus, system (2) has the two equilibrium solutions $x(t) = 0$, $y(t) = 0$ and $x(t) = c/d$, $y(t) = a/b$. Of course, only the second of these is of interest in this application.

Let us compute the trajectories of the solutions of (2). Clearly, $x(t) = 0$, $y(t) = y(0)e^{-ct}$ is a solution of (2) with trajectory the positive y axis. Also, $y(t) = 0$, $x(t) = x(0)e^{at}$ is another solution of (2), with trajectory the positive x axis. Because of the uniqueness of solutions, it follows that every solution of (2) which at $t = 0$ starts in the first quadrant cannot cross the x or the y axis, and therefore should remain in the first quadrant forever. The remaining trajectories of the solutions of (2) satisfy the differential equation

$$\frac{dy}{dx} = \frac{-cy + dxy}{ax - bxy} = \frac{(-c + dx)y}{(a - by)x},$$

which is a separable differential equation. Separating the variables x and y, we obtain

$$\frac{a - by}{y}\,dy = \frac{-c + dx}{x}\,dx$$

or

$$\left(\frac{a}{y} - b\right) dy = \left(-\frac{c}{x} + d\right) dx. \tag{3}$$

Integrating both sides of (3), we obtain the general solution of (3),

$$a \ln y - by = -c \ln x + dx + k, \tag{4}$$

where k is an arbitrary constant. Equation (4) can be rewritten as follows:

$$\begin{aligned} &\ln y^a + \ln x^c = by + dx + k \\ \Rightarrow\quad &\ln y^a x^c = by + dx + k \\ \Rightarrow\quad &y^a x^c = e^{by+dx+k} \\ \Rightarrow\quad &\frac{y^a}{e^{by}} \cdot \frac{x^c}{e^{dx}} = K, \end{aligned} \tag{5}$$

where K is e^k. If we allow K to be equal to zero, Eq. (5) gives all the trajectories of system (2).

It can be shown that for each $K > 0$, the trajectory (5) is a closed curve, and therefore each solution $(x(t), y(t))$ of (2) with initial value $(x(0), y(0))$ in the first quadrant is a periodic function of time t (see Figure 8.11). If T is the period of a solution $(x(t), y(t))$, that is, if $(x(t + T), y(t + T)) = (x(t), y(t))$ for all $t \geq 0$, then the average values of the populations $x(t)$ and $y(t)$ are, by definition, given by the integrals

$$\bar{x} = \frac{1}{T}\int_0^T x(t)dt, \qquad \bar{y} = \frac{1}{T}\int_0^T y(t)dt.$$

Surprisingly, these average values can be computed directly for system (2) without knowing explicitly the solution and its period. In fact, from the second equation in (2), we obtain

$$\frac{\dot{y}}{y} = -c + dx.$$

Integrating both sides from 0 to T, we find that

$$\ln y(T) - \ln y(0) = -cT + d\int_0^T x(t)dt.$$

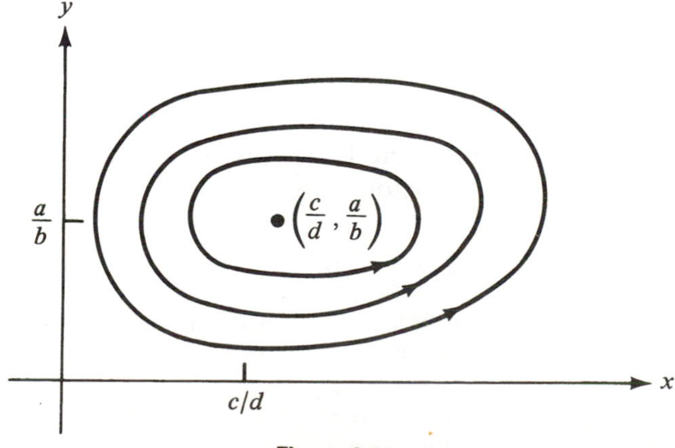

Figure 8.11

Since $y(T) = y(0)$, it follows that

$$-cT + d\int_0^T x(t)dt = 0 \qquad \text{or} \qquad \frac{1}{T}\int_0^T x(t)dt = \frac{c}{d}.$$

Therefore,

$$\bar{x} = \frac{c}{d}. \tag{6}$$

Similarly, from the first equation in (2), we obtain

$$\frac{\dot{x}}{x} = a - by.$$

Integrating both sides from 0 to T and using the fact that $x(T) = x(0)$, we find

$$\bar{y} = \frac{a}{b}. \tag{7}$$

From (6) and (7) we can make the interesting prediction that the average sizes of two populations $x(t)$ and $y(t)$ which interact according to the mathematical model described by system (2) will be exactly their equilibrium values $x = c/d$ and $y = a/b$.

We can utilize this observation to make another interesting prediction. In addition to the hypotheses about the prey and predator populations which lead to the mathematical model described by system (2), assume that the prey population $x(t)$ is harvested in "moderate amounts." Then both the prey and predator populations will decrease at rates, say, $\epsilon x(t)$ and $\epsilon y(t)$, respectively. In this case, system (2) should be replaced by the system

$$\dot{x} = ax - bxy - \epsilon x, \qquad \dot{y} = -cy + dxy - \epsilon y$$

or

$$\dot{x} = (a - \epsilon)x - bxy, \qquad \dot{y} = -(c + \epsilon)y + dxy. \tag{8}$$

Applying equations (6) and (7) to system (8), we conclude, with surprise, that if the harvesting of the preys is such that $a > \epsilon$, the average size of the prey population will be

$$\bar{x} = \frac{c + \epsilon}{d}, \tag{6$'$}$$

which is somewhat higher than before there was any harvesting. On the other hand, the average size of the predator population will be

$$\bar{y} = \frac{a - \epsilon}{b}, \tag{7$'$}$$

which is somewhat smaller than before there was any harvesting.[1]

[1]For more details on this and other related applications, the reader is referred to U. D'Ancona, *The Struggle for Existence* (Leiden: Brill, 1954).

Nonlinear Mechanics
The Motion of a Simple Pendulum

■ A simple pendulum consists of a bob B of mass m at the end of a very light and rigid rod of length L, pivoted at the top so that the system can swing in a vertical plane (see Figure 8.12). We pull the bob to one side and release it from rest at time $t = 0$. Let $\theta(t)$, in radians, be the angular displacement of the rod from its equilibrium position OA at time t. The angle $\theta(t)$ is taken to be positive when the bob is to the right of the equilibrium and negative when it is to the left. We want to study $\theta(t)$ as the bob swings back and forth along the circular arc CC'. From this information we know already that

$$\theta(0) = \theta_0 \quad \text{and} \quad \dot{\theta}(0) = 0, \tag{9}$$

where θ_0 is the initial angular displacement of the rod and $\dot{\theta}(0) = 0$ because the bob was released from rest. There are two forces acting upon the bob B at any time t. One is its weight $-mg$, where m is the mass of the bob and g is the acceleration of gravity. The other force is the tension T from the rod.

The force $-mg$ is resolved into the two components $-mg \cos \theta$ and $-mg \sin \theta$, as shown in Figure 8.12. The force $-mg \cos \theta$ balances the tension T in the rod, while the force $-mg \sin \theta$ moves the bob along the circular arc BA. By Newton's second law of motion, we obtain

$$m\frac{d^2s}{dt^2} = -mg \sin \theta, \tag{10}$$

where s is the length of the arc AB, and d^2s/dt^2 is the acceleration along the arc. Since L is the length of the rod, we have $s = L\theta$ and therefore

$$\frac{d^2s}{dt^2} = L\frac{d^2\theta}{dt^2}. \tag{11}$$

Using Eq. (11) in (10) and simplifying the resulting equation, we obtain

$$\ddot{\theta} + \frac{g}{L}\sin \theta = 0. \tag{12}$$

The nonlinear differential equation (12), together with the initial conditions (9), describes completely the motion (in a vacuum) of the simple pendulum.

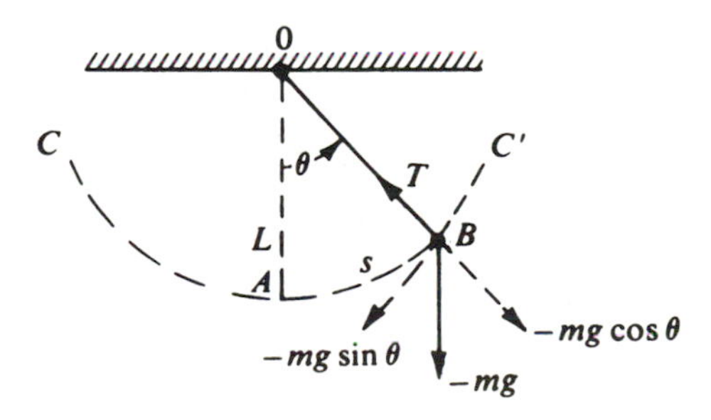

Figure 8.12

Let us denote for simplicity the constant g/L by ω^2. If we now set $\theta = x$ and $\dot{\theta} = y$, the differential equation (12) is equivalent to the system

$$\dot{x} = y, \qquad \dot{y} = -\omega^2 \sin x. \tag{13}$$

In what follows we present a phase-plane analysis of system (13).

The critical points of system (13) are the solutions of the simultaneous equations

$$y = 0 \quad \text{and} \quad -\omega^2 \sin x = 0.$$

Since $\sin x = 0$ for $x = 0, \pm\pi, \pm 2\pi, \ldots$, the critical points are

$$(0, 0), (\pi, 0), (-\pi, 0), (2\pi, 0), (-2\pi, 0), \ldots. \tag{14}$$

Clearly, all of them are located on the x axis. Since $x = \theta$ is the angular displacement of the rod from the equilibrium position OA (see Figure 8.12), we conclude that when $x = 0, 2\pi, -2\pi, \ldots$, the pendulum is pointing vertically downward, and when $x = \pi, -\pi, 3\pi, -3\pi, \ldots$, the pendulum is pointing vertically upward. Because of the periodicity of the function $\sin x$ it suffices to study the nature of the critical points $(0, 0)$ and $(\pi, 0)$ only.

First, we study the critical point $(0, 0)$. Since

$$\sin x = x - \frac{x^3}{3!} + \frac{x^5}{5!} - \cdots,$$

we can approximate (13) by the linear system

$$\dot{x} = y, \qquad \dot{y} = -\omega^2 x. \tag{15}$$

The characteristic equation of (15) is $\lambda^2 + \omega^2 = 0$ with characteristic roots $\lambda = \pm\omega i$ (pure imaginary). Hence, the critical point $(0, 0)$ is stable and a center for system (15) (see Figure 8.13). The arrows on the trajectories point in the clockwise direction because, as we see from the first equation in (15), x increases

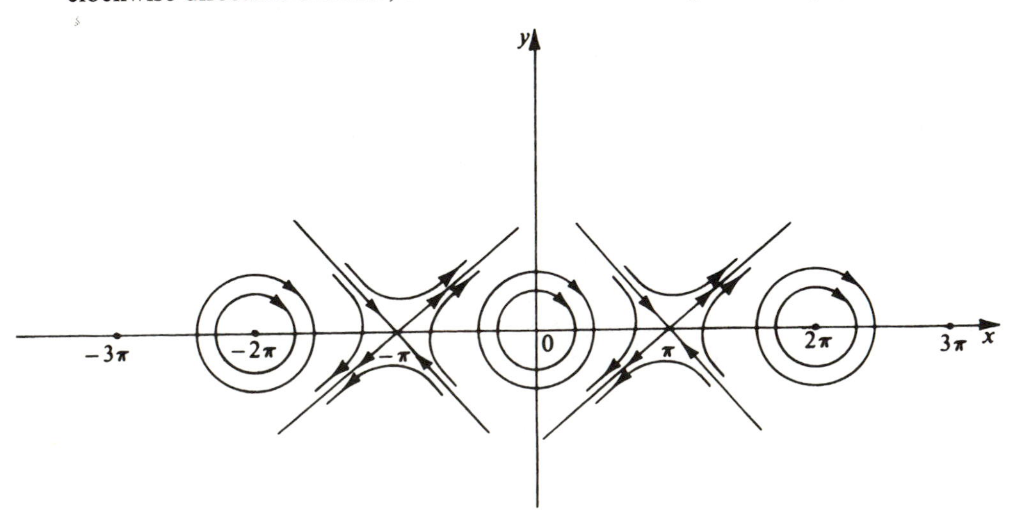

Figure 8.13

when y is positive. By Theorem 1(d) of Section 8.5 (whose hypotheses are satisfied here), it follows that (0, 0) is either a center or a focus point for the nonlinear system (13). In this case, however, it can be shown that it is a center (the simple pendulum is a conservative system, and hence there is no gain or loss of energy as in the case of a focus). System (13) also has a center at each of the critical points $(2\pi n, 0)$ for $n = \pm 1, \pm 2, \ldots$. The closed trajectories in Figure 8.14 correspond to the oscillatory motions of the pendulum about the stable points $(2\pi n, 0)$, $n = 0, \pm 1, \pm 2, \ldots$.

Next we study the critical point $(\pi, 0)$. The Taylor-series expansion of $\sin x$ about $x = \pi$ is given by

$$\sin x = -(x - \pi) + \frac{(x - \pi)^3}{3!} - \frac{(x - \pi)^5}{5!} + \cdots.$$

Hence, the linearized system is

$$\dot{x} = y, \qquad \dot{y} = \omega^2(x - \pi) \tag{16}$$

and has a critical point at $(\pi, 0)$. We can transform the critical point $(\pi, 0)$ to (0, 0) by setting $v = x - \pi$ in (16). Then (16) becomes

$$\dot{v} = y, \qquad \dot{y} = \omega^2 v. \tag{17}$$

The characteristic equation of (17) is $\lambda^2 - \omega^2 = 0$ with characteristic roots $\pm\omega$. Since the roots are real with opposite sign, it follows that the critical point (0, 0) of (17) is a saddle point and hence an unstable equilibrium point. Thus, $(\pi, 0)$ is also a saddle point and hence an unstable equilibrium point of the linearized system (16) (see Figure 8.13). Now, from Theorem 1(b) of Section 8.5, we conclude that $(\pi, 0)$ is a saddle point and therefore an unstable equilibrium point for system (13). System (13) also has a saddle point at each of the critical points $((2n + 1)\pi, 0)$ for $n = \pm 1, \pm 2, \ldots$. The trajectories in this case are the heavy loops shown in Figure 8.14.

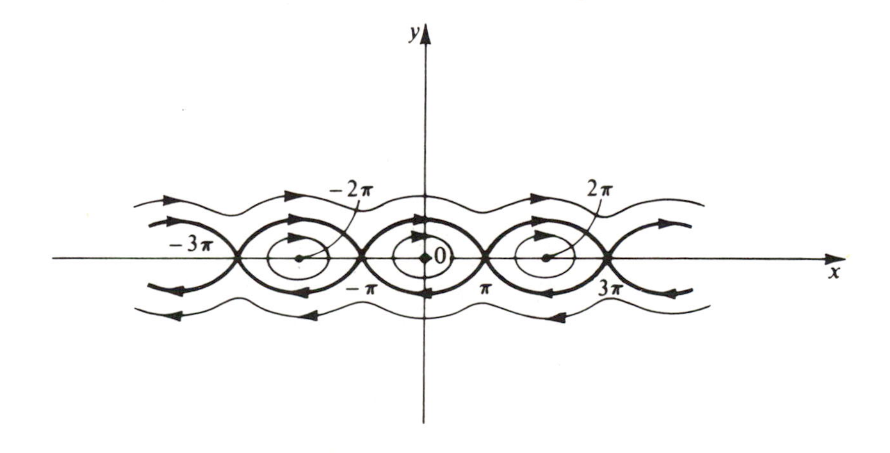

Figure 8.14

The equations of the trajectories of the solutions of system (13) can be found explicitly. In fact, from (13) we see that the trajectories satisfy the separable differential equation

$$\frac{dy}{dx} = -\omega^2 \frac{\sin x}{y}.$$

Separating the variables x and y and integrating, we obtain the family of trajectories

$$\tfrac{1}{2}y^2 - \omega^2 \cos x = c, \tag{18}$$

where c is the integration constant. For the sake of describing the phase portrait, it is convenient to express c in terms of initial conditions. Assume that $y = y_0$ when $x = 0$. Then from (18) we find that

$$c = \tfrac{1}{2}y_0^2 - \omega^2$$

$$\tfrac{1}{2}y^2 - \omega^2 \cos x = \tfrac{1}{2}y_0^2 - \omega^2$$

$$y^2 + 2\omega^2(1 - \cos x) = y_0^2$$

$$y^2 + 4\omega^2 \sin^2 \frac{x}{2} = y_0^2. \tag{19}$$

The three types of curves shown in Figure 8.14 correspond to the following three cases.

CASE 1 $|y_0| < 2\omega$ Then the maximum value of the angle x (remember that $x = \theta$) is attained when $y = 0$ and

$$x_{\max} = 2 \arcsin \frac{y_0}{2\omega} < \pi.$$

In this case the pendulum oscillates between the extreme angles $\pm x_{\max}$. The trajectories in this case are the closed curves shown in Figure 8.14.

CASE 2 $|y_0| > 2\omega$ In this case the pendulum makes complete revolutions. The trajectories in this case are the wavy curves on the top and bottom of Figure 8.14.

CASE 3 $|y_0| = 2\omega$ In this case the trajectories are the heavy loops that separate the closed and wavy trajectories described in the previous cases. The equations of these trajectories can be found from (19) if we substitute y_0^2 by $4\omega^2$, and they are given by

$$y = \pm 2\omega \cos \frac{x}{2}.$$

EXERCISES

1. Linearize system (2) in the biology application in the neighborhood of the critical point $(c/d, a/b)$ and study the nature and stability of the critical point $(0, 0)$ of the linearized system. [*Hint:* Set $x = X + c/d$, $y = Y + a/b$.]

2. Biology In the predator–prey interaction that we discussed in the text, we assumed that the prey population has an abundant supply of food. Assume now that the prey has a limited supply of food available. In this case it is reasonable to assume that the predator–prey interaction is described by a system of the form

$$\dot{x} = ax - bxy - \epsilon_1 x^2, \qquad \dot{y} = -cy + dxy - \epsilon_2 y^2,$$

where a, b, c, and d are positive constants and ϵ_1 and ϵ_2 are nonnegative constants. [For $\epsilon_1 = \epsilon_2 = 0$, we obtain system (2) in the biology application.] Find the critical points of the system (negative populations are not permissible). In the special case

$$\dot{x} = 3x - xy - 2x^2, \qquad \dot{y} = -y + 2xy - y^2,$$

where x and y are measured in hundreds of organisms, study the stability of each critical point and determine whether it is a node, saddle, or focus.

3. The differential equation

$$\ddot{\theta} + k\dot{\theta} + \omega^2 \sin\theta = 0, \qquad k > 0,$$

describes the motion of a simple pendulum under the influence of a frictional force proportional to the angular velocity $\dot{\theta}$. Transform the differential equation into an equivalent system by setting $x = \theta$ and $y = \dot{\theta}$. Show that $(n\pi, 0)$ for $n = 0, \pm 1, \pm 2, \ldots$ are the only critical points of the system. In each of the following cases, study the stability of the system at $(0, 0)$ and determine whether $(0, 0)$ is a node, saddle, or focus.

(a) $k < 2\omega$ (b) $k = 2\omega$ (c) $k > 2\omega$

4. Electric Circuits The differential equation

$$\ddot{x} + \mu(x^2 - 1)\dot{x} + x = 0, \qquad \mu > 0$$

is called the *van der Pol equation* and governs certain electric circuits that contain vacuum tubes. Transform the differential equation into an equivalent system by setting $\dot{x} = y$. Show that $(0, 0)$ is the only critical point of the system. Study the stability of the system at the point $(0, 0)$ when $\mu < 2$ and when $\mu > 2$. In each case, determine whether $(0, 0)$ is a node, saddle, or focus.

REVIEW EXERCISES

1. Show that the IVP

$$\dot{x} = 2x^2 - y + t$$
$$\dot{y} = x + 2y^2 - e^t$$
$$x(0) = 1; \qquad y(0) = 0$$

has a unique solution defined in some interval about the point $t_0 = 0$.

2. Show that $(0, 0)$ is the only critical point of the system

$$\dot{x} = 2x^2 - xy, \qquad \dot{y} = x + 2y.$$

3. Find the critical points of the system

$$\dot{x} = 2x^2 - xy, \qquad \dot{y} = -2x + y.$$

4. Find the critical points and the equations of the trajectories of the solutions of the system

$$\dot{x} = x - 4y, \qquad \dot{y} = 2x - y.$$

5. Find the critical points and the equations of the trajectories of the solutions of the system

$$\dot{x} = x, \qquad \dot{y} = -x^2 + 2y.$$

In particular, draw the trajectory corresponding to the solution with initial conditions

$$x(0) = 1, \qquad y(0) = 0$$

and indicate the direction of increasing t.

In Exercises 6 through 12, determine whether the critical point $(0, 0)$ of the system is stable, asymptotically stable, or unstable.

6. $\dot{x} = -3x + 2y,\ \dot{y} = 4x - y$ **7.** $\dot{x} = 2x - 8y,\ \dot{y} = x - 2y$

8. $\dot{x} = 3x + 8y,\ \dot{y} = -x - y$ **9.** $\dot{x} = 4x - 9y,\ \dot{y} = 5x - 10y$

10. $\dot{x} = -3x + 2y + x^2 - y^2,\ \dot{y} = 4x - y + xy$

11. $\dot{x} = 3x + 8y - xy,\ \dot{y} = -x - y + xy$

12. $\dot{x} = 4x - 9y + (x^2 + y^2)^{3/2},\ \dot{y} = 5x - 10y - (x^2 + y^2)$

In Exercises 13 through 16, determine whether the critical point $(0, 0)$ is a node, saddle point, focus, or center, and graph a few trajectories.

13. $\dot{x} = -3x + 2y,\ \dot{y} = 4x - y$ **14.** $\dot{x} = 3x + 8y,\ \dot{y} = -x - y$

15. $\dot{x} = 2x - 8y,\ \dot{y} = x - 2y$ **16.** $\dot{x} = 4x - 9y,\ \dot{y} = 5x - 10y$

17. Determine the nature of the critical point $(0, 0)$ of the system

$$\dot{x} = 4x - 9y + (x^2 + y^2)^2, \qquad \dot{y} = 5x - 10y - (x^2 + y^2)^{3/2}.$$

18. Determine[2] the stability of the linear time-invariant network shown in Figure 8.15. Assume that the initial voltage on the capacitor is v_0. (The state variables are $x_1 = I$ and $x_2 = Q$.)

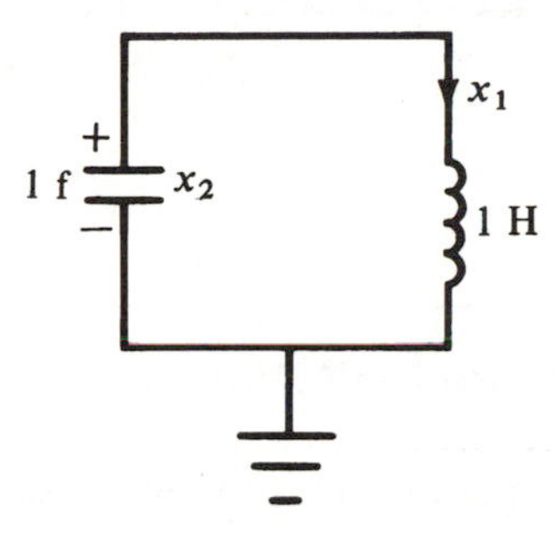

Figure 8.15

19. Consider[3] the linear time-invariant *RLC* network shown in Figure 8.16. Let the initial current through the inductor be i_0. Determine the stability of this network. (The state variables are $x_1 = -Q$ and $x_2 = I$.)

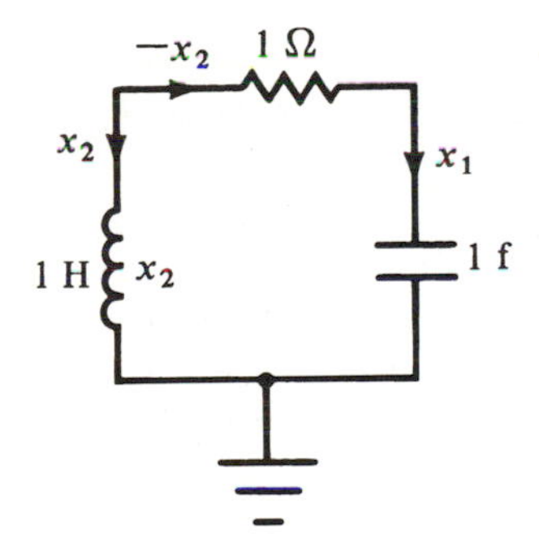

Figure 8.16

[2]This is Example 11.2-2 in Behrouz Peikari, *Fundamentals of Network Analysis and Synthesis* (Englewood Cliffs, N.J.: Prentice-Hall, 1974), © 1974. Reprinted by permission of Prentice-Hall, Inc.
[3]This is Example 11.2-3, ibid., p. 455. © 1974. Reprinted by permission of Prentice-Hall, Inc.

CHAPTER 9

Difference Equations

9.1 INTRODUCTION AND DEFINITIONS

Difference equations are equations that involve discrete changes or differences of the unknown function. This is in contrast to differential equations, which involve instantaneous rates of changes, or derivatives, of the unknown function. Difference equations are the discrete analogs of differential equations; they appear as mathematical models in situations where the variable takes or is assumed to take only a discrete set of values. As we will see in this chapter, the theory and solutions of difference equations in many ways parallel the theory and solutions of differential equations.

DEFINITION 1

A difference equation over the set of k-values 0, 1, 2, . . . is an equation of the form

$$F(k, y_k, y_{k+1}, \ldots, y_{k+n}) = 0, \tag{1}$$

where F is a given function, n is some positive integer, and $k = 0, 1, 2, \ldots$.

The following are examples of difference equations over the set of k-values $0, 1, 2, \ldots$.

$$3y_{k+1} - 8y_k = 0$$

$$y_{k+2} + 5y_{k+1} + 6y_k = 0$$

$$y_{k+3} - 3y_{k+2} + 6y_{k+1} - 4y_k = -2k + 5$$

$$y_{k+1}^2 - 2y_{k+1}y_k = k^3.$$

DEFINITION 2

A solution of the difference equation

$$F(k, y_k, y_{k+1}, \ldots, y_{k+n}) = 0 \tag{1}$$

is a sequence y_k *which satisfies (1) for* $k = 0, 1, 2, \ldots$.

For example, the sequence

$$y_k = 2^k, k = 0, 1, 2, \ldots$$

is a solution of the difference equation

$$y_{k+1} - 2y_k = 0.$$

In fact,

$$y_{k+1} - 2y_k = 2^{k+1} - 2 \cdot 2^k = 2 \cdot 2^k - 2 \cdot 2^k = 0$$

for $k = 0, 1, 2, \ldots$.

DEFINITION 3

A difference equation over the set of k-values 0, 1, 2, . . . is said to be linear if it can be written in the form

$$a_n(k)y_{k+n} + a_{n-1}(k)y_{k+n-1} + \ldots + a_1(k)y_{k+1} + a_0(k)y_k = f(k), \tag{2}$$

where n is some positive integer and the coefficients a_n, a_{n-1}, . . . , a_0, together with the function f, are given functions of k defined for $k = 0, 1, 2, \ldots$.

The following are examples of linear difference equations over the set of k-values $0, 1, 2, \ldots$.

$$5y_{k+1} - 3y_k = 0 \tag{3}$$

$$y_{k+2} + 2y_{k+1} + y_k = k^2 \tag{4}$$

$$y_{k+1} - (1-k)y_k = 1 \tag{5}$$

$$ky_{k+3} - 4y_{k+2} + y_k = 0 \tag{6}$$

$$y_{k+4} + y_k = k \tag{7}$$

On the other hand, none of the following difference equations can be written in the form of Eq. (2).

$$y_{k+2} - y_k^2 = 0$$

$$y_k\, y_{k+3} - y_{k+1} = 3k - 2$$

$$\frac{1}{2 + y_{k+1}^2} + y_{k+2} = 1.$$

Such equations are called *nonlinear*.

As in the case of ordinary differential equations, when the function f is identically zero, Eq. (2) is called *homogeneous*. When f is not identically zero, Eq. (2) is called *nonhomogeneous*.

For example, Eqs. (3) and (6) are homogeneous, but Eqs. (4), (5), and (7) are nonhomogeneous.

If all the coefficients $a_n, a_{n-1}, \ldots, a_0$ are constants, we speak of Eq. (2) as a linear difference equation with *constant coefficients*; otherwise, it is a linear difference equation with *variable coefficients*.

For example, Eqs. (3), (4), and (7) are linear difference equations with constant coefficients, but Eqs. (5) and (6) have variable coefficients.

For reasons that will become clear in subsequent sections (see Remark 1 in Section 9.2 and Remark 1 in Section 9.3), in order to define the order of the linear difference equation (2), we must make the assumption that the coefficients $a_n(k)$ and $a_0(k)$ are nonzero for all $k = 0, 1, 2, \ldots$. In this case, Eq. (2) is said to be of *order n*.

For example, Eq. (3) is of order 1, Eq. (4) is of order 2, and Eq. (7) is of order 4. But Eq. (5) is not of order 1 since the coefficient $a_0(k)$ vanishes when $k = 1$. We say that $k = 1$ is a *singular* point for Eq. (5). Also, Eq. (6) is not of order 3 because the leading coefficient $a_3(k)$ vanishes when $k = 0$. For Eq. (6), $k = 0$ is a singular point. We should remark here that over a set of k-values different from $0, 1, 2, \ldots$, Eqs. (5) and (6) may not have any singular points. For example, Eq. (5) has no singular points over the set of k-values $2, 3, 4, \ldots$; its order over this set is 1. Also over the set of k-values $1, 2, 3, \ldots$, Eq. (6) has no singular point and its order is 3.

APPLICATIONS 9.1.1

Difference equations appear as mathematical models in situations where the variable under study takes or is assumed to take only a discrete set of values. As we have already seen in Chapter 7, difference equations are useful in connection with the numerical solution of differential equations. The basic idea there is to approximate the differential equation by a difference equation. Difference equations were also encountered in Chapter 5 as recurrence formulas, which enabled us to compute the coefficients $a_2, a_3, \ldots$ of a power series solution $y(x) = \sum_{n=0}^{\infty} a_n x^n$ in terms of the coefficients a_0 and a_1.

In this section we see how difference equations appear as mathematical models describing realistic situations in biology, electric circuits, and optics. Further applications on these topics and applications to business, probability theory, and Fibonacci sequences will be given in subsequent sections.

Biology

■ The Malthusian law of population growth (see the biology application of Section 1.1.1) assumes that the rate of change of a population is proportional to the population present. That is, if $N(t)$ denotes the size of the population at time t, then $\dot{N}(t) = kN(t)$, where k is the constant of proportionality. Here $N(t)$ is assumed to be a differentiable function of time. For some populations, however, it is more realistic to assume that the size N of the population is a step function. For example, this is the case in some insect populations, where one generation dies out before the next generation hatches. A simple model for such a population will be to assume that the increase in size, from one generation to the next, is proportional to the size of the former generation. Let N_k denote the size of the population of the kth generation. Then

$$N_{k+1} - N_k = \alpha N_k, \tag{8}$$

where α is the constant of proportionality. Equation (8) can be written in the form

$$N_{k+1} - (1+\alpha)N_k = 0, \tag{9}$$

which is a linear homogeneous difference equation of order 1. Its solution is (see Section 9.4)

$$N_k = N_0(1+\alpha)^k,$$

where N_0 is the initial size of the population.

Equation (9) assumes that the size of the population depends on the population in the previous generation. In some cases it may be more realistic to assume that

$$N_{k+2} + pN_{k+1} + qN_k = 0, \tag{10}$$

which is the mathematical formulation of the assumption that the size of the population of the $(k+2)$nd generation depends (linearly) on the population of the previous two generations.

Electric Circuits

■ Consider the electric circuit shown in Figure 9.1. Assume that V_0 is a given voltage, $V_n = 0$, and the shaded region indicates the ground where the voltage is zero. Each resistance in the horizontal branch is equal to R and in the vertical branches equal to $6R$. We want to find the voltage V_k for $k = 1, 2, \ldots, n-1$. We will show that the voltage V_k for $k = 0, 1, 2, \ldots, n$ satisfies a linear homogeneous difference equation of order 2 which, together with the boundary conditions V_0 equal to a given constant and $V_n = 0$, will give the value of V_k for $k = 1, 2, \ldots, n-1$. In fact, according to Kirchhoff's current law, the sum of the currents flowing into a junction point is equal to the sum of the currents flowing away from the junction point. Applying this law at the junction point corresponding to the voltage V_{k+1}, we have

$$I_{k+1} = I_{k+2} + i_{k+1}.$$

Using Ohm's law, $I = V/R$, the above equation can be replaced by

$$\frac{V_k - V_{k+1}}{R} = \frac{V_{k+1} - V_{k+2}}{R} + \frac{V_{k+1} - 0}{6R}. \tag{11}$$

Equation (11) can be simplified and reduced to

$$6V_{k+2} - 13V_{k+1} + 6V_k = 0.$$

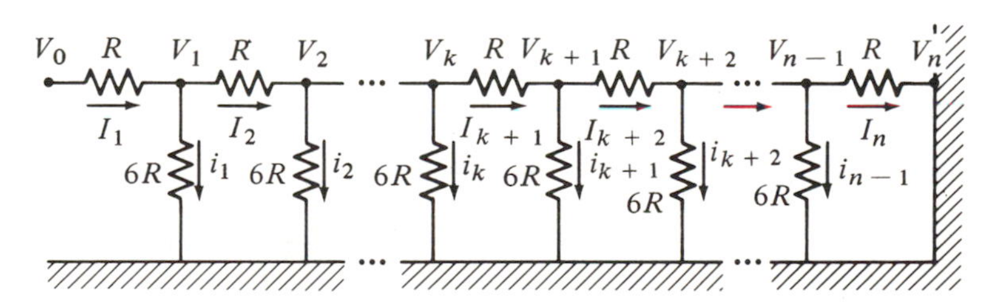

Figure 9.1

Optics: Rays Guided by Lenses

■ Consider the converging thin lens L shown in Figure 9.2.
A ray PQ passing through L will be refracted by the lens in the direction QR, which can be found as follows: From the center 0 of the lens, draw a line parallel

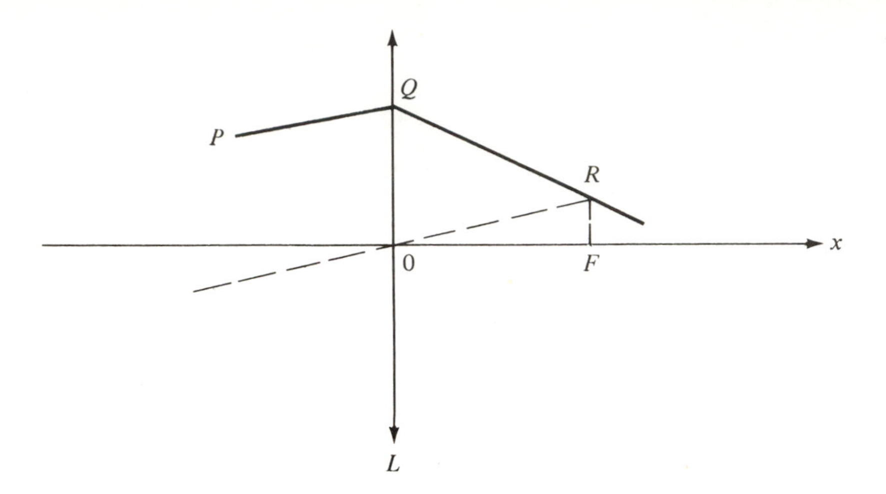

Figure 9.2

to the ray PQ; from the focal point F of the lens, draw a line parallel to the lens. The two lines meet at the point R, through which the refracted ray passes. Now consider a system of identical converging thin lenses, each of focal length f and spaced d units apart on the x-axis, as shown in Figure 9.3. We want to study the path $Q_0Q_1Q_2 \ldots$ of the ray PQ_0 through the system of lenses. That is, we want to find the y-coordinate y_k of the point Q_k, with respect to the optical axis x. The result, as we are about to prove, is that the displacement y_k of the ray, as it traverses the lenses, satisfies the linear difference equation

$$fy_{k+2} - (2f-d)y_{k+1} + fy_k = 0, \tag{12}$$

where f is the focal length of each lens and d the distance between lenses. Let us denote by m_k the slope of the segment $Q_{k-1}Q_k$. (See Figure 9.4.) Then the line $0A$ also has slope m_k, and its equation is $y = m_kx$. Thus the coordinates of the point A are (f,m_kf). Since Q_k has coordinates $(0,y_k)$, it follows that the slope m_{k+1} of the line segment Q_kQ_{k+1} is given by

$$m_{k+1} = \frac{m_kf - y_k}{f} = -\frac{1}{f}y_k + m_k. \tag{13}$$

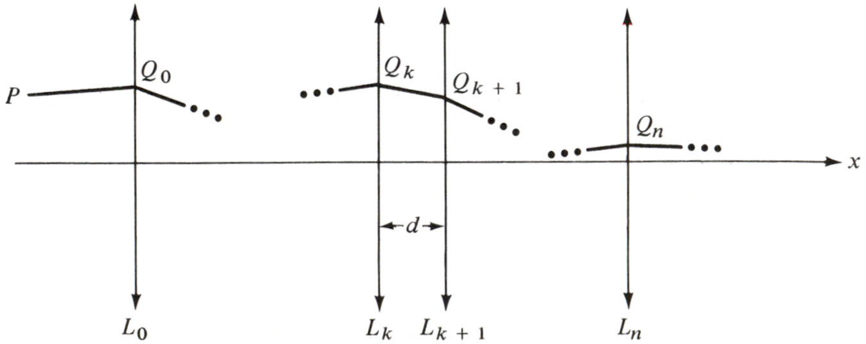

P, Q_0, Q_k, Q_{k+1}, Q_n, x, d, L_0, L_k, L_{k+1}, L_n

Figure 9.3

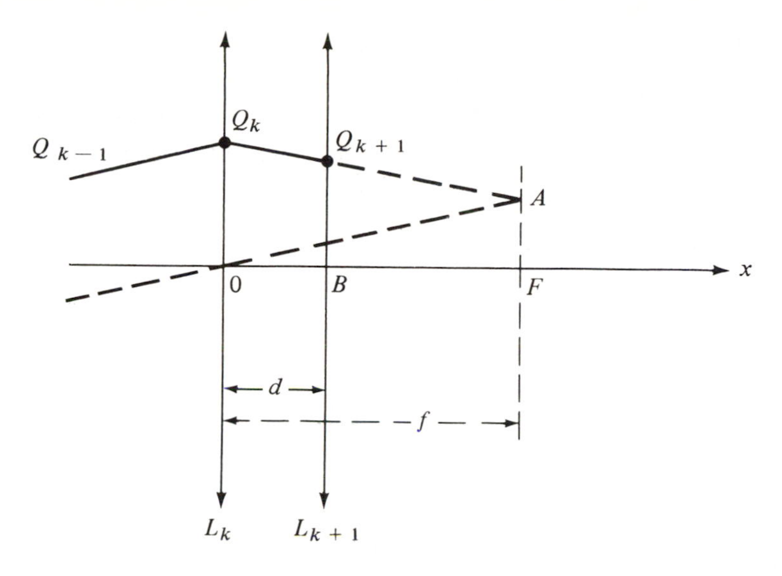

Figure 9.4

On the other hand, the equation of the line through the points Q_k and A is

$$y = y_k + \frac{m_k f - y_k}{f - 0} x.$$

For $x = d$, we find

$$y_{k+1} = y_k + \left(m_k - \frac{1}{f} y_k\right) d,$$

or

$$y_{k+1} = \left(1 - \frac{d}{f}\right) y_k + d m_k. \tag{14}$$

Equations (14) and (13) form a system of first-order linear difference equations in y_k and m_k. Eliminating m_k gives rise to Eq. (12). In fact, from Eqs. (14) and (13) we have

$$\begin{aligned} y_{k+2} &= \left(1 - \frac{d}{f}\right) y_{k+1} + d m_{k+1} \\ &= \left(1 - \frac{d}{f}\right) y_{k+1} + d\left(-\frac{1}{f} y_k + m_k\right) \\ &= \left(1 - \frac{d}{f}\right) y_{k+1} - \frac{d}{f} y_k + d m_k \\ &= \left(1 - \frac{d}{f}\right) y_{k+1} - \frac{d}{f} y_k + y_{k+1} - \left(1 - \frac{d}{f}\right) y_k \\ &= \left(2 - \frac{d}{f}\right) y_{k+1} - y_k, \end{aligned}$$

from which Eq. (12) follows.

EXERCISES

In Exercises 1 through 16, classify the difference equation as being linear or nonlinear. Furthermore, classify the linear ones as homogeneous or nonhomogeneous, with constant coefficients or with variable coefficients, and state the order.

1. $y_{k+2} + (k+1)y_k = 0$

2. $y_{k+3} + (k^2+1)y_k = 2^k$

3. $y_{k+4} + y_k y_{k+2} = 3$

4. $y_{k+2} - 4y_k = 0$

5. $y_{k+5} - 2y_{k+4} + y_k = 2k^2-3$

6. $(1+k^2)y_{k+2} - ky_{k+1} + y_k = 0$

7. $y_{k+1} - 3^k y_k = -2$

8. $2y_{k+3} + 3y_{k+2} - 4y_{k+1} + y_k = 0$

9. $3y_{k+1} - 2^k y_k - 3 = 0$

10. $y_{k+1} - 4y_k = 5k - 2$

11. $y_{k+2} - \left(\sin \frac{k\pi}{2}\right)y_{k+1} + y_k = k$

12. $y_k - 3y_{k+2} + 4y_{k+1} = 3 \cdot 2^k$

13. $y_{k+3} - y_k + 2 = 0$

14. $(y_{k+1})^2 - 3y_k = 0$

15. $y_{k+2} - 8y_{k+1} + 15y_k = 0$

16. $y_{k+1}y_k = 3$

In Exercises 17 through 24, answer true or false.

17. $y_k = 3^k$ is a solution of $y_{k+2} - 8y_{k+1} + 15y_k = 0$.

18. $y_k = 5^k$ is a solution of $y_{k+1} + 5y_k = 0$.

19. $k = 0$ is a singular point for $y_{k+2} - ky_{k+1} + 3y_k = 3^k$.

20. $k = 0$ is a singular point for $y_{k+2} - 3y_{k+1} + ky_k = 3^k$.

21. The constant sequence $y_k = -1$ is a solution of $y_{k+2} - y_k^2 = 0$.

22. The constant sequence $y_k = 0$ is a solution of any linear difference equation.

23. The constant sequence $y_k = 0$ is a solution of any linear homogeneous difference equation.

24. $y_k = k^2$ is a solution of $(2k+1)y_{k+2} - 4(k+1)y_{k+1} + (2k+3)y_k = 0$.

9.2 EXISTENCE AND UNIQUENESS OF SOLUTIONS

Here and in Section 9.3 we develop the basic theory of solutions of linear difference equations. There are great similarities between this theory and the corresponding theory of linear differential equations developed in Chapter 2. For simplicity we state and prove the theorems for second-order linear difference equations. But the results can be extended to equations of any order.

Consider the second-order linear difference equation

$$a_2(k)y_{k+2} + a_1(k)y_{k+1} + a_0(k)y_k = f(k), \tag{1}$$

where the coefficients a_2, a_1, a_0 and the function f are given functions of k defined for $k = 0, 1, 2, \ldots$. Since Eq. (1) is assumed to be of order 2, the coefficients $a_2(k)$ and $a_0(k)$ are nonzero for all $k = 0, 1, 2, \ldots$. Dividing both sides of Eq. (1) by $a_2(k)$ and setting

$$\frac{a_1(k)}{a_2(k)} = p_k, \quad \frac{a_0(k)}{a_2(k)} = q_k, \quad \frac{f(k)}{a_2(k)} = r_k \quad \text{for} \quad k = 0, 1, 2, \ldots,$$

we get

$$y_{k+2} + p_k y_{k+1} + q_k y_k = r_k, \tag{2}$$

where p_k, q_k, and r_k are given sequences of numbers, and $q_k \neq 0$ for all $k = 0, 1, 2, \ldots$. A solution of Eq. (2) is a sequence y_k which satisfies Eq. (2) for all $k = 0, 1, 2, \ldots$.

The proof of the existence and uniqueness theorem for differential equations (see Appendix D) is so involved that it is almost never proved at the elementary level. On the other hand, the proof of the corresponding theorem for difference equations is almost trivial.

THEOREM 1 EXISTENCE AND UNIQUENESS

The initial value problem

$$y_{k+2} + p_k y_{k+1} + q_k y_k = r_k, \tag{3}$$

$$y_0 = A, \; y_1 = B, \tag{4}$$

where p_k, q_k, and r_k are given sequences of numbers with $q_k \neq 0$ for all $k = 0, 1, 2, \ldots$ and A, B are given constants, has exactly one solution.

Proof For $k = 0$, Eq. (3) becomes

$$y_2 + p_0 y_1 + q_0 y_0 = r_0.$$

Using the initial conditions (4) we find

$$y_2 = r_0 - p_0 B - q_0 A.$$

Hence the value of y_2 is uniquely determined in terms of known quantities. Next, setting $k = 1$ in Eq. (3) and using the values of y_1 and y_2, we find that y_3 is uniquely determined, and so forth. For a rigorous proof that every term y_k of the solution is uniquely determined, we should continue by mathematical induction as follows: Assume that the terms $y_0, y_1, \ldots, y_{n+1}$ of the solution y_k are uniquely determined. We must prove that y_{n+2} is also uniquely determined. In fact, from Eq. (3) with $k = n$, we find

$$y_{n+2} = -p_n y_{n+1} - q_n y_n,$$

and the result follows.

EXAMPLE 1 Solve the IVP

$$y_{k+2} - 3y_{k+1} + 2y_k = 0, \tag{5}$$

$$y_0 = 1, y_1 = 2. \tag{6}$$

Solution The hypotheses of Theorem 1 are satisfied, so the IVP (5)–(6) has a unique solution. Setting $k = 0, 1, 2, \ldots$ into the difference equation (5), we find

for $k = 0$, $y_2 - 3y_1 + 2y_0 = 0 \Rightarrow y_2 = 3y_1 - 2y_0 = 6 - 2 = 4 = 2^2$,

for $k = 1$, $y_3 - 3y_2 + 2y_1 = 0 \Rightarrow y_3 = 3y_2 - 2y_1 = 12 - 4 = 8 = 2^3$,

for $k = 2$, $y_4 - 3y_3 + 2y_2 = 0 \Rightarrow y_4 = 3y_3 - 2y_2 = 24 - 8 = 16 = 2^4$,

and so forth. From the above results we conjecture that the unique solution of the IVP(5)–(6) is $y_k = 2^k$. To prove it we need only demonstrate that y_k satisfies the difference equation (5) and the initial conditions (6). Since $y_0 = 2^0 = 1$ and $y_1 = 2^1 = 2$, the initial conditions (6) are satisfied. Also, Eq. (5) is satisfied because

$$\begin{aligned} y_{k+2} - 3y_{k+1} + 2y_k &= 2^{k+2} - 3 \cdot 2^{k+1} + 2 \cdot 2^k = 2^2 \cdot 2^k - 3 \cdot 2 \cdot 2^k + 2 \cdot 2^k \\ &= 4 \cdot 2^k - 6 \cdot 2^k + 2 \cdot 2^k = 0. \end{aligned}$$

REMARK 1 The hypothesis that the coefficient $a_2(k)$ of the difference equation (1) never vanishes is crucial for existence and uniqueness of solutions. When $a_2(k)$ vanishes for some value of k, the IVP consisting of the difference equation (1) and the initial conditions (4) may not have any solution at all or it may have infinitely many solutions. The following two examples substantiate these comments:

(a) The IVP

$$\left.\begin{aligned} ky_{k+2} - y_k &= 0 \\ y_0 = 1, y_1 &= 0 \end{aligned}\right\} \tag{7}$$

has no solution. In fact, for $k = 0$ the difference equation gives $y_0 = 0$, which violates the first of the initial conditions in (7).

(b) The IVP

$$ky_{k+2} - y_k = 0, \tag{8}$$

$$y_0 = 0, y_1 = 0 \tag{9}$$

has infinitely many solutions. In fact, setting successively $k = 0, 1, 2, \ldots$ into Eq. (8) and using (9), we find (putting the result into compact form)

$$y_k = \begin{cases} 0 \quad \text{for} \quad k = 0 \quad \text{and} \quad k = 1, 3, 5, \ldots \\ \dfrac{c}{2^{[(k/2) - 1]} \left(\dfrac{k}{2} - 1\right)!} \quad \text{for} \quad k = 2, 4, 6, \ldots \end{cases} \tag{10}$$

where c is an arbitrary constant. In Exercise 14 the reader is asked to verify directly that (10) is a solution of the IVP (8)–(9).

EXERCISES

For each of the initial value problems in Exercises 1 through 4, state whether or not the hypotheses of Theorem 1 are satisfied.

1. $y_{k+2} - 8y_{k+1} + 15y_k = 3^k,\ y_0 = 0,\ y_1 = 2$

2. $y_{k+2} - 8y_{k+1} + 15y_k = 0,\ y_0 = 0,\ y_1 = 2$

3. $y_{k+2} - 8y_{k+1} + (k-2)y_k = 3^k,\ y_0 = 0,\ y_1 = 0$

4. $y_{k+2} - 8y_{k+1} + ky_k = 3^k,\ y_0 = 0,\ y_1 = 2$

In Exercises 5 through 9, verify that the unique solution of the initial value problem is the given sequence.

5. $y_{k+2} - 2y_{k+1} + y_k = 4,\ y_0 = 3,\ y_1 = 5;\ y_k = 3 + 2k^2$

6. $y_{k+2} + y_k = -2\cos\dfrac{k\pi}{2},\ y_0 = 0,\ y_1 = 1;\ y_k = \sin\dfrac{k\pi}{2} - k\cos\dfrac{k\pi}{2}$

7. $y_{k+2} - 3(k^2+1)y_k = 0,\ y_0 = y_1 = 0;\ y_k = 0$

8. $y_{k+2} - 8y_{k+1} + 15y_k = 6 \cdot 2^k,\ y_0 = 2,\ y_1 = 4;\ y_k = 2^{k+1}$

9. $(2k+1)y_{k+2} - 4(k+1)y_{k+1} + (2k+3)y_k = 0;\ y_0 = y_1 = 1;\ y_k = 1$

In Exercises 10 through 13, find (by iteration as in Example 1) the unique solution of the initial value problem.

10. $y_{k+2} - 8y_{k+1} + 15y_k = 0,\ y_0 = 1,\ y_1 = 3$

11. $y_{k+2} - 4y_{k+1} + 4y_k = 0,\ y_0 = 0,\ y_1 = 2$

12. $y_{k+2} - y_k = 0,\ y_0 = 1,\ y_1 = -1$

13. $4y_{k+2} - y_k = 0,\ y_0 = 1,\ y_1 = \dfrac{1}{2}$

14. Verify that the sequence

$$y_k = \begin{cases} 0 & \text{for} \quad k = 0 \quad \text{and} \quad k = 1, 3, 5, \ldots \\ \dfrac{c}{2^{[(k/2)-1]} \cdot \left(\dfrac{k}{2} - 1\right)!} & \text{for} \quad k = 2, 4, 6, \ldots \end{cases} \tag{10}$$

is a solution of the IVP (8)–(9) for any choice of the constant c. Why doesn't this contradict Theorem 1?

15. Extend the existence and uniqueness theorem (Theorem 1) to linear equations of any order.

Fibonacci Sequence In the sequence

$$1, 1, 2, 3, 5, 8, 13, \ldots, \tag{11}$$

called the *Fibonacci sequence*, each term after the second is the sum of the two preceding terms. Therefore, if y_k denotes the kth term of (11), then the Fibonacci sequence is the unique solution of the IVP

$$y_{k+2} = y_{k+1} + y_k, \quad k = 1, 2, 3, \ldots, \tag{12}$$

$$y_1 = 1, \quad y_2 = 1. \tag{13}$$

In Exercise 17, the reader is asked to show that the unique solution of the IVP (12)–(13) is given by

$$y_k = \frac{1}{\sqrt{5}}\left[\left(\frac{1+\sqrt{5}}{2}\right)^k - \left(\frac{1-\sqrt{5}}{2}\right)^k\right]. \tag{14}$$

The numbers in the Fibonacci sequence, called the *Fibonacci numbers*, have many fascinating properties and appear mysteriously in many diverse situations. They were discovered in the famous Fibonacci rabbit problem.[1]

16. **The Fibonacci Rabbit Problem** A pair of rabbits gives birth to a new pair of rabbits every month after the pair is two months old. Every new pair of rabbits does exactly the same. How many pairs will there be after k months if we begin with a pair of newborn rabbits?

17. Show that the unique solution of the IVP (12)–(13) is given by (14).

9.3 LINEAR INDEPENDENCE AND THE GENERAL SOLUTION

The following definition of linear dependence and independence of sequences is similar to the corresponding definition for functions. (See Definition 1 of Section 2.2.)

DEFINITION 1

Two sequences a_k and b_k, $k = 0, 1, 2, \ldots$ are said to be linearly dependent if there exist constants A and B, not both zero, such that

$$Aa_k + Bb_k = 0 \quad \text{for} \quad k = 0, 1, 2, \ldots. \tag{1}$$

Otherwise the sequences are said to be linearly independent.

For example, the sequences

$$a_k = 3k + \frac{12}{5} \quad \text{and} \quad b_k = 5k + 4, k = 0, 1, 2, \ldots \tag{2}$$

[1]The rabbit problem first appeared in Leonardo "Fibonacci" da Pisa, *Liber abaci* (1202).

are linearly dependent because (1) is satisfied with $A = 5$ and $B = -3$. On the other hand, the sequences

$$a_k = 2^k \quad \text{and} \quad b_k = 3^k, k = 0, 1, 2, \ldots \tag{3}$$

are linearly independent. Otherwise, they are linearly dependent, which means there exist constants A and B, not both zero, such that (1) holds. That is,

$$A2^k + B3^k = 0 \quad \text{for} \quad k = 0, 1, 2, \ldots. \tag{4}$$

Setting $k = 0$ and $k = 1$ in (4), we find

$$A + B = 0 \quad \text{and} \quad 2A + 3B = 0. \tag{5}$$

But the only solution of system (5) is $A = B = 0$. This contradicts the hypothesis and establishes our claim that the two sequences in (3) are linearly independent.

Before we present a simple test for linear dependence and independence of solutions of linear homogeneous difference equations, we need the following definition.

DEFINITION 2

The Casoratian of two sequences a_k and b_k, $k = 0, 1, 2, \ldots$ is denoted by $C(a_k, b_k)$ and is defined to be the determinant

$$C(a_k, b_k) = \begin{vmatrix} a_k & b_k \\ a_{k+1} & b_{k+1} \end{vmatrix}. \tag{6}$$

As we will see, the Casoratian plays for difference equations the role that the Wronskian plays for differential equations. Let us compute the Casoratians of the sequences in (2) and (3). We have

$$\begin{aligned} C\left(3k + \frac{12}{5}, 5k + 4\right) &= \begin{vmatrix} 3k + \frac{12}{5} & 5k + 4 \\ 3(k+1) + \frac{12}{5} & 5(k+1) + 4 \end{vmatrix} \\ &= \left(3k + \frac{12}{5}\right)(5k + 9) - (5k + 4)\left(3k + \frac{27}{5}\right) \\ &= 0 \qquad \text{for all } k = 0, 1, 2, \ldots \end{aligned}$$

and

$$\begin{aligned} C(2^k, 3^k) &= \begin{vmatrix} 2^k & 3^k \\ 2^{k+1} & 3^{k+1} \end{vmatrix} \\ &= 2^k 3^{k+1} - 2^{k+1} 3^k = 3(2^k 3^k) - 2(2^k 3^k) \\ &= 2^k 3^k = 6^k \neq 0 \qquad \text{for all } k = 0, 1, 2, \ldots. \end{aligned}$$

THEOREM 1
Let $y_k^{(1)}$ and $y_k^{(2)}$ be two solutions of the linear homogeneous difference equation

$$y_{k+2} + p_k y_{k+1} + q_k y_k = 0, \tag{7}$$

where $q_k \neq 0$ for $k = 0, 1, 2, \ldots$. Then their Casoratian $C(y_k^{(1)}, y_k^{(2)})$ is either never zero or vanishes for all $k = 0, 1, 2, \ldots$.

Proof Set

$$C_k = C(y_k^{(1)}, y_k^{(2)}).$$

We will show that C_k satisfies a linear first-order difference equation, namely,

$$C_{k+1} = q_k C_k. \tag{8}$$

In fact, using Eq. (6) and the fact that $y_k^{(1)}$ and $y_k^{(2)}$ are solutions of the difference equation (7), we have

$$\begin{aligned} C_{k+1} = \begin{vmatrix} y_{k+1}^{(1)} & y_{k+1}^{(2)} \\ y_{k+2}^{(1)} & y_{k+2}^{(2)} \end{vmatrix} &= y_{k+1}^{(1)} y_{k+2}^{(2)} - y_{k+1}^{(2)} y_{k+2}^{(1)} \\ &= y_{k+1}^{(1)}(-p_k y_{k+1}^{(2)} - q_k y_k^{(2)}) - y_{k+1}^{(2)}(-p_k y_{k+1}^{(1)} - q_k y_k^{(1)}) \\ &= q_k(y_k^{(1)} y_{k+1}^{(2)} - y_k^{(2)} y_{k+1}^{(1)}) \\ &= q_k C_k. \end{aligned}$$

Now Eq. (8) can be solved explicitly. In fact, from (8) and for $k = 0, 1, 2, \ldots,$ we find

$$C_1 = q_0 C_0,\ C_2 = q_1 C_1 = q_1 q_0 C_0,\ C_3 = q_2 C_2 = q_2 q_1 q_0 C_0,$$

and, in general,

$$C_k = \left(\prod_{i=0}^{k-1} q_i\right) C_0, \tag{9}$$

where the symbol $\prod_{i=0}^{k-1} q_i$ denotes the product $q_0 q_1 \cdots q_{k-1}$. Finally, from (9), and thanks to the hypothesis that $q_i \neq 0$ for $i = 0, 1, 2, \ldots,$ it follows that C_k is either always zero or never zero, depending on whether C_0 is zero or not. The proof is complete.

REMARK 1 The hypothesis $q_k \neq 0$ for $k = 0, 1, 2, \ldots$ is necessary for Theorem 1 to be true. For example, the conclusion of Theorem 1 fails to be true for the difference equation

$$y_{k+2} - k y_k = 0$$

considered over the set of k-values 0, 1, 2, ... In fact,

$$y_k^{(1)} : 1, 1, 0, 1, 0, 3, 0, 15, \ldots$$

and

$$y_k^{(2)} : 0, 2, 0, 2, 0, 6, 0, 30, \ldots$$

are solutions. Their Casoratian is

$$C(y_k^{(1)}, y_k^{(2)}) = \begin{cases} 1 & \text{if} \quad k = 0 \\ 0 & \text{if} \quad k \neq 0 \end{cases}.$$

On the other hand, on the basis of Definition 1, the solutions $y_k^{(1)}$ and $y_k^{(2)}$ are linearly independent, although

$$y_k^{(2)} = 2y_k^{(1)} \qquad \text{for} \qquad k = 1, 2, 3, \ldots.$$

THEOREM 2

Two solutions, $y_k^{(1)}$ and $y_k^{(2)}$, of the linear homogeneous difference equation

$$y_{k+2} + p_k\, y_{k+1} + q_k\, y_k = 0, \tag{10}$$

where $q_k \neq 0$ for $k = 0, 1, 2, \ldots$, are linearly independent if and only if their Casoratian

$$C(y_k^{(1)}, y_k^{(2)}) = \begin{vmatrix} y_k^{(1)} & y_k^{(2)} \\ y_{k+1}^{(1)} & y_{k+1}^{(2)} \end{vmatrix}$$

is different from zero.

Proof Assume that the solutions $y_k^{(1)}$ and $y_k^{(2)}$ are linearly independent. We should prove that

$$C(y_k^{(1)}, y_k^{(2)}) \neq 0 \qquad \text{for} \qquad k = 0, 1, 2, \ldots.$$

Otherwise the Casoratian is identically zero, and, in particular,

$$\begin{vmatrix} y_0^{(1)} & y_0^{(2)} \\ y_1^{(1)} & y_1^{(2)} \end{vmatrix} = 0. \tag{11}$$

Consider the homogeneous system

$$\left.\begin{aligned} Ay_0^{(1)} + By_0^{(2)} &= 0 \qquad (12) \\ Ay_1^{(1)} + By_1^{(2)} &= 0 \qquad (13) \end{aligned}\right\}.$$

Since, by (11), the determinant of the coefficients of the system is zero, it follows, from Appendix A, that the system has a nontrivial solution (A, B).

Multiplying both sides of (12) by q_0 and both sides of (13) by p_0 and adding the results, we find

$$q_0(Ay_0^{(1)} + By_0^{(2)}) + p_0(Ay_1^{(1)} + By_1^{(2)}) = 0$$

or

$$A(p_0\, y_1^{(1)} + q_0\, y_0^{(1)}) + B(p_0\, y_1^{(2)} + q_0\, y_0^{(2)}) = 0. \tag{14}$$

Now, using the fact that $y_k^{(1)}$ and $y_k^{(2)}$ are solutions of (10), we have

$$y_2^{(1)} = -(p_0\, y_1^{(1)} + q_0\, y_0^{(1)}) \qquad \text{and} \qquad y_2^{(2)} = -(p_0\, y_1^{(2)} + q_0\, y_0^{(2)}),$$

and (14) becomes

$$Ay_2^{(1)} + By_2^{(2)} = 0. \tag{15}$$

One can show by induction that

$$Ay_k^{(1)} + By_k^{(2)} = 0 \qquad \text{for} \qquad k = 0, 1, 2, \ldots .$$

Since A and B are not both zero, it follows that the solutions $y_k^{(1)}$ and $y_k^{(2)}$ are linearly dependent. This contradiction proves the claim that

$$C(y_k^{(1)}, y_k^{(2)}) \neq 0.$$

Conversely, assume that

$$C(y_k^{(1)}, y_k^{(2)}) \neq 0.$$

We must prove that $y_k^{(1)}$ and $y_k^{(2)}$ are linearly independent. Otherwise they are linearly dependent which, by Definition 1, means there are numbers A and B, not both zero, such that

$$Ay_k^{(1)} + By_k^{(2)} = 0 \qquad \text{for} \qquad k = 0, 1, 2, \ldots . \tag{16}$$

Assume $A \neq 0$. The case $B \neq 0$ is treated in a similar fashion. Then, from (16),

$$y_k^{(1)} = -\frac{B}{A} y_k^{(2)},$$

and so

$$\begin{aligned} C(y_k^{(1)}, y_k^{(2)}) &= y_k^{(1)} y_{k+1}^{(2)} - y_k^{(2)} y_{k+1}^{(1)} \\ &= -\frac{B}{A} y_k^{(2)} y_{k+1}^{(2)} - y_k^{(2)} \left(-\frac{B}{A} y_{k+1}^{(2)} \right) \\ &= 0. \end{aligned}$$

This contradicts our assumption and the proof is complete.

As in the case of differential equations, we have the following theorem for the general solution of a linear homogeneous and a linear nonhomogeneous difference equation of order 2. The proofs are straightforward. See Exercises 9 and 10.

THEOREM 3

Let $y_k^{(1)}$ and $y_k^{(2)}$ be two linearly independent solutions of the linear homogeneous difference equation

$$y_{k+2} + p_k y_{k+1} + q_k y_k = 0, \tag{17}$$

where $q_k \neq 0$ for $k = 0, 1, 2, \ldots$. The general solution of (17) is $c_1 y_k^{(1)} + c_2 y_k^{(2)}$, where c_1 and c_2 are arbitrary constants.

THEOREM 4

Let $y_k^{(1)}$ and $y_k^{(2)}$ be two linearly independent solutions of the linear homogeneous difference equation

$$y_{k+2} + p_k y_{k+1} + q_k y_k = 0, \tag{18}$$

where $q_k \neq 0$ for $0, 1, 2, \ldots$, and let y_k^p be a particular solution of the nonhomogeneous equation

$$y_{k+2} + p_k y_{k+1} + q_k y_k = r_k. \tag{19}$$

Then the general solution of Eq. (19) is $c_1 y_k^{(1)} + c_2 y_k^{(2)} + y_k^p$, where c_1 and c_2 are arbitrary constants.

EXAMPLE 1 Consider the difference equations

$$y_{k+2} - 8y_{k+1} + 15y_k = 16 \tag{20}$$

and

$$y_{k+2} - 8y_{k+1} + 15y_k = 0. \tag{21}$$

(a) Show that $y_k^{(1)} = 3^k$ and $y_k^{(2)} = 5^k$ are linearly independent solutions of Eq. (21) and that $y_k^p = 2$ is a particular solution of Eq. (20).
(b) Find the general solutions of Eq. (20) and (21).

Solution (a) By direct substitution into Eq. (21), we see that $y_k^{(1)}$ and $y_k^{(2)}$ are indeed solutions of Eq. (21). They are linearly independent because their Casoratian

$$\begin{aligned} C(3^k, 5^k) &= \begin{vmatrix} 3^k & 5^k \\ 3^{k+1} & 5^{k+1} \end{vmatrix} \\ &= 3^k \cdot 5^{k+1} - 5^k \cdot 3^{k+1} \\ &= 5(3^k \cdot 5^k) - 3(5^k \cdot 3^k) \\ &= 2 \cdot 3^k \cdot 5^k \end{aligned}$$

is different from zero for all $k = 0, 1, 2, \ldots$. Also by direct substitution, y_k^p is a solution of Eq. (20).

(b) From (a) and Theorem 3, the general solution of Eq. (21) is

$$y_k^h = c_1 3^k + c_2 5^k,$$

where c_1 and c_2 are arbitrary constants. From (a) and Theorem 4, the general solution of Eq. (20) is

$$y_k = c_1 3^k + c_2 5^k + 2,$$

where c_1 and c_2 are arbitrary constants.

EXERCISES

In Exercises 1 though 5, a second-order linear homogeneous difference equation and two solutions $y_k^{(1)}$ and $y_k^{(2)}$ are given. Test whether the solutions are linearly dependent or independent. If they are linearly independent, give the general solution of the equation.

1. $y_{k+2} - 8y_{k+1} + 15y_k = 0;\ y_k^{(1)} = 3^k,\ y_k^{(2)} = 5^k$

2. $y_{k+2} - 8y_{k+1} + 15y_k = 0;\ y_k^{(1)} = 3^{k+1},\ y_k^{(2)} = 5^k$

3. $y_{k+2} - 8y_{k+1} + 15y_k = 0;\ y_k^{(1)} = 3^k,\ y_k^{(2)} = 3^{k+1}$

4. $y_{k+2} - 2y_{k+1} + y_k = 0;\ y_k^{(1)} = 1,\ y_k^{(2)} = k$

5. $y_{k+2} - 2y_{k+1} + y_k = 0;\ y_k^{(1)} = k,\ y_k^{(2)} = 2k$

In Exercises 6 though 8, a second-order linear nonhomogeneous difference equation is given together with two solutions $y_k^{(1)}$ and $y_k^{(2)}$ of the corresponding homogeneous equation and a particular solution y_k^p of the nonhomogeneous equation. Determine whether $y_k^{(1)}$ and $y_k^{(2)}$ are linearly independent; if they are, give the general solution of the nonhomogeneous equation.

6. $y_{k+2} - 8y_{k+1} + 15y_k = 3 \cdot 2^k - 5 \cdot 3^k$

$y_k^{(1)} = 3^k,\ y_k^{(2)} = 5^k;\ y_k^p = 2^k + \frac{5}{6} k3^k$

7. $y_{k+2} - y_k = 12k^2$

$y_k^{(1)} = 1,\ y_k^{(2)} = (-1)^k;\ y_k^p = 2k^3 - 6k^2 + 4k$

8. $4y_{k+2} - y_k = 3 \cos \dfrac{k\pi}{2}$

$y_k^{(1)} = 2^{-k},\ y_k^{(2)} = \dfrac{1}{2^{k+1}};\ y_k^p = -\dfrac{3}{5} \cos \dfrac{k\pi}{2}$

9. Prove Theorem 3.

10. Prove Theorem 4.

11. Extend Definition 2 to n sequences $y_k^{(1)}, y_k^{(2)}, \ldots, y_k^{(n)}$.

12. Extend Theorem 2 to linear equations of order n.

13. Extend Theorem 3 to linear equations of order n.

14. Extend Theorem 4 to linear equations of order n.

In Exercises 15 through 19, answer true or false.

15. The sequences

$$y_k^{(1)} = 1, y_k^{(2)} = 2^k, y_k^{(3)} = 3^k$$

are linearly independent solutions of the equation

$$y_{k+3} - 6y_{k+2} + 11y_{k+1} - 6y_k = 0.$$

16. The general solution of $y_{k+1} - 2y_k = 1$ is $y_k = c2^k - 1$, where c is an arbitrary constant.

17. The general solution of $y_{k+1} - y_k = 2$ is $y_k = c + 2$, where c is an arbitrary constant.

18. The general solution of $y_{k+4} - y_k = 4$ is

$$y_k = c_1 + c_2(-1)^k + c_3 \cos\frac{k\pi}{2} + c_4 \sin\frac{k\pi}{2} + k.$$

19. The general solution of

$$y_{k+3} - 6y_{k+2} + 11y_{k+1} - 6y_k = 0$$

is

$$y_k = c_1 2^k + c_2 3^k.$$

20. First-Order Linear Difference Equations Show that the general solution of the first-order linear difference equation with variable coefficients

$$y_{k+1} - q_k y_k = r_k$$

is given by

$$y_k = \left(\prod_{i=0}^{k-1} q_i\right) y_0 + \sum_{n=0}^{k-1} \left(\prod_{i=n+1}^{k-1} q_i\right) r_n,$$

where the first term y_0 of the solution is an arbitrary constant. [*Hint:* Use the procedure used in Theorem 1 to solve the homogeneous equation (8).]

21. Find the general solution of the equation $y_{k+1} - y_k = r_k$. [*Hint:* Use Exercise 20 or sum up both sides of the equation for $k = 0, 1, \ldots, k - 1$.]

In Exercises 22 through 27, find the general solution.

22. $y_{k+1} - y_k = 2k + 1$

23. $y_{k+1} - y_k = k$

24. $y_{k+1} - (k + 1)y_k = 0$

25. $y_{k+1} + 3y_k = -1$

26. $y_{k+1} - 3y_k = 1$

27. $y_{k+1} - \dfrac{2k+3}{2k+1} y_k = 0$

28. A certain colony of bacteria undergoes a 10 percent relative increase in size per hour. If the initial size of the colony is 2000, what will its size be after 24 hours?

9.4 HOMOGENEOUS EQUATIONS WITH CONSTANT COEFFICIENTS

In this section we show how to find the general solution of a linear homogeneous difference equation with constant coefficients. We present the main features of the method by means of examples.

EXAMPLE 1 Find the general solution of the linear homogeneous equation

$$y_{k+2} - 8y_{k+1} + 15y_k = 0. \tag{1}$$

Solution With differential equations, we looked for a solution of the form $y(x) = e^{\lambda x}$. Here the idea is to look for a solution of the form

$$y_k = \lambda^k, \tag{2}$$

where $\lambda \neq 0$. Substituting (2) into Eq. (1), we find

$$\lambda^{k+2} - 8\lambda^{k+1} - 15\lambda^k = 0,$$

and dividing by λ^k,

$$\lambda^2 - 8\lambda + 15 = 0. \tag{3}$$

Equation (3) is called the *characteristic equation* of the difference equation (1). The roots of the characteristic equation (3), called the *characteristic roots*, are $\lambda_1 = 3$ and $\lambda_2 = 5$. Hence

$$y_k^{(1)} = 3^k \qquad \text{and} \qquad y_k^{(2)} = 5^k$$

are two solutions of Eq. (1). They are linearly independent because their Casoratian

$$\begin{vmatrix} 3^k & 5^k \\ 3^{k+1} & 5^{k+1} \end{vmatrix} = 3^k5^{k+1} - 3^{k+1}5^k = 5 \cdot 3^k \cdot 5^k - 3 \cdot 3^k \cdot 5^k = 2 \cdot 3^k \cdot 5^k$$

is never zero for $k = 0, 1, 2, \ldots$. Hence, by Theorem 3 of Section 9.3, the general solution of Eq. (1) is

$$y_k = c_1 3^k + c_2 5^k,$$

where c_1 and c_2 are arbitrary constants.

EXAMPLE 2 Find the general solution of the difference equation

$$3y_{k+2} - 5y_{k+1} - 2y_k = 0.$$

Solution Characteristic equation: $3\lambda^2 - 5\lambda - 2 = 0$.

Characteristic roots:

$$\lambda = \frac{5 \pm \sqrt{25+24}}{6} = \frac{5 \pm 7}{6} = \begin{cases} 2 \\ -1/3 \end{cases}.$$

General solution: $y_k = c_1 2^k + c_2\left(-\frac{1}{3}\right)^k.$

EXAMPLE 3 Find the general solution of

$$y_{k+2} - 6y_{k+1} + 9y_k = 0. \tag{4}$$

Solution Looking for a solution of the form $y_k = \lambda^k$, as in Example 1, we find that λ must be a root of the characteristic equation

$$\lambda^2 - 6\lambda + 9 = 0.$$

But here the characteristic roots are equal, $\lambda_1 = \lambda_2 = 3$, and so Eq. (4) has the solution

$$y_k^{(1)} = 3^k \tag{5}$$

(and all its multiples); but it does not have two linearly independent solutions of the form λ^k. However, in the case of equal roots, the product $ky_k^{(1)}$ gives always another solution which, together with $y_k^{(1)}$, are linearly independent. That is, in addition to (5), Eq. (4) has the solution

$$y_k^{(2)} = k3^k, \tag{6}$$

and $y_k^{(1)}$, $y_k^{(2)}$ are linearly independent. (Show it.)

Hence, the general solution of Eq. (4) is

$$y_k = c_1 3^k + c_2 k 3^k.$$

EXAMPLE 4 Find the general solution of the equation

$$y_{k+2} + 2y_{k+1} + 2y_k = 0. \tag{7}$$

Solution Characteristic equation: $\lambda^2 + 2\lambda + 2 = 0$.
Characteristic roots: $\lambda = -1 \pm i$.
General solution:

$$y_k = c_1(-1+i)^k + c_2(-1-i)^k. \tag{8}$$

REMARK 1 We will show here how to write the general solution (8) of Eq. (7) as a linear combination of two real and linearly independent solutions. In general, consider the difference equation

$$a_2 y_{k+2} + a_1 y_{k+1} + a_0 y_k = 0, \tag{9}$$

with real coefficients, and assume that the roots of the characteristic equation

$$a_2\lambda^2 + a_1\lambda + a_0 = 0 \tag{10}$$

are complex numbers. Then they must be complex conjugate. That is, the roots of Eq. (10) are of the form $a \pm ib$, where a and b are real numbers. Now the crucial step is to write the complex number $a + ib$ in *complex polar form*. That is, in the form

$$a + ib = r(\cos\theta + i\sin\theta), \tag{11}$$

where the *modulus* r of $a + ib$ is given by

$$r = \sqrt{a^2 + b^2}, \tag{12}$$

and the *argument* θ satisfies the equations

$$\cos\theta = \frac{a}{r}, \quad \sin\theta = \frac{b}{r}, \quad \text{with} \quad -\pi < \theta \leq \pi. \tag{13}$$

From (11), using Euler's identity, we find

$$a + ib = re^{i\theta}.$$

Taking complex conjugates, we also have

$$a - ib = re^{-i\theta}.$$

Thus, the general solution of the difference equation (9) is

$$\begin{aligned} y_k &= c_1(a+ib)^k + c_2(a-ib)^k = c_1(re^{i\theta})^k + c_2(re^{-i\theta})^k \\ &= c_1r^ke^{ik\theta} + c_2r^ke^{-ik\theta} \\ &= c_1r^k(\cos k\theta + i\sin k\theta) + c_2r^k(\cos k\theta - i\sin k\theta) \\ &= r^k(c_1+c_2)\cos k\theta + r^k(ic_1 - ic_2)\sin k\theta \\ &= C_1r^k\cos k\theta + C_2r^k\sin k\theta, \end{aligned}$$

where $C_1 = c_1 + c_2$ and $C_2 = ic_1 - ic_2$ are arbitrary constants which, for economy in notation, we can still denote by c_1 and c_2. In summary, the general solution of Eq. (9) is

$$y_k = c_1r^k\cos k\theta + c_2r^k\sin k\theta,$$

where r and θ are determined by Eqs. (12) and (13). In Exercise 38, you are asked to show directly that the sequences

$$y_k^{(1)} = r^k\cos k\theta \qquad \text{and} \qquad y_k^{(2)} = r^k\sin k\theta \tag{14}$$

are two linearly independent solutions of Eq. (9).

Let us apply the above results to Eq. (7). The characteristic roots are $-1 \pm i$. In polar form,

$$-1+i = r(\cos\theta + i\sin\theta),$$

where

$$r = \sqrt{(-1)^2 + 1^2} = \sqrt{2}$$

and

$$\cos\theta = \frac{-1}{\sqrt{2}}, \quad \sin\theta = \frac{1}{\sqrt{2}}, \quad -\pi < \theta \le \pi.$$

Therefore, $\theta = \frac{3\pi}{4}$. And the general solution of Eq. (7) is

$$y_k = c_1(\sqrt{2})^k \cos\frac{3k\pi}{4} + c_2(\sqrt{2})^k \sin\frac{3k\pi}{4}.$$

Examples 1 through 4 are special cases of the following general result.

THEOREM 1

Consider the second-order linear homogeneous difference equation

$$a_2 y_{k+2} + a_1 y_{k+1} + a_0 y_k = 0, \tag{15}$$

with constant and real coefficients and $a_2 a_0 \neq 0$. *Let* λ_1 *and* λ_2 *be the two roots of the characteristic equation*

$$a_2\lambda^2 + a_1\lambda + a_0 = 0.$$

Then the form of the general solution y_k *of Eq.* (15) *depends on the characteristic roots as follows:*

CASE 1 Real and distinct roots *That is,* $\lambda_1 \neq \lambda_2$:

$$y_k = c_1\lambda_1^k + c_2\lambda_2^k.$$

CASE 2 Equal roots *That is,* $\lambda_1 = \lambda_2$:

$$y_k = c_1\lambda_1^k + c_2 k\lambda_1^k.$$

CASE 3 Complex conjugate roots *That is,* $\lambda_{1,2} = a \pm ib$:

$$y_k = c_1 r^k \cos k\theta + i r^k \sin k\theta,$$

where r and θ *are found by writing* $a + ib$ *in polar form:*

$$a + ib = r(\cos\theta + i\sin\theta),$$

$r = \sqrt{a^2 + b^2}$ *and* $\cos\theta = \frac{a}{r}$, $\sin\theta = \frac{b}{r}$, $-\pi < \theta \le \pi$.

Higher- and lower-order linear homogeneous difference equations can be solved in a similar fashion.

EXAMPLE 5 Solve

$$y_{k+5} - 3y_{k+4} + 4y_{k+3} - 4y_{k+2} + 3y_{k+1} - y_k = 0.$$

Solution Characteristic equation: $\lambda^5 - 3\lambda^4 + 4\lambda^3 - 4\lambda^2 + 3\lambda - 1 = 0$ or $(\lambda - 1)^3(\lambda^2 + 1) = 0$.
Characteristic roots: $\lambda_1 = \lambda_2 = \lambda_3 = 1$, $\lambda_{4,5} = \pm i$.

Since $i = 1\left(\cos\frac{\pi}{2} + i\sin\frac{\pi}{2}\right)$, we have $r = 1$ and $\theta = \pi/2$.

General solution:

$$y_k = c_1 1^k + c_2 k 1^k + c_3 k^2 1^k + c_4 1^k \cos\frac{k\pi}{2} + c_5 1^k \sin\frac{k\pi}{2}$$

or

$$y_k = c_1 + c_2 k + c_3 k^2 + c_4 \cos\frac{k\pi}{2} + c_5 \sin\frac{k\pi}{2}.$$

EXAMPLE 6 Solve the first-order linear homogeneous difference equation

$$y_{k+1} + Ay_k = 0,$$

where A is a given constant.

Solution Characteristic equation: $\lambda + A = 0$.
Characteristic root: $\lambda = -A$.
General solution: $y_k = c(-A)^k$, where c is an arbitrary constant.

EXERCISES

In Exercises 1 through 28, find the general solution of the difference equation.

1. $y_{k+2} - 7y_{k+1} + 10y_k = 0$

2. $4y_{k+2} - y_k = 0$

3. $y_{k+2} + 6y_{k+1} + 9y_k = 0$

4. $y_{k+2} - 2y_{k+1} + y_k = 0$

5. $y_{k+2} + 6y_{k+1} + 18y_k = 0$

6. $y_{k+2} + y_{k+1} + y_k = 0$

7. $3y_{k+1} - 2y_k = 0$

8. $y_{k+1} + y_k = 0$

9. $y_{k+2} - 4y_{k+1} + 4y_k = 0$

10. $4y_{k+2} + 9y_k = 0$

11. $9y_{k+2} - 16y_k = 0$

12. $y_{k+2} + 5y_{k+1} - 6y_k = 0$

13. $y_{k+2} - y_k = 0$

14. $y_{k+2} + y_k = 0$

15. $y_{k+1} + 0.1y_k = 0$

16. $y_{k+3} - y_k = 0$

17. $y_{k+4} + 4y_k = 0$

18. $y_{k+2} = y_{k+1} + y_k$

19. $y_{k+3} - y_{k+2} + y_{k+1} - y_k = 0$

20. $y_{k+4} - 16y_k = 0$

21. $6y_{k+2} + 5y_{k+1} + y_k = 0$

22. $4y_{k+1} - 3y_k = 0$

23. $y_{k+3} + y_k = 0$

24. $y_{k+2} + \sqrt{3}y_{k+1} + y_k = 0$

25. $y_{k+3} - 6y_{k+2} + 12y_{k+1} - 8y_k = 0$

26. $y_{k+3} + 3y_{k+2} + 3y_{k+1} + y_k = 0$

27. $y_{k+4} + y_{k+3} - 3y_{k+2} - y_{k+1} + 2y_k = 0$

28. $y_{k+5} - 3y_{k+4} + 4y_{k+3} - 4y_{k+2} + 3y_{k+1} - y_k = 0$

In Exercises 29 through 37, solve the initial value problem.

29. $y_{k+2} - 7y_{k+1} + 10y_k = 0;\ y_0 = 3,\ y_1 = 15$

30. $y_{k+2} + 6y_{k+1} + 9y_k = 0;\ y_0 = 1,\ y_1 = 3$

31. $y_{k+2} + 6y_{k+1} + 18y_k = 0;\ y_0 = 0,\ y_1 = 1$

32. $3y_{k+1} - 2y_k = 0;\ y_0 = -3$

33. $y_{k+2} = y_{k+1} + y_k;\ y_0 = 0,\ y_1 = 1$

34. $y_{k+4} + 4y_k = 0;\ y_0 = 0,\ y_1 = 1,\ y_2 = 2,\ y_3 = 2$

35. $y_{k+3} - y_k = 0;\ y_0 = 1,\ y_1 = \dfrac{4 + \sqrt{3}}{4},\ y_2 = \dfrac{4 - \sqrt{3}}{4}$

36. $y_{k+3} - y_{k+2} + y_{k+1} - y_k = 0;\ y_0 = y_1 = y_2 = 0$

37. $y_{k+5} - 3y_{k+4} + 4y_{k+3} - 4y_{k+2} + 3y_{k+1} - y_k = 0;$
$y_0 = 0,\ y_1 = 1,\ y_2 = 2,\ y_3 = 1,\ y_4 = 0$

38. Show that

$$y_k^{(1)} = r^k \cos k\theta \qquad \text{and} \qquad y_k^{(2)} = r^k \sin k\theta$$

are linearly independent solutions of Eq. (9).

39. Let λ_1 and λ_2 be the characteristic roots of the difference equation

$$a_2 y_{k+2} + a_1 y_{k+1} + a_0 y_k = 0,$$

and set

$$L = \max(|\lambda_1|, |\lambda_2|).$$

Show that $L < 1$ is a necessary and sufficient condition for all solutions of the equation to approach zero as $k \to \infty$. [*Note:* $|a + ib| = \sqrt{a^2 + b^2}$.]

40. Show that the conditions

$$q - p + 1 > 0,\ q + p + 1 > 0, \qquad \text{and} \qquad 1 - q > 0$$

are necessary and sufficient for all solutions of the difference equation

$$y_{k+2} + py_{k+1} + qy_k = 0, q \neq 0$$

to approach zero as $k \to \infty$.

41. In the optics application [see Eq. (12), Section 9.1.1], assume that $d = 4f$, $y_0 = 2$, and $y_1 = 1$. Investigate the displacement y_k of the ray as it traverses the lenses.

42. In the electric circuit application (see Section 9.1.1), find the voltage V_k for $k = 1, 2, \ldots, n - 1$.

43. It has been estimated that the student enrollment in a college undergoes a 5 percent relative decrease in size per year. If the college currently has 20,000 students, what will the enrollment be after 10 years?

44. A certain colony of bacteria undergoes a 5 percent relative increase in size per hour. If the initial size of the colony is 1000, what will its size be after 12 hours?

45. Assume that the resistances in the electric circuit application (see Section 9.1.1) are all equal to R. Find the voltage V_k for $k = 1, 2, \ldots, n - 1$.

46. Reduction of Order First-order linear difference equations with variable coefficients can always be solved explicitly (see Exercise 20, Section 9.3). But there exists no general method for solving linear equations with variable coefficients of order higher than one. However, as with differential equations, if we know a nontrivial solution of the homogeneous equation, we can use a transformation to obtain an equation of lower order.

Consider the second-order linear equation

$$y_{k+2} + p_k y_{k+1} + q_k y_k = 0,$$

where $q_k \neq 0$ for $k = 0, 1, 2, \ldots$. Assume that x_k is a solution and $x_k \neq 0$ for $k = 0, 1, 2, \ldots$. Show that the transformation

$$y_k = x_k u_k$$

produces a first-order linear difference equation for u_k. [*Hint:*

$$u_{k+1} - u_k = \frac{y_{k+1}}{x_{k+1}} - \frac{y_k}{x_k} = \frac{C(x_k, y_k)}{x_k x_{k+1}}.$$

Use Eq. (9) of Section 9.3 with $C_0 = 1$.]

47. Given that $y_k = 1$ is a solution of the equation

$$(2k + 1)y_{k+2} - 4(k + 1)y_{k+1} + (2k + 3)y_k = 0,$$

find the general solution.

9.5 NONHOMOGENEOUS EQUATIONS WITH CONSTANT COEFFICIENTS

Consider the linear nonhomogeneous difference equation with constant coefficients

$$a_2 y_{k+2} + a_1 y_{k+1} + a_0 y_k = r_k \tag{1}$$

and the corresponding homogeneous equation

$$a_2 y_{k+2} + a_1 y_{k+1} + a_0 y_k = 0, \tag{2}$$

where $a_2 a_0 \neq 0$. From Theorem 4 of Section 9.3, we know that the general solution of Eq. (1) is the sum of the general solution of Eq. (2) and a particular solution of Eq. (1). That is,

$$y_k^g = y_k^h + y_k^p, \tag{3}$$

where y_k^g is the general solution of Eq. (1), y_k^h is the general solution of the homogeneous equation (2), and y_k^p is a particular solution of the nonhomogeneous equation (1). In Section 9.4 we explained how to obtain y_k^h. Hence y_k^g will be completely known if we show how to obtain a particular solution of Eq. (1).

As in the case of differential equations, there are two methods which can be used to find a particular solution of Eq. (1): the method of *undetermined coefficients* and the method of *variation of parameters*. The method of undetermined coefficients requires that the coefficients a_2, a_1, and a_0 of Eq. (1) be constants and that the nonhomogeneous term r_k be of a special form. (See Section 9.5.1.) On the other hand, the method of variation of parameters imposes no restriction on r_k; it applies to equations with constant coefficients always, and to equations with variable coefficients only if we know the general solution of the corresponding homogeneous equation. It should be remarked, however, that when we have a choice, the method of undetermined coefficients usually requires less labor.

9.5.1 UNDETERMINED COEFFICIENTS

This method applies successfully when the nonhomogeneous term r_k in Eq. (1) is a linear combination of sequences of the following types:

1. k^α, where α is a positive integer or zero.
2. β^k, where β is a nonzero constant.
3. $\cos \gamma k$, where γ is a nonzero constant.
4. $\sin \delta k$, where δ is a nonzero constant.
5. A (finite) product of two or more sequences of types 1–4.

The sequences k^α, β^k, $\cos \gamma k$, $\sin \delta k$ correspond to the functions x^α, $e^{\beta x}$, $\cos \gamma x$, $\sin \delta x$ listed under types 1–4 in Section 2.11. With this correspondence in mind, one can find the form of a particular solution of a difference equation in a manner similar to that used in Section 2.11 for a particular solution of a differential equation.

EXAMPLE 1 Find a particular solution of the nonhomogeneous difference equation

$$y_{k+2} - 8y_{k+1} + 15y_k = 3 \cdot 2^k - 5 \cdot 3^k. \tag{4}$$

Solution As with differential equations, first we find the general solution of the corresponding homogeneous equation

$$y_{k+2} - 8y_{k+1} + 15y_k = 0, \tag{5}$$

since the solution of Eq. (5) can influence the form of a particular solution of Eq. (4).

Characteristic equation: $\lambda^2 - 8\lambda + 15 = 0$.

Characteristic roots: $\lambda_1 = 3, \lambda_2 = 5$.

Homogeneous solution: $y_k^h = c_1 3^k + c_2 5^k$. (6)

Here r_k is a linear combination of two sequences of type 2. Using notation similar to that of Section 2.11 and treating terms of the form β^k much in the way we treated $e^{\beta x}$ in Section 2.11, we have

$$3 \cdot 2^k \rightarrow \{2^k\} \tag{7}$$

$$-5 \cdot 3^k \rightarrow \{3^k\} \rightarrow \{k \cdot 3^k\}, \tag{8}$$

where in (8) we multiplied 3^k by k because 3^k is a solution of Eq. (5). The rule is to multiply by k^n, where n is the smallest positive integer such that $k^n 3^k$ is not a solution of Eq. (5). Therefore,

$$y_k^p = A2^k + Bk3^k \tag{9}$$

is a particular solution of Eq. (4). The *undetermined coefficients A* and *B* will be determined by substituting (9) into Eq. (4) and equating coefficients of similar terms. We now have

$$\left[A2^{k+2} + B(k+2)3^{k+2}\right] - 8\left[A2^{k+1} + B(k+1)3^{k+1}\right] + 15(A2^k + Bk3^k) = 3 \cdot 2^k - 5 \cdot 3^k$$

$$\left[4A2^k + 9B(k+2)3^k\right] - 8\left[2A2^k + 3B(k+1)\right] + 15(A2^k + Bk3^k) = 3 \cdot 2^k - 5 \cdot 3^k$$

$$3A2^k - 6B3^k = 3 \cdot 2^k - 5 \cdot 3^k$$

$$3A = 3 \quad \text{and} \quad -6B = -5$$

$$A = 1 \quad \text{and} \quad B = 5/6.$$

Hence,

$$y_k^p = 2^k + \frac{5}{6}k3^k$$

is a particular solution of Eq. (4).

EXAMPLE 2 Find a particular solution of

$$y_{k+2} - y_k = 12k^2. \tag{10}$$

Solution Homogeneous solution $y_k^h = c_1 \cdot 1 + c_2(-1)^k$.
Treating k^2 in the same way x^2 was treated in Section 2.11, we have $12k^2 \to \{k^2, k, 1\} \to \{k^3, k^2, k\}$. (We multiplied by k because 1 is a solution of the corresponding homogeneous equation. As in Example 1, the rule is to multiply all terms in the bracket by the lowest positive integral power of k so that no term in the last bracket is a solution of the homogeneous equation.) Hence,

$$y_k^p = Ak^3 + Bk^2 + Ck. \tag{11}$$

Substituting (11) into Eq. (10) and equating coefficients of similar terms, we find $A = 2$, $B = -6$, and $C = 4$. Therefore,

$$y_k^p = 2k^3 - 6k^2 + 4k.$$

EXAMPLE 3 Find a particular solution of the equation

$$4y_{k+2} - y_k = 3\cos\frac{k\pi}{2}. \tag{12}$$

Solution Homogeneous solution: $y_k^h = c_1\left(\frac{1}{2}\right)^k + c_2\left(-\frac{1}{2}\right)^k$.
Treating $\cos(k\pi/2)$ in the way $\cos(x\pi/2)$ was treated in Section 2.11, we have

$$3\cos\frac{k\pi}{2} \to \left\{\cos\frac{k\pi}{2}, \sin\frac{k\pi}{2}\right\}.$$

Hence,

$$y_k^p = A\cos\frac{k\pi}{2} + B\sin\frac{k\pi}{2}. \tag{13}$$

Substituting (13) into Eq. (12), we find $A = -\frac{3}{5}$, $B = 0$. Therefore,

$$y_k^p = -\frac{3}{5}\cos\frac{k\pi}{2}.$$

Particular solutions of lower- or higher-order difference equations are found in a similar fashion.

EXAMPLE 4 Find the general solution of the first-order difference equation

$$y_{k+1} - ay_k = b, \tag{14}$$

where a and b are given constants.

Solution Homogeneous solution: $y_k^h = ca^k$, (15)
where c is an arbitrary constant. We have

$$b \to \{1\} \to \begin{cases} \{k\} & \text{if} \quad a = 1 \\ \{1\} & \text{if} \quad a \neq 1 \end{cases},$$

because, when $a = 1$, the sequence $\{1\}$ is a solution of the homogeneous equation. Hence,

$$y_k^p = \begin{cases} Ak & \text{if} \quad a = 1 \\ A & \text{if} \quad a \neq 1 \end{cases}. \tag{16}$$

Substituting (16) into Eq. (14), we find $A = b$ when $a = 1$, and $A = b/(1-a)$ when $a \neq 1$. Therefore, the general solution of Eq. (14) is

$$y_k^g = y_k^h + y_k^p = \begin{cases} c + bk & \text{if} \quad a = 1 \\ ca^k + \dfrac{b}{1-a} & \text{if} \quad a \neq 1 \end{cases}.$$

EXAMPLE 5 Find the form of a particular solution of the third-order difference equation

$$y_{k+3} - 19y_{k+1} - 30y_k = -3k^2 5^k + 2k \sin\frac{k\pi}{3} - 2^k \cos\frac{k\pi}{4}.$$

Solution Characteristic equation:

$$\lambda^3 - 19\lambda - 30 = 0 \text{ or } (\lambda+3)(\lambda+2)(\lambda-5) = 0$$

Characteristic roots: $\lambda_1 = -3$, $\lambda_2 = -2$, $\lambda_3 = 5$

Homogeneous solution: $y_k^h = c_1(-3)^k + c_2(-2)^k + c_3 5^k$.

We have

$$-3k^2 5^k \to \{k^2 5^k, k5^k, 5^k\} \to \{k^3 5^k, k^2 5^k, k5^k\}$$

$$2k \sin\frac{k\pi}{3} \to \left\{k \sin\frac{k\pi}{3}, k\cos\frac{k\pi}{3}, \sin\frac{k\pi}{3}, \cos\frac{k\pi}{3}\right\}$$

$$2^k \cos\frac{k\pi}{4} \to \left\{2^k \cos\frac{k\pi}{4}, 2^k \sin\frac{k\pi}{4}\right\}.$$

Hence,

$$y_k^p = A_1k^3 5^k + A_2k^2 5^k + A_3 k5^k + A_4 k \sin\frac{k\pi}{3} + A_5 k \cos\frac{k\pi}{3} + A_6 \sin\frac{k\pi}{3} + A_7 \cos\frac{k\pi}{3} + A_8 2^k \cos\frac{k\pi}{4} + A_9 2^k \sin\frac{k\pi}{4}.$$

9.5.2 VARIATION OF PARAMETERS

Here we will exhibit a particular solution of Eq. (1) under no restriction whatsoever on r_k. The same method also applies successfully to difference equations with variable coefficients, provided we know the general solution of the corresponding homogeneous equation. More precisely, we prove the following theorem.

THEOREM 1

Consider the difference equation

$$y_{k+2} + p_k y_{k+1} + q_k y_k = r_k, \tag{17}$$

where p_k, q_k, and r_k are given sequences, and $q_k \neq 0$ for $k = 0, 1, 2, \ldots$. Let $y_k^{(1)}$ and $y_k^{(2)}$ be two linearly independent solutions of the homogeneous equation corresponding to Eq. (17). A particular solution of Eq. (17) is given by

$$y_k^p = u_k^{(1)} y_k^{(1)} + u_k^{(2)} y_k^{(2)}, \tag{18}$$

where the sequences $u_k^{(1)}$ and $u_k^{(2)}$ satisfy the system of equations

$$y_{k+1}^{(1)} \Delta u_k^{(1)} + y_{k+1}^{(2)} \Delta u_k^{(2)} = 0 \tag{19}$$

$$y_{k+2}^{(1)} \Delta u_k^{(1)} + y_{k+2}^{(2)} \Delta u_k^{(2)} = r_k \tag{20}$$

and the symbol $\Delta u_k^{(j)} = u_{k+1}^{(i)} - u_k^{(i)}, i = 1, 2.$

Proof From (18) we find

$$\begin{aligned} y_{k+1}^p &= u_{k+1}^{(1)} y_{k+1}^{(1)} + u_{k+1}^{(2)} y_{k+1}^{(2)} \\ &= u_k^{(1)} y_{k+1}^{(1)} + y_{k+1}^{(1)}(u_{k+1}^{(1)} - u_k^{(1)}) + u_k^{(2)} y_{k+1}^{(2)} + y_{k+1}^{(2)}(u_{k+1}^{(2)} - u_k^{(2)}) \\ &= u_k^{(1)} y_{k+1}^{(1)} + u_k^{(2)} y_{k+1}^{(2)} + \underbrace{y_{k+1}^{(1)} \Delta u_k^{(1)} + y_{k+1}^{(2)} \Delta u_k^{(2)}}_{\text{equal to zero by (19)}} \\ &= u_k^{(1)} y_{k+1}^{(1)} + u_k^{(2)} y_{k+1}^{(2)} \end{aligned} \tag{21}$$

and

$$\begin{aligned} y^p_{k+2} &= u^{(1)}_{k+1} y^{(1)}_{k+2} + u^{(2)}_{k+1} y^{(2)}_{k+2} \\ &= u^{(1)}_k y^{(1)}_{k+2} + u^{(2)}_k y^{(2)}_{k+2} + \underbrace{y^{(1)}_{k+2}\Delta u^{(1)}_k + y^{(2)}_{k+2}\Delta u^{(2)}_k}_{\text{equal to } r_k \text{ by (20)}} \\ &= u^{(1)}_k y^{(1)}_{k+2} + u^{(2)}_k y^{(2)}_{k+2} + r_k. \end{aligned} \tag{22}$$

From (22), (21), and (18) we have

$$\begin{aligned} y^p_{k+2} + p_k y^p_{k+1} + q_k y^p_k &= u^{(1)}_k y^{(1)}_{k+2} + u^{(2)}_k y^{(2)}_{k+2} + r_k \\ &\quad + p_k u^{(1)}_k y^{(1)}_{k+1} + p_k u^{(2)}_k y^{(2)}_{k+1} \\ &\quad + q_k u^{(1)}_k y^{(1)}_k + q_k u^{(2)}_k y^{(2)}_k \\ &= u^{(1)}_k (y^{(1)}_{k+2} + p_k y^{(1)}_{k+1} + q_k y^{(1)}_k) \\ &\quad + u^{(2)}_k (y^{(2)}_{k+2} + p_k y^{(2)}_{k+1} + q_k y^{(2)}_k) \\ &\quad + r_k \\ &= r_k \end{aligned}$$

because $y^{(1)}_k$ and $y^{(2)}_k$ are solutions of the homogeneous equation corresponding to Eq. (17). Thus, (18) is a solution of Eq. (17). The proof is complete.

We will now solve the system (19)–(20) and find an explicit formula for the particular solution (18). We have

$$\Delta u^{(1)}_k = \frac{\begin{vmatrix} 0 & y^{(2)}_{k+1} \\ r_k & y^{(2)}_{k+2} \end{vmatrix}}{\begin{vmatrix} y^{(1)}_{k+1} & y^{(2)}_{k+1} \\ y^{(1)}_{k+2} & y^{(2)}_{k+2} \end{vmatrix}} = -\frac{r_k y^{(2)}_{k+1}}{C_{k+1}} \tag{23}$$

and

$$\Delta u^{(2)}_k = \frac{\begin{vmatrix} y^{(1)}_{k+1} & 0 \\ y^{(1)}_{k+2} & r_k \end{vmatrix}}{C_{k+1}} = \frac{r_k y^{(1)}_{k+1}}{C_{k+1}} \tag{24}$$

where, by Theorem 2 of Section 9.3, the Casoratian C_{k+1} is never zero. Writing Eq. (23) in the equivalent form

$$u^{(1)}_{n+1} - u^{(1)}_n = -\frac{r_n y^{(2)}_{n+1}}{C_{n+1}},$$

and summing up from $n = 0$ to $k - 1$, we find

$$u_k^{(1)} = u_0^{(1)} - \sum_{n=0}^{k-1} \frac{r_n y_{n+1}^{(2)}}{C_{n+1}}. \tag{25}$$

Similarly, from (24), we obtain

$$u_k^{(2)} = u_0^{(2)} + \sum_{n=0}^{k-1} \frac{r_n y_{n+1}^{(1)}}{C_{n+1}}. \tag{26}$$

Substituting (25) and (26) into (18) and neglecting the term

$$u_0^{(1)} y_k^{(1)} + u_0^{(2)} y_k^{(2)},$$

which is a solution of the homogeneous equation, we find

$$y_k^p = -y_k^{(1)} \sum_{n=0}^{k-1} \frac{r_n y_{n+1}^{(2)}}{C_{n+1}} + y_k^{(2)} \sum_{n=0}^{k-1} \frac{r_n y_{n+1}^{(1)}}{C_{n+1}}.$$

Combining the two series and using determinants, we obtain

$$y_k^p = \sum_{n=0}^{k-1} r_n \frac{\begin{vmatrix} y_{n+1}^{(1)} & y_{n+1}^{(2)} \\ y_k^{(1)} & y_k^{(2)} \end{vmatrix}}{\begin{vmatrix} y_{n+1}^{(1)} & y_{n+1}^{(2)} \\ y_{n+2}^{(1)} & y_{n+2}^{(2)} \end{vmatrix}}. \tag{27}$$

Formula (27) gives explicitly a particular solution of a second-order linear nonhomogeneous difference equation in terms of the nonhomogeneous term r_k and two linearly independent solutions of the corresponding homogeneous equation.

EXAMPLE 6 Verify that the sequences

$$y_k^{(1)} = 1 \quad \text{and} \quad y_k^{(2)} = k^2, k = 0, 1, 2, \ldots \tag{28}$$

are linearly independent solutions of the difference equation

$$(2k + 1)y_{k+2} - 4(k + 1)y_{k+1} + (2k + 3)y_k = 0. \tag{29}$$

Then solve the IVP

$$(2k + 1)y_{k+2} - 4(k + 1)y_{k+1} + (2k + 3)y_k = (2k + 1)(2k + 3). \tag{30}$$

$$y_0 = 0, y_1 = \frac{1}{2}. \tag{31}$$

Solution The reader can verify by direct substitution that the sequences (28) are solutions of Eq. (29). Since the coefficients of Eq. (29) are different from zero for $k = 0, 1, 2, \ldots$, the theory that we have developed so far applies to

Eqs. (29) and (30). The two sequences in (28) are linearly independent solutions of Eq. (29) because they are solutions, as we just proved, and their Casoratian is

$$\begin{vmatrix} 1 & k^2 \\ 1 & (k+1)^2 \end{vmatrix} = (k+1)^2 - k^2 = 2k + 1 \neq 0 \qquad \text{for} \qquad k = 0, 1, 2, \ldots.$$

Thus, the general solution of Eq. (29) is

$$y_k^h = c_1 \cdot 1 + c_2 k^2,$$

where c_1 and c_2 are arbitrary constants. Next we will find a particular solution of Eq. (30). Here we have no choice of method. We must use variation of parameters because the coefficients of the equation are variables. Dividing both sides of Eq. (30) by the leading coefficient $(2k + 1)$, we find that

$$r_k = 2k + 3.$$

Now, using formula (27), we find

$$y_k^p = \sum_{n=0}^{k-1} (2n+3) \frac{\begin{vmatrix} 1 & (n+1)^2 \\ 1 & k^2 \end{vmatrix}}{\begin{vmatrix} 1 & (n+1)^2 \\ 1 & (n+2)^2 \end{vmatrix}} = \sum_{n=0}^{k-1} (2n+3) \frac{k^2 - (n+1)^2}{(n+2)^2 - (n+1)^2}$$

$$= \sum_{n=0}^{k-1} [k^2 - (n+1)^2] = k^3 - \frac{k(k+1)(2k+1)}{6} = \frac{4k^3 - 3k^2 - k}{6},$$

where we used the fact that

$$\sum_{n=0}^{k-1} (n+1)^2 = 1^2 + 2^2 + \cdots + k^2 = \frac{k(k+1)(2k+1)}{6}.$$

Hence, the general solution of Eq. (30) is

$$y_k^g = y_k^h + y_k^p = c_1 + c_2 k^2 + \frac{4k^3 - 3k^2 - k}{6}.$$

Using the initial conditions (30), we find $c_1 = 0$ and $c_2 = 1/2$. Thus, the unique solution of the IVP (30)–(31) is

$$y_k = \frac{k^2}{2} + \frac{4k^3 - 3k^2 - k}{6} = \frac{4k^3 - k}{6}.$$

9.5.3 APPLICATIONS

Business: Mathematics of Finance

■ Problems in compound interest often lead to simple difference equations. First let us recall the meaning of simple and compound interest. Suppose we invest \$500 *principal* at 8 percent *simple interest*. At the end of the first year the bank will add (0.08) 500 = \$40 to the account; at the end of the second year it will add another \$40, bringing the total to \$580, and so forth. That is, we earn interest only from the principal investment of \$500. On the other hand, suppose we invest \$500 at 8 percent compound interest, with interest compounded quarterly. Interest rates are usually quoted as annual rates, so 8 percent interest compounded quarterly really means 2 percent per quarter. At the end of the first quarter (each quarter is 3 months), the bank will add (0.02) 500 = \$10 to the account. This interest is now compounded with the principal and from now on will also earn interest. Hence, at the end of the second quarter the bank will add (0.02) (510) = \$10.20 to the account, bringing the total to 510 + 10.20 = \$520.20. Continuing this process, after 2 years the account will be worth \$585.82, compared to \$580 in the case of simple interest. (See Example 7).

Theorem 2 is a basic model for compound interest.

THEOREM 2

Assume that the amount P is deposited in a savings account and, together with the interest, is kept there for k time periods, at the interest rate of i per period. Show that the value A_k of the account at the end of the kth period will be

$$A_k = P(1+i)^k. \tag{32}$$

Proof Since A_k denotes the value of the account at the end of the kth period, and iA_k is the interest during the next period, it follows that

$$A_{k+1} = A_k + iA_k, \; k = 0, 1, 2, \ldots$$

or

$$A_{k+1} - (1+i)A_k = 0.$$

Also

$$A_0 = P.$$

The solution of this IVP is given by (32).

EXAMPLE 7 I invested \$500 at 8 percent annual interest compounded quarterly. What will the amount be after 2 years?

Solution $P = 500$, $k = 4 \cdot 2 = 8$ periods, and the interest is $\frac{8}{4} = 2$ percent per quarter, that is, $i = 0.02$. Hence, from (32), we find

$$A_8 = 500(1 + 0.02)^8 = 500(1.02)^8 = \$585.82.$$

Theorem 3 gives the formula for periodic payments of a fixed amount, or *annuities*.

THEOREM 3

Assume that, at the beginning of each time period for the next k periods, we deposit in a savings account the fixed amount P, at the interest rate of i (compounded) per period. Show that the value of the account immediately after the kth payment will be

$$A_k = P\frac{(1+i)^k - 1}{i}. \tag{33}$$

Proof The assumptions of the theorem lead to the IVP

$$A_{k+1} = A_k + iA_k + P,\ k = 1, 2, 3, \ldots$$

and

$$A_1 = P$$

or

$$A_{k+1} - (1+i)\,A_k = P \tag{34}$$

$$A_1 = P. \tag{35}$$

Characteristic equation: $\lambda - (1+i) = 0$.

Characteristic root: $\lambda = 1 + i$.

Homogeneous solution: $A_k^h = c(1+i)^k$.

Particular solution: $A_k^p = -\dfrac{P}{i}$.

General solution: $A_k = c(1+i)^k - \dfrac{P}{i}$.

Using the initial condition (35), we find

$$c(1+i) - \frac{P}{i} = P$$

and so

$$c = \frac{P}{i}.$$

Hence,

$$A_k = \frac{P}{i}(1+i)^k - \frac{P}{i} = \frac{P}{i}[(1+i)^k - 1],$$

which is the desired result.

Theorem 4 derives the formula for the periodic payment P necessary to pay off a loan A.

THEOREM 4

The periodic payment P necessary to pay off a loan A in N periods at an interest rate of i per period is

$$P = A\frac{i}{1-(1+i)^{-N}}. \tag{36}$$

Proof Let y_k be the amount still owed immediately after the kth payment. Then,

$$y_{k+1} = y_k + iy_k - P \tag{37}$$

and

$$y_0 = A,\ y_N = 0. \tag{38}$$

The general solution of Eq. (37) is

$$y_k^g = c(1+i)^k + \frac{P}{i}.$$

Using the boundary conditions (38), we find

$$c + \frac{P}{i} = A \qquad \text{and} \qquad c(1+i)^N + \frac{P}{i} = 0. \tag{39}$$

Subtracting the two equations in (39), we find

$$c = \frac{A}{1-(1+i)^N}.$$

Substituting this value of c into the second equation in (39), and solving for P, we obtain (36).

EXAMPLE 8 I have a 25-year mortgage on my house. I borrowed \$25,700 at 7 percent compounded monthly. What is the monthly payment?

Solution

$$A = 25{,}700,\ i = \frac{0.07}{12} = 0.00583, \text{ and } N = 25 \cdot 12 = 300.$$

Hence, the monthly payment is

$$P = 25{,}700\frac{0.00583}{1-(1.00583)^{-300}} = \$181.64.$$

EXERCISES

In Exercises 1 through 16, find a particular solution of the nonhomogeneous difference equation by the method of undetermined coefficients.

1. $y_{k+2} - 7y_{k+1} + 10y_k = 8 \cdot 3^k$ **2.** $y_{k+2} - 7y_{k+1} + 10y_k = 3 \cdot 2^k$

3. $y_{k+2} + 6y_{k+1} + 9y_k = 2 - 4 \cdot 3^k$

4. $y_{k+2} + 6y_{k+1} + 9y_k = -4 \sin \dfrac{k\pi}{2}$

5. $y_{k+2} - y_k = 3$

6. $y_{k+2} + y_k = -2 \cos \dfrac{k\pi}{2}$

7. $4y_{k+2} + 9y_k = -3k2^k$

8. $y_{k+1} + y_k = -1$

9. $3y_{k+1} - 2y_k = \sin \dfrac{k\pi}{4}$

10. $y_{k+2} - 4y_{k+1} + 4y_k = 2$

11. $y_{k+2} - 2y_{k+1} + y_k = 6k$

12. $y_{k+2} - 8y_{k+1} + 15y_k = 3 \cdot 2^k - 4 \cdot 3^k$

13. $y_{k+3} - y_k = -k \cos \dfrac{k\pi}{3}$

14. $y_{k+3} + y_k = 2$

15. $y_{k+4} + 4y_k = 2^k \cos \dfrac{3k\pi}{4}$

16. $y_{k+5} - 3y_{k+4} + 4y_{k+3} - 4y_{k+2} + 3y_{k+1} - y_k = 10 \cdot 2^{-k}$

In Exercises 17 through 34, find the general solution of the equation.

17. $y_{k+2} - 7y_{k+1} + 10y_k = 8 \cdot 3^k$

18. $y_{k+2} - 7y_{k+1} + 10y_k = 3 \cdot 2^k$

19. $y_{k+2} + 6y_{k+1} + 9y_k = 2 - 4 \cdot 3^k$

20. $y_{k+2} + 6y_{k+1} + 9y_k = -4 \sin \dfrac{k\pi}{2}$

21. $y_{k+2} - y_k = 3$

22. $y_{k+2} + y_k = -2 \cos \dfrac{k\pi}{2}$

23. $4y_{k+2} + 9y_k = -3k2^k$

24. $y_{k+1} + y_k = -1$

25. $y_{k+1} + y_k = \sin \dfrac{k\pi}{4}$

26. $y_{k+2} - 4y_{k+1} + 4y_k = 2$

27. $y_{k+2} - 2y_{k+1} + y_k = 6k$

28. $y_{k+2} - 8y_{k+1} + 15y_k = 3 \cdot 2^k - 4 \cdot 3^k$

29. $4y_{k+2} - y_k = 3^k \sin \dfrac{k\pi}{4}$

30. $y_{k+2} + 5y_{k+1} - 6y_k = \cos \dfrac{k\pi}{2}$

31. $y_{k+3} - y_k = -k \cos \dfrac{k\pi}{3}$

32. $y_{k+3} + y_k = 2$

33. $y_{k+4} + 4y_k = 2^k \cos \dfrac{3k\pi}{4}$

34. $y_{k+5} - 3y_{k+4} + 4y_{k+3} - 4y_{k+2} + 3y_{k+1} - y_k = 10 \cdot 2^{-k}$

In Exercises 35 through 44, solve the initial value problem.

35. $y_{k+2} - 7y_{k+1} + 10y_k = 8 \cdot 3^k;\ y_0 = -1,\ y_1 = -6$

36. $y_{k+2} + 6y_{k+1} + 9y_k = 2 - 4 \cdot 3^k;\ y_0 = 0,\ y_1 = -\dfrac{1}{6}$

37. $y_{k+2} - y_k = 3;\ y_0 = 1,\ y_1 = 5/2$

38. $4y_{k+2} + 9y_k = -3k2^k;\ y_0 = 1,\ y_1 = 0$

39. $y_{k+1} + y_k = \sin \dfrac{k\pi}{4};\ y_0 = 1$

40. $y_{k+2} - 2y_{k+1} + y_k = 6k;\ y_0 = 1,\ y_1 = 2$

41. $4y_{k+2} - y_k = 3^k \sin \dfrac{k\pi}{4};\ y_0 = 0,\ y_1 = 1$

42. $y_{k+3} - y_k = -k \cos \dfrac{k\pi}{3};\ y_0 = \dfrac{1}{4}, y_1 = \dfrac{7}{8}, y_2 = \dfrac{7}{8}$

43. $y_{k+3} + y_k = 2;\ y_0 = 1,\ y_1 = -\dfrac{1}{2}, y_2 = \dfrac{5}{2}$

44. $y_{k+5} - 3y_{k+4} + 4y_{k+3} - 4y_{k+2} + 3y_{k+1} - y_k = 10 \cdot 2^{-k}$

$y_0 = -64,\ y_1 = -32,\ y_2 = -16,\ y_3 = -8,\ y_4 = -4$

In Exercises 45 through 49, use the method of undetermined coefficients to find the form of a particular solution of the difference equation.

45. $y_{k+2} - 7y_{k+1} + 10y_k = 4 \cdot 5^k$

46. $y_{k+2} - 2y_{k+1} + y_k = 3k^2 - 1$

47. $4y_{k+2} + 9y_k = 2 \sin \dfrac{k\pi}{2} - \left(\dfrac{3}{2}\right)^k \cos \dfrac{k\pi}{2}$

48. $y_{k+2} - 4y_{k+1} + 4y_k = 5k2^k + 3 \cdot (-2)^k$

49. $y_{k+1} - y_k = 4 - 2k^2 \sin \dfrac{k\pi}{3}$

In Exercises 50 through 54, find a particular solution of the difference equation by the method of variation of parameters.

50. $y_{k+2} - y_k = 3$

51. $y_{k+2} - 7y_{k+1} + 10y_k = 8 \cdot 3^k$

52. $y_{k+2} - 7y_{k+1} + 10y_k = 3 \cdot 2^k$

53. $y_{k+2} - 2y_{k+1} + y_k = 6k$

54. $y_{k+2} + y_k = -2 \cos \dfrac{k\pi}{2}$

55. Verify that the sequences

$$y_k^{(1)} = 1 \quad \text{and} \quad y_k^{(2)} = k^3$$

are linearly independent solutions of the difference equation

$$y_{k+2} - \frac{(k+2)^3 - k^3}{(k+1)^3 - k^3} y_{k+1} + \frac{(k+2)^3 - (k+1)^3}{(k+1)^3 - k^3} y_k = 0.$$

Then use the method of variation of parameters to find a particular solution of the equation

$$y_{k+2} - \frac{(k+2)^3 - k^3}{(k+1)^3 - k^3} y_{k+1} + \frac{(k+2)^3 - (k+1)^3}{(k+1)^3 - k^3} y_k = (k+2)^3 - (k+1)^3.$$

56. Linear Systems with Constant Coefficients: Method of Elimination Linear systems with constant coefficients of the form

$$\left.\begin{aligned} x_{k+1} &= a_{11}x_k + a_{12}y_k + r_k \\ y_{k+1} &= a_{21}x_k + a_{22}y_k + t_k \end{aligned}\right\}$$

can be solved by the method of elimination. Write the first equation in the form

$$x_{k+2} = a_{11}x_{k+1} + a_{12}y_{k+1} + r_{k+1},$$

then eliminate y by using the second equation and then the first (exactly as in the case of differential equations in Section 3.2).

Show that, after we eliminate y, we obtain the second-order linear equation with constant coefficients

$$\begin{aligned} x_{k+2} - (a_{11} + a_{22})x_{k+1} &+ (a_{11}a_{22} - a_{12}a_{21})x_k \\ &= a_{12}t_k - a_{22}r_k + r_{k+1}. \end{aligned}$$

After we find x_k we compute y_k from the first or second equation.

In Exercises 57 through 60, find the general solution of the system by the method of elimination (see Exercise 56).

57. $x_{k+1} = -3x_k + 4y_k$
$y_{k+1} = -2x_k + 3y_k$

58. $x_{k+1} = 4x_k - y_k$
$y_{k+1} = 2x_k + y_k$

59. $x_{k+1} = 5x_k - 6y_k + 1$
$y_{k+1} = 6x_k - 7y_k + 1$

60. $x_{k+1} = 3x_k - 2y_k + 2k$
$y_{k+1} = 5x_k - y_k - 1$

61. Equilibrium and Stability A constant E is called an *equilibrium* value for the difference equation

$$y_{k+2} + py_{k+1} + qy_k = r, \; q \neq 0,$$

if $y_k = E$ is a solution. The equilibrium value E is *stable* if every solution of the equation approaches E as $k \to \infty$. Show that the value

$$E = \frac{r}{1 + p + q}$$

is a stable equilibrium of the above difference equation if and only if the following conditions are satisfied:

$$q - p + 1 > 0, \; q + p + 1 > 0, \qquad \text{and} \qquad 1 - q > 0.$$

62. Boundary Value Problems Show that the BVP

$$y_{k+2} + py_{k+1} + qy_k = r_k,\ q \neq 0$$
$$y_0 = A,\ y_N = B$$

has a unique solution y_k for $k = 0, 1, \ldots, N$, provided that the roots λ_1 and λ_2 of the characteristic equation are either equal or distinct and $\lambda_1^N \neq \lambda_2^N$.

63. How long does it take to become a millionaire if you invest \$100 at 8 percent compounded quarterly?

64. At the beginning of each month for the next 5 years I plan to put \$400 in the bank. If the bank pays 8 percent interest compounded monthly, how much will my account be worth immediately after the last payment?

65. I borrowed \$756 at 15.55 percent compounded monthly, and I agreed to pay it in 24 equal successive monthly payments. What is the monthly payment?

66. I have a 30-year mortgage on a house. I borrowed \$40,000 at 12 percent compounded monthly. What is the monthly payment?

REVIEW EXERCISES

In Exercises 1 through 5, find the general solution of the difference equation.

1. $y_{k+2} - 7y_{k+1} + 10y_k = 8 \cdot 3^k + 3 \cdot 2^k$

2. $y_{k+2} + y_k = 2\left(2 - \cos\dfrac{k\pi}{2}\right)$

3. $y_{k+2} + 6y_{k+1} + 9y_k = 144 - 288 \cdot 3^k$

4. $3y_{k+1} + 2y_k = 5$

5. $y_{k+3} + y_k = 3$

In Exercises 6 though 10, solve the initial value problem.

6. $y_{k+2} - 7y_{k+1} + 10y_k = 8 \cdot 3^k + 3 \cdot 2^k$
$y_0 = -3,\ y_1 = -8$

7. $y_{k+2} + y_k = 4 - 2\cos\dfrac{k\pi}{2}$
$y_0 = 0,\ y_1 = 2$

8. $y_{k+2} + 6y_{k+1} + 9y_k = 144 - 288 \cdot 3^k$
$y_0 = 1,\ y_1 = -15$

9. $3y_{k+1} + 2y_k = 5$
$y_0 = 0$

10. $y_{k+3} + y_k = 3$
$y_0 = 5,\ y_1 = 1,\ y_2 = 2$

11. Principle of Superposition Show that if $y_k^{(1)}$ is a solution of

$$y_{k+2} + p_k y_{k+1} + q_k y_k = r_k^{(1)},$$

and $y_k^{(2)}$ is a solution of

$$y_{k+2} + p_k y_{k+1} + q_k y_k = r_k^{(2)},$$

then $ay_k^{(1)} + by_k^{(2)}$ is a solution of

$$y_{k+2} + p_k y_{k+1} + q_k y_k = ar_k^{(1)} + br_k^{(2)}.$$

12. Use the principle of superposition (see Exercise 11) to find a particular solution of the equation

$$y_{k+2} - 8y_{k+1} + 15y_k = 3 \cdot 2^k - 4 \cdot 3^k.$$

13. Given that $y_k = 1$ is a solution of the homogeneous equation

$$y_{k+2} - \frac{4(k+1)}{2k+1} y_{k+1} + \frac{2k+3}{2k+1} y_k = 0,$$

solve the IVP

$$y_{k+2} - \frac{4(k+1)}{2k+1} y_{k+1} + \frac{2k+3}{2k+1} y_k = 2k + 3$$
$$y_0 = 1,\ y_1 = 0.$$

In Exercises 14 and 15, solve the IVP.

14. $x_{k+1} = -3x_k - 4y_k$
$y_{k+1} = 2x_k + 3y_k$
$x_0 = 1,\ y_0 = -1$

15. $x_{k+1} = 5x_k + 6y_k + 1$
$y_{k+1} = -6x_k - 7y_k - 1$
$x_0 = \frac{3}{2},\ y_0 = -\frac{3}{2}$

16. A colony of bacteria undergoes a 10 percent relative decrease in size per day. If the initial size of the colony is 1000, what will its size be after 5 days?

17. In the electric circuit application of Section 9.1.1 (see also Figure 9.1), assume that each resistance in the horizontal branch is equal to 1 Ω and each resistance in the vertical branch is equal to 2 Ω. If $V_0 = 10$ volts and $V_{10} = 0$ volts, find the voltage V_5.

18. Probability: The Gambler's Ruin Two people, A and B, play a game (cards, tennis, chess) in which their skills or chances of winning are rated as p to q. That is, the probability that A wins is p, and the probability that B wins is q, where $p + q = 1$. On each play of the game, the winner wins one dollar. There are no ties, and the outcomes of the successive plays are independent. If A starts with a dollars and B with b dollars, and if they continue to play until one of them is ruined, what is the probability that A will ruin B by winning all of B's money?

19. Mary has a 25-year mortgage on her house. She borrowed $25,700 at 9 percent compounded monthly. What is the monthly payment?

In Exercises 20 through 23, verify the summation property of Fibonacci numbers.

20. $y_1 + y_2 + \cdots + y_k = y_{k+2} - 1$

21. $y_1 + y_3 + \cdots + y_{2k+1} = y_{2k}$

22. $y_2 + y_4 + \cdots + y_{2k} = y_{2k+1} - 1$

23. $y_1 - y_2 + y_3 - y_4 + \cdots + (-1)^{k+1} y_k = (-1)^{k+1} y_{k+1} + 1$

24. An optical system[2] of identical converging thin lenses is constructed with $d = 5$ cm and $f = 16$ cm. If the ray enters the first lens at height 2 cm and at an angle of 35°, plot the ray for the distance of five lenses.

25. Suppose an investor[3] wishes to double her money in six years by loaning it out at a rate of r compounded yearly. The borrower is to pay the loan back in one lump payment at the completion of six years. Find the necessary value of r.

26. In the network shown in the figure,[4] assume that all resistances except R_L on the end are of the same value R. Show that the current i_n in the nth loop satisfies the difference equation

$$i_{n+2} - 3i_{n+1} + i_n = 0,$$

and solve for i_n.

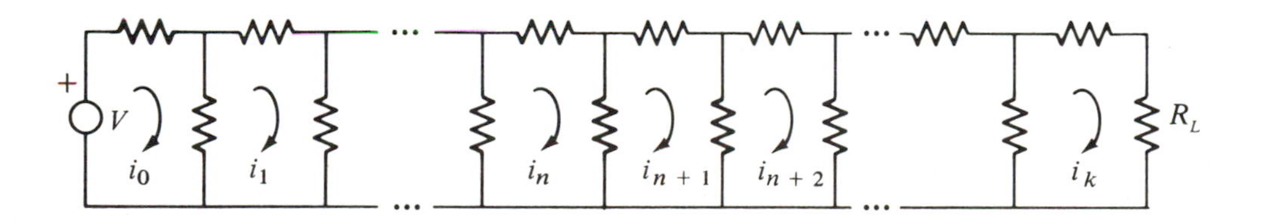

[2]This is Exercise 11–9.2 in S. Seely and A.D. Poularikas, *Electromagnetics: Classical and Modern Theory and Applications* (New York: Marcel Dekker, 1979), p. 533. Reprinted by permission of Marcel Dekker, Inc. N.Y.

[3]This is Exercise 2.4–2 in J.A. Cadzow, *Discrete-Time Systems: An Introduction with Interdisciplinary Applications* (Englewood Cliffs, N.J.: Prentice-Hall, 1973), p. 83. Reprinted by permission of Prentice-Hall, Inc., Englewood Cliffs, New Jersey.

[4]See Section 16–1 in J.A. Aseltine, *Transform Method in Linear System Analysis* (New York: McGraw-Hill Book Co., 1958), p. 246. Used with the permission of McGraw-Hill Book Company.

Fourier Series

10.1 INTRODUCTION

In connection with the solution of the heat equation in Section 6.2.1, we still have to show how to choose constants b_n for $n = 1, 2, 3, \ldots$ in such a way that a given function f can be expressed as a trigonometric series of the form

$$f(x) = \sum_{n=1}^{\infty} b_n \sin \frac{n\pi x}{l}. \tag{1}$$

This, and the more general problem of expressing a given function f as a series of the form

$$f(x) = \frac{a_0}{2} + \sum_{n=1}^{\infty} \left(a_n \cos \frac{n\pi x}{l} + b_n \sin \frac{n\pi x}{l}\right), \tag{2}$$

will be the subject matter of this chapter.

Series like the ones which appear in the right-hand sides of (1) and (2) are called *trigonometric series* or *Fourier series* in honor of the French scientist J. B. Fourier. Fourier discovered an ingenious method for computing the coefficients a_n and b_n of (2) and made systematic use of such series in connection with his work on heat conduction in 1807 and 1811. However, Fourier's work lacked mathematical rigor, thus leading him to the false conclusion that any arbitrary function can be expressed as a series of the form (2). This is not true, as we will see in Section 10.4. Consequently, when Fourier published his results, they were considered to be nonsense by many of his contemporaries. Since then, the theory of Fourier series has been developed on a rigorous basis and has become an indispensable tool in many areas of scientific work.

10.2 PERIODICITY AND ORTHOGONALITY OF SINES AND COSINES

In this section we study the periodic character and orthogonality properties of the functions

$$\cos \frac{n\pi x}{l} \quad \text{and} \quad \sin \frac{n\pi x}{l} \quad \text{for} \quad n = 1, 2, 3, \ldots; l > 0, \tag{1}$$

which are the building blocks of Fourier series.

Let us recall that a function f is called *periodic with period* $T > 0$ if for all x in the domain of the function

$$f(x+T) = f(x). \tag{2}$$

Geometrically, this means that the graph of f repeats itself in successive intervals of length T. The functions $\sin x$ and $\cos x$ are simple examples of periodic functions with period 2π. A constant function is also periodic with period any positive number. Other examples of periodic functions are the functions in (1) with period $2l$ (see Remark 1) and the functions shown graphically in Figure 10.1. The function in Figure 10.1(a) has period 2 and that in Figure 10.1(b) has period 1.

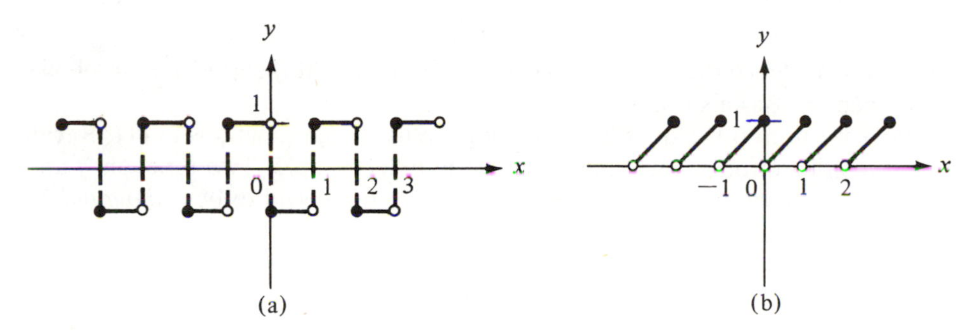

Figure 10.1

Periodic functions appear in a variety of real life situations. Alternating currents, the vibrations of a spring, sound waves, and the motion of a pendulum are examples of periodic functions.

Periodic functions have many periods. For example, $2\pi, 4\pi, 6\pi, \ldots$ are periods of $\sin x$ and $\cos x$. More generally, it follows from (2) that if f has period T, then

$$f(x) = f(x+T) = f(x+2T) = f(x+3T) = \ldots,$$

which means that $2T, 3T, \ldots$ are also periods of f. The smallest positive number T for which Eq. (2) holds, if it exists, is called the *fundamental period* of f. For example, the fundamental period of $\sin x$ and $\cos x$ is 2π, and in general (see Theorem 1) the fundamental period of each function in (1) is $2l/n$. On the other hand, a constant function has no fundamental period. This is because any positive number, no matter how small, is a period.

THEOREM 1

The functions $\cos px$ and $\sin px$, $p > 0$, are periodic with fundamental period $2\pi/p$. In particular, the functions

$$\cos\frac{n\pi x}{l} \quad \text{and} \quad \sin\frac{n\pi x}{l}, \quad n = 1, 2, 3, \ldots,$$

where l is a positive number, are periodic with fundamental period $T = 2l/n$.

Proof We give the proof for the function $\cos px$. The proof for $\sin px$ is similar. Assume that T is a period of $f(x) = \cos px$. Then the statement $f(x+T) = f(x)$ for all x is equivalent to $\cos(px+pT) = \cos px$ or, after expanding the cosine of the sum $px + pT$,

$$\cos px \cos pT - \sin px \sin pT = \cos px, \text{ for all } x. \tag{3}$$

But (3) is true for all x if and only if $\cos pT = 1$ and $\sin pT = 0$. That is, $pT = 2n\pi$, with $n = 1, 2, 3, \ldots$ (recall that p and T are positive). Thus $T = 2n\pi/p$, $n = 1, 2, \ldots$, and so the fundamental period (the least positive) of $\cos px$ is $2\pi/p$. Clearly, each of the functions listed in (1) is periodic with fundamental period $2\pi/(n\pi/l) = 2l/n$.

REMARK 1 Since the functions in (1) have (fundamental) period $2l/n$, it follows that each one has also period $2l$.

Next we turn to the question of orthogonality of the functions in (1). Recall the definition of orthogonal functions given in Section 6.2. Two functions f and g defined and continuous in an interval $a \leq x \leq b$ are said to be *orthogonal* on $a \leq x \leq b$ if

$$\int_a^b f(x)g(x)dx = 0.$$

The following result establishes the fact that the functions listed in (1) are *mutually orthogonal* in the interval $-l \leq x \leq l$. This means that any two distinct functions from (1) are orthogonal in $-l \leq x \leq l$. More precisely, we have Theorem 2.

THEOREM 2

The functions

$$\cos\frac{n\pi x}{l} \quad \text{and} \quad \sin\frac{n\pi x}{l} \quad \text{for} \quad n = 1, 2, 3, \ldots; l > 0$$

satisfy the following orthogonality properties in the interval $-l \leq x \leq l$.

$$\int_{-l}^{l} \cos\frac{n\pi x}{l}\cos\frac{k\pi x}{l}dx = \begin{cases} 0 & \text{if } n \neq k \\ l & \text{if } n = k. \end{cases} \tag{4}$$

$$\int_{-l}^{l} \sin\frac{n\pi x}{l}\sin\frac{k\pi x}{l}dx = \begin{cases} 0 & \text{if } n \neq k \\ l & \text{if } n = k. \end{cases} \tag{5}$$

$$\int_{-l}^{l} \cos\frac{n\pi x}{l}\sin\frac{k\pi x}{l}dx = 0 \qquad \text{for all } n, k. \tag{6}$$

Proof We can (and do) immediately establish (6) by using the fact that the integrand is an odd function and so its integral over the interval $-l \leq x \leq l$ (which is symmetric with respect to the origin) is zero. (For a review of odd and even functions, see Section 10.5.)

Next we prove (4). For $n = k$ we have, using the identity $\cos^2 x = \dfrac{1 + \cos 2x}{2}$,

$$\int_{-l}^{l} \cos\frac{n\pi x}{l}\cos\frac{k\pi x}{l}\,dx = \int_{-l}^{l}\left(\cos\frac{n\pi x}{l}\right)^2 dx$$

$$= \int_{-l}^{l} \tfrac{1}{2}\left[1 + \cos\frac{2n\pi x}{l}\right]dx = \tfrac{1}{2}\left[x + \frac{l}{2n\pi}\sin\frac{2n\pi x}{l}\right]\Big|_{-l}^{l} = l.$$

Now for $n \neq k$ we find, using the identity

$$\cos x \cdot \cos y = \tfrac{1}{2}[\cos(x+y) + \cos(x-y)],$$

$$\int_{-l}^{l} \cos\frac{n\pi x}{l}\cos\frac{k\pi x}{l}\,dx = \int_{-l}^{l} \tfrac{1}{2}\left[\cos\frac{(n+k)\pi x}{l} + \cos\frac{(n-k)\pi x}{l}\right]dx$$

$$= \left[\tfrac{1}{2}\frac{l}{(n+k)\pi}\sin\frac{(n+k)\pi x}{l} + \tfrac{1}{2}\frac{l}{(n-k)\pi}\sin\frac{(n-k)\pi x}{l}\right]\Big|_{-l}^{l} = 0.$$

The proof of (5) is similar to that of (4). In the case $n = k$ we need the identity $\sin^2 x = (1 - \cos 2x)/2$, and in the case $n \neq k$ we need the identity

$$\sin x \sin y = \tfrac{1}{2}[\cos(x-y) - \cos(x+y)].$$

EXERCISES

In Exercises 1 through 12, answer true or false.

1. The functions $3\sin\dfrac{x}{2}$ and $2\cos\dfrac{x}{2}$ have fundamental period equal to 4π.

2. The function $2 + 3\sin x + 4\cos x$ has period 3π.

3. The function $x\cos x$ is periodic.

4. The function $\sin^2 x \cos x$ has period 2π.

5. The functions $\sin x$ and x^2 are orthogonal in the interval $-1 \leq x \leq 1$.

6. The functions $\cos\dfrac{n\pi x}{l}$, $n = 1, 2, 3, \ldots$ are mutually orthogonal in the interval $0 \leq x \leq l$.

7. $\displaystyle\int_{-l}^{l} \sin\frac{n\pi x}{l}\,dx = \int_{0}^{2l} \sin\frac{n\pi x}{l}\,dx = 2\int_{0}^{l} \sin\frac{n\pi x}{l}\,dx.$

8. $\displaystyle\int_{-l}^{l} \cos\frac{n\pi x}{l}\,dx = \int_{0}^{2l} \cos\frac{n\pi x}{l}\,dx = 2\int_{0}^{l} \cos\frac{n\pi x}{l}\,dx.$

9. If the series

$$f(x) = \frac{a_0}{2} + \sum_{n=1}^{\infty} \left(a_n \cos \frac{n\pi x}{l} + b_n \sin \frac{n\pi x}{l}\right)$$

converges, the sum would be a periodic function with period $2l/n$.

10. Any solution of the differential equation $y''' + y' = 0$ has fundamental period 2π.

11. Any solution of the differential equation $y''' + y' = 0$ has period 2π.

12. The differential equation $y'' - y = \sin x$ has periodic solutions.

In Exercises 13 through 16, graph the given function.

13. $f(x) = \begin{cases} 1, & -1 \le x < 0 \\ -1, & 0 \le x < 1 \end{cases}; \quad f(x+2) = f(x)$

14. $f(x) = x, \quad -\pi \le x < \pi; \quad f(x+2\pi) = f(x)$

15. $f(x) = x, \quad 0 < x \le 1; \quad f(x+1) = f(x)$

16. $f(x) = \begin{cases} -3, & -2 \le x < -1 \\ 0, & -1 \le x \le 1 \\ 3, & 1 < x < 2 \end{cases}; \quad f(x+4) = f(x)$

17. Assume that the functions f and g are defined for all x and that they are periodic with common period T. Show that for any constants a and b, the functions $af + bg$ and fg are also periodic with period T.

18. Assume that the function f is defined for all x and is periodic with period T. Show that if f is integrable in the interval $0 \le x \le T$, then for any constant c,

$$\int_0^T f(x)\,dx = \int_c^{c+T} f(x)\,dx.$$

19. Find a necessary and sufficient condition for all solutions of the differential equation $y'' + py = 0$, with p constant to be periodic. What is the period?

20. Find a necessary and sufficient condition for all solutions of the system with constant coefficients

$$\begin{aligned} \dot{x} &= ax + by \\ \dot{y} &= cx + dy \end{aligned}$$

to be periodic.

10.3 FOURIER SERIES

Let us begin by assuming that a given function f, defined in the interval

$-l \le x \le l$ and outside of this interval by $f(x + 2l) = f(x)$, so that f has period $2l$, can be expressed as a trigonometric series of the form

$$f(x) = \frac{a_0}{2} + \sum_{n=1}^{\infty} \left(a_n \cos \frac{n\pi x}{l} + b_n \sin \frac{n\pi x}{l} \right). \tag{1}$$

We want to compute the coefficients a_n, $n = 0, 1, 2, \ldots$, and b_n, $n = 1, 2, 3, \ldots$, of (1). Consider the orthogonality properties of the functions $\cos(n\pi x/l)$ and $\sin(n\pi x/l)$ for $n = 1, 2, 3, \ldots$ in the interval $-l \le x \le l$.

(i) To compute the coefficients a_n for $n = 1, 2, 3, \ldots$, multiply both sides of (1) by $\cos(k\pi x/l)$, with k a positive integer, then integrate from $-l$ to l. For the moment we assume that the integrals exist and that it is legal to integrate the series term by term. Then, using (4) and (6) from Section 10.2, we find

$$\int_{-l}^{l} f(x) \cos \frac{k\pi x}{l}\, dx = \frac{a_0}{2} \underbrace{\int_{-l}^{l} \cos \frac{k\pi x}{l}\, dx}_{0}$$

$$+ \sum_{n=1}^{\infty} \left[a_n \underbrace{\int_{-l}^{l} \cos \frac{n\pi x}{l} \cos \frac{k\pi x}{l}\, dx}_{\substack{0 \text{ if } n \neq k \\ l \text{ if } n = k}} + b_n \underbrace{\int_{-l}^{l} \sin \frac{n\pi x}{l} \cos \frac{k\pi x}{l}\, dx}_{0} \right] = a_k l.$$

Thus,

$$a_k = \frac{1}{l} \int_{-l}^{l} f(x) \cos \frac{k\pi x}{l}\, dx, \qquad k = 1, 2, 3, \ldots$$

or, replacing k by n,

$$a_n = \frac{1}{l} \int_{-l}^{l} f(x) \cos \frac{n\pi x}{l}\, dx, \qquad n = 1, 2, 3, \ldots. \tag{2}$$

(ii) To compute a_0, integrate both sides of (1) from $-l$ to l. Then

$$\int_{-l}^{l} f(x) dx = \frac{a_0}{2} \underbrace{\int_{-l}^{l} dx}_{2l} + \sum_{n=1}^{\infty} \left[a_n \underbrace{\int_{-l}^{l} \cos \frac{n\pi x}{l}\, dx}_{0} + b_n \underbrace{\int_{-l}^{l} \sin \frac{n\pi x}{l}\, dx}_{0} \right]$$

$$= a_0 l.$$

Hence,

$$a_0 = \frac{1}{l} \int_{-l}^{l} f(x) dx. \tag{3}$$

That is, a_0 is twice the average value of the function f over the interval $-l \le x \le l$. Note that the value of a_0 can be obtained from formula (2) for $n = 0$. Of course, if the constant a_0 in (1) were not divided by 2, we would need a separate formula for a_0. As it is, all a_n are given by a single formula, namely,

$$a_n = \frac{1}{l} \int_{-l}^{l} f(x) \cos \frac{n\pi x}{l}\, dx, \qquad n = 0, 1, 2, \ldots. \tag{4}$$

(iii) Finally, to compute b_n for $n = 1, 2, 3, \ldots$, multiply both sides of (1) by $\sin(k\pi x/l)$, with k a positive integer; then integrate from $-l$ to l. Using (6) and (5) from Section 10.2, we find

$$\int_{-l}^{l} f(x)\sin\frac{k\pi x}{l}\,dx = \frac{a_0}{2}\underbrace{\int_{-l}^{l}\sin\frac{k\pi x}{l}\,dx}_{0}$$

$$+\sum_{n=1}^{\infty}\left[a_n\underbrace{\int_{-l}^{l}\cos\frac{n\pi x}{l}\sin\frac{k\pi x}{l}\,dx}_{0} + b_n\underbrace{\int_{-l}^{l}\sin\frac{n\pi x}{l}\sin\frac{k\pi x}{l}\,dx}_{\substack{0 \text{ if } n \neq k \\ l \text{ if } n = k}}\right] = b_k l.$$

Thus, replacing k by n, we find

$$b_n = \frac{1}{l}\int_{-l}^{l} f(x)\sin\frac{n\pi x}{l}\,dx, \qquad n = 1, 2, 3, \ldots. \tag{5}$$

When the coefficients a_n and b_n of (1) are given by the formulas (4) and (5) above, then the right-hand side of (1) is called the *Fourier series of the function* f over the interval of definition of the function. The formulas (4) and (5) are known as the *Euler-Fourier formulas,* and the numbers a_n and b_n are called the *Fourier coefficients* of f. We will write

$$f(x) \sim \frac{a_0}{2} + \sum_{n=1}^{\infty}\left(a_n\cos\frac{n\pi x}{l} + b_n\sin\frac{n\pi x}{l}\right) \tag{6}$$

to indicate that the right-hand side of (6) is the Fourier series of the function f.

Before we present any examples, the following remarks are in order.

REMARK 1 So far we proved that, if the right-hand side of (1) converges and has sum $f(x)$, if f is integrable in the interval $-l \le x \le l$, and if the term by term integrations could be justified, then the coefficients a_n and b_n in (1) must be given by the formulas (4) and (5) respectively. On the other hand, if a function f is given and if we formally write down its Fourier series, there is no guarantee that the series converges. Even if the Fourier series converges, there is no guarantee that its sum is equal to $f(x)$. The convergence of the Fourier series and how its sum is related to $f(x)$ will be investigated in the next section.

REMARK 2 To compute the Fourier coefficients a_n and b_n, we *only* need the values of f in the interval $-l \le x \le l$ and the assumption that f is integrable there. It is a fact, however, that an integral is not affected by changing the values of the integrand at a finite number of points. In particular we can compute the Fourier coefficients a_n and b_n if f is integrable in $-l \le x \le l$, although the function may not be defined, or may be discontinuous at a finite number of points in that interval. Of course the interval does not have to be closed; it may be open or closed at one end and open at the other.

REMARK 3 When the series in (1) converges for all x, its sum must be a periodic function of period $2l$. This is because every term of the series is periodic with period $2l$. For this reason Fourier series is an indispensable tool for the study of periodic phenomena. Assume that a function f is not periodic and is only defined in the interval $-l \le x < l$ (or $-l < x \le l$, or $-l < x < l$). We can write its Fourier series in $-l \le x < l$. We also have the choice of extending f outside of this interval as a periodic function with period $2l$. The periodic extension of f, F, agrees with f in the interval $-l \le x < l$. Therefore, a function f, defined in $-l \le x < l$, and its periodic extension, which is defined for all x, have identical Fourier series. (See Example 4.) Finally we should mention that if f is defined in a closed interval $-l \le x \le l$ and if $f(-l) \ne f(l)$, then f cannot be extended periodically. In such a case we can either ignore (as we do in this book) or modify the values of f at $\pm l$ and proceed with the periodic extension.

REMARK 4 When f is periodic with period $2l$, the Fourier coefficients of f can be determined from formulas (4) and (5) or, equivalently, from

$$a_n = \frac{1}{l}\int_c^{c+2l} f(x)\cos\frac{n\pi x}{l}\,dx, \quad n = 0, 1, 2, \ldots \tag{4'}$$

and

$$b_n = \frac{1}{l}\int_c^{c+2l} f(x)\sin\frac{n\pi x}{l}\,dx, \quad n = 1, 2, 3, \ldots, \tag{5'}$$

where c is any real number. This follows immediately from Exercise 18 of Section 10.2, with $T = 2l$ and $c = -l$. Observe that for $c = -l$, formulas (4′) and (5′) reduce to (4) and (5) respectively.

EXAMPLE 1 Compute the Fourier series of the function

$$f(x) = \begin{cases} 0, & -\pi \le x < \dfrac{\pi}{2} \\ 1, & \dfrac{\pi}{2} \le x < \pi \end{cases}; \qquad f(x + 2\pi) = f(x).$$

Solution Since the period of f is 2π, we have $2l = 2\pi$, $l = \pi$. Hence, the Fourier series of f is

$$f(x) \sim \frac{a_0}{2} + \sum_{n=1}^{\infty} (a_n \cos nx + b_n \sin nx),$$

with

$$a_n = \frac{1}{\pi}\int_{-\pi}^{\pi} f(x)\cos nx\,dx = \frac{1}{\pi}\int_{\pi/2}^{\pi} \cos nx\,dx, \qquad n = 0, 1, 2, \ldots \tag{7}$$

and

$$b_n = \frac{1}{\pi}\int_{-\pi}^{\pi} f(x)\sin nx\,dx = \frac{1}{\pi}\int_{\pi/2}^{\pi} \sin nx\,dx, \qquad n = 1, 2, 3, \ldots . \tag{8}$$

To evaluate the integral in (7), we have to distinguish between the two cases $n = 0$ and $n \neq 0$. For $n = 0$, the integrand is $\cos 0 = 1$, and so

$$a_0 = \frac{1}{\pi}\int_{\pi/2}^{\pi} dx = \frac{1}{\pi}x\Big|_{\pi/2}^{\pi} = \frac{1}{\pi}\left(\pi - \frac{\pi}{2}\right) = \frac{1}{\pi}\frac{\pi}{2} = \frac{1}{2}.$$

On the other hand, for $n \neq 0$, and so for $n = 1, 2, 3, \ldots$, we find

$$a_n = \frac{1}{\pi}\frac{\sin nx}{n}\Big|_{\pi/2}^{\pi} = \frac{1}{n\pi}\left(\sin n\pi - \sin\frac{n\pi}{2}\right) = -\frac{\sin(n\pi/2)}{n\pi}, \qquad n = 1, 2, 3, \ldots .$$

From (8) we find, for $n = 1, 2, 3, \ldots$,

$$\begin{aligned} b_n &= \frac{1}{\pi}\left(-\frac{\cos nx}{n}\right)\Big|_{\pi/2}^{\pi} = -\frac{1}{n\pi}[\cos n\pi - \cos(n\pi/2)] \\ &= \frac{\cos(n\pi/2) - (-1)^n}{n\pi}, \qquad n = 1, 2, 3, \ldots . \end{aligned}$$

Hence, the Fourier series of f is

$$\begin{aligned} f(x) &\sim \frac{1}{4} + \sum_{n=1}^{\infty}\left[-\frac{\sin(n\pi/2)}{n\pi}\cos nx + \frac{\cos(n\pi/2) - (-1)^n}{n\pi}\sin nx\right] \\ &= \frac{1}{4} - \frac{1}{\pi}\sum_{n=1}^{\infty}\frac{1}{n}\{\sin(n\pi/2)\cos nx + [(-1)^n - \cos(n\pi/2)]\sin nx\}. \end{aligned}$$

EXAMPLE 2 Find the Fourier series of the function

$$f(x) = x^2, \qquad -1 < x \le 1; \qquad f(x + 2) = f(x).$$

Solution Since the period of f is 2, we have $2l = 2$, $l = 1$. Hence the Fourier series of f is

$$f(x) \sim \frac{a_0}{2} + \sum_{n=1}^{\infty}(a_n \cos n\pi x + b_n \sin n\pi x),$$

with

$$a_n = \int_{-1}^{1} x^2 \cos n\pi x\,dx, \qquad n = 0, 1, 2, \ldots . \tag{9}$$

and

$$b_n = \int_{-1}^{1} x^2 \sin n\pi x\,dx, \qquad n = 1, 2, 3, \ldots . \tag{10}$$

From (9) we find, for $n = 0$,

$$a_0 = \int_{-1}^{1} x^2 dx = \frac{x^3}{3}\bigg|_{-1}^{1} = \frac{2}{3}.$$

On the other hand, for $n = 1, 2, 3, \ldots$, and integrating by parts twice, or using the integral tables in the book, we find

$$a_n = \left(\frac{x^2}{n\pi}\sin n\pi x + \frac{2x}{n^2\pi^2}\cos n\pi x - \frac{2}{n^3\pi^3}\sin n\pi x\right)\bigg|_{-1}^{1}$$

$$= \frac{4\cos n\pi}{n^2\pi^2} = \frac{4(-1)^n}{n^2\pi^2}, \qquad n = 1, 2, 3, \ldots .$$

Again from the tables in the book, or integrating by parts, or using the fact that the integrand in (10) in an odd function of x, we find

$$b_n = 0, \qquad n = 1, 2, 3, \ldots .$$

Hence, the Fourier series of f is

$$f(x) \sim \frac{1}{3} + \sum_{n=1}^{\infty} \frac{4(-1)^n}{n^2\pi^2}\cos n\pi x = \frac{1}{3} + \frac{4}{\pi^2}\sum_{n=1}^{\infty}\frac{(-1)^n}{n^2}\cos n\pi x.$$

EXAMPLE 3 Determine the Fourier series of the function

$$f(x) = x, \qquad -\pi < x \leq \pi.$$

Solution Here f is only defined in the interval $-\pi < x \leq \pi$. Hence $2l = 2\pi$ or $l = \pi$. Its Fourier series in this interval is

$$f(x) \sim \frac{a_0}{2} + \sum_{n=1}^{\infty} (a_n \cos nx + b_n \sin nx),$$

with

$$a_n = \frac{1}{\pi}\int_{-\pi}^{\pi} x \cos nx\, dx, \qquad n = 0, 1, 2, \ldots \tag{11}$$

and

$$b_n = \frac{1}{\pi}\int_{-\pi}^{\pi} x \sin nx\, dx, \qquad n = 1, 2, 3, \ldots . \tag{12}$$

Since the integrand in (11) is an odd function of x for all n, we have

$$a_n = 0, \qquad n = 0, 1, 2, \ldots .$$

Integrating by parts or using the tables in the book, we find

$$b_n = \frac{1}{\pi}\left(\frac{1}{n^2}\sin nx - \frac{x}{n}\cos nx\right)\bigg|_{-\pi}^{\pi} = -\frac{2}{n}(-1)^n, \qquad n = 1, 2, 3, \ldots .$$

Hence, the Fourier series of $f(x) = x$ in the interval $-\pi < x \leq \pi$ is

$$x \sim \sum_{n=1}^{\infty} -\frac{2}{n}(-1)^n \sin nx = 2\left(\sin x - \frac{1}{2}\sin 2x + \frac{1}{3}\sin 3x - \cdots\right).$$

EXAMPLE 4 Find the Fourier series of each of the following functions:

(i)	$f(x) = x,$	$-\pi \leq x < \pi$	
(ii)	$f(x) = x,$	$-\pi < x < \pi$	
(iii)	$f(x) = x,$	$-\pi < x \leq \pi;$	$f(x + 2\pi) = f(x)$
(iv)	$f(x) = x,$	$-\pi \leq x < \pi;$	$f(x + 2\pi) = f(x)$
(v)	$f(x) = x,$	$-\pi \leq x \leq \pi;$	$f(x + 2\pi) = f(x).$

Solution As we explained in Remark 3, or as we can see directly from formulas (11) and (12), all the above functions and the function of Example 3 have identical Fourier series, namely,

$$f(x) \sim \sum_{n=1}^{\infty} -\frac{2}{n}(-1)^n \sin nx.$$

EXERCISES

In Exercises 1 through 20, find the Fourier series of the given function.

1. $f(x) = x, \ -1 \leq x \leq 1; f(x + 2) = f(x)$

2. $f(x) = |x|, \ -\pi \leq x < \pi; f(x + 2\pi) = f(x)$

3. $f(x) = \begin{cases} 0, & -\pi \leq x < 0 \\ 1, & 0 \leq x < \pi \end{cases}; f(x + 2\pi) = f(x)$

4. $f(x) = \begin{cases} -1, & -\pi < x < 0 \\ 1, & 0 < x < \pi \end{cases}$

5. $f(x) = x, \ 0 < x \leq 2\pi; f(x + 2\pi) = f(x)$
[*Hint:* Use formulas (4′) and (5′) with $c = 0$.]

6. $f(x) = x^2, \ -\pi \leq x \leq \pi$

7. $f(x) = x^2, \ 0 \leq x < 2\pi; f(x + 2\pi) = f(x)$

8. $f(x) = x^2, \ -l < x \leq l; f(x + 2l) = f(x)$

9. $f(x) = 2\cos^2 x, \ -\pi \leq x \leq \pi; f(x + 2\pi) = f(x)$

10. $f(x) = 2\sin^2 x, \ -\pi < x \leq \pi$

11. $f(x) = \sin 2x, \ -\frac{\pi}{2} \leq x \leq \frac{\pi}{2}$

12. $f(x) = \cos 2x, \ -\frac{\pi}{2} < x < \frac{\pi}{2}$

13. $f(x) = x^3,\ -\pi < x \leq \pi;\ f(x + 2\pi) = f(x)$

14. $f(x) = x^3,\ -1 \leq x \leq 1$

15. $f(x) = \cos\dfrac{x}{2},\ -\pi \leq x < \pi$

16. $f(x) = \cos px,\ -\pi < x < \pi\ (p \neq \text{integer})$

17. $f(x) = \begin{cases} 0, & -2 < x \leq -1 \\ 1, & -1 < x \leq 1 \\ 0, & 1 < x \leq 2 \end{cases};\qquad f(x + 4) = f(x)$

18. $f(x) = \begin{cases} -1, & -2 \leq x < -1 \\ 1, & -1 \leq x < 1 \\ -1, & 1 \leq x < 2 \end{cases};\qquad f(x + 4) = f(x)$

19. $f(x) = \begin{cases} 0, & -l < x < 0 \\ 1, & 0 \leq x < l \end{cases};\qquad f(x + 2l) = f(x)$

20. $f(x) = e^x,\ -1 \leq x < 1;\qquad f(x + 2) = f(x)$

10.4 CONVERGENCE OF FOURIER SERIES

Assume that a function f is defined in the interval $-l \leq x < l$ and outside this interval by $f(x + 2l) = f(x)$, so that f has period $2l$. In Section 10.3 we defined the Fourier series of f,

$$f(x) \sim \frac{a_0}{2} + \sum_{n=1}^{\infty} \left(a_n \cos\frac{n\pi x}{l} + b_n \sin\frac{n\pi x}{l} \right), \tag{1}$$

where the Fourier coefficients a_n and b_n of f are given by the Euler-Fourier formulas

$$a_n = \frac{1}{l}\int_{-l}^{l} f(x)\cos\frac{n\pi x}{l}\,dx,\ n = 0, 1, 2, \ldots \tag{2}$$

and

$$b_n = \frac{1}{l}\int_{-l}^{l} f(x)\sin\frac{n\pi x}{l}\,dx,\ n = 1, 2, 3, \ldots. \tag{3}$$

When Fourier announced his famous theorem to the Paris Academy in 1807, he claimed that *any* function f could be represented by a series of the form

$$f(x) = \frac{a_0}{2} + \sum_{n=1}^{\infty} \left(a_n \cos\frac{n\pi x}{l} + b_n \sin\frac{n\pi x}{l} \right), \tag{4}$$

where the coefficients a_n and b_n are given by (2) and (3). Fourier was wrong in asserting that (4) is true without any restrictions on the function f. As we will see in Theorem 1, there is a huge class of functions for which (4) fails at the points of discontinuities of the functions. Examples are also known of functions whose Fourier series diverge at "almost" every point. Sufficient conditions for (4) to be true were given by Dirichlet in 1829. However, necessary and sufficient conditions for (4) to hold have not been discovered.

In this section we state conditions which are sufficient to insure that the Fourier series converges for all x and furthermore that the sum of the series is equal to the value $f(x)$ at each point where f is continuous. These conditions, although not the most general sufficient conditions known today, are, nevertheless, generally satisfied in practice.

DEFINITION 1

A function f is said to be piecewise continuous on an interval I if I can be subdivided into a finite number of subintervals, in each of which f is continuous and has finite left- and right-hand limits. An example of a piecewise continuous function is shown graphically in Figure 10.2. *Clearly, a piecewise continuous function on an interval I has a finite number of discontinuities on I. Such discontinuities (where the left- and right-hand limits exist but are unequal) are called jump discontinuities. The notation $f(c-)$ denotes the limit of $f(x)$ as $x \to c$ from the left. That is,*

$$f(c-) = \lim_{h \to 0-} f(x+h).$$

Similarly we write $f(c+)$ to denote the limit of $f(x)$ as $x \to c$ from the right. If f is continuous at c, then

$$f(c-) = f(c+) = f(c). \tag{5}$$

THEOREM 1

Assume that f is a periodic function with period $2l$ and such that f and f' are piecewise continuous on the interval $-l \le x \le l$. Then the Fourier series of f converges to the value $f(x)$ at each point x where f is continuous, and to the average $[f(x-) + f(x+)]/2$ of the left- and right-hand limits at each point x where f is discontinuous.

The hypotheses of the above theorem[1] are known by the name *Dirichlet conditions*. Hence, if f satisfies the Dirichlet conditions, then

$$\frac{a_0}{2} + \sum_{n=1}^{\infty}\left(a_n \cos\frac{n\pi x}{l} + b_n \sin\frac{n\pi x}{l}\right) = \begin{cases} f(x) & \text{if } x \text{ is point of continuity of } f \\ \dfrac{f(x-)+f(x+)}{2} & \text{if } x \text{ is point of discontinuity} \end{cases} \tag{6}$$

[1]For a proof of Theorem 1 see, for example, W. Kaplan, *Advanced Calculus* (Reading, Mass.: Addison-Wesley Publishing Co., 1973).

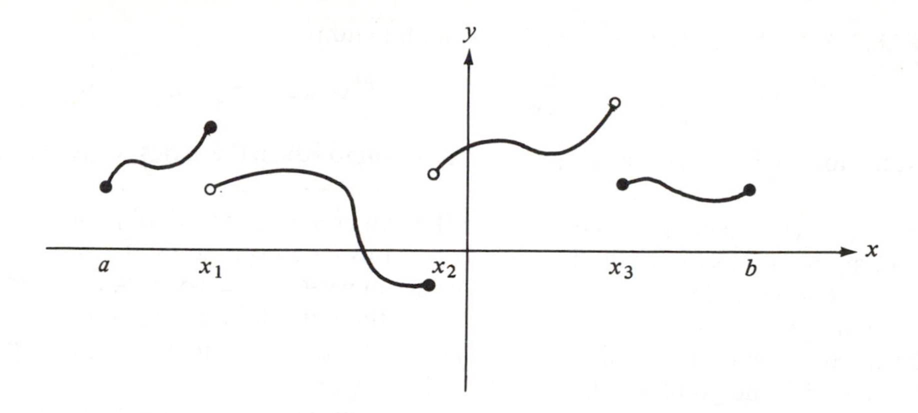

Figure 10.2

where a_n and b_n are given by (2) and (3). It follows from (6) that (4) is, in general, false at the point where f is discontinuous. On the other hand, if f is continuous everywhere and satisfies the Dirichlet conditions, then (4) is true for all x. Unless (4) is true for all x, we will continue using the symbol $\sim$ to indicate that the right-hand side is the Fourier series of the function to the left.

Using (5) we can write (6) in the form

$$\frac{f(x-) + f(x+)}{2} = \frac{a_0}{2} + \sum_{n=1}^{\infty} \left(a_n \cos \frac{n\pi x}{l} + b_n \sin \frac{n\pi x}{l} \right) \tag{6'}$$

which is true for all x. In fact, if x is a point of discontinuity of f, (6′) agrees with (6); if x is a point of continuity, we have $f(x-) = f(x+) = f(x)$, and the left-hand side of (6′) reduces to $f(x)$.

The Remarks 2, 3, and 4 of the last section are relevant to this section as well.

The conclusion of Theorem 1 is also true for functions f which are only defined on an interval I with endpoints $-l$ and $+l$, provided that f and f' are piecewise continuous on I. Then the periodic extension of f, which agrees with f on I, satisfies the Dirichlet conditions. Furthermore, the Fourier series of f and its periodic extension are identical. The periodic extension of f can also be utilized in finding the sum of the Fourier series of f at the endpoints $\pm l$. In fact, if F denotes the periodic extension of f, then, from (6′), the sum of the Fourier series of f at l is

$$\frac{F(l-) + F(l+)}{2} = \frac{f(l-) + f(-l+)}{2}$$

and at $-l$ is

$$\frac{F(-l-) + F(-l+)}{2} = \frac{f(l-) + f(-l+)}{2}.$$

Thus, the sum of the Fourier series of f at each of the endpoints $\pm l$ is $\frac{1}{2}[f(l-) + f(-l+)]$.

EXAMPLE 1 Find the Fourier series of the function

$$f(x) = \begin{cases} -1, & -\pi < x < \pi \\ 1, & 0 < x < \pi \end{cases}; \qquad f(x + 2\pi) = f(x).$$

Sketch for a few periods the graph of the function to which the series converges.

Solution The function f, whose graph is known as a *square wave* of period 2π and amplitude 1, satisfies the Dirichlet conditions (the hypotheses of Theorem 1) with $l = \pi$. In fact, the only points in the interval $-\pi \le x \le \pi$ where f or f' is not continuous are $x = 0$, and $x = \pm\pi$; the left- and right-hand limits at these points exist and are finite. [*Note:* $f'(x) = 0$ in $-\pi < x < 0$ and $0 < x < \pi$.] The graph of f is sketched in Figure 10.3.

The function f is continuous everywhere except at the points $0, \pm\pi, \pm 2\pi, \ldots$, where f is not even defined. From Theorem 1, the Fourier series of f converges to $f(x)$ at each point except $0, \pm\pi, \pm 2\pi, \ldots$. At each of the points $0, \pm\pi, \pm 2\pi, \ldots$, the Fourier series converges to the average of the left- and right-hand limit, which in this case is 0. Therefore, the graph of the function to which the Fourier series of f converges is now completely known. It is identical to f everywhere except at the discontinuities of f, where the value of the Fourier series is zero. See Figure 10.4.

Next we compute the Fourier series of f. Here $l = \pi$ and

$$f(x) \sim \frac{a_0}{2} + \sum_{n=1}^{\infty} (a_n \cos nx + b_n \sin nx),$$

with

$$a_n = \frac{1}{\pi}\int_{-\pi}^{\pi} f(x)\cos nx\,dx = -\frac{1}{\pi}\int_{-\pi}^{0} \cos nx\,dx + \frac{1}{\pi}\int_{0}^{\pi} \cos nx\,dx = 0,$$
$$n = 0, 1, 2, \ldots$$

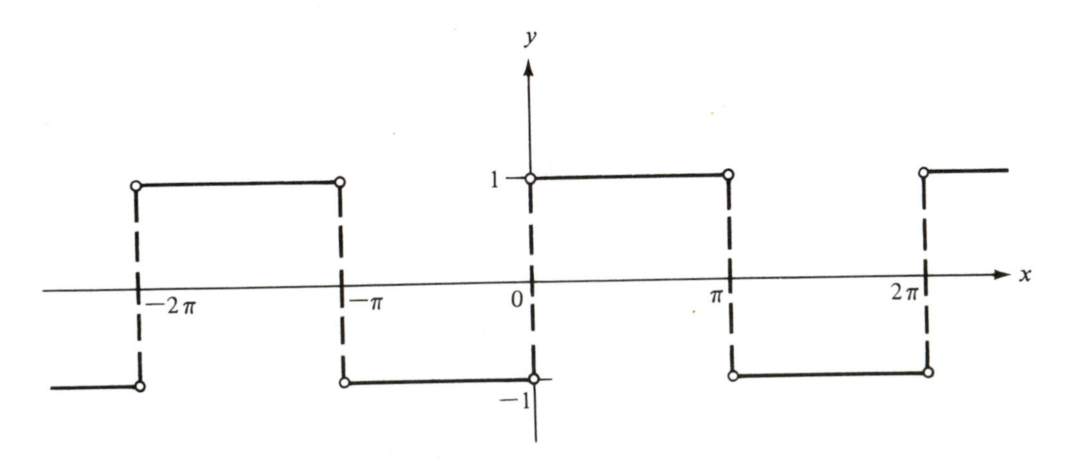

Figure 10.3

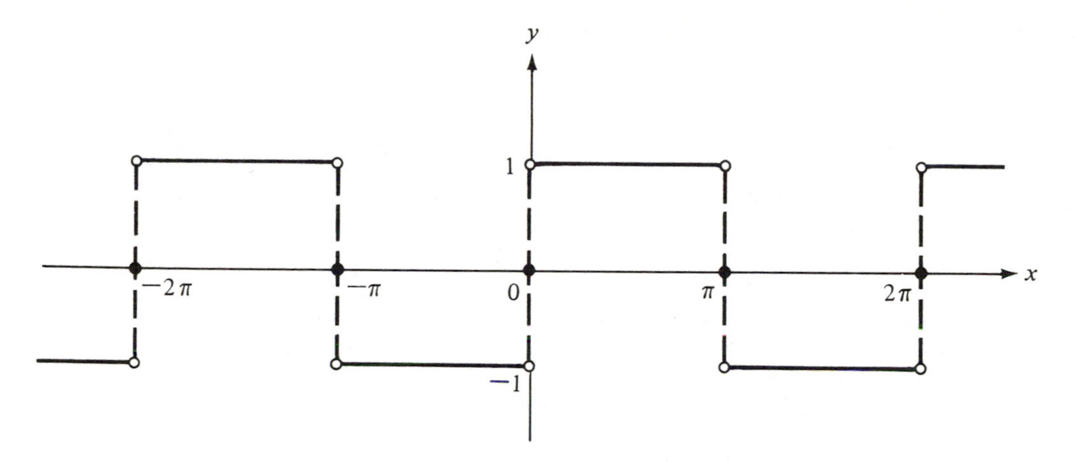

Figure 10.4

and

$$b_n = \frac{1}{\pi}\int_{-\pi}^{\pi} f(x)\sin nx\,dx = -\frac{1}{\pi}\int_{-\pi}^{0}\sin nx\,dx + \frac{1}{\pi}\int_{0}^{\pi}\sin nx\,dx$$

$$= \frac{2[1-(-1)^n]}{n\pi}, \qquad n = 1, 2, 3, \ldots .$$

Hence,

$$f(x) \sim \frac{2}{\pi}\sum_{n=1}^{\infty}\frac{1-(-1)^n}{n}\sin nx = \frac{4}{\pi}\left(\sin x + \frac{1}{3}\sin 3x + \frac{1}{5}\sin 5x + \ldots\right).$$

The symbol $\sim$ can be replaced by the equality sign everywhere except for $x = 0, \pm\pi, \pm 2\pi, \ldots$. Applying this idea to Fourier series often leads to interesting results. For example, in the above series, f is continuous and equal to 1 for all x in the interval $0 < x < \pi$. This leads to the trigonometric identity

$$\sin x + \frac{1}{3}\sin 3x + \frac{1}{5}\sin 5x + \ldots = \frac{\pi}{4}, \qquad 0 < x < \pi,$$

from which, for $x = \pi/2$, we find

$$\frac{\pi}{4} = 1 - \frac{1}{3} + \frac{1}{5} - \frac{1}{7} + \ldots .$$

EXAMPLE 2 Find the Fourier series of the function

$$f(x) = x, \qquad 0 \le x < 2\pi; \qquad f(x+2\pi) = f(x).$$

Sketch for a few periods the graph of the function to which the series converges.

Solution The graph of f is shown in Figure 10.5.

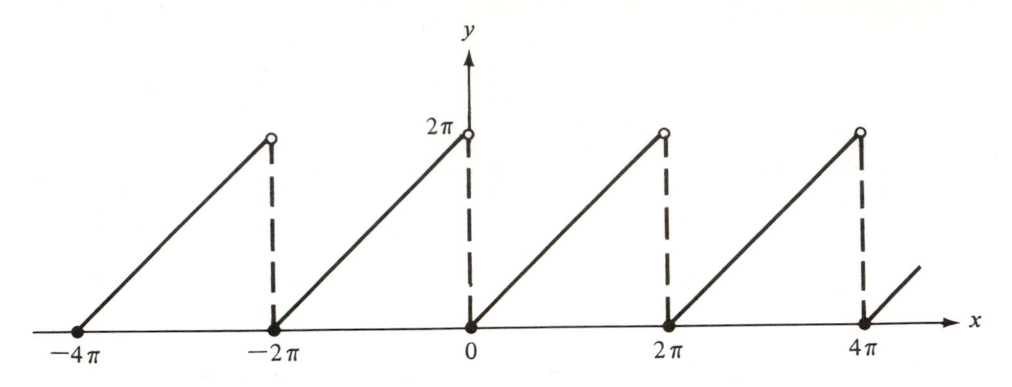

Figure 10.5

The function in this example is different from the function

$$g(x) = x, \qquad -\pi \leq x < \pi; \qquad g(x+2\pi) = g(x).$$

In the interval $0 \leq x < 2\pi$, the functions f and f' are piecewise continuous with jump discontinuities only at the points 0 and 2π. Therefore, Theorem 1 applies and Figure 10.6 shows the graph of the function to which the Fourier series of f converges. At the points $0, \pm 2\pi, \pm 4\pi, \ldots$, where f is discontinuous, the Fourier series converges to the value π, which is the average value at the jumps. (See Figure 10.6.) At all other points the graphs of f and the function to which its Fourier series converges are identical.

Next, we compute the Fourier series of f. Since the interval $0 \leq x < 2\pi$ is not symmetric with respect to the origin, it is advisable to use formulas (4′) and (5′) of Section 10.3 with $c = 0$ and $l = \pi$. The Fourier series of f is

$$f(x) \sim \frac{a_0}{2} + \sum_{n=1}^{\infty} (a_n \cos nx + b_n \sin nx),$$

with

$$a_n = \frac{1}{\pi}\int_0^{2\pi} f(x) \cos nx \, dx = \frac{1}{\pi}\int_0^{2\pi} x \cos nx \, dx = \begin{cases} 2\pi, & n = 0 \\ 0, & n = 1, 2, 3, \ldots \end{cases}$$

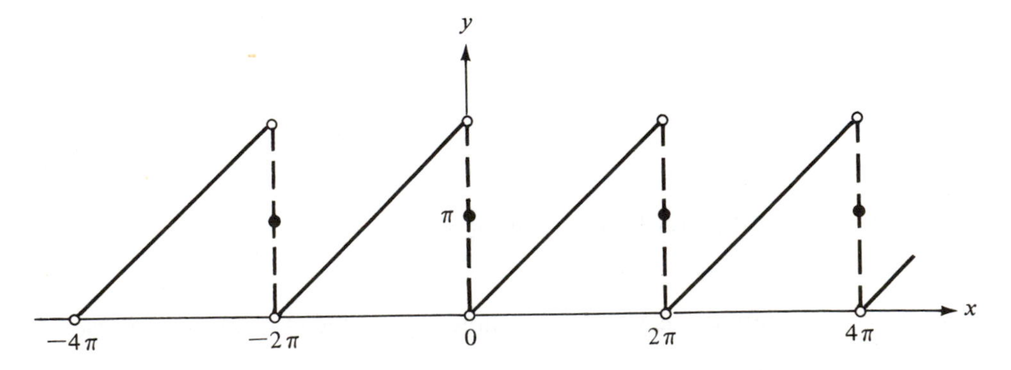

Figure 10.6

and

$$b_n = \frac{1}{\pi}\int_0^{2\pi} f(x)\sin nx\,dx = \frac{1}{\pi}\int_0^{2\pi} x\sin nx\,dx = -\frac{2}{n}, \qquad n = 1, 2, 3, \ldots .$$

Hence,

$$f(x) \sim \pi - \sum_{n=1}^{\infty} \frac{2}{n}\sin nx = \pi - 2(\sin x + \frac{1}{2}\sin 2x + \frac{1}{3}\sin 3x + \ldots).$$

The symbol $\sim$ can be replaced by the equality sign everywhere except for $x = 0, \pm 2\pi, \pm 4\pi, \ldots$. In particular, in the interval $0 < x < 2\pi$ we obtain the trigonometric identity

$$\sin x + \frac{1}{2}\sin 2x + \frac{1}{3}\sin 3x + \cdots = \frac{\pi - x}{2},$$

from which, for $x = \pi/2$, we find again

$$\frac{\pi}{4} = 1 - \frac{1}{3} + \frac{1}{5} - \frac{1}{7} + \cdots .$$

EXERCISES

In Exercises 1 through 16, find the Fourier series of the given function. Sketch for a few periods the graph of the function to which the Fourier series converges.

1. $f(x) = x, \ -\pi \le x < \pi; \ f(x+2\pi) = f(x)$

2. $f(x) = \begin{cases} 0, & -\pi < x < 0 \\ 1, & 0 \le x < \pi \end{cases}; \ f(x+2\pi) = f(x)$

3. $f(x) = x^2, \ -1 < x \le 1$

4. $f(x) = x^2, \ 0 \le x < 2\pi; \ f(x+2\pi) = f(x)$

5. $f(x) = \begin{cases} 1 + x, & -2 \le x < 0 \\ 1 - x, & 0 \le x \le 2 \end{cases}; \ f(x+4) = f(x)$

6. $f(x) = |x|, \ -1 \le x < 1; \ f(x+2) = f(x)$

7. $f(x) = \begin{cases} 0, & -\pi < x \le 0 \\ \sin x, & 0 < x \le \pi \end{cases}; f(x+2\pi) = f(x)$

8. $f(x) = \begin{cases} 0, & -1 < x < 0 \\ x, & 0 \le x < 1 \end{cases}$

9. $f(x) = \begin{cases} 0, & -2 < x \le -1 \\ 1, & -1 < x \le 1 \\ 0, & 1 < x \le 2 \end{cases}; \ f(x+4) = f(x)$

10. $f(x) = \begin{cases} -1, & -2 \le x < -1 \\ 0, & -1 \le x < 1 \\ -1, & 1 \le x < 2 \end{cases}; f(x+4) = f(x)$

11. $f(x) = \sin^2 x,\ 0 \le x \le 2\pi;\ f(x+2\pi) = f(x)$

12. $f(x) = \sin^2 x,\ -\pi \le x \le \pi;\ f(x+2\pi) = f(x)$

13. $f(x) = \cos^2 x,\ 0 \le x \le 2\pi;\ f(x+2\pi) = f(x)$

14. $f(x) = \cos^2 x,\ -\pi \le x \le \pi;\ f(x+2\pi) = f(x)$

15. $f(x) = \cos 2x,\ 0 \le x \le \pi$

16. $f(x) = \sin 2x,\ 0 \le x \le \pi$

17. Show that

$$\sin x - \frac{1}{2}\sin 2x + \frac{1}{3}\sin 3x - \cdots = \frac{x}{2}, \quad -\pi < x < \pi.$$

18. Show that

$$\frac{1}{3} + \frac{4}{\pi^2}\sum_{n=1}^{\infty}\frac{(-1)^n}{n^2}\cos n\pi x = x^2, \quad -1 \le x \le 1.$$

[*Hint:* Use the result of Exercise 3.]

19. Utilize the Fourier series of the function $f(x) = x^2$, $-1 \le x \le 1$, to establish the following results:

$$\frac{\pi^2}{6} = \frac{1}{1^2} + \frac{1}{2^2} + \frac{1}{3^2} + \cdots \quad \text{and} \quad \frac{\pi^2}{12} = \frac{1}{1^2} - \frac{1}{2^2} + \frac{1}{3^2} - \cdots.$$

20. Utilize the Fourier series of the function $f(x) = |x|$, $-1 \le x \le 1$, to obtain the following result:

$$\frac{\pi^2}{8} = \frac{1}{1^2} + \frac{1}{3^2} + \frac{1}{5^2} + \cdots.$$

In Exercises 21 through 30, answer true or false.

21. The function $f(x) = \dfrac{1}{x}$ is piecewise continuous in the interval $-\pi \le x \le \pi$.

22. The function $f(x) = \dfrac{1}{x}$ is piecewise continuous in the interval $0 \le x \le \pi$.

23. The Fourier series of the function

$$f(x) = \begin{cases} -3, & -2 < x < 0 \\ 5, & 0 < x < 2 \end{cases}$$

converges to 1 at the points $x = 0$, 2, and -2.

24. The Fourier series of the function

$$f(x) = |x|, -\pi \leq x < \pi; f(x+2\pi) = f(x)$$

converges to $f(x)$ everywhere.

25. The Fourier series of the function

$$f(x) = x, -1 \leq x < 1; f(x+2) = f(x)$$

converges to $f(x)$ everywhere.

26. The function

$$f(x) = \sqrt{|x|}, -1 \leq x \leq 1; f(x+2) = f(x)$$

is continuous everywhere.

27. The function

$$f(x) = \sqrt{|x|}, -1 \leq x \leq 1; f(x+2) = f(x)$$

satisfies the hypotheses of Theorem 1.

28. The Fourier series of the function

$$f(x) = x^2, 0 \leq x < 2\pi; f(x+2\pi) = f(x)$$

does not involve any sine term.

29. The Fourier series of the function

$$f(x) = x^2, -\pi \leq x \leq \pi; f(x+2\pi) = f(x)$$

does not involve any sine terms.

30. The Fourier series of the function

$$f(x) = x^2, -\pi \leq x \leq \pi; f(x+2\pi) = f(x)$$

is equal to $f(x)$ everywhere.

10.5 FOURIER SINE AND FOURIER COSINE SERIES

As we saw in Section 6.2.1 in connection with the solution of the heat equation, sometimes it is necessary to express a given function f as a Fourier series of the form

$$f(x) = \sum_{n=1}^{\infty} b_n \sin\frac{n\pi x}{l}. \tag{1}$$

In other cases it is necessary to express f as a series of the form

$$f(x) = \frac{a_0}{2} + \sum_{n=1}^{\infty} a_n \cos\frac{n\pi x}{l}. \tag{2}$$

A Fourier series of the form (1) is called a *Fourier sine series*; a Fourier series of the form (2) is called a *Fourier cosine series*. In this section we will show that if a function f is defined in the interval $0 < x < l$, and if f and f' are piecewise continuous there, then in the interval $0 < x < l$ we have the choice to represent f as a Fourier sine series or a Fourier cosine series.

Before we establish the above claim, we will review the concepts of odd and even functions and see how, for such functions, the labor of computing the Fourier coefficients is reduced.

DEFINITION 1

A function f, whose domain is symmetric with respect to the origin, is called even *if $f(x) = f(-x)$ for each x in the domain of f and* odd *if $f(x) = -f(-x)$ for each x in the domain of f.*

For example, the functions $\cos ax$, 1, x^2, $|x|$, $3 - x^2 + x^4 \cos 2x$ are even and $\sin ax$, x, $5x - x^2 \sin 4x$ are odd. Geometrically speaking, a function is even if its graph is symmetric with respect to the y-axis and odd if its graph is symmetric with respect to the origin. The functions whose graphs are sketched in Figure 10.7 are even, and those in Figure 10.8 are odd.

With respect to the operations of addition and multiplication, even and odd functions have the following properties:

(i) even + even = even; (ii) odd + odd = odd;
(iii) even × even = even; (iv) odd × odd = even;
(v) even × odd = odd.

Let us prove, for example, (v). (The others are proved in a similar fashion.) Assume f is even and g is odd. Set $F = fg$. Then

$$F(-x) = f(-x)g(-x) = f(x)\,(-g(x)) = -f(x)g(x) = -F(x),$$

which proves that F is odd.

With respect to integration, even and odd functions have the following useful properties:

$$\int_{-l}^{l} (\text{even})\, dx = 2\int_{0}^{l} (\text{even})\, dx \tag{3}$$

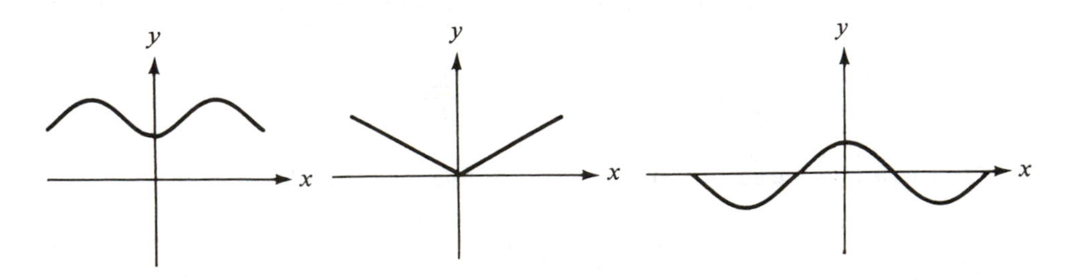

Figure 10.7

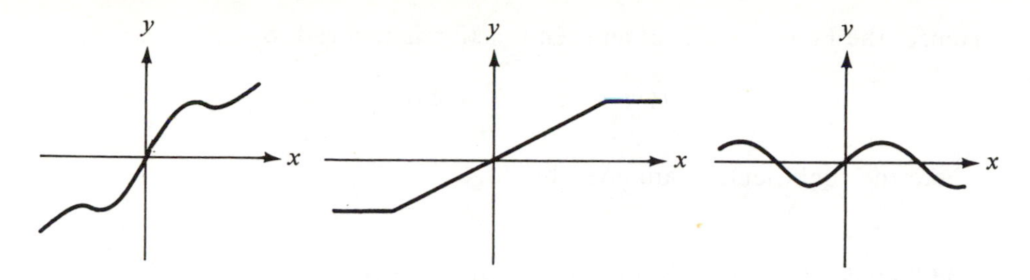

Figure 10.8

and

$$\int_{-l}^{l} (\text{odd})\, dx = 0. \tag{4}$$

We will prove (4). (The proof of (3) is similar.) Assume that f is an odd function. Then

$$\int_{-l}^{l} f(x)\, dx = \int_{-l}^{0} f(x)\, dx + \int_{0}^{l} f(x) dx. \tag{5}$$

Setting $x = -t$, we find

$$\int_{-l}^{0} f(x)\, dx = \int_{l}^{0} f(-t)\,(-dt) = \int_{l}^{0} -f(-t)\, dt = \int_{l}^{0} f(t)\, dt$$

$$= -\int_{0}^{l} f(t)\, dt.$$

This proves that the right-hand side of (5) is zero, and the proof is complete.

Using properties (3) and (4), the evaluation of the Fourier coefficients

$$a_n = \frac{1}{l}\int_{-l}^{l} f(x) \cos\frac{n\pi x}{l}\, dx, \qquad n = 0, 1, 2, \ldots \tag{6}$$

and

$$b_n = \frac{1}{l}\int_{-l}^{l} f(x) \sin\frac{n\pi x}{l}\, dx, \qquad n = 1, 2, 3, \ldots \tag{7}$$

is considerably simplified in the case of even or odd functions.

Even Functions When f is even, the integrand in (6) is even and in (7) is odd. Then, from (3) and (4), we find

$$a_n = \frac{2}{l}\int_{0}^{l} f(x) \cos\frac{n\pi x}{l}\, dx, \qquad n = 0, 1, 2, \ldots \tag{8}$$

and

$$b_n = 0, \qquad n = 1, 2, 3, \ldots .$$

Hence, the Fourier series of an even function is reduced to

$$f(x) \sim \frac{a_0}{2} + \sum_{n=1}^{\infty} a_n \cos \frac{n\pi x}{l}, \tag{9}$$

where the coefficients a_n are given by (8).

Odd Functions When f is odd, the integrand in (6) is odd and in (7) is even. Then, from (3) and (4), we find

$$a_n = 0, \qquad n = 0, 1, 2, \ldots$$

and

$$b_n = \frac{2}{l} \int_0^l f(x) \sin \frac{n\pi x}{l}\, dx, \qquad n = 1, 2, 3, \ldots. \tag{10}$$

Hence, the Fourier series of an odd function is reduced to

$$f(x) \sim \sum_{n=1}^{\infty} b_n \sin \frac{n\pi x}{l}, \tag{11}$$

where the coefficients b_n are given by (10).

It should be remarked that, for even or odd functions, the formulas for the Fourier coefficients use the values of the function in the interval $0 < x < l$ *only*.

If a function f is defined *only* in the interval $0 < x < l$, we define its *even periodic extension* by

$$g(x) = \begin{cases} f(x), & 0 < x < l \\ f(-x), & -l < x < 0 \end{cases}; \qquad g(x + 2l) = g(x) \tag{12}$$

and its *odd periodic extension* by

$$h(x) = \begin{cases} f(x), & 0 < x < l \\ -f(-x), & -l < x < 0 \end{cases}; \qquad h(x + 2l) = h(x). \tag{13}$$

Note that the functions f, g, and h agree in the interval $0 < x < l$. Furthermore, if f and f' are piecewise continuous on $0 < x < l$, then g and g' and also h and h' are piecewise continuous on $-l < x < l$. Therefore, the hypotheses of Theorem 1 of Section 10.4 are satisfied for the functions g and h. Since g is even, it has a Fourier cosine series which converges to $g(x) = f(x)$ at each point x in the interval $0 < x < l$ where f is continuous; and it converges to the average

$$\frac{g(x-) + g(x+)}{2} = \frac{f(x-) + f(x+)}{2}$$

at each point in $0 < x < l$ where f is discontinuous. Similarly, since h is odd, it has a Fourier sine series which converges to $h(x) = f(x)$ at each point x in $0 < x < l$ where f is continuous, and to the average

$$\frac{h(x-) + h(x+)}{2} = \frac{f(x-) + f(x+)}{2}$$

at each point in $0 < x < l$ where f is discontinuous.

In summary, if a function f is defined *only* in the interval $0 < x < l$, and if f and f' are piecewise continuous there, then f has a Fourier cosine series of the form

$$f(x) \sim \frac{a_0}{2} + \sum_{n=1}^{\infty} a_n \cos\frac{n\pi x}{l} \qquad \text{with } a_n = \frac{2}{l}\int_0^l f(x)\cos\frac{n\pi x}{l}\,dx, \qquad n = 0, 1, 2, \ldots \tag{14}$$

and a Fourier sine series of the form

$$f(x) \sim \sum_{n=1}^{\infty} b_n \sin\frac{n\pi x}{l} \qquad \text{with } b_n = \frac{2}{l}\int_0^l f(x)\sin\frac{n\pi x}{l}\,dx, \qquad n = 1, 2, 3, \ldots. \tag{15}$$

Furthermore, the convergence of the series in (14) and (15) is as described by Theorem 1 of Section 10.4 for the even and odd periodic extensions of f, respectively.

EXAMPLE 1 Compute the Fourier series of each function:

(a) $f(x) = x^2,\ -\pi < x \le \pi;\ f(x + 2\pi) = f(x)$

(b) $f(x) = x,\ -1 \le x < 1;\ f(x + 2) = f(x)$.

Solution (a) f is an even function and $l = \pi$. Hence, from Equations (9) and (8), we have

$$f(x) \sim \frac{a_0}{2} + \sum_{n=1}^{\infty} a_n \cos nx,$$

with

$$a_n = \frac{2}{\pi}\int_0^{\pi} x^2 \cos nx\,dx = \begin{cases} \dfrac{2\pi^2}{3}, & n = 0 \\[2mm] \dfrac{4(-1)^n}{n^2}, & n = 1, 2, 3, \ldots. \end{cases}$$

Thus,

$$f(x) \sim \frac{\pi^2}{3} + 4\sum_{n=1}^{\infty} \frac{(-1)^n}{n^2}\cos nx.$$

(b) f is an odd function and $l = 1$. Hence, from Eqs. (11) and (10), we have

$$f(x) \sim \sum_{n=1}^{\infty} b_n \sin n\pi x,$$

with

$$b_n = 2\int_0^1 x \sin \pi x\,dx = -\frac{2(-1)^n}{n\pi}, \qquad n = 1, 2, 3, \ldots.$$

Thus,

$$f(x) \sim -\frac{2}{\pi}\sum_{n=1}^{\infty} \frac{(-1)^n}{n}\sin n\pi x.$$

EXAMPLE 2 Sketch the even and the odd periodic extension of the function

$$f(x) = \begin{cases} x, 0 < x < 1 \\ 0, 1 < x < 2 \end{cases}.$$

Solution The even periodic extension of f is sketched in Figure 10.9, and the odd periodic extension is sketched in Figure 10.10.

EXAMPLE 3 Compute the Fourier cosine and the Fourier sine series of the function

$$f(x) = \begin{cases} x, & 0 < x < 1 \\ 0, & 1 < x < 2 \end{cases}.$$

Sketch the graph of the function to which the Fourier cosine series converges and the graph of the function to which the Fourier sine series converges.

Solution $l = 2$. From (14), the Fourier cosine series of f is

$$f(x) \sim \frac{a_0}{2} + \sum_{n=1}^{\infty} a_n \cos\frac{n\pi x}{2},$$

with

$$\begin{aligned} a_n &= \int_0^2 f(x)\cos\frac{n\pi x}{2}\,dx = \int_0^1 x\cos\frac{n\pi x}{2}\,dx \\ &= \begin{cases} \dfrac{1}{2}, & n = 0 \\ \dfrac{4}{n^2\pi^2}\left(\cos\dfrac{n\pi}{2} - 1\right) + \dfrac{2}{n\pi}\sin\dfrac{n\pi}{2}, & n = 1, 2, 3, \ldots. \end{cases} \end{aligned}$$

Hence,

$$f(x) \sim \frac{1}{4} + \sum_{n=1}^{\infty}\left[\frac{4}{n^2\pi^2}\left(\cos\frac{n\pi}{2} - 1\right) + \frac{2}{n\pi}\sin\frac{n\pi}{2}\right]\cos\frac{n\pi x}{2}.$$

The even periodic extension of f, referred to here as g, is shown graphically in Figure 10.9. From this graph we see that Theorem 1 of Section 10.4 is applicable. The only discontinuities of g are at the points $0, \pm 2, \pm 4, \pm 6, \ldots$ and ± 1,

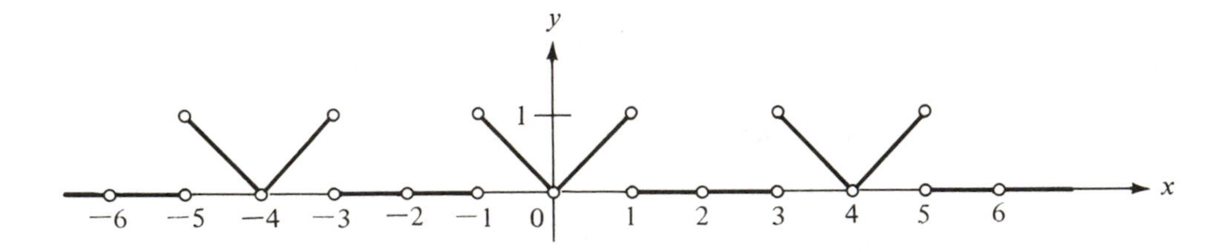

Figure 10.9

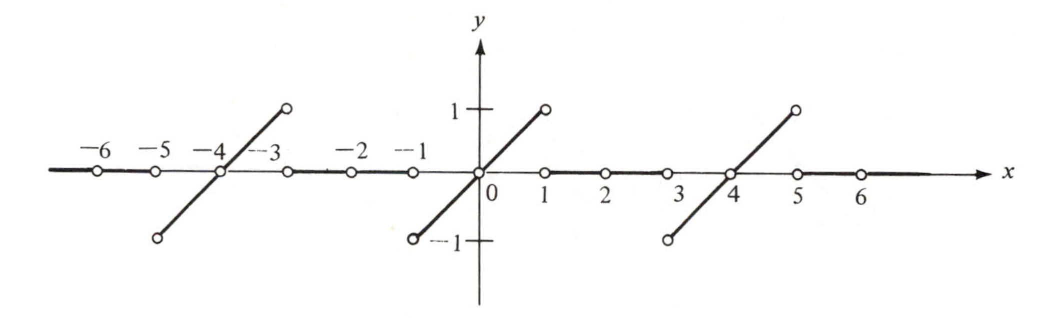

Figure 10.10

$\pm 3, \pm 5, \ldots$. From Figure 10.9 and from Theorem 1 of Section 10.4, we conclude that the Fourier series converges to 0 at the points $\pm 0, \pm 2, \pm 4, \ldots$ and converges to $\frac{1}{2}$ at the points $\pm 1, \pm 2, \pm 3, \ldots$. At every other point x the series converges to $g(x)$. The graph of the function to which the series converges is shown in Figure 10.11 (which is easily constructed from Figure 10.9).

Next we compute the Fourier sine series of f. From (15), with $l = 2$, we have

$$f(x) \sim \sum_{n=1}^{\infty} b_n \sin\frac{n\pi x}{2},$$

with

$$b_n = \int_0^2 f(x)\sin\frac{n\pi x}{2}\,dx = \int_0^1 x\sin\frac{n\pi x}{2}\,dx = \frac{4}{n^2\pi^2}\sin\frac{n\pi}{2} - \frac{2}{n\pi}\cos\frac{n\pi}{2}.$$

Hence,

$$f(x) \sim \sum_{n=1}^{\infty}\left(\frac{4}{n^2\pi^2}\sin\frac{n\pi}{2} - \frac{2}{n\pi}\cos\frac{n\pi}{2}\right)\sin\frac{n\pi x}{2}.$$

The odd periodic extension of f, referred to here as h, is shown graphically in Figure 10.10. From this graph we see that Theorem 1 of Section 10.4 is applicable. The only discontinuities of h are at the points $0, \pm 1, \pm 2, \pm 3, \ldots$. The graph of the function to which the series converges is sketched in Figure 10.12.

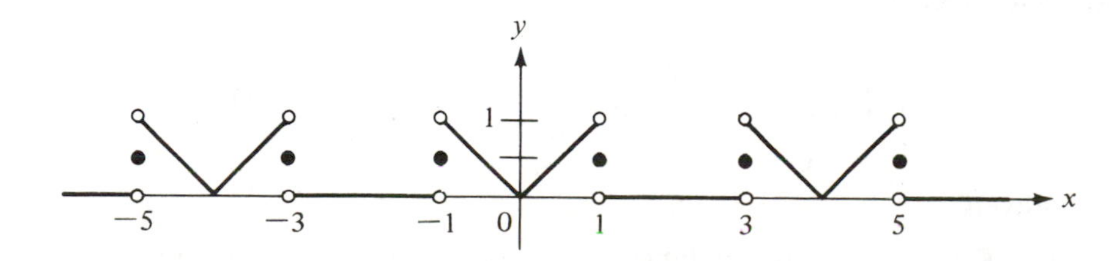

Figure 10.11

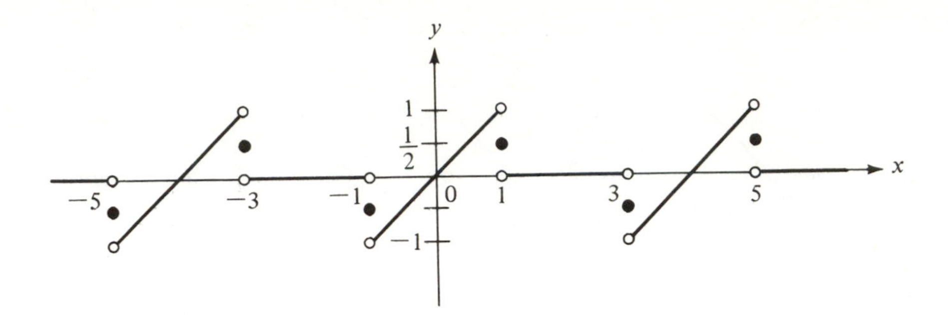

Figure 10.12

EXERCISES

In Exercises 1 through 10, answer true or false.

1. $x^2 \cos x + 2|x|$ is an even function in the interval $-\infty < x < +\infty$.

2. The function

$$F(x) = \begin{cases} x, & 0 < x < 2 \\ -x, & -2 < x < 0 \end{cases}$$

is the odd periodic extension of the function $f(x) = x$, $0 < x < 2$.

3. The function

$$f(x) = x^2, \qquad 0 \le x \le 2; \qquad f(x + 2) = f(x)$$

is even.

4. If a function f is odd and is defined at $x = 0$, then $f(0) = 0$.

5. The function

$$f(x) = \begin{cases} 0, & -2 < x < -1 \\ 1, & -1 < x < 1 \\ 0, & 1 < x < 2 \end{cases}; \qquad f(x + 4) = f(x)$$

is odd.

6. The function

$$f(x) = \begin{cases} -1, & -2 < x < -1 \\ 1, & -1 < x < 1 \\ -1, & 1 < x < 2 \end{cases}; \qquad f(x + 4) = f(x)$$

is even.

7. The Fourier series of the function in Exercise 5 does not contain any sine terms.

8. The Fourier series of the function

$$f(x) = \begin{cases} -1, & -3 < x < 0 \\ 1, & 0 < x < 3 \end{cases}; \quad f(x+6) = f(x)$$

is of the form $\sum_{n=1}^{\infty} b_n \sin \frac{n\pi x}{3}$ with $b_n = \frac{2}{3}\int_0^3 \sin \frac{n\pi x}{3} dx.$

9. The Fourier series of the function

$$f(x) = \begin{cases} -\sin x, & -\pi < x < 0 \\ \sin x, & 0 < x < \pi \end{cases}; \quad f(x+2\pi) = f(x)$$

contains only sine terms.

10. The Fourier series of the function

$$f(x) = \begin{cases} -x, & -2 < x < 0 \\ x, & 0 \le x < 2 \end{cases}$$

is of the form

$$\frac{a_0}{2} + \sum_{n=1}^{\infty} a_n \cos \frac{n\pi x}{2} \quad \text{with} \quad a_n = \int_0^2 x \cos \frac{n\pi x}{2} dx, \quad n = 0, 1, 2, \ldots.$$

In Exercises 11 through 18, determine whether the given function is even or odd and utilize this information to compute its Fourier series.

11. $f(x) = |x|, \ -1 \le x < 1; f(x+2) = f(x)$

12. $f(x) = x, \ -\pi \le x < \pi; f(x+2\pi) = f(x)$

13. $f(x) = \begin{cases} -1, & -\pi < x < 0 \\ 1, & 0 < x < \pi \end{cases}$

14. $f(x) = x^2, \ -1 < x \le 1; f(x+2) = f(x)$

15. $f(x) = x^3, \ -1 \le x \le 1$

16. $f(x) = \begin{cases} -1, & -2 \le x < -1 \\ 1, & -1 \le x < 1 \\ -1, & 1 \le x < 2 \end{cases}; f(x+4) = f(x)$

17. $f(x) = \sin 2x, \ -\frac{\pi}{2} \le x \le \frac{\pi}{2}$

18. $f(x) = \begin{cases} 1+x, & -2 \le x < 0 \\ 1-x, & 0 \le x \le 2 \end{cases}; \quad f(x+4) = f(x)$

In Exercises 19 through 28, compute, as indicated, the Fourier sine or Fourier cosine series [see Eqs. (14) and (15)] for the given function. Sketch for a few periods the graph of the function to which the series converges.

19. $f(x) = 1,\ 0 < x < \pi;$ Fourier sine series

20. $f(x) = 1,\ 0 < x < \pi;$ Fourier cosine series

21. $f(x) = x,\ 0 < x < \pi;$ Fourier sine series

22. $f(x) = x,\ 0 < x < \pi;$ Fourier cosine series

23. $f(x) = x^2,\ 0 < x < 1;$ Fourier sine series

24. $f(x) = x^2,\ 0 < x < 1;$ Fourier cosine series

25. $f(x) = \begin{cases} 1, & 0 < x < \dfrac{\pi}{2} \\ 0, & \dfrac{\pi}{2} < x < \pi \end{cases};$ Fourier sine series

26. $f(x) = \begin{cases} 1, & 0 < x < \dfrac{\pi}{2} \\ 0, & \dfrac{\pi}{2} < x < \pi \end{cases};$ Fourier cosine series

27. $f(x) = \begin{cases} x, & 0 < x < \dfrac{1}{2} \\ 0, & \dfrac{1}{2} < x < 1 \end{cases};$ Fourier sine series

28. $f(x) = \begin{cases} x, & 0 < x < \dfrac{1}{2} \\ 0, & \dfrac{1}{2} < x < 1 \end{cases};$ Fourier cosine series

REVIEW EXERCISES

In Exercises 1 through 6, compute the Fourier series of the function.

1. $f(x) = -x^2,\ -3 < x < 3$

2. $f(x) = x,\ -l \le x \le l;\ f(x+2l) = f(x)$

3. $f(x) = \cos\dfrac{x}{3},\ -\pi < x < \pi;\ f(x+2\pi) = f(x)$

4. $f(x) = x^3,\ -l \le x \le l;\ f(x+2l) = f(x)$

5. $f(x) = 1 + e^x,\ -1 \le x < 1$

6. $f(x) = x + |x|,\ -\pi < x \le \pi;\ f(x + 2\pi) = f(x)$

7. The pressure in a fluid passage[2] has been found to vary, as shown in the figure. In other words, there are regular pulses consisting of a sudden surge followed by an exponential decay occurring at the rate of 100/sec. Find the Fourier series for this periodic wave of pressure versus time.

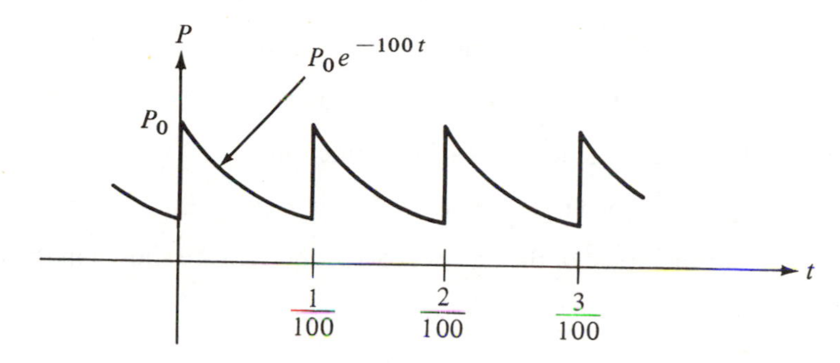

8. An underdamped harmonic oscillator[3] is subject to a force given by

$$F(t) = \begin{cases} 0, & -\frac{\tau}{2} \le t < 0 \\ F_0, & 0 < t \le \frac{\tau}{2}. \end{cases}$$

Solve for the motion, using the method of Fourier series. [*Hint:* Assume a particular solution of the form

$$y_p(t) = \frac{1}{2}a_0 + \sum_{n=1}^{\infty}\left(a_n \cos\frac{2n\pi t}{\tau} + b_n \sin\frac{2n\pi t}{\tau}\right).\Bigg]$$

In Exercises 9 through 12, find the Fourier series of the given function. Sketch for a few periods the graph of the function to which the Fourier series converges.

9. $f(x) = \begin{cases} 1, & 0 \le x < \pi \\ -1, & \pi \le x < 2\pi \end{cases};\ f(x + 2\pi) = f(x)$

10. $f(x) = 1 + |x|,\ -1 \le x < 1;\ f(x + 2) = f(x)$

11. $f(x) = 2\cos^2\frac{x}{2},\ -\pi \le x \le \pi$

12. $f(x) = \sin 3x,\ 0 \le x \le \frac{2\pi}{3};\ f\left(x + \frac{2\pi}{3}\right) = f(x)$

[2]This is Exercise 8 in J. L. Shearer, A. T. Murphy, and H. H. Richardson, *Introduction to Systems Dynamics* (Reading, Mass.: Addison-Wesley Publishing Co., 1971), p. 279. Reprinted with permission of Addison-Wesley Publishing Company.

[3]This is Exercise 3-3.8 in J. Norwood, Jr., *Intermediate Classical Mechanics* (Englewood Cliffs, N.J.: Prentice-Hall, 1979), p. 102. Reprinted by permission of Prentice-Hall, Inc., Englewood Cliffs, New Jersey.

13. Show that, for $-1 \leq x < 1$,

$$\sum_{n=1}^{\infty} \frac{(-1)^n - 1}{n^2} \cos n\pi x = \frac{(2|x| - 1)\pi^2}{4}.$$

14. Use the result in Exercise 13 to prove that

$$\frac{1}{1^2} + \frac{1}{3^2} + \frac{1}{5^2} + \cdots = \frac{\pi^2}{8}.$$

15. Compute the Fourier cosine series of the function

$$f(x) = 1 - x, \qquad 0 < x < \pi.$$

16. Compute the Fourier sine series of the function

$$f\mathrm{x}(x) = 1 + x, \qquad 0 < x < \pi.$$

17. Find the Fourier series[4] of the half-wave–rectified cosine shown in the figure.

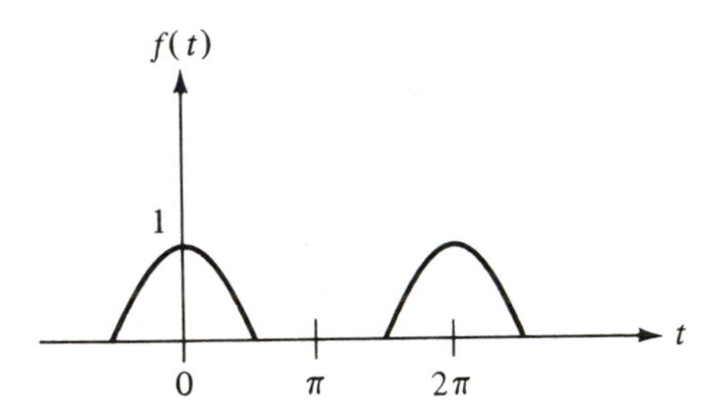

18. Find the Fourier series expansion[5] of the pulse train shown in the figure. Each pulse has a height H, a duration of W seconds, and a repetition period of T seconds.

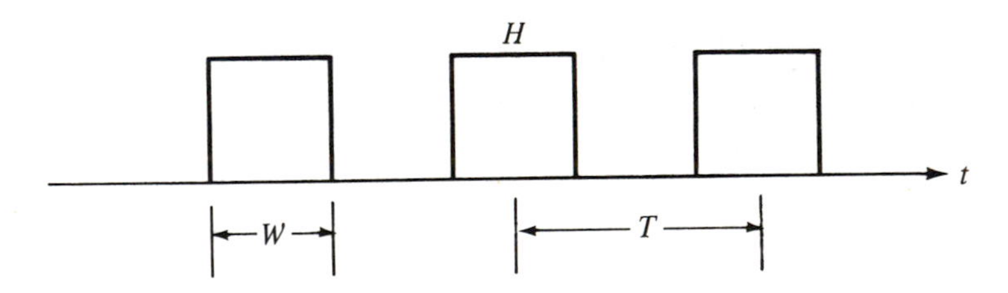

[4]This is Exercise c6.4 in R. B. Kerr, *Electrical Network Science* (Englewood Cliffs, N.J.: Prentice-Hall, 1977), p. 335. Reprinted by permission of Prentice-Hall, Inc., Englewood Cliffs, New Jersey.
[5]This is Example 6.4 in ibid., p. 195. Reprinted by permission of Prentice-Hall, Inc., Englewood Cliffs, New Jersey.

CHAPTER 11

An Introduction To Partial Differential Equations

11.1 INTRODUCTION

Partial differential equations are equations that involve partial derivatives of an unknown function. In the case of ordinary differential equations, the unknown function depends on a single independent variable. In contrast, for partial differential equations the unknown function depends on two or more independent variables. In this chapter we present an elementary treatment of partial differential equations. The general theory of partial differential equations is beyond the scope of this book, and we make no attempt to develop it here. Our treatment will focus on some simple cases of first-order equations and some special cases of second-order equations that occur frequently in applications. In this latter category we concentrate on the classical equations of mathematical physics—the heat equation, the potential (Laplace) equation, and the wave equation. The reader interested in the more practical aspects of this subject should concentrate on Sections 11.3, 11.4, and 11.6 through 11.10.

At first glance, a partial differential equation seems to differ from an ordinary differential equation only in that there are more independent variables. However, although the study of partial differential equations frequently utilizes known facts about ordinary differential equations, in general it is quite different. For instance, one of the simplest types of ordinary differential equation is the first-order linear equation (see Section 1.4). The general solution of this ordinary differential equation contains one arbitrary constant; hence this general solution can be interpreted geometrically as a set of two-dimensional curves, each coming about by assigning a different value to the arbitrary constant. Uniqueness of solution (in other words, isolating a specific curve in the set of two-dimensional curves) is accomplished by specifying a two-dimensional point (an initial value $y = y_0$ when $x = x_0$) that the general solution must contain. That is to say, the initial value fixes the value of the arbitrary constant in the general solution. If one considers a first-order linear partial differential equation (to be defined in Section 11.2), the results are very different. Let us assume that the independent variables are x and y and that the unknown function is u. The general solution can be interpreted geometrically as a collection of three-dimensional surfaces. Unfortunately, no *simple* condition serves to isolate (uniquely determine) a specific surface from this collection. Example 1 illustrates this fact.

EXAMPLE 1 Show that the following functions are solutions of the differential equation

$$\frac{\partial u}{\partial x} + \frac{\partial u}{\partial y} = 0.$$

(a) $u = f(x - y)$, where f is *any* function having a continuous derivative; and

(b) $u = (x - y)^n$, where n is any positive integer.

Solution (a) $u = f(x - y) \Rightarrow \dfrac{\partial u}{\partial x} = [f'(x - y)](1),$

$$\frac{\partial u}{\partial y} = [f'(x - y)](-1).$$

Thus,

$$\frac{\partial u}{\partial x} + \frac{\partial u}{\partial y} = f'(x - y) - f'(x - y) = 0.$$

(b) $$u = (x - y)^n \Rightarrow \frac{\partial u}{\partial x} = n(x - y)^{n-1}(1),$$

$$\frac{\partial u}{\partial y} = n(x - y)^{n-1}(-1).$$

Thus,

$$\frac{\partial u}{\partial x} + \frac{\partial u}{\partial y} = n(x - y)^{n-1} - n(x - y)^{n-1} = 0.$$

The solution in part (a) demonstrates that we cannot expect to obtain uniqueness of solution by specifying a single point that the general solution must contain. One point cannot uniquely determine the arbitrary function f. More specifically, suppose that $f(x - y) \equiv c(x - y)$, where c is an arbitrary constant. If we specify that the solution surface contain the origin (in other words, $u = 0$ when $x = y = 0$), there is an infinity of solution surfaces (corresponding to the values of c) that contain this point. The solution in part (b), on the other hand, demonstrates that we cannot expect (in general) to obtain uniqueness of solution by specifying a particular curve that the general solution must contain. In particular, if it is specified that the solution surface contain the curve $y = x$ in the xy-plane (in other words, $u = 0$ when $y = x$), then there is an infinity of solution surfaces (corresponding to the values of n) which contain this curve.

Example 1 illustrates that, for partial differential equations, uniqueness of solutions is not (in general) accomplished by simply specifying a point or a curve that the general solution must contain. Appropriate conditions that produce uniqueness of solutions usually depend on the form of the partial differential equation. We will not attempt to investigate all of the ramifications of this issue, but rather we will discuss particular situations.

In spite of the above differences, there are some similarities and analogies between ordinary differential equations and partial differential equations. In Section 11.2 we will discuss a few of them.

EXERCISES

1. Show that (a) $u = f(x + y)$, where f is any function possessing a continuous derivative, and (b) $u = (x + y)^n$, where n is a positive integer are solutions of the partial differential equation $\dfrac{\partial u}{\partial x} - \dfrac{\partial u}{\partial y} = 0$. Give an argument to support the statement that there is an infinity of solutions which contain the point (0, 0, 0).

2. Show that $u = f(2x - y)$, where f is any function possessing a continuous derivative, is a solution of the partial differential equation $\dfrac{\partial u}{\partial x} + 2\dfrac{\partial u}{\partial y} = 0$.

3. If g is any function such that $g(0) = 0$, and if g possesses a continuous derivative, show that $u = g(3x + 2y)$ is a solution of the partial differential equation $2\dfrac{\partial u}{\partial x} - 3\dfrac{\partial u}{\partial y} = 0$. Give an argument to support the statement that there is an infinity of solutions which contain the curve $y = -\frac{3}{2}x$ in the xy-plane.

11.2 DEFINITIONS AND GENERAL COMMENTS

To simplify the notation, we restrict our attention to the case of one unknown function, denoted by u, and no more than three independent variables, denoted by x, y, z. With these constraints we can define a partial differential equation to be a relation of the form

$$F(x, y, z, u, u_x, u_y, u_z, u_{xx}, u_{xy}, u_{xz}, u_{yy}, u_{yz}, u_{zz}, u_{xxx}, \ldots) = 0. \tag{1}$$

In Eq. (1) we have used the subscript notation for partial differentiation, that is,

$$u_z = \frac{\partial u}{\partial z}, \; u_{xy} = \frac{\partial^2 u}{\partial x \partial y}, \; u_{xxx} = \frac{\partial^3 u}{\partial x^3}, \text{ and so on.}$$

We always assume that the unknown function u is "sufficiently well behaved" so that all necessary partial derivatives exist and corresponding mixed partial derivatives are equal; for example,

$$u_{xy} = u_{yx};\; u_{xzx} = u_{xxz} = u_{zxx};\text{ and so on.}$$

Just as in the case of an ordinary differential equation, we define the *order* of the partial differential equation (1) to be the order of the partial derivative of highest order appearing in the equation. Furthermore, we define the partial differential equation (1) to be *linear* if F is linear as a function of the variables $u, u_x, u_y, u_z, u_{xx}, \ldots$; that is, F is a linear combination of the unknown function and its derivatives. Equation (1) is said to be *quasilinear* if F is linear as a function of the highest-order derivatives.

The following are examples of partial differential equations:

$$u_x + u_y = 3u_z - 2x^2 - 3z \tag{2}$$

$$u_{xx} = x^2 \tag{3}$$

$$5xyu_{xy} - 3zu_y + 2u = 0, \tag{4}$$

$$7u_x + 8u_y + 3u = 2xe^y, \tag{5}$$

$$u_{xz} + uu_y - 4z = 0, \tag{6}$$

$$5u_{xx} - 8u_{xy} + 9u_{yy} + 4u_x - 3u_y + 2u = 0. \tag{7}$$

Equations (2) and (5) are first order and the rest are second order. Equation (6) is quasilinear, and the rest are linear.

With very few exceptions, we will limit our discussion to linear partial differential equations of order one or two. Thus our most general partial differential equation can be written in the form

$$\begin{aligned} &a_1(x, y, z)u_{xx} + a_2(x, y, z)u_{xy} + a_3(x, y, z)u_{xz} \\ &+ a_4(x, y, z)u_{yy} + a_5(x, y, z)u_{yz} + a_6(x, y, z)u_{zz} \\ &+ a_7(x, y, z)u_x + a_8(x, y, z)u_y + a_9(x, y, z)u_z \\ &+ a_{10}(x, y, z)u = f(x, y, z). \end{aligned} \tag{8}$$

In Eq. (8) it is understood that the function f and the coefficients a_i are known, and u is unknown. By a *solution* of Eq. (8) we mean a continuous function u, of the independent variables x, y, z, with continuous first- and second-order partial derivatives, which, when substituted in (8), reduces Eq. (8) to an identity. Thus, in Example 1 of Section 11.1, we demonstrated by direct substitution that $u = f(x - y)$ was a solution of $u_x + u_y = 0$.

If $f(x, y, z) \equiv 0$, the partial differential equation (8) is called *homogeneous*; otherwise it is called *nonhomogeneous* (or inhomogeneous). Note that Eqs. (4) and (7) are homogeneous, and Eqs. (2), (3), and (5) are nonhomogeneous. Equation (6) is nonlinear because it is not of the form of Eq. (8).

If every one of the coefficients a_i is a constant, Eq. (8) is called a partial differential equation with *constant coefficients.* If at least one of the a_i is not a constant, Eq. (8) is called a partial differential equation with *variable coefficients.* Equations (2), (3), (5), and (7) have constant coefficients, and Eq. (4) has variable coefficients.

EXAMPLE 1 Find a solution $u = u(x, y)$ of the partial differential equation

$$u_x = x + y. \tag{9}$$

Solution We begin by integrating Eq. (9) "partially with respect to x" (in other words, we integrate with respect to x, treating the variable y as if it were a constant) to obtain

$$u = \tfrac{1}{2}x^2 + xy + c. \tag{10}$$

We note that the "constant of integration" is denoted by c. In order to verify that u, as given in Eq. (10), is a solution of Eq. (9), we need only substitute this expression for u in Eq. (9). When verifying this, notice that even if c were not a constant but a function of the variable y, such as $f(y)$, then u as given by Eq. (10), with c replaced by $f(y)$, would still be a solution of the partial differential equation (9), since $\dfrac{\partial f(y)}{\partial x} = 0$. We conclude that Eq. (10) is not the most general result possible unless we emphasize that c is to be replaced by an arbitrary function of y. Since we seek the most general form possible for the solution, we write

$$u = \tfrac{1}{2}x^2 + xy + f(y), \tag{11}$$

where f is an arbitrary function of y.

This example illustrates another strong contrast between partial differential equations and ordinary differential equations in that solution (11) contains an *arbitrary function* rather than an arbitrary constant.

EXAMPLE 2 Find a solution $u = u(x, y, z)$ of the partial differential equation

$$u_{xy} = z + x. \tag{12}$$

Solution First we integrate partially with respect to y (treating x and z as constants) to obtain

$$u_x = yz + xy + f_1(x, z),$$

where f_1 is an arbitrary function of the variables x and z. Next we integrate partially with respect to x (treating y and z as constants) to obtain

$$u = xyz + \tfrac{1}{2}x^2y + \int^x f_1(s, z)ds + f_2(y, z),$$

where f_2 is an arbitrary function of the variables y and z. If we set

$$f(x, z) \equiv \int^x f_1(s, z)ds, \qquad g(y, z) \equiv f_2(y, z),$$

then our solution takes the form

$$u = xyz + \tfrac{1}{2}x^2y + f(x, z) + g(y, z), \tag{13}$$

where f is an arbitrary function of x and z, and g is an arbitrary function of y and z. f and g are to have continuous first and second partial derivatives with respect to their arguments.

As in the case of ordinary differential equations, we call the solutions (11) and (13) the *general solution* of Eqs. (9) and (12) respectively. Each specific assignment of the arbitrary function(s) in the general solution gives rise to a *particular solution* of the corresponding partial differential equation. Thus,

$$u = \tfrac{1}{2}x^2 + xy + e^y$$

is a particular solution of Eq. (9) [for the particular choice $f(y) = e^y$], and

$$u = xyz + \tfrac{1}{2}x^2y + z\cos x + ye^z$$

is a particular solution of Eq. (12) [for the particular choices $f(x, z) = z\cos x$ and $g(y, z) = ye^z$].

Even though it is difficult (if not impossible) to make all-inclusive general statements about partial differential equations, the following is a reasonable claim for the partial differential equations to be treated in this text.

The general solution to a linear partial differential equation of order n, for an unknown function depending on s independent variables, involves n arbitrary functions, each of which depends on $(s - 1)$ *variables.* (Not necessarily the same set of $(s - 1)$ variables applies to each arbitrary function.)

Thus, in Example 1, $n = 1$, $s = 2$, and the general solution contains the arbitrary function f, which depends on the single variable y. In Example 2, $n = 2$, $s = 3$, and the general solution contains the two arbitrary functions f and g, f depending on the variables x and z, and g depending on the variables y and z. In Example 1 of Section 11.1, $n = 1$, $s = 2$, and the general solution [given as solution (a)] contains the arbitrary function f, which depends on the *single* variable $x - y$.

EXERCISES

For each of the linear partial differential equations in Exercises 1 through 10, state its order, determine whether it is homogeneous or nonhomogeneous, and determine whether it has constant or variable coefficients.

1. $u_{xyz} - 3u_{yy} + 2zu = 3u_x$

2. $u_y + u_{xx} - u_{yy} = 3x^2 - y^2$

3. $u_{xxyz} + 17u = 0$

4. $u_z + u_{zz} - 5u = 0$

5. $3u_{xx} - 5u_{xy} + 4u_{yy} = 0$

6. $u_x - u_y = 3u$

7. $u_y - 3zu + u_z - u_y = 27y + z^2$

8. $u_{xxyyz} + 5xu_y = 0$

9. $u_z - u_x + u_y = \cos z$

10. $u_y - 4u_{xx} = 0$

In Exercises 11 through 20, assume that u is a function of the two independent variables x and y. Integrate each equation to obtain the general solution.

11. $u_y = 0$

12. $u_x = 0$

13. $u_x = 3x^2 + 4y$

14. $u_y = \sin x - \sin y$

15. $u_{xy} = 0$

16. $u_{xx} = 0$

17. $u_{yy} = 0$

18. $u_{yy} = \cos y + e^x$

19. $u_{xy} = \cos y + e^x$

20. $u_x = \sec^2 y$

In Exercises 21 through 30, assume that u is a function of the three independent variables x, y, and z. Integrate each equation to obtain the general solution.

21. $u_{zx} = y$

22. $u_{yz} = 0$

23. $u_{xxy} = 0$

24. $u_{yyz} = \sec^2 x$

25. $u_{xxyz} = 0$

26. $u_x = x^2 + z$

27. $u_z = y$

28. $u_{yxz} = 2$

29. $u_{zz} = y + 3x$

30. $u_{xx} = \sec^2 x$

31. Verify that the general solution of Exercise 21 conforms to the claim about the form of the general solution made at the end of this section.

32. Verify that the general solution of Exercise 23 conforms to the claim about the form of the general solution made at the end of this section.

33. Verify that the general solution of Exercise 25 conforms to the claim about the form of the general solution made at the end of this section.

34. **Elasticity.** The shearing stress and the normal stress in an elastic body are obtainable from *Airy's stress function,* ϕ, where ϕ is a solution of the partial differential equation

$$\phi_{xxxx} + 2\phi_{xxyy} + \phi_{yyyy} = 0.$$

Classify this equation utilizing all definitions of this section that are appropriate.

11.3 THE PRINCIPLE OF SUPERPOSITION

In this section and the next two sections we outline some general ideas and methods of solution. In some instances the methods will be further illustrated in subsequent sections.

Our partial differential equation has the appearance of Eq. (8), Section 11.2. For simplicity, we write Eq. (8) in the abbreviated form

$$A[u] = f. \tag{1}$$

We speak of the symbol A as an *operator,* and the manner in which the operator A "operates" on the function u (denoted by $A[u]$) is defined by the left-hand side of Eq. (8). Thus, for the operator of Eq. (8) we write,

$$A[u] \equiv a_1(x, y, z)u_{xx} + \cdots + a_9(x, y, z)u_z + a_{10}(x, y, z)u.$$

DEFINITION 1

An operator A is called a linear operator if $A[c_1u_1 + c_2u_2] = c_1A[u_1] + c_2A[u_2]$ for every choice of the constants c_1, c_2 and for every permissable (in other words, $A[u_1]$, $A[u_2]$, and $A[c_1u_1 + c_2u_2]$ make sense) choice of functions u_1 and u_2.

DEFINITION 2

Given the nonhomogeneous partial differential equation (1), *the equation* $A[u] = 0$ *is called the associated homogeneous equation.*

A very important property of linear partial differential equations is contained in the following principle.

Principle of Superposition *Let $f_1, f_2, \ldots, f_m$ be any functions and let $c_1, c_2, \ldots, c_m$ be any constants. If A is a linear operator and if $u_1, u_2, \ldots, u_m$ are, respectively, solutions of the equations $A[u_1] = f_1$, $A[u_2] = f_2, \ldots, A[u_m] = f_m$, then $u = c_1u_1 + c_2u_2 + \cdots + c_mu_m$ is a solution of the equation*

$$A[u] = c_1f_1 + c_2f_2 + \cdots + c_mf_m.$$

Proof Using the linearity of A we have

$$\begin{aligned} A[u] &= A[c_1u_1 + c_2u_2 + \cdots + c_mu_m] \\ &= c_1A[u_1] + c_2A[u_2] + \cdots + c_mA[u_m] \\ &= c_1f_1 + c_2f_2 + \cdots + c_mf_m, \end{aligned}$$

and the proof is complete.

Two important consequences of the principle of superposition are as follows: (i) If $u_1, u_2, \ldots, u_m$ are solutions of $A[u] = 0$ and $c_1, c_2, \ldots, c_m$ are any constants, then $\sum_{i=1}^{m} c_iu_i$ is also a solution of $A[u] = 0$; that is, any linear combination of solutions is a solution. (ii) If u_h is a solution of $A[u] = 0$ and if u_p is a particular solution of $A[u] = f$, then $u = u_h + u_p$ is a solution of $A[u] = f$; that is, the sum of a solution of the homogeneous equation and a particular solution is also a solution. As in the case of ordinary differential equations, if u_h is the general solution of the homogeneous equation $A[u] = 0$, and u_p is a particular solution of $A[u] = f$, then $u = u_h + u_p$ is the general solution of $A[u] = f$.

These two consequences are similar to the manner in which we generated the general solution for a nonhomogeneous linear ordinary differential equation (see Chapter 2). The same approach is sometimes useful in trying to solve a nonhomogeneous linear partial differential equation, but the methods of solution are not as concrete or systematic for partial differential equations as for ordinary differential equations.

EXAMPLE 1 Suppose that u is a function depending on the variables, x, y, and z, and that u satisfies the partial differential equation

$$u_{xy} = z + x. \tag{2}$$

Show that (i) the general solution of the associated homogeneous equation $u_{xy} = 0$ is given by $u_h = f(x, z) + g(y, z)$, where f and g are arbitrary functions; (ii) $u_1 = xyz$ is a particular solution of the equation $u_{xy} = f_1$, where

$f_1(x, y, z) = z$; (iii) $u_2 = \frac{1}{2}x^2y$ is a solution of $u_{xy} = f_2$, where $f_2(x, y, z) = x$; and (iv) $u = u_h + u_1 + u_2$ is the general solution of (2).

Solution

$$\text{(i)}\ u_h = f(x, z) + g(y, z) \Rightarrow (u_h)_x = \frac{\partial f(x, z)}{\partial x} \equiv h(x, z)$$
$$\Rightarrow (u_h)_{xy} = 0,$$

(ii) $u_1 = xyz \Rightarrow (u_1)_x = yz \Rightarrow (u_1)_{xy} = z$,

(iii) $u_2 = \frac{1}{2}x^2y \Rightarrow (u_2)_x = xy \Rightarrow (u_2)_{xy} = x$.

(iv) Note that Eq. (2) is of the form $A[u] = f_1 + f_2$,

where $A[u] = u_{xy} \equiv \dfrac{\partial^2 u}{\partial x \partial y}$. Now

$$\begin{aligned} A[c_1u_1 + c_2u_2] &= \frac{\partial^2}{\partial x \partial y}[c_1u_1 + c_2u_2] \\ &= \frac{\partial^2}{\partial x \partial y}[c_1u_1] + \frac{\partial^2}{\partial x \partial y}[c_2u_2] \\ &= c_1\frac{\partial^2}{\partial x \partial y}[u_1] + c_2\frac{\partial^2}{\partial x \partial y}[u_2] \\ &= c_1A[u_1] + c_2A[u_2]. \end{aligned}$$

Thus A is a linear operator. By the principle of superposition (with $c_1 = c_2 = 1$), we have that

$$u_p = u_1 + u_2 = xyz + \frac{1}{2}x^2y$$

is a particular solution of Eq. (2). Therefore, the general solution of (2) (compare with Example 2, Section 11.2) is given by

$$u = f(x,z) + g(y,z) + xyz + \frac{1}{2}x^2y. \tag{3}$$

EXERCISES

In Exercises 1 through 13, show that the operator A is linear.

1. $A[u] = u_{xx}$

2. $A[u] = u_y$

3. $A[u] = u_{yy}$

4. $A[u] = u_x$

5. $A[u] = u_{xy}$

6. $A[u] = u$

7. $A[u] = u_{xx} + u_{yy}$

8. $A[u] = u_y - cu_{xx}$, c a constant

9. $A[u] = u_{yy} - c^2u_{xx}$, c a constant

10. $A[u] = 3u_x + 4u_y - 7u$

11. $A[u] = u_{xx} + 5u_{xy} - 2u_{yy}$

12. $A[u] = au_{xx} + 2bu_{xy} + cu_{yy}$, a, b, c constants

13. $A[u] = au_{xx} + 2bu_{xy} + cu_{yy} + du_x + eu_y + fu$, a, b, c, d, e, f constants

In Exercises 14 through 23, assume that u is a function of the two independent variables x and y. Verify that the given u_h is the general solution of the associated homogeneous equation (f, g, and k denote arbitrary functions of sufficient differentiability for the solution to be meaningful).

14. $u_x = y$; $u_h = f(y)$

15. $u_y = xy$; $u_h = f(x)$

16. $u_{xy} = yx$; $u_h = f(x) + g(y)$

17. $u_{yy} = 2 \cos x$; $u_h = yf(x) + g(x)$

18. $u_{xx} = 0$; $u_h = xf(y) + g(y)$

19. $u_{xxy} = x$; $u_h = f(x) + xg(y) + k(y)$

20. $u_{xyy} = xy$; $u_h = yf(x) + g(x) + k(y)$

21. $u_x = y + x^2$; $u_h = f(y)$

22. $u_y = \sin x - e^y$; $u_h = f(x)$

23. $u_{xx} = 2x - 3y$; $u_h = xf(y) + g(y)$

24. Show that $u_p = xy$ is a particular solution of $u_x = y$. Write the general solution for this equation.

25. Show that $u_p = \frac{1}{2}xy^2$ is a particular solution of $u_y = xy$. Write the general solution for this equation.

26. Show that $u_p = \frac{1}{4}x^2y^2$ is a particular solution of $u_{xy} = yx$. Write the general solution of this equation.

27. Show that $u_p = 0$ is a particular solution of $u_{xx} = 0$. Write the general solution of this equation.

28. Show that $u_p = \frac{1}{6}x^3y$ is a particular solution of $u_{xxy} = x$. Write the general solution of this equation.

29. Show that $u_p = \frac{1}{12}x^2y^3$ is a particular solution of $u_{xyy} = xy$. Write the general solution of this equation.

30. Show that $u_p = xy$ is a particular solution of $u_x = y$, and $u_p = \frac{1}{3}x^3$ is a particular solution of $u_x = x^2$. Write the general solution of $u_x = y + x^2$.

31. Show that $u_p = y \sin x$ is a particular solution of $u_y = \sin x$, and $u_p = -e^y$ is a particular solution of $u_y = -e^y$. Write the general solution of $u_y = \sin x - e^y$.

32. Show that $u_p = \frac{1}{3}x^3$ is a particular solution of $u_{xx} = 2x$, and $u_p = -\frac{3}{2}x^2y$ is a particular solution of $u_{xx} = -3y$. Write the general solution of $u_{xx} = 2x - 3y$.

Linear Homogeneous Partial Differential Equations with Constant Coefficients In Section 2.4 we saw that for linear homogeneous ordinary differential equations with constant coefficients, the trial solution $e^{\lambda x}$, λ a constant, led to the requirement that λ be a root of an algebraic polynomial, namely the characteristic polynomial. If we have a linear homogeneous partial differential equation which has x and y as independent variables, we are motivated to assume a trial solution of the form $e^{\lambda x+\mu y}$, where λ and μ are constants to be determined. Thus we substitute $u = e^{\lambda x+\mu y}$ into the equation

$$a_1u_{xx} + a_2u_{xy} + a_3u_{yy} + a_4u_x + a_5u_y + a_6u = 0 \tag{4}$$

to obtain

$$e^{\lambda x+\mu y}[a_1\lambda^2 + a_2\lambda\mu + a_3\mu^2 + a_4\lambda + a_5\mu + a_6] = 0.$$

Consequently $e^{\lambda x+\mu y}$ will be a solution of Eq. (4) if and only if λ, μ satisfy

$$a_1\lambda^2 + a_2\lambda\mu + a_3\mu^2 + a_4\lambda + a_5\mu + a_6 = 0. \tag{5}$$

Note that Eq. (5) is a single algebraic equation[1] in two unknowns, and in general there will be an infinity of solutions (λ, μ).[2] In particular, if (λ_1, μ_1) is a specific solution of Eq. (5), then $e^{\lambda_1 x+\mu_1 y}$ is a particular solution of Eq. (4). By the principle of superposition, it follows that if $(\lambda_1, \mu_1), (\lambda_2, \mu_2), \ldots, (\lambda_n, \mu_n)$ are n different solutions of Eq. (5), and if $c_1, c_2, \ldots, c_n$ are arbitrary constants, then $\sum_{i=1}^{n} c_i e^{\lambda_i x+\mu_i y}$ is a solution of Eq. (4).

In Exercises 33 through 48, find a particular solution in the form $u = e^{\lambda x+\mu y}$.

33. $3u_x + u_y + u = 0$

34. $u_x - 2u_y + 5u = 0$

35. $2u_x + 3u_y - 8u = 0$

36. $u_x + u_y - u = 0$

37. $5u_x - 3u_y - 2u = 0$

38. $u_x - u_y + 5u = 0$

39. $u_{xx} + 2u_{xy} + u_{yy} + 3u_x + 3u_y + 2u = 0$

40. $u_{xx} - 2u_{xy} + u_{yy} - 2u_x + 2u_y - 3u = 0$

41. $u_{xx} - 2u_{xy} + u_{yy} + 5u_x - 5u_y + 4u = 0$

42. $u_{xx} + 2u_{xy} + u_{yy} - 5u_x - 5u_y + 6u = 0$

43. $4u_{xx} - 4u_{xy} + u_{yy} + 4u_x - 2u_y - 3u = 0$

44. $u_{xx} + 4u_{xy} + 4u_{yy} + 6u_x + 12u_y + 8u = 0$

[1]We recognize Eq. (5) as representing a conic section in the $\lambda\mu$-plane. Thus the graph is either a hyperbola, a parabola, an ellipse, or a degeneracy of one of these curves.
[2]For motivation see Footnote 1.

45. $4u_{xx} + 12u_{xy} + 9u_{yy} - 2u_x - 3u_y - 6u = 0$

46. $u_{xx} - 10u_{xy} + 25u_{yy} - 3u_x + 15u_y - 10u = 0$

47. $u_{xx} - u_{xy} + u_{yy} - u_x + 2u_y - 2u = 0$

48. $2u_{xx} - u_{xy} - 2u_{yy} - u_x + 2u_y + 2u = 0$

49. To find particular solutions of linear partial differential equations, one can sometimes use an approach similar to that of undetermined coefficients for ordinary differential equations. Thus, if the partial differential equation is of the form

$$A[u] = \sum_{i=1}^{n} f_i,$$

with each f_i a "simple" function, then for each f_i we can find a particular solution by undetermined coefficients and then use the principle of superposition to obtain a particular solution. Find a particular solution for each of the following partial differential equations.

(a) $u_x + u_y - u = 3x^2$ (b) $u_x + u_y + u = 2\cos y$

(c) $u_{xx} + u_{yy} + u_x - u_y + u = 2x^2 - 3y^2$

(d) $u_{xx} - 3u_{xy} + 5u_{yy} = 10e^{3x+4y}$ [*Hint:* Set $u_p = Ae^{3x+4y}$.]

50. Laplacian The operator Δ defined by

$$\Delta[u] = u_{xx} + u_{yy} + u_{zz}$$

is referred to as the three-dimensional *Laplacian operator*. If the term u_{zz} is removed, the resulting operator is called two-dimensional, and if u_{zz} and u_{yy} are removed, the resulting operator is called one-dimensional.

(a) Show that the one-, two-, and three-dimensional Laplacian operators are linear operators.

(b) If we change variables from rectangular coordinates to polar coordinates by the formulas $x = r\cos\theta$, $y = r\sin\theta$, show that the two-dimensional Laplacian operator has the form

$$\Delta[u] = \frac{1}{r}[(ru_r)_r + u_{\theta\theta}/r].$$

[*Hint:* Make repeated use of the chain rule; for example, $u_r = u_x x_r + u_y y_r$.]

(c) If we change variables from rectangular coordinates to spherical coordinates by the formulas $x = r\sin\phi\cos\theta$, $y = r\sin\phi\sin\theta$, $z = r\cos\phi$, show that the three-dimensional Laplacian operator has the form

$$\Delta[u] = \frac{1}{r^2}(r^2u_r)_r + \frac{1}{r^2\sin\phi}(\sin\phi\, u_\phi)_\phi + \frac{1}{r^2\sin^2\phi}u_{\theta\theta}.$$

[*Hint:* Make repeated use of the chain rule; for example, $u_\phi = u_x x_\phi + u_y y_\phi + u_z z_\phi$.]

Remark The classical equations of mathematical physics are

$$u_{tt} = c^2\Delta[u] \qquad \text{(the wave equation);}$$

$$u_t = a\Delta[u] \qquad \text{(the heat equation);}$$

$$0 = \Delta[u] \qquad \text{(the potential or Laplace equation).}$$

In each case the equation is called one-, two-, or three-dimensional, depending on whether the Laplacian operator is one-, two-, or three-dimensional. The one-dimensional wave equation is discussed in Section 11.6; the one-dimensional heat equation is discussed in Section 11.7; and the two-dimensional Laplace equation is discussed in Section 11.8.

11.4 SEPARATION OF VARIABLES

A frequently used method for finding solutions to linear homogeneous partial differential equations is known as *separation of variables*. In this method we try to write the solution as a product of functions, each of which depends on exactly one of the independent variables. For example, we would try to write the solution of the partial differential equation

$$u_{xx} + u_{yy} + u_{zz} = 0$$

in the form $u = X(x)Y(y)Z(z)$, where the functions X, Y, and Z are to be determined. For the partial differential equation

$$u_{xx} - u_{yy} = 0,$$

we would try to write the solution in the form $u = X(x)Y(y)$, where the functions X, Y are to be determined. The basic ideas and manipulations involved in the method are illustrated in the following examples.

EXAMPLE 1 For the partial differential equation

$$u_{xx} - u_{yy} = 0, \tag{1}$$

find a solution in the form $u = X(x)Y(y)$.

Solution If

$$u = X(x)Y(y),$$

then

$$u_x = X'(x)Y(y),\ u_{xx} = X''(x)Y(y),\ u_y = X(x)Y'(y),\ u_{yy} = X(x)Y''(y).$$

Substitution of these results in Eq. (1) leads to

$$X''(x)Y(y) - X(x)Y''(y) = 0.$$

Divide this latter equation by $u = X(x)Y(y)$ (assuming that $u \neq 0$) to obtain

$$\frac{X''(x)}{X(x)} - \frac{Y''(y)}{Y(y)} = 0$$

or

$$\frac{X''(x)}{X(x)} = \frac{Y''(y)}{Y(y)}. \tag{2}$$

Since $\frac{X''(x)}{X(x)}$ does not contain the variable y, we note that changes in y will not have any effect on the expression $\frac{X''(x)}{X(x)}$. Thus, if (2) is to be an equality, it must happen that changes in the variable y do not affect the expression $\frac{Y''(y)}{Y(y)}$ either. Similarly, changes in x should not affect the expression $\frac{X''(x)}{X(x)}$. The net conclusion is that in order for (2) to be an equality, the expressions $\frac{X''(x)}{X(x)}$ and $\frac{Y''(y)}{Y(y)}$ must be constants. In fact, they must be the same constant. If the constant is denoted by λ, we can write

$$\frac{X''(x)}{X(x)} = \lambda$$

and

$$\frac{Y''(y)}{Y(y)} = \lambda.$$

Thus,

$$X''(x) - \lambda X(x) = 0 \tag{3}$$

and

$$Y''(y) - \lambda Y(y) = 0. \tag{4}$$

Equations (3) and (4) are ordinary differential equations with constant coefficients and can be solved by the methods of Section 2.5 to yield.

$$X(x) = \begin{cases} c_1 e^{\sqrt{\lambda}x} + c_2 e^{-\sqrt{\lambda}x}, & \lambda > 0 \\ c_1 + c_2 x, & \lambda = 0 \\ c_1 \cos\sqrt{-\lambda}x + c_2 \sin\sqrt{-\lambda}x, & \lambda < 0\,; \end{cases}$$

$$Y(y) = \begin{cases} c_3 e^{\sqrt{\lambda}y} + c_4 e^{-\sqrt{\lambda}y}, & \lambda > 0 \\ c_3 + c_4 y, & \lambda = 0 \\ c_3 \cos\sqrt{-\lambda}y + c_4 \sin\sqrt{-\lambda}y, & \lambda < 0\,. \end{cases}$$

Thus,

$$u = X(x)Y(y) = \begin{cases} (c_1e^{\sqrt{\lambda}x} + c_2e^{-\sqrt{\lambda}x})(c_3e^{\sqrt{\lambda}y} + c_4e^{-\sqrt{\lambda}y}), & \lambda > 0 \\ (c_1 + c_2x)(c_3 + c_4y), & \lambda = 0 \\ (c_1 \cos\sqrt{-\lambda}x + c_2 \sin\sqrt{-\lambda}x)\cdot \\ \quad (c_3 \cos\sqrt{-\lambda}y + c_4 \sin\sqrt{-\lambda}y), & \lambda < 0. \end{cases}$$

Without further information we have no way of knowing the value of λ; hence we cannot specify the form of the solution. In many practical problems there are other conditions that the solution must satisfy; these conditions usually dictate the value of λ and the form of the solution (see Sections 11.6–11.10).

EXAMPLE 2 For the partial differential equation

$$3u_x - 2u_y - 5u_z = 0,$$

find a solution in the form $u = X(x)Y(y)Z(z)$.

Solution If

$$u = X(x)Y(y)Z(z)$$

then

$$u_x = X'(x)Y(y)Z(z), \quad u_y = X(x)Y'(y)Z(z), \quad \text{and} \quad u_z = X(x)Y(y)Z'(z).$$

Substitution into the partial differential equation yields

$$3X'(x)Y(y)Z(z) - 2X(x)Y'(y)Z(z) - 5X(x)Y(y)Z'(z) = 0.$$

Dividing by $u = X(x)Y(y)Z(z)$ (assuming $u \neq 0$), we have

$$\frac{3X'(x)}{X(x)} - \frac{2Y'(y)}{Y(y)} - \frac{5Z'(z)}{Z(z)} = 0$$

or

$$\frac{3X'(x)}{X(x)} = \frac{2Y'(y)}{Y(y)} + \frac{5Z'(z)}{Z(z)}. \tag{5}$$

Using the same type of argument as in Example 1, we conclude that the only way that Eq. (5) can be an equality is that both sides of the equation equal a constant, say λ. Thus

$$\frac{3X'(x)}{X(x)} = \lambda \tag{6}$$

and

$$\frac{2Y'(y)}{Y(y)} + \frac{5Z'(z)}{Z(z)} = \lambda. \tag{7}$$

Equation (6) has as general solution $X(x) = c_1e^{(\lambda/3)x}$. Equation (7) can be rewritten as

$$\frac{2Y'(y)}{Y(y)} = \lambda - \frac{5Z'(z)}{Z(z)}. \tag{8}$$

Once more we argue that in order for (8) to be an equality, we must have

$$\frac{2Y'(y)}{Y(y)} = \mu, \tag{9}$$

and

$$\lambda - \frac{5Z'(z)}{Z(z)} = \mu, \tag{10}$$

where μ is a constant. The solution to Eq. (9) is $Y(y) = ce^{(\mu/2)y}$, and the solution to Eq. (10) is $Z(z) = c_3e^{[(\lambda-\mu)/5]z}$. Therefore a solution to the partial differential is

$$u = (c_1e^{(\lambda/3)x})(c_2e^{(\mu/2)y})(c_3e^{[(\lambda-\mu)/5]z}) = ke^{(\lambda/3)x+(\mu/2)y+[(\lambda-\mu)/5]z}.$$

In Examples 1 and 2, the partial differential equations have constant coefficients, but the method can also be applied to equations with variable coefficients. It is not the case, however, that every linear homogeneous partial differential equation can be solved by the method of separation of variables. There are many equations for which the method does not apply. Example 3 is a case in point.

EXAMPLE 3 Show that the variables "do not separate" for the partial differential equation

$$u_{xy} + u_{xx} + u = 0.$$

Solution We try a solution in the form $u = X(x)Y(y)$. Then

$$u_x = X'(x)Y(y),\ u_{xy} = X'(x)Y'(y),\ u_{xx} = X''(x)Y(y).$$

Substitution of these results into the partial differential equation leads to

$$X'(x)Y'(y) + X''(x)Y(y) + X(x)Y(y) = 0.$$

It is not possible to algebraically manipulate this latter equation to a form $P(x) = Q(y)$, therefore we conclude that the method does not work for this partial differential equation; that is, the variables "do not separate."

EXERCISES

In Exercises 1 through 22, assume a solution in the form $u = X(x)Y(y)$. Show that the equation "separates," and find the differential equations that X and Y must satisfy.

1. $u_x - 3u_y = 0$

2. $4u_x + 3u_y = 0$

3. $2u_x + 5u_y = 0$

4. $7u_x - 6u_y = 0$

5. $u_x + u_y + u = 0$

6. $2u_x - 3u_y - u = 0$

7. $u_y - u_{xx} = 0$

8. $u_y - cu_{xx} = 0$, c a constant

9. $u_{yy} - u_{xx} = 0$

10. $u_{yy} - c^2u_{xx} = 0$, c a constant

11. $u_{xx} + u_{yy} + u_x + u_y = 0$

12. $u_{xx} + u_{yy} + u_x - u_y = 0$

13. $u_{xx} - u_{yy} - u_x + u_y = 0$

14. $u_{xx} - u_{yy} - u_x - u_y = 0$

15. $u_y - u_{xx} + u_x = 0$

16. $u_{xx} + u_{yy} + u_x + u_y + u = 0$

17. $3u_y - 5u_{xx} - 6u_x = 0$

18. $5u_{xx} - 6u_{yy} + u_x - 3u_y + u = 0$

19. $2u_{xx} + 3u_{yy} + u = 0$

20. $au_{xx} + cu_{yy} + du_x + eu_y = 0$, a, c, d, e, constants

21. $au_{xx} + cu_{yy} + du_x + eu_y + fu = 0$, a, c, d, e, f, constants

22. $a(x)u_{xx} + c(y)u_{yy} + d(x)u_x + e(y)u_y + fu = 0$, $a(x)$, $d(x)$ continuous functions of x; $c(y)$, $e(y)$ continuous functions of y, f a constant

In Exercises 23 through 32, assume a solution in the form $u = X(x)Y(y)Z(z)$. Show that the equation "separates" and find the differential equations that X, Y, and Z must satisfy.

23. $u_x - u_y + u_z = 0$

24. $2u_x + 3u_y + 4u_z = 0$

25. $u_x + 2u_y - 2u_z = 0$

26. $u_x + u_y + u_z + u = 0$

27. $u_{xx} + u_{yy} + u_{zz} = 0$

28. $u_y - u_{xx} - u_{zz} = 0$

29. $u_{yy} - u_{xx} - u_{zz} = 0$

30. $u_{yy} - 3u_{xx} - 3u_{zz} = 0$

31. $2u_{xx} - u_{yy} + u_{zz} + u = 0$

32. $u_{xx} - u_{yy} + u_z - u = 0$

In Exercises 33 through 39, assume a solution in the form $u = X(x)Y(y)$. Show that the equation does not "separate."

33. $(y + x)u_x + u_y = 0$

34. $u_{xy} + (2x + 3y)u_y = 0$

35. $u_x + f(x, y)u_y = 0$, where $\dfrac{\partial f}{\partial x} \neq 0$, $\dfrac{\partial f}{\partial y} \neq 0, f(x, y) \neq f_1(x) \cdot f_2(y)$

36. $u_{xx} + u_{xy} - 2u = 0$

37. $u_{xx} + xu_{yy} - u_y = 0$

38. $u_{yy} + xu_y - u = 0$

39. $u_{xx} - u_{yy} + 3xu_x - 3xu_y = 0$

40. Show that if f is an arbitrary twice differentiable function, and λ is a constant, then $u = f(y + \lambda x)$ is a solution of

$$a_1u_{xx} + a_2u_{xy} + a_3u_{yy} = 0, \tag{11}$$

if and only if λ is a solution of

$$a_1\lambda^2 + a_2\lambda + a_3 = 0. \tag{12}$$

Thus, if λ_1 and λ_2 ($\lambda_1 \neq \lambda_2$) are the roots of Eq. (12), then the general solution of Eq. (11) is given by $u = f_1(y + \lambda_1 x) + f_2(y + \lambda_2 x)$, where f_1 and f_2 are arbitrary twice differentiable functions. If $\lambda_1 = \lambda_2$, then the general solution is of the form

$$u = f_1(y + \lambda_1 x) + x f_2(y + \lambda_1 x),$$

which can be verified by direct substitution.

DEFINITION

The partial differential equation

$$au_{xx} + 2bu_{xy} + cu_{yy} + du_x + eu_y + fu = 0$$

is called hyperbolic *if* $b^2 - ac > 0$, parabolic *if* $b^2 - ac = 0$, elliptic *if* $b^2 - ac < 0$.[2]

In Exercises 41 through 60, classify the second-order partial differential equation as hyperbolic, parabolic, or elliptic. Find the general solution in each case. Refer to Exercise 40.

41. $u_{xx} + 6u_{xy} + 12u_{yy} = 0$

42. $u_{xx} + 20u_{xy} + 64u_{yy} = 0$

43. $5u_{xx} + 10u_{xy} + 20u_{yy} = 0$

44. $6u_{xx} + 4u_{xy} + u_{yy} = 0$

45. $u_{xx} + 9u_{xy} + 4u_{yy} = 0$

46. $u_{xx} + 2u_{xy} + u_{yy} = 0$

47. $u_{xx} + 5u_{xy} + u_{yy} = 0$

48. $4u_{xx} + 8u_{xy} + 4u_{yy} = 0$

49. $u_{xx} + 8u_{xy} + 16u_{yy} = 0$

50. $u_{xx} + u_{xy} + u_{yy} = 0$

51. $6u_{xx} - u_{xy} - u_{yy} = 0$

52. $36u_{xx} + 13u_{xy} + u_{yy} = 0$

53. $21u_{xx} - 10u_{xy} + u_{yy} = 0$

54. $u_{xx} + 2u_{xy} + 5u_{yy} = 0$

55. $u_{xx} + 4u_{xy} + 5u_{yy} = 0$

56. $4u_{xx} - 4u_{xy} + u_{yy} = 0$

57. $u_{xx} - 3u_{xy} - 4u_{yy} = 0$

58. $10u_{xx} + 7u_{xy} + u_{yy} = 0$

59. $u_{xx} + 6u_{xy} + 9u_{yy} = 0$

60. $u_{xx} - u_{yy} = 0$

61. Quantum Mechanics: Helmholtz's Equation In Section 2.8.1 we introduced the Schrödinger wave equation of quantum mechanics, namely

$$\frac{ih}{2\pi}\phi_t = -\frac{h^2}{8m\pi^2}(\phi_{xx} + \phi_{yy} + \phi_{zz}) + V(x, y, z)\phi.$$

(a) Set $\phi(x, y, z, t) = e^{-(i2\pi Et)/h}\, u(x, y, z)$, where E is a constant. Show that u satisfies

$$u_{xx} + u_{yy} + u_{zz} + \frac{8m\pi^2}{h}[E - V(x, y, z)]u = 0. \tag{13}$$

(b) If we set $V \equiv 0$ in Eq. (13), the resulting equation is called *Helmholtz's* equation. For Helmholtz's equation set $u = X(x)Y(y)Z(z)$ and determine the differential equations for X, Y, and Z.

[2]See Footnote 1, page 431.

62. Acoustics The nonlinear partial differential equation

$$(u_x)^{n+1}u_{tt} = a^2u_{xx}$$

occurs in the study of the propagation of sound in a medium. a is a constant and represents the velocity of sound in the medium. Set $u = X(x)T(t)$ and determine the differential equations for X and T.

63. Supersonic Fluid Flow In the study of the supersonic flow of an ideal compressible fluid past an obstacle, the velocity potential satisfies the equation

$$(M^2 - 1)u_{xx} - u_{yy} = 0,$$

where $M(>1)$ is a constant known as the *Mach number* of the flow. Set $u = X(x)Y(y)$ and determine the differential equations for X and Y.

64. Isentropic Fluid Flow The second-order linear partial differential equation

$$u_{xy} - \frac{\alpha}{x + y}(u_x + u_y) = 0$$

occurs in the one-dimensional isentropic flow of a compressible fluid.[3] α is a constant which depends on the fluid.

(a) Set $u = X(x)Y(y)$ and determine the differential equations for X and Y.

(b) Solve the differential equations of part (a).

65. Given

$$u_{tt} + a^2u = u_{xx},$$

where a^2 is a constant. Set $u = X(x)T(t)$ and determine the differential equations for X and T. Solve these differential equations.

66. Acoustics In the study of the transmission of sound through a moving fluid, one considers the velocity potential $u(x, y, z, t)$. (The term *potential* is usually used in physics to describe a quantity whose gradient furnishes a field of force, where gradient F is the vector $[F_x, F_y, F_z]$. In this case the gradient of u yields the velocity of the flow.) It can be shown that u satisfies the three-dimensional wave equation (see the remark following Exercise 50, Section 11.3)

$$u_{xx} + u_{yy} + u_{zz} = \frac{1}{c^2}u_{tt},$$

where the constant c represents the velocity of sound in the medium. Set $u = X(x)Y(y)Z(z)T(t)$ and determine the differential equations for X, Y, Z, and T.

67. Separation of variables for partial differential equations normally relates to the method described in this section. There are other techniques that may also be called separation of variables. To illustrate, let us assume a solution of the equation

[3]R. Courant and D. Hilbert, *Methods of Mathematical Physics*, vol. 2 (New York: Interscience Publishers, 1962), p. 459.

$$u_x + u_y = 0$$

in the form

$$u(x, y) = X(x) + Y(y). \tag{14}$$

X and Y satisfy the differential equations $X' = \lambda$ and $Y' = -\lambda$, respectively, therefore $u = \lambda(x - y) + c$. Find a solution of the form (14) for the partial differential equation

$$(u_x)^2 + (u_y)^2 = 1,$$

where u_x and u_y are assumed to be positive.

68. Magnetic Field Intensity in a Solenoid When a long copper rod is wound with a coil of wire and excited by a current, the following partial differential equation results[4] for the magnetic field intensity H:

$$rH_{rr} + H_r = \frac{4\pi r}{\rho} H_t,$$

where t is time, r is measured from the axis of the copper rod, and ρ is a constant known as the *resistivity* of copper. (a) Set $H = R(r)T(t)$ and determine the differential equations for R and T (call the separation constant $-\lambda$ instead of λ). (b) Find T. (c) Show that the differential equation for R is a special case of Bessel's equation (Section 5.1). [*Hint:* Make the change of variable $r = \dfrac{1}{\sqrt{\lambda}} x$.]

11.5 INITIAL-BOUNDARY VALUE PROBLEMS: AN OVERVIEW

In Sections 11.6 through 11.10 we consider special cases of the following general problem.

PROBLEM 1

Consider the second-order linear partial differential equation

$$a_1(x, y)u_{xx} + a_2(x, y)u_{xy} + a_3(x, y)u_{yy} + a_4(x, y)u_x + a_5(x, y)u_y$$
$$+ a_6(x, y)u = F(x, y), \qquad 0 < x < l, \qquad 0 < y < m \tag{1}$$

and the conditions

$$A_1u(0, y) + A_2u_x(0, y) = f_1(y), \qquad 0 < y < m \tag{2}$$

$$A_3u(l, y) + A_4u_x(l, y) = f_2(y), \qquad 0 < y < m \tag{3}$$

$$A_5u(x, 0) = f_3(x), \qquad 0 < x < l \tag{4}$$

$$A_6u_y(x, 0) = f_4(x), \qquad 0 < x < l, \tag{5}$$

[4] H. W. Reddick and F. H. Miller, *Advanced Mathematics for Engineers*, 3rd ed. (New York: John Wiley & Sons, 1960), p. 307.

where $a_1(x, y), \ldots, a_6(x, y), f_1(y), f_2(y), f_3(x), f_4(x)$, and $F(x, y)$ are assumed to be known functions and $A_1, \ldots, A_6$ are known constants. l or m, or both, may be infinite.

Is there a function u of two variables x and y that satisfies the partial differential equation (1) and each of the conditions (2)–(5)? Such a problem is referred to as an *initial-boundary value problem* (I-BVP). Equations (2) and (3) are called the *boundary conditions*, and Eqs. (4) and (5) are called the *initial conditions*. For a given problem, one or more of the boundary conditions, or one or more of the initial conditions may be missing.

In Problem 1, if we set F, f_1, f_2, f_3, and f_4 each equal to zero, we obtain the *associated homogeneous problem*. In Sections 11.6–11.8 we demonstrate, for specific cases, how to solve Problem 1 with $F(x, y) = 0$. In Sections 11.9 and 11.10 we consider special cases of Problem 1 with $F(x, y) \neq 0$. While we do not solve the general case of Problem 1, we do provide the necessary ingredients for the solution of many problems that fall into the category of Problem 1. The main prerequisite for utilizing the method we discuss is that the method of separation of variables be applicable to the partial differential equation.

11.6 THE HOMOGENEOUS ONE-DIMENSIONAL WAVE EQUATION: SEPARATION OF VARIABLES

Consider that we have a string that is perfectly flexible (that is to say, the string is capable of transmitting tension but will not transmit bending or shearing forces) and that its mass per unit length is a constant. The string is to be stretched and attached to two fixed points on the x-axis, $x = 0$ and $x = l$. The string is then given an initial displacement and/or an initial velocity parallel to the y-axis, thus setting it in motion. The distance along the string will be denoted by s and as usual

$$ds = \sqrt{(dx)^2 + (dy)^2}.$$

Thus,

$$\frac{\partial s}{\partial x} = \sqrt{1 + \left(\frac{\partial y}{\partial x}\right)^2}.$$

If we assume that the displacement, y, is small enough so that $\left(\frac{\partial y}{\partial x}\right)^2$ is a very small quantity (in comparison to 1), then approximately $\frac{\partial s}{\partial x} = 1$; that is, the length of the string is approximately unchanged (since $s \approx x$). Consequently, the tension in the string is approximately constant. If no other forces are acting on the string, we have the situation illustrated in Figure 11.1, where T represents the (constant) force due to tension.

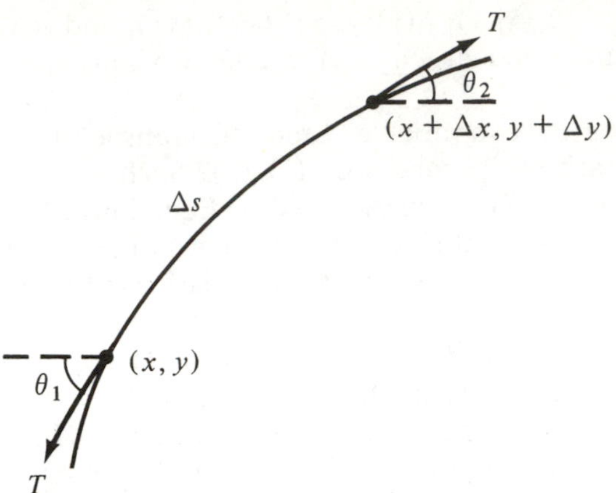

Figure 11.1

The vertical component of the force T at the point (x, y) is given by $-T \sin \theta_1 = -T \dfrac{\partial y}{\partial s}$; however,

$$-T\frac{\partial y}{\partial s} = -T\frac{\partial y}{\partial x}\frac{\partial s}{\partial x} = -T\frac{\partial y}{\partial x}.$$

At the point $(x + \Delta x, y + \Delta y)$, the vertical component of the force T is $T \sin \theta_2$. Since the displacement is considered to be small, $\sin \theta_2 \approx \tan \theta_2 = y'(x + \Delta x)$. Expanding $y'(x + \Delta x)$ in a Taylor's series, the vertical component is

$$T\frac{\partial y}{\partial x} + T\frac{\partial^2 y}{\partial x^2}\Delta x + R,$$

where the remainder, R, has the property

$$\lim_{\Delta x \to 0} \frac{1}{\Delta x} \cdot R = 0.$$

If ρ represents the mass per unit length, then the mass of the portion of string in Figure 11.1 is given by $\rho\Delta s$, which is approximately $\rho\Delta x$. Thus, for motion in the vertical direction, we have by Newton's Second Law of Motion

$$\begin{aligned}\rho\Delta x\frac{\partial^2 y}{\partial t^2} &= T\frac{\partial y}{\partial x} + T\frac{\partial^2 y}{\partial x^2}\Delta x + R + \left(-T\frac{\partial y}{\partial x}\right)\\ &= T\frac{\partial^2 y}{\partial x^2}\Delta x + R.\end{aligned}$$

Dividing by Δx and taking the limit as Δx tends to zero, we obtain the equation of motion of the vertical displacement of the string, namely,

$$\rho\frac{\partial^2 y}{\partial t^2} = T\frac{\partial^2 y}{\partial x^2}.$$

Conforming now to our convention of denoting the unknown function by u, we rewrite the equation of motion in the form

$$u_{tt} - c^2 u_{xx} = 0, \qquad 0 < x < l, \qquad t > 0, \tag{1}$$

where $c^2 = T/\rho$ is a constant according to our assumptions.

For obvious reasons, Eq. (1) is frequently referred to as the equation of the vibrating string. It is also customary to call Eq. (1) the *one-dimensional wave equation.*

The wave equation is one of three partial differential equations known as the classical equations of mathematical physics. The other two, the potential (Laplace) equation and the heat equation, are discussed in Sections 11.8 and 11.7 respectively. Note that the wave equation is an example of a *hyperbolic* partial differential equation.

A typical initial-boundary value problem for the one-dimensional wave equation is the following.

$$u_{tt} - c^2 u_{xx} = 0, \quad 0 < x < l, \quad t > 0 \tag{2}$$

$$u(x, 0) = f(x), \quad 0 < x < l \tag{3}$$

$$u_t(x, 0) = g(x), \quad 0 < x < l \tag{4}$$

$$u(0, t) = 0, \quad t \geq 0 \tag{5}$$

$$u(l, t) = 0, \quad t \geq 0. \tag{6}$$

It can be shown that if f and g satisfy the Dirichlet conditions (see Theorem 1, Section 10.4), the initial-boundary value problem (2)–(6) has a unique solution.

Conditions (5) and (6) can be interpreted as indicating that the ends of the string are attached ("tied") to the x-axis for all time. Condition (3) represents the initial position of the string and condition (4) is the initial velocity.

A standard approach to solving this initial-boundary value problem is to use the separation of variables method of Section 11.4 and Fourier series (Chapter 10). To this end we assume a solution of Eq. (2) to be of the form

$$u(x, t) = X(x)T(t), \tag{7}$$

where X, T are unknown functions to be determined. Substitution of (7) into Eq. (2) yields

$$XT'' - c^2 X''T = 0.$$

Thus,

$$\frac{T''}{T} = \frac{c^2 X''}{X} = \lambda,$$

with λ a constant. Consequently we have two separate problems:

$$\frac{T''}{T} = \lambda \tag{8}$$

and

$$\frac{c^2X''}{X} = \lambda. \tag{9}$$

The boundary condition (5) demands that $X(0)T(t) = 0$ for all $t \geq 0$; thus, $X(0) = 0$. Similarly, the boundary condition (6) indicates that $X(l) = 0$. The function X, then, is to be a solution of the eigenvalue problem

$$X'' - \frac{\lambda}{c^2}X = 0,$$

$$X(0) = X(l) = 0.$$

This problem can be solved by the methods of Section 6.2, to yield the eigenvalues

$$\lambda_n = -\frac{n^2\pi^2c^2}{l^2}, \qquad n = 1, 2, 3, \ldots \tag{10}$$

and the corresponding eigenfunctions

$$X_n(x) = \sin\frac{n\pi x}{l}, \qquad n = 1, 2, 3, \ldots. \tag{11}$$

With λ given by Eq. (10), Eq. (8) takes the form

$$T'' + \frac{n^2\pi^2c^2}{l^2}T = 0.$$

Hence

$$T_n(t) = a_n \cos\frac{n\pi c}{l}t + b_n \sin\frac{n\pi c}{l}t, \qquad n = 1, 2, 3, \ldots, \tag{12}$$

where a_n and b_n are the integration constants in the general solution.

We conclude that for each specific value of n ($n = 1, 2, 3, \ldots$), the function $X_n(x)T_n(t)$ is a solution of Eq. (2) that satisfies conditions (5) and (6). What about conditions (3) and (4)? Let us investigate condition (3) when $u(x, t) = X_n(x)T_n(t)$, with n not specified but otherwise considered fixed.

$$u(x, 0) = f(x) \Rightarrow X_n(x)T_n(0) = f(x) \Rightarrow \left(\sin\frac{n\pi x}{l}\right)a_n = f(x). \tag{13}$$

The only way that (13) can be satisfied is that $f(x)$ be restricted to be of the form $A \sin\frac{n\pi x}{l}$, where A is a constant. If f is of this form, condition (4) demands that

$$\left(\sin\frac{n\pi x}{l}\right)\left(\frac{n\pi c}{l}b_n\right) = g(x). \tag{14}$$

This, too, restricts g to be of the form $B \sin\frac{n\pi x}{l}$, where B is a constant.

Conditions (13) and (14) place too great a restriction on the permissible forms for f and g; therefore, we consider an alternate approach.

Since $X_n(x)T_n(t)$ is a solution of Eq. (2) for each value of n ($n = 1, 2, 3, \ldots$), and since Eq. (2) is a linear partial differential equation, it seems reasonable to expect that $\sum_{n=1}^{\infty} X_n(x)T_n(t)$ is a solution of Eq. (2). Naturally there is the question of whether or not this infinite series converges. We will not investigate this question here but rather emphasize the method of solution. We consider

$$u(x, t) = \sum_{n=1}^{\infty} X_n(x)T_n(t), \tag{15}$$

with X_n given by (11) and T_n given by (12), to be a solution of Eq. (2). $u(x, t)$ satisfies conditions (5) and (6), as is easily verified. $u(x, t)$ will satisfy condition (3) provided that

$$\sum_{n=1}^{\infty} a_n \sin \frac{n\pi x}{l} = f(x),$$

that is, that the Fourier sine series for $f(x)$ in the interval $0 \le x \le l$ be $\sum_{n=1}^{\infty} a_n \sin \frac{n\pi x}{l}$. Consequently, a_n is given by (see Section 10.5)

$$a_n = \frac{2}{l} \int_0^l f(x) \sin \frac{n\pi x}{l}\, dx. \tag{16}$$

Likewise, condition (4) will be satisfied provided that

$$\sum_{n=1}^{\infty} \left(\frac{n\pi c}{l} b_n \right) \sin \frac{n\pi x}{l} = g(x).$$

In other words, b_n is given by

$$b_n = \frac{2}{n\pi c} \int_0^l g(x) \sin \frac{n\pi x}{l}\, dx. \tag{17}$$

We conclude that the solution of the initial-boundary value problem (2)–(6) is given by

$$u(x, t) = \sum_{n=1}^{\infty} \left(a_n \cos \frac{n\pi c}{l} t + b_n \sin \frac{n\pi c}{l} t \right) \sin \frac{n\pi x}{l}, \tag{18}$$

where a_n and b_n are given by Eqs. (16) and (17) respectively.

EXAMPLE 1 Solve the following initial-boundary value problem.

$$\begin{aligned}
u_{tt} - 4u_{xx} &= 0, && 0 < x < \pi, \quad t > 0,\\
u(x, 0) &= x(\pi - x), && 0 < x < \pi,\\
u_t(x, 0) &= 0, && 0 < x < \pi,\\
u(0, t) &= 0, && t \ge 0,\\
u(\pi, t) &= 0, && t \ge 0.
\end{aligned}$$

Solution The above problem is the initial-boundary value problem (2)–(6) with $c = 2$, $l = \pi$, $f(x) = x(\pi - x)$, $g(x) = 0$. From formula (17) we have $b_n = 0$; applying formula (16) we have (using 49 and 51 in the integral tables)

$$\begin{aligned}
a_n &= \frac{2}{\pi}\int_0^\pi x(\pi - x)\sin nx\,dx = \frac{2}{\pi}\left[\pi\int_0^\pi x\sin nx\,dx - \int_0^\pi x^2\sin nx\,dx\right] \\
&= \frac{2}{\pi}\left\{\pi\left[-\frac{x}{n}\cos nx\Big|_0^\pi + \frac{1}{n}\int_0^\pi \cos nx\,dx\right] - \left[-\frac{x^2}{n}\cos nx\Big|_0^\pi \right.\right. \\
&\quad \left.\left. + \frac{2}{n}\int_0^\pi x\cos nx\,dx\right]\right\} \\
&= \frac{2}{\pi}\left\{\pi\left[\frac{(-1)^{n+1}\pi}{n} + \frac{1}{n^2}\sin nx\Big|_0^\pi\right] - \left[\frac{(-1)^{n+1}\pi^2}{n} + \right.\right. \\
&\quad \left.\left. \frac{2}{\pi}\left(\frac{x}{n}\sin nx\Big|_0^\pi - \frac{1}{n}\int_0^\pi \sin nx\,dx\right)\right]\right\} \\
&= \frac{2}{\pi}\left\{\frac{(-1)^{n+1}\pi^2}{n} - \frac{(-1)^{n+1}\pi^2}{n} + \frac{2}{n^2}\left[-\frac{\cos nx}{n}\Big|_0^\pi\right]\right\} \\
&= \frac{2}{\pi}\left\{\frac{2}{n^2}\left[\frac{1 - (-1)^n}{n}\right]\right\} = \frac{4}{\pi n^3}[1 - (-1)^n].
\end{aligned}$$

Thus, the solution given by Eq. (18) is

$$u(x, t) = \sum_{n=1}^{\infty} \frac{4}{\pi n^3}[1 - (-1)^n]\cos 2nt \sin nx$$

or

$$u(x, t) = \sum_{n=0}^{\infty} \frac{8}{\pi(2n + 1)^3}\cos 2(2n + 1)t \sin (2n + 1)x.$$

EXAMPLE 2 Solve the following initial-boundary value problem.

$$\begin{aligned}
u_{tt} - 25u_{xx} &= 0, & 0 < x < \pi, \quad t > 0, \\
u(x, 0) &= \sin 3x, & 0 < x < \pi, \\
u_t(x, 0) &= 4, & 0 < x < \pi, \\
u(0, t) &= 0, & t \geq 0, \\
u(\pi, t) &= 0, & t \geq 0.
\end{aligned}$$

Solution Applying formula (16) we have

$$a_n = \frac{2}{\pi}\int_0^\pi \sin 3x \sin nx\,dx.$$

Using the orthogonality property of the functions sin nx, we obtain

$$a_n = \begin{cases} 0, & n \neq 3 \\ \dfrac{2}{\pi}\left[\dfrac{x}{2} - \dfrac{\sin 2(3)x}{4(3)}\right]_0^\pi = 1, & n = 3. \end{cases}$$

From formula (17) we have

$$b_n = \frac{2}{5\pi n}\int_0^\pi 4 \sin nx\, dx = \frac{8}{5\pi n}\left[\frac{-\cos nx}{n}\Bigg|_0^\pi\right] = \frac{8}{5\pi n^2}[1 - (-1)^n].$$

Thus, the solution of the initial-boundary value problem is

$$\begin{aligned} u(x, t) &= \sum_{n=1}^{\infty} [a_n \cos 5nt + b_n \sin 5nt] \sin nx, \\ &= \cos 15t \sin 3x + \sum_{n=1}^{\infty} \frac{8}{5\pi n^2}[1 - (-1)^n] \sin 5nt \sin nx, \\ &= \cos 15t \sin 3x + \sum_{n=0}^{\infty} \frac{16}{5(2n+1)^2\pi} \sin 5(2n+1)t \sin (2n+1)x. \end{aligned}$$

The method introduced in this section is applicable to many problems that come under the classification of Problem 1, Section 11.5. Other types of equations are presented in Sections 11.7, 11.8, 11.9, and 11.10. We conclude this section with another example involving the wave equation but with different boundary conditions.

EXAMPLE 3 Solve the initial-boundary value problem

$$u_{tt} - c^2 u_{xx} = 0, \qquad 0 < x < l, \quad t > 0 \tag{19}$$

$$u(x, 0) = f(x), \quad 0 < x < l, \tag{20}$$

$$u_t(x, 0) = g(x), \quad 0 < x < l, \tag{21}$$

$$u_x(0, t) = 0, \qquad t > 0, \tag{22}$$

$$u_x(l, t) = 0, \qquad t > 0. \tag{23}$$

Solution Conditions (22) and (23) are different than those imposed for the problem (2)–(6); therefore, if we assume a solution in the form $u(x, t) = X(x)T(t)$, X must satisfy the eigenvalue problem

$$X'' - \frac{\lambda}{c^2} X = 0,$$

$$X'(0) = X'(l) = 0.$$

Except for a slight change in symbolism, this eigenvalue problem is that of Example 2, Section 6.2. Therefore, we have

$$\lambda_n = -\frac{n^2\pi^2c^2}{l^2}, \qquad n = 1, 2, 3, \ldots$$

and

$$X_n(x) = \cos\frac{n\pi x}{l}, \qquad n = 1, 2, 3, \ldots.$$

Thus we can repeat the development of the solution as we did in the beginning of this section. The solution is

$$u(x, t) = \sum_{n=1}^{\infty}\left(a_n \cos\frac{n\pi c}{l}t + b_n \sin\frac{n\pi c}{l}t\right)\cos\frac{n\pi x}{l},$$

where

$$a_n = \frac{2}{l}\int_0^l f(x)\cos\frac{n\pi x}{l}\,dx \tag{24}$$

and

$$b_n = \frac{2}{n\pi c}\int_0^l g(x)\cos\frac{n\pi x}{l}\,dx. \tag{25}$$

That is, a_n and $\left(\frac{n\pi c}{l}\right) b_n$ are the coefficients of the Fourier cosine series of the functions f and g respectively.

EXERCISES

In Exercises 1 through 14, solve the initial-boundary value problem (2)–(6) for the conditions given.

2. $c = 1, l = 1, f(x) = x(1 - x), g(x) = 0$

3. $c = 1, l = \pi, f(x) = x^2(\pi - x), g(x) = 0$

4. $c = 1, l = \pi, f(x) = x(\pi - x)^2, g(x) = 0$

5. $c = 1, l = \pi, f(x) = 0, g(x) = 3$

6. $c = 1, l = \pi, f(x) = 0, g(x) = \pi$

7. $c = 1, l = 1, f(x) = 0, g(x) = A$, A a constant

8. $c = 1, l = \pi, f(x) = 0, g(x) = A$, A a constant

9. $c = 1, l = \pi, f(x) = x(\pi - x), g(x) = 3$

10. $c = 1, l = \pi, f(x) = x^2(\pi - x), g(x) = 3$

11. $c = 1, l = \pi, f(x) = x(\pi - x)^2, g(x) = \pi$

12. $c = 2, l = \pi, f(x) = \sin x, g(x) = \cos x$

13. $c = 4$, $l = \pi$, $f(x) = \sin 5x$, $g(x) = \cos 2x$

14. $c = 3$, $l = \pi$, $f(x) = \sin x$, $g(x) = 0$

In Exercises 15 through 28, solve the initial-boundary value problem (19)–(23) for the conditions given.

15. The same as Exercise 1.

16. The same as Exercise 2.

17. The same as Exercise 3.

18. The same as Exercise 4.

19. The same as Exercise 5.

20. The same as Exercise 6.

21. The same as Exercise 7.

22. The same as Exercise 8.

23. The same as Exercise 9.

24. The same as Exercise 10.

25. The same as Exercise 11.

26. The same as Exercise 12.

27. The same as Exercise 13.

28. The same as Exercise 14.

In Exercises 29 and 30, solve the initial-boundary value problem given $0 \leq x \leq l$, $t \geq 0$.

29.
$$\begin{aligned} &u_{tt} - c^2 u_{xx} = 0 \\ &u(x, 0) = f(x) \\ &u_t(x, 0) = g(x) \\ &u(0, t) = 0 \\ &u_x(l, t) = 0. \end{aligned}$$

30.
$$\begin{aligned} &u_{tt} - c^2 u_{xx} = 0 \\ &u(x, 0) = f(x) \\ &u_t(x, 0) = g(x) \\ &u_x(0, t) = 0 \\ &u(l, t) = 0. \end{aligned}$$

31. Suppose that u is a solution of the initial-boundary value problem

$$u_{tt} - c^2 u_{xx} = 0, \qquad 0 < x < l, \quad t > 0, \tag{26}$$

$$u(x, 0) = f(x), \quad 0 < x < l, \tag{27}$$

$$u_t(x, 0) = g(x), \quad 0 < x < l, \tag{28}$$

$$u(0, t) = A, \qquad t > 0, \tag{29}$$

$$u(l, t) = B, \qquad t > 0, \tag{30}$$

where A and B are constants. Show that if

$$v(x, t) = u(x, t) + \left(\frac{x - l}{l}\right) A - \frac{x}{l} B, \tag{31}$$

then v is a solution of the initial-boundary value problem

$$\begin{aligned} &v_{tt} - c^2 v_{xx} = 0, && 0 < x < l, \quad t > 0, \\ &v(x, 0) = f(x) + \left(\frac{x - l}{l}\right) A - \frac{x}{l} B, && 0 < x < l, \\ &v_t(x, 0) = g(x), && 0 < x < l, \\ &v(0, t) = 0, && t > 0, \\ &v(l, t) = 0, && t > 0. \end{aligned}$$

In Exercises 32 through 45, solve the initial-boundary value problem (26)–(30) of Exercise 31 for the conditions given. [*Hint:* Find v, then determine u from Eq. (31).]

32. $A = 3$, $B = 0$ and the conditions of Exercise 2.

33. $A = 0$, $B = 3$ and the conditions of Exercise 1.

34. $A = -3$, $B = 2$ and the conditions of Exercise 4.

35. $A = 2$, $B = 2$ and the conditions of Exercise 3.

36. $A = 0$, $B = -2$ and the conditions of Exercise 6.

37. $A = 10$, $B = \pi$ and the conditions of Exercise 5.

38. $A = 8\pi$, $B = \pi$ and the conditions of Exercise 8.

39. $A = 7$, $B = 2$ and the conditions of Exercise 7.

40. $A = 4$, $B = 0$ and the conditions of Exercise 10.

41. $A = 0$, $B = 5$ and the conditions of Exercise 9.

42. $A = 9$, $B = 5$ and the conditions of Exercise 12.

43. $A = 0$, $B = 8$ and the conditions of Exercise 11.

44. $A = 13$, $B = -3$ and the conditions of Exercise 14.

45. $A = 6$, $B = 0$ and the conditions of Exercise 13.

46. Give a physical interpretation of the initial conditions of Example 1 (assume that the problem relates to a "string").

47. Give a physical interpretation to the initial conditions of Example 2 (consider that the problem relates to a "string").

48. A tightly stretched string 3 feet long weighs 0.9 lb. and is under a constant tension of 10 lb. The string is initially straight and is set into motion by imparting to each of its points an initial velocity of 1 ft/sec. (a) Find the displacement u as a function of x and t. (b) Find an expression for the displacement of the midpoint one minute after the motion has begun.

49. Wave Equation with Damping If there is a damping force present, for example, air resistance, the equation of the vibrating string becomes

$$u_{tt} + 2\alpha u_t - c^2 u_{xx} = 0, \tag{32}$$

where α is a positive constant known as the *damping factor.*
(a) Set $u(x, t) = e^{-\alpha t}v(x, t)$ and show that v satisfies

$$v_{tt} - \alpha^2 v - c^2 v_{xx} = 0. \tag{33}$$

(b) Set $v(x, t) = w(x)e^{i\beta t}$ in Eq. (33) and show that w satisfies $w'' + \gamma w = 0$, where $\gamma = (\alpha^2 + \beta^2)/c^2$.
(c) Set $v(x, t) = w(x)e^{-i\beta t}$ in Eq. (33) and show that w satisfies $w'' + \gamma w = 0$, where $\gamma = (\alpha^2 + \beta^2)/c^2$.

(d) Utilizing parts (b) and (c), verify that Eq. (32) possesses particular solutions u_1, u_2 of the form

$$u_1 = Ae^{-\alpha t}e^{i(\sqrt{\gamma}\,x - \beta t)}, \; u_2 = Be^{-\alpha t}e^{-i(\sqrt{\gamma}x - \beta t)}.$$

(e) Describe u_1, u_2 as "waves" and determine their speed (see Exercise 61b).

50. A taut string of length 1m and $c = 1$ is subjected to air resistance damping for which $\alpha = 1$ (see Exercise 49). Using the method of separation of variables and Fourier series, find the displacement as a function of t and x if the initial displacement is zero and the initial velocity is 1m/sec.

51. **Torsional Vibration in Shafts** A shaft (rod) of circular cross section has its axis along the x-axis, the ends coinciding with $x = 0$ and $x = l$. The shaft is subjected to a twisting action and then released. $\theta(x, t)$ denotes the *angular* displacement undergone by the mass in the circular cross section located at the position x and time t (that is, the mass of a very thin disc located there). It can be shown that θ satisfies the one-dimensional wave equation with $c = G/\rho$, where G (a constant) is known as the *shear modulus* of the shaft and ρ (a constant) is the density of the shaft. Furthermore, it is known in the theory of elasticity that $\theta_x = \tau/G\mu$, where τ is the *twisting moment* (torque) and μ is the *polar moment* of inertia. Two types of end conditions are common. They are a *fixed end* (for which $\theta = 0$) and a *free end* (for which $\theta_x = 0$, since $\tau = 0$).

Set up and solve the initial-boundary value problem for the angular displacement of a rod that is fixed at the end $x = 0$, free at the end $x = l$ (take $l = 1$), whose initial velocity (θ_t) is zero, and whose initial displacement is given by $3x$. Take $c = 1$.

52. Repeat Exercise 51 by considering all the information to be the same except that the end at $x = 0$ is free instead of fixed, and the end at $x = 1$ is fixed instead of free.

53. **Plucked String** When the initial displacement of the string is of the form

$$f(x) = \begin{cases} mx, & 0 \le x \le x_0 \\ \dfrac{mx_0}{l - x_0}(l - x), & x_0 \le x \le l \end{cases},$$

one can say that the string has been "pinched" at the point $x = x_0$, lifted (in case $m > 0$) to the height mx_0, and then released. This action is described as *plucking* the string. Find the displacement of the string that is plucked at its midpoint and released from rest. Take $c = l = 1$, $m = \frac{1}{4}$.

Remark It is reasonable to think of a guitar string as being plucked, since a guitar pick or a person's fingernail can be thought of as acting at a point on the string. The same would be true for a harp string except that the plucking very often occurs at two or more positions on the string. On the other hand, piano strings are set into motion when struck by a hammer, which does not act at a certain point on the string, but rather on a segment of the string. In this case, it is reasonable to think of the string as being in an initial horizontal position; the hammer imparts an initial velocity to the portion of the string it strikes, and the rest of the string has zero initial velocity.

54. Piano String Find the displacement of a 2-foot piano string that is struck by a 2-inch hammer having an initial velocity of 1 ft/sec, if it is known that the center of the hammer strikes the center of the string. Take $c = 1$. [*Hint:* See the preceding remark.]

55. A 2-inch-wide acorn travelling at 3 in/sec. strikes a taut spider's web consisting of a single horizontal thread 6 inches long. Find the displacement of the web if it is known that the center of the acorn strikes the web at a point 2 inches from its left end. Take $c = 1$. [*Hint:* See the preceding remark.]

56. Harmonics The solution given in Eq. (18) can be written in the form

$$u(x, t) = \sum_{n=1}^{\infty} u_n(x, t),$$

where

$$u_n(x, t) = A_n g(t)h(x)$$

and

$$A_n = (a_n^2 + b_n^2)^{1/2}, \quad h(x) = \sin\frac{n\pi x}{l}, \qquad \text{and} \qquad g(t) = \cos\left[\frac{n\pi c}{l}t - \beta\right].$$

(Recall from trigonometry that $A\cos\alpha t + B\sin\alpha t = C\cos(\alpha t - \beta)$, with $C = (A^2 + B^2)^{1/2}$ and $\cos\beta = A/C$.) By itself, u_n is a possible motion of the string and is called the nth *normal* (or *natural*) *mode of vibration* or the nth *harmonic;* u_1 is called the *fundamental mode* or the *fundamental harmonic.* If we consider x to be fixed, then u_n is a simple harmonic motion of period

$$p_n = \frac{2l}{nc}$$

and frequency

$$\nu_n = \frac{nc}{2l}.$$

ν_n is called the nth *natural frequency,* and ν_1 is called the *fundamental frequency* of vibration. If $u_n \equiv 0$, then ν_n is taken to be zero. Find the first four harmonics and the first four natural frequencies for the string of Example 1 of this section.

57. A clothesline 10 feet in length is to be considered as a taut flexible string. A boy strikes the clothesline with a paddle that is one foot wide, so that the paddle is travelling with speed 3 ft/sec at the time of contact. The point of contact of the center of the paddle is at the point 3.5 feet from the left end of the clothesline. Find the displacement of the clothesline. Take $c = 1$.

58. The transverse displacement, $u(x, t)$, of a uniform beam satisfies the partial differential equation $u_{tt} + c^2 u_{xxxx} = 0$, where $c^2 = EI/\rho S$, with E a constant (*Young's modulus*), I a constant (moment of inertia of the cross section area), ρ a constant (the density), and S a constant (the area of cross section).[5] An end of the beam is said to be *hinged* if $u = 0$ and $u_{xx} = 0$ at that end. Set up and solve the initial-boundary value problem for the transverse displacement of a beam of length l that is hinged at both ends, having initial displacement $f(x)$ and initial velocity zero. [*Hint:* Use the method of separation of variables and Fourier series.]

59. If the beam of Exercise 58 is subjected to axial forces, $F(t)$, applied at its ends, the transverse displacement, $u(x, t)$, satisfies[6] the partial differential equation

$$EIu_{xxxx} - F(t)u_{xx} + \rho S u_{tt} = 0.$$

Assume that the beam is of length l and has hinged ends. This equation does not "separate," and we cannot use the method of this section. Set $u(x, t) = \sum_{n=1}^{\infty} T_n(t) \sin \frac{n\pi x}{l}$.

(a) Show that u satisfies the boundary conditions.
(b) Determine a differential equation that $T_n(t)$ must satisfy if u is to be a solution of the partial differential equation.

60. The Telegraph Equation If $u(x, t)$ and $i(x, t)$ represent the voltage and current, respectively, in a cable where t denotes time and x denotes the position in the cable measured from a fixed initial position, the governing equations[7] are

$$Cu_t + Gu + i_x = 0$$

$$Li_t + Ri + u_x = 0.$$

The constants C, G, L, and R represent electrostatic capacity per unit length, leakage conductance per unit length, inductance per unit length, and resistance per unit length, respectively.

(a) Eliminate i from the above system to show that u satisfies the *telegraph equation*

$$LCu_{tt} + (LG + RC)u_t + RGu = u_{xx}. \tag{34}$$

[*Hint:* Differentiate the first equation with respect to t and the second with respect to x.]

(b) Eliminate u from the above system and show that i satisfies Eq. (34).

[5] S. Timoshenko, *Vibration Problems in Engineering* (Princeton, N.J.: Van Nostrand, 1928), p. 221.
[6] Ibid., p. 374.
[7] R. Courant and D. Hilbert, *Methods of Mathematical Physics,* vol. 2 (New York: Interscience Publishers, 1962), p. 193.

61. Travelling Waves Divide by LC and rewrite the telegraph equation, Eq. (34), in the form

$$u_{tt} + (\alpha + \beta)u_t + \alpha\beta u = c^2 u_{xx},$$

with $c^2 = 1/LC$, $\alpha = G/C$, $\beta = R/L$.

(a) Set $u(x, t) = e^{-1/2(\alpha+\beta)t}v(x, t)$, and show that if u is a solution of the telegraph equation, then v satisfies

$$v_{tt} - \left(\frac{\alpha - \beta}{2}\right)^2 v = c^2 v_{xx}.$$

(b) If $\alpha = \beta$, that is, $GL = RC$, then v satisfies the one-dimensional wave equation; although we do not have initial-boundary conditions, we can still discuss the solution. Using the method of Exercise 40, Section 11.4, show that v has the form

$$v(x, t) = f(x + ct) + g(x - ct).$$

The expression $f(x + ct)$ can be thought of as a "wave"[8] travelling to the left with speed c, and $g(x - ct)$ as a "wave" travelling to the right with speed c. Thus the solution u of part (a) can be described as a voltage (or current) subjected to damping that is travelling in both directions in the cable. Thus, for appropriate properties of the cable ($\alpha = \beta$), signals can be transmitted along the cable in a "relatively" undistorted form yet damped in time.

62. Maxwell's Equation for the Electromagnetic Field Intensity in a Homogeneous Medium The partial differential equation

$$(\underset{\sim}{E})_{tt} = \frac{\delta^2}{\mu\epsilon}\Delta[\underset{\sim}{E}] - \frac{4\pi\sigma}{\epsilon}(\underset{\sim}{E})_t \tag{35}$$

is one of *Maxwell's equations* that occur in electrodynamics.[9] $\underset{\sim}{E}$ is a vector representing the *electromagnetic field intensity,* and σ (*conductivity*), ϵ (*dielectric constant*), and μ (*permeability*) are constants associated with the medium; δ is a constant associated with the conversion of units. The one-dimensional version of Eq. (35) can be written in the form (32) with $c = \delta/\sqrt{\mu\epsilon}$, $\alpha = 2\pi\sigma/\epsilon$. Repeat the procedure outlined in Exercise 49. In this application the factor $e^{-\alpha t}$ is called the *attenuation factor.*

63. Determine the tension in a string of length 100 cm. and density 1.5 gram per meter, so that the fundamental frequency of the string is 256 cycles per second (256 cps is middle C on the musical scale).

[8]See, for example, J. M. Pearson, *A Theory of Waves* (Boston: Allyn and Bacon, 1966), p. 2.
[9]Pearson, *Theory of Waves,* p. 29.

11.7 THE ONE-DIMENSIONAL HEAT EQUATION

In the investigation of the flow of heat in a conducting body, the following three laws have been deduced from experimentation.

LAW 1 Heat will flow from a region of higher temperature to a region of lower temperature.

LAW 2 The amount of heat in the body is proportional to the temperature of the body and to the mass of the body.

LAW 3 Heat flows across an area at a rate proportional to the area and to the temperature gradient (that is, the rate of change of temperature with respect to distance where the distance is taken perpendicular to the area).

We consider a rod of length l and constant cross sectional area A. The rod is assumed to be made of material that conducts heat uniformly. The lateral surface of the rod is insulated so that the streamlines of heat are straight lines perpendicular to the cross sectional area A. The x-axis is taken parallel to and in the same direction as the flow of heat. The point $x = 0$ is at one end of the rod and the point $x = l$ is at the other end. ρ denotes the density of the material, and c (a constant) denotes the *specific heat* of the material. (Specific heat is the amount of heat energy required to raise a unit mass of the material one unit of temperature change.) $u(x, t)$ denotes the temperature at time t (> 0) in a cross sectional area A, x units from the end $x = 0$. Consider a small portion of the rod of thickness Δx that is between x and $x + \Delta x$. The amount of heat in this portion is, by Law 2, $c\rho A\Delta xu$. Thus, the time rate of change of this quantity of heat is $c\rho A\Delta x \dfrac{\partial u}{\partial t}$. Thus,

$$c\rho A\Delta x \frac{\partial u}{\partial t} = (\text{rate into this portion}) - (\text{rate out of this portion}).$$

From Law 3 we have

$$\text{rate in} = -kA \left.\frac{\partial u}{\partial x}\right|_{x},$$

$$\text{rate out} = -kA \left.\frac{\partial u}{\partial x}\right|_{x + \Delta x},$$

where the minus sign is a consequence of Law (1) and our assumption regarding the orientation of the x-axis. The constant of proportionality k is known as the *thermal conductivity*. Thus,

$$c\rho A\Delta x \frac{\partial u}{\partial t} = -kA \left.\frac{\partial u}{\partial x}\right|_{x} + kA \left.\frac{\partial u}{\partial x}\right|_{x + \Delta x}$$

or

$$\frac{\partial u}{\partial t} = \frac{k}{c\rho}\left[\frac{\left.\frac{\partial u}{\partial x}\right|_{x+\Delta x} - \left.\frac{\partial u}{\partial x}\right|_{x}}{\Delta x}\right].$$

Taking the limit as Δx tends to zero, we obtain the *one-dimensional heat equation*

$$u_t - au_{xx} = 0, \tag{1}$$

where $a = k/c\rho$ is known as the *diffusivity*. Note that the heat equation is a *parabolic* partial differential equation.

EXAMPLE 1 Solve[10] the initial-boundary value problem

$$u_t - au_{xx} = 0, \qquad 0 < x < l, \quad t > 0, \tag{2}$$

$$u(0, t) = 0, \qquad t > 0, \tag{3}$$

$$u(l, t) = 0, \qquad t > 0, \tag{4}$$

$$u(x, 0) = f(x), \quad 0 < x < l. \tag{5}$$

Solution We note that conditions (3) and (4) indicate that the ends of the rod are in contact with a heat reservoir of constant temperature zero; (5) gives the initial distribution of temperature. We seek a solution in the form $u(x, t) = X(x)T(t)$. Using the method of separation of variables, we find that X is to be a solution of the eigenvalue problem

$$X'' - \frac{\lambda}{a}X = 0, \tag{6}$$

$$X(0) = 0, X(l) = 0, \tag{7}$$

and T is to satisfy the differential equation

$$T' - \lambda T = 0. \tag{8}$$

The eigenvalues and eigenfunctions of problem (6)–(7) are, respectively,

$$\lambda_n = -\frac{n^2\pi^2 a}{l^2}, \quad n = 1, 2, 3, \ldots, \tag{9}$$

$$X_n(x) = \sin\frac{n\pi x}{l}, \quad n = 1, 2, 3, \ldots. \tag{10}$$

For λ as given in (9), the general solution of Eq. (8) is

$$T_n(t) = c_n e^{-[(n^2\pi^2 a)/l^2]t}.$$

Thus,

$$u(x, t) = \sum_{n=1}^{\infty} X_n(x)T_n(t)$$

[10]It can be shown that if f satisfies the Dirichlet conditions (Theorem 1, Section 10.4), problem (2)–(5) has a unique solution.

is a solution of Eq. (2) that satisfies conditions (3) and (4) and will satisfy condition (5) if

$$\sum_{n=1}^{\infty} c_n \sin \frac{n\pi x}{l} = f(x).$$

Thus, the constants c_n are the Fourier sine coefficients of f; that is,

$$c_n = \frac{2}{l}\int_0^l f(x)\sin\frac{n\pi x}{l}\,dx. \tag{11}$$

The solution of the initial-boundary value problem (2)–(5) is

$$u(x, t) = \sum_{n=1}^{\infty} c_n e^{-[(n^2\pi^2 a)/l^2]t} \sin\frac{n\pi x}{l},$$

where c_n is given by Eq. (11).

EXAMPLE 2 If the ends of the rod are in contact with heat reservoirs of constant temperatures A at $x = 0$ and B at $x = l$, the corresponding initial-boundary value problem is

$$u_t - au_{xx} = 0, \quad 0 < x < l, \quad t > 0, \tag{12}$$

$$u(0, t) = A, \quad t > 0, \tag{13}$$

$$u(l, t) = B, \quad t > 0, \tag{14}$$

$$u(x, 0) = f(x), \quad 0 < x < l. \tag{15}$$

Solve the initial-boundary value problem (12)–(15).

Solution If u is a solution of the initial-boundary value problem (12)–(15), then

$$v(x, t) = u(x, t) + \left(\frac{x - l}{l}\right)A - \frac{x}{l}B \tag{16}$$

is a solution of the initial-boundary value problem

$$v_t - av_{xx} = 0, \quad 0 < x < l, \quad t > 0,$$

$$v(0, t) = 0, \quad t > 0,$$

$$v(l, t) = 0, \quad t > 0,$$

$$v(x, 0) = f(x) + \left(\frac{x - l}{l}\right)A - \frac{x}{l}B, \quad 0 < x < l.$$

Thus the solution is obtained by first finding v using the method of Example 1, and then determining u from Eq. (16).

EXERCISES

In Exercises 1 though 16, solve the initial-boundary value problem (2)–(5) for the given conditions.

1. $a = 1, l = 1, f(x) = x$

2. $a = 1, l = \pi, f(x) = x$

3. $a = 4, l = \pi, f(x) = x^2$

4. $a = 4, l = \pi, f(x) = \sin x$

5. $a = 2, l = 1, f(x) = \cos \dot{x} + 3x$

6. $a = 1, l = 3, f(x) = x + 3$

7. $a = 5, l = 1, f(x) = e^x$

8. $a = 3, l = \pi, f(x) = \sin 3x$

9. $a = 1, l = \pi, f(x) = x + \sin x$

10. $a = 2, l = 1, f(x) = \sin^2 x$

11. $a = 1, l = 1, f(x) = x(1 - x)$

12. $a = 1, l = 1, f(x) = x^2(1 - x)$

13. $a = 2, l = \pi, f(x) = x(\pi - x)^2$

14. $a = 1, l = 1, f(x) = x^3$

15. $a = 1, l = 1, f(x) = x + e^x$

16. $a = 1, l = 1, f(x) = x^2 + e^x$

In Exercises 17 through 25, solve the initial-boundary value problem (12)–(15) for the given conditions.

17. $a = 5, l = \pi, f(x) = x, A = 10, B = 0$

18. $a = 3, l = \pi, f(x) = x, A = 0, B = -5$

19. $a = 1, l = 1, f(x) = x(1 - x), A = 7, B = 3$

20. $a = 2, l = 1, f(x) = x^2, A = 5, B = 5$

21. $a = 1, l = 1, f(x) = \sin^2 x, A = 7, B = 2$

22. $a = 1, l = 1, f(x) = x + 3, A = 6, B = 4$

23. $a = 3, l = 1, f(x) = e^x, A = 7, B = 3$

24. $a = 1, l = 1, f(x) = x + e^x, A = 25, B = 15$

25. $a = 5, l = 1, f(x) = x(1 - x)^2, A = 10, B = 10$

Remark In general the heat equation can be written in the form

$$u_t = a\Delta[u], \tag{17}$$

where Δ is the Laplacian operator introduced in Exercise 50, Section 11.3.

Consequently one can consider heat diffusion problems in two and three dimensions also. Some of these higher-dimensional problems can be handled in precisely the same manner as in this section; however, the analysis is complicated by the fact that the corresponding eigenvalue problems involve Bessel and Legendre functions. These functions are considerably more difficult to deal with than the sine and cosine functions heretofore encountered. For this reason we have restricted our attention to the one-dimensional heat equation. There are, however, some special types of two- and three-dimensional problems that lead to relatively simple solutions (see Exercises 26, and 32 through 36).

26. If the temperature is a function of the radius only, show that the two-dimensional heat equation can be written in the form

$$u_t = \frac{a}{r}(ru_r)_r. \tag{18}$$

[*Hint:* See Exercise 50, Section 11.3.]

27. Heat Equation: Source Terms If internal sources of heat are present in the rod and if the rate of production of heat is the same throughout any given cross section of area, then the equation[11] of heat flow is

$$u_t - au_{xx} = q(x, t, u), \tag{19}$$

where $h(x, t, u) = c\rho q(x, t, u)$ is the rate of production of heat energy per unit volume per unit time. The following special cases are of interest.

(a) $q(x, t, u) = F(x, t)$. This situation corresponds to heating caused by an electric current through the rod and gives rise to a nonhomogeneous problem. See Exercise 24, Section 11.9 for a specific example.

(b) $q(x, t, u) = r(x, t)u$. This case can be thought of as heating of the rod caused by a chemical reaction that is proportional to the local temperature. If r is positive, the rod is receiving heat (thus, a heat *source* exists); if r is negative, the rod is giving off heat (thus, a heat *loss* exists).

Set up and solve the initial-boundary value problem consisting of Eq. (19) with $q = -\alpha u$, $\alpha(> 0)$ a constant, $u(x, 0) = x(1 - x)$, $u(0, t) = u(l, t) = 0$, $l = a = 1$. [*Hint:* Set $u = e^{-\alpha t} v(x, t)$; the initial-boundary value problem for v is solvable by the methods of this section.]

28. Automatic Heat Control The initial-boundary value problem

$$\begin{aligned} u_t &= au_{xx}, && 0 < x < l, \quad t > 0 \\ u_x(0, t) &= 0, && t > 0, \\ u(l, t) &= Au(0, t), && t > 0, \\ u(x, 0) &= f(x), && 0 < x < l, \end{aligned}$$

[11] P. W. Berg and J. L. McGregor, *Elementary Partial Differential Equations* (San Francisco: Holden-Day, 1966), p. 32.

where A is a constant ($A \neq 1$), describes the temperature in a rod whose left end ($x = 0$) is thermally insulated (that is, no heat is lost through this end, hence $u_x = 0$), and whose right end is equipped with an *automatic heat control* which keeps the temperature at this end proportional to the temperature at the left end. The initial distribution of temperature in the rod is $f(x)$.

(a) If $-1 < A < 1$, solve this initial-boundary value problem.

(b) Show that there are no real eigenvalues in the case $A < -1$.

29. Solve the initial-boundary value problem (2)–(5) when

$$f(x) = \begin{cases} A, & 0 < x < \dfrac{l}{2}, \\ 0, & \dfrac{l}{2} < x < l, \end{cases}$$

where A is a constant.

30. Two rods of the same type of material, each 10 cm. long, are placed face to face in perfect contact. The outer faces are maintained at 0°C. Just prior to contact, one of the rods has constant temperature 100°C; the other has constant temperature 0°C throughout. (a) Find the temperature distribution as a function of x and t after contact is made. [*Hint:* See Exercise 29.] (b) If $a = 0.2$ for the material, find to the nearest degree the temperature at a point on the common face and at points 5 cm. from this common face twenty minutes after contact is made.

31. Three rods of the same material, $a = 1.02$, are each 10 cm. long. The rods each have constant temperature, two at 0°C and one at 100°C. The rods are placed end to end, with the hot rod in the middle. Find the temperature at a point on the middle (center) cross section of the "new" rod if the ends of the "new" rod are maintained at 0°C.

32. (a) Referring to Exercise 50(c) of Section 11.3 and the Remark which follows that exercise, show that the heat equation in spherical coordinates for a temperature distribution that depends only on the radius r and the time t is

$$u_t = \frac{a}{r^2}(r^2 u_r)_r.$$

(b) Show that $\dfrac{1}{r^2}(r^2 u_r)_r = \dfrac{1}{r}(ru)_{rr}$.

33. Temperature Distribution in a Sphere The surface of a homogeneous solid sphere of radius R is maintained at the constant temperature 0°C and has an initial temperature distribution given by $g(r)$. If $u(r, t)$ denotes the temperature in the sphere as a function of the radius r and the time t only, then $u(r, t)$ satisfies the problem (see Exercise 32)

$$u_t = \frac{a}{r}(ru)_{rr}, \quad 0 < r < R, \quad t > 0$$

$$u(R, t) = 0, \qquad t > 0,$$

$$u(r, 0) = g(r), \qquad 0 \le r < R.$$

It is reasonable to assume that u is bounded at $r = 0$; therefore, if we set $v(r, t) = ru(r, t)$, we would have $v(0, t) = 0$. (a) Set up and solve the corresponding initial-boundary value problem for v. (b) Find u.

34. Repeat Exercise 33 for $g(r) = A$, a constant.

35. Repeat Exercise 33 for $g(r) = R^2 - r^2$.

36. A solid sphere of radius 10 cm. is made of material for which $a = 0.2$. Initially the temperature is 100°C throughout the sphere, and the sphere is cooled by keeping the surface of the sphere at 0°C. Find the temperature (to the nearest degree) at the center of the sphere 20 minutes after the cooling begins. [*Hint:* See Exercise 33.]

11.8 THE POTENTIAL (LAPLACE) EQUATION

We begin this section with a consideration of heat flow in two dimensions. Refer to Section 11.7 for the appropriate definitions and the experimental laws associated with heat flow.

The conducting material is in the shape of a rectangular sheet of constant thickness D, and consider the flow of heat in the differential portion of the sheet depicted in Figure 11.2. Using the same arguments as in Section 11.7, we determine that the gain of heat in the horizontal direction for this portion is

$$kD\Delta y\left[\left.\frac{\partial u}{\partial x}\right|_{x+\Delta x} - \left.\frac{\partial u}{\partial x}\right|_{x}\right],$$

and in the vertical direction it is

$$kD\Delta x\left[\left.\frac{\partial u}{\partial y}\right|_{y+\Delta y} - \left.\frac{\partial u}{\partial y}\right|_{y}\right].$$

Thus, the total gain of heat for this portion of the sheet is

$$kD\left\{\Delta y\left[\left.\frac{\partial u}{\partial x}\right|_{x+\Delta x} - \left.\frac{\partial u}{\partial x}\right|_{x}\right] + \Delta x\left[\left.\frac{\partial u}{\partial y}\right|_{y+\Delta y} - \left.\frac{\partial u}{\partial y}\right|_{y}\right]\right\},$$

or

$$kD\Delta x\Delta y\left[\frac{\left.\frac{\partial u}{\partial x}\right|_{x+\Delta x} - \left.\frac{\partial u}{\partial x}\right|_{x}}{\Delta x} + \frac{\left.\frac{\partial u}{\partial y}\right|_{y+\Delta y} - \left.\frac{\partial u}{\partial y}\right|_{y}}{\Delta y}\right].$$

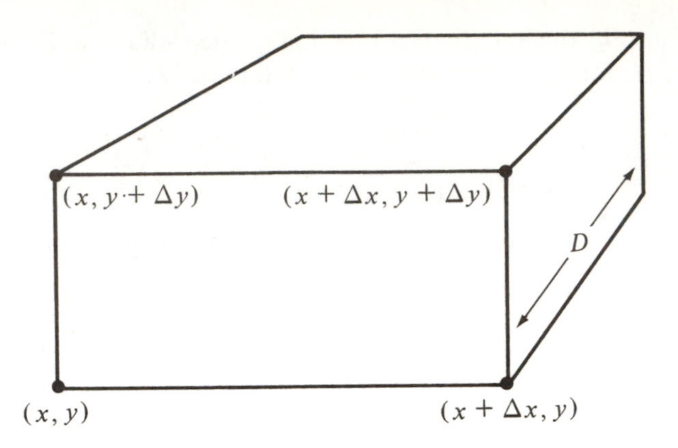

Figure 11.2

By Law 2 of Section 11.7, this rate of gain of heat is also given by $c\rho D\Delta x\Delta y \dfrac{\partial u}{\partial t}$. Hence,

$$c\rho D\Delta x\Delta y \frac{\partial u}{\partial t} = kD\Delta x\Delta y \left[\frac{\left.\dfrac{\partial u}{\partial x}\right|_{x+\Delta x} - \left.\dfrac{\partial u}{\partial x}\right|_{x}}{\Delta x} + \frac{\left.\dfrac{\partial u}{\partial y}\right|_{y+\Delta y} - \left.\dfrac{\partial u}{\partial y}\right|_{y}}{\Delta y} \right].$$

Simplifying and taking the limit as both Δx and Δy tend to zero, we obtain the two-dimensional heat equation

$$u_t = a(u_{xx} + u_{yy}),$$

where $a = k/\rho c$.

If at a future point in time, the temperature u is a function of x and y only (that is, u is independent of time and $u_t = 0$), one says that *steady state* conditions have been achieved. Consequently, the partial differential equation for the steady state temperature for two-dimensional heat flow is

$$u_{xx} + u_{yy} = 0. \tag{1}$$

Equation (1) is called the *two-dimensional potential (Laplace) equation.* Note that the potential equation is an *elliptic* partial differential equation.

In addition to the steady state temperature for two-dimensional heat flow, there are many other physical applications in which the potential equation occurs. One application is that u represents the potential of a two-dimensional electrostatic field. In this case and for that of two-dimensional heat flow (as well as others), a common situation is that u is to satisfy Eq. (1) in a fixed bounded region, say R, of the xy plane, and that the values of u are known on the boundaries of this region. Thus, a typical problem associated with the potential equation is a boundary value problem, commonly referred to as *Dirichlet's problem,* for R. It is the geometry of R and the nature of the boundary conditions that dictate whether or not it is easy (or even possible) to solve the Dirichlet problem.

EXAMPLE 1 Solve[12] the following Dirichlet problem on the rectangle $0 < x < l, \quad 0 < y < m$.

$$u_{xx} + u_{yy} = 0, \qquad 0<x<l, \quad 0<y<m, \tag{2}$$

$$u(x, 0) = f(x), \quad 0<x<l, \tag{3}$$

$$u(x, m) = g(x), \quad 0<x<l, \tag{4}$$

$$u(0, y) = 0, \qquad 0<y<m, \tag{5}$$

$$u(l, y) = 0, \qquad 0<y<m. \tag{6}$$

Solution We seek a solution of Eq. (2) of the form $u(x, y) = X(x)Y(y)$. Thus, by the separation of variables method, we find that X must satisfy the eigenvalue problem

$$X'' + \lambda X = 0, \tag{7}$$

$$X(0) = 0, X(l) = 0, \tag{8}$$

and Y must satisfy the differential equation

$$Y'' - \lambda Y = 0. \tag{9}$$

The eigenvalues and eigenfunctions of problem (7)–(8) are given respectively by

$$\lambda_n = \frac{n^2\pi^2}{l^2}, \qquad n = 1, 2, 3, \ldots, \tag{10}$$

$$X_n(x) = \sin\frac{n\pi x}{l}, \quad n = 1, 2, 3, \ldots. \tag{11}$$

For λ as given in (10), the general solution of the differential equation (9) is

$$Y_n(y) = a_n \cosh\frac{n\pi y}{l} + b_n \sinh\frac{n\pi y}{l}. \tag{12}$$

Thus we seek a solution of the boundary value problem (2)–(6) in the form

$$u(x, y) = \sum_{n=1}^{\infty} X_n(x)Y_n(y). \tag{13}$$

Now (13) satisfies each of (2), (5), and (6) and will satisfy (3) if

$$\sum_{n=1}^{\infty} Y_n(0) \sin\frac{n\pi x}{l} = f(x).$$

Also, (13) will satisfy condition (4) if

$$\sum_{n=1}^{\infty} Y_n(m) \sin\frac{n\pi x}{l} = g(x).$$

[12]It can be shown that if f and g satisfy the Dirichlet conditions (Theorem 1, Section 10.4), problem (2)–(6) has a unique solution. Also, for the potential equation, we assume that $l \neq \infty$, $m \neq \infty$.

We conclude that $Y_n(0)$ should be the Fourier sine coefficients of f, and $Y_n(m)$ should be the Fourier sine coefficients of g; in other words,

$$Y_n(0) = \frac{2}{l}\int_0^l f(x)\sin\frac{n\pi x}{l}\,dx,$$

and

$$Y_n(m) = \frac{2}{l}\int_0^l g(x)\sin\frac{n\pi x}{l}\,dx.$$

Now

$$Y_n(0) = a_n$$

and

$$Y_n(m) = a_n\cosh\frac{n\pi m}{l} + b_n\sinh\frac{n\pi m}{l}.$$

Hence,

$$a_n = \frac{2}{l}\int_0^l f(x)\sin\frac{n\pi x}{l}\,dx \tag{14}$$

and

$$b_n = \frac{1}{\sinh\dfrac{n\pi m}{l}}\left[\frac{2}{l}\int_0^l g(x)\sin\frac{n\pi x}{l}\,dx - \left(\cosh\frac{n\pi m}{l}\right)\frac{2}{l}\int_0^l f(x)\sin\frac{n\pi x}{l}\,dx\right]. \tag{15}$$

We conclude that the solution of the boundary value problem (2)–(6) is given by

$$u(x, y) = \sum_{n=1}^{\infty}\left[a_n\cosh\frac{n\pi y}{l} + b_n\sinh\frac{n\pi y}{l}\right]\sin\frac{n\pi x}{l}, \tag{16}$$

with a_n and b_n given by Eqs. (14) and (15) respectively.

EXAMPLE 2 Solve the following Dirichlet problem for the rectangle $0 < x < l,\quad 0 < y < m$.

$$u_{xx} + u_{yy} = 0, \qquad 0<x<l, \quad 0<y<m, \tag{17}$$

$$u(x, 0) = 0, \qquad 0<x<l, \tag{18}$$

$$u(x, m) = 0, \qquad 0<x<l, \tag{19}$$

$$u(0, y) = h(y), \quad 0<y<m, \tag{20}$$

$$u(l, y) = k(y), \quad 0<y<m. \tag{21}$$

Solution Using the method of separation of variables, we find that Y must satisfy the eigenvalue problem

$$Y'' + \lambda Y = 0,$$
$$Y(0) = 0,\ Y(m) = 0,$$

and X must satisfy the differential equation

$$X'' - \lambda X = 0.$$

Note that the problems we have to solve are analogous to those of Example 1 except that the roles of X and Y have interchanged. Consequently we can borrow the results of Example 1 to conclude that the solution of the boundary value problem (17)–(21) is given by

$$u(x, y) = \sum_{n=1}^{\infty} \left[\alpha_n \cosh\frac{n\pi x}{m} + \beta_n \sinh\frac{n\pi x}{m}\right] \sin\frac{n\pi y}{m}, \tag{22}$$

where

$$\alpha_n = \frac{2}{m}\int_0^m h(y) \sin\frac{n\pi y}{m}\, dy$$

and

$$\beta_n = \frac{1}{\sinh\dfrac{n\pi l}{m}}\left[\frac{2}{m}\int_0^m k(y)\sin\frac{n\pi y}{m}\,dy - \left(\cosh\frac{n\pi l}{m}\right)\frac{2}{m}\int_0^m h(y)\sin\frac{n\pi y}{m}\,dy\right].$$

EXAMPLE 3 Solve the following Dirichlet problem for the rectangle $0 < x < l,\ \ 0 < y < m$.

$$u_{xx} + u_{yy} = 0, \qquad 0 < x < l,\ \ 0 < y < m, \tag{23}$$
$$u(x, 0) = f(x), \quad 0 < x < l, \tag{24}$$
$$u(x, m) = g(x), \quad 0 < x < l. \tag{25}$$
$$u(0, y) = h(y), \quad 0 < y < m, \tag{26}$$
$$u(l, y) = k(y), \quad 0 < y < m. \tag{27}$$

Solution We consider the two separate problems

$$u_{xx} + u_{yy} = 0, \qquad 0 < x < l,\ \ 0 < y < m,$$
$$u(x, 0) = f(x), \quad 0 < x < l,$$
$$u(x, m) = g(x), \quad 0 < x < l,$$
$$u(0, y) = 0, \qquad 0 < y < m,$$
$$u(l, y) = 0, \qquad 0 < y < m,$$

and

$$
\begin{aligned}
u_{xx} + u_{yy} &= 0, && 0 < x < l, \quad 0 < y < m \\
u(x, 0) &= 0, && 0 < x < l, \\
u(x, m) &= 0, && 0 < x < l, \\
u(0, y) &= h(y), && 0 < y < m, \\
u(l, y) &= k(y), && 0 < y < m.
\end{aligned}
$$

The solution of the first problem will be denoted by $u_1(x, y)$ and is given by formula (16). The solution of the second problem will be denoted by $u_2(x, y)$ and is given by formula (22). Since Eq. (23) is a homogeneous linear partial differential equation, we can use the principle of superposition to conclude that the solution of the boundary value problem (23)–(27) is

$$u(x, y) = u_1(x, y) + u_2(x, y).$$

EXERCISES

In Exercises 1 through 16, certain conditions are given. In each case find the solution of the boundary value problem (2)–(6).

1. $l = 1, m = 1, f(x) = x, g(x) = 0$

2. $l = 1, m = 1, f(x) = 0, g(x) = x$

3. $l = 1, m = 1, f(x) = x^2, g(x) = 0$

4. $l = 1, m = 1, f(x) = 0, g(x) = x^2$

5. $l = 1, m = 1, f(x) = \sin \pi x, g(x) = 0$

6. $l = 1, m = 1, f(x) = 0, g(x) = \sin \pi x$

7. $l = 1, m = 2, f(x) = 0, g(x) = \cos x$

8. $l = 1, m = 2, f(x) = \cos x, g(x) = 0$

9. $l = 2, m = 1, f(x) = 0, g(x) = e^x$

10. $l = 2, m = 1, f(x) = e^x, g(x) = 0$

11. $l = 1, m = 1, f(x) = x, g(x) = \sin \pi x$

12. $l = 1, m = 1, f(x) = \sin \pi x, g(x) = x^2$

13. $l = 2, m = 2, f(x) = e^x, g(x) = \cos x$

14. $l = 2, m = 2, f(x) = \cos x, g(x) = e^x$

15. $l = 1, m = 2, f(x) = x^2, g(x) = e^x$

16. $l = 1, m = 3, f(x) = \sin x, g(x) = x^2$

In Exercises 17 through 32, certain conditions are given. In each case find the solution of the boundary value problem (17)–(21).

17. $l = 1, m = 1, h(y) = \sin y, k(y) = 0$

18. $l = 1, m = 1, h(y) = 0, k(y) = \sin y$

19. $l = 1, m = 1, h(y) = 0, k(y) = y$

20. $l = 1, m = 1, h(y) = y, k(y) = 0$

21. $l = 1, m = 1, h(y) = \cos y, k(y) = 0$

22. $l = 1, m = 1, h(y) = 0, k(y) = \cos y$

23. $l = 1, m = 2, h(y) = 0, k(y) = y^2$

24. $l = 1, m = 2, h(y) = y^2, k(y) = 0$

25. $l = 2, m = 1, h(y) = e^y, k(y) = 0$

26. $l = 2, m = 1, h(y) = 0, k(y) = e^y$

27. $l = 1, m = 1, h(y) = e^{3y}, k(y) = 3y$

28. $l = 1, m = 1, h(y) = \cos y, k(y) = e^y$

29. $l = 2, m = 2, h(y) = y + 2, k(y) = 0$

30. $l = 2, m = 2, h(y) = e^y, k(y) = \cos y$

31. $l = 3, m = 2, h(y) = 1, k(y) = 3$

32. $l = 2, m = 3, h(y) = \sin y, k(y) = 0$

In Exercises 33 through 40, certain conditions are given. In each case solve the boundary value problem (23)–(27).

33. $l = 1, m = 1, f(x) = x, g(x) = 0, h(y) = \sin y, k(y) = 0$

34. $l = 1, m = 1, f(x) = 0, g(x) = x^2, h(y) = 0, k(y) = \cos y$

35. $l = 2, m = 1, f(x) = 0, g(x) = e^x, h(y) = e^y, k(y) = 0$

36. $l = 1, m = 1, f(x) = \sin \pi x, g(x) = 0, h(y) = \cos y, k(y) = 0$

37. $l = 1, m = 1, f(x) = \sin \pi x, g(x) = x^2, h(y) = \sin y, k(y) = 0$

38. $l = 2, m = 1, f(x) = e^x, g(x) = 0, h(y) = 0, k(y) = e^y$

39. $l = 2, m = 2, f(x) = \cos x, g(x) = e^x, h(y) = e^y, k(y) = \cos y$

40. $l = 1, m = 1, f(x) = 0, g(x) = \sin \pi x, h(y) = 0, k(y) = \cos y$

41. A rectangular homogeneous thermally conducting sheet of material (of small thickness) of width 10 cm. and height 15 cm. lies in the first quadrant with one vertex at the origin and one side on the vertical axis. Find the steady state temperature in the sheet if the edge $x = 0$ is maintained at 100°C and the other three edges are maintained at 0°C.

42. Laplace's equation in polar coordinates is (Exercise 50(b), Section 11.3)

$$u_{rr} + \frac{1}{r}u_r + \frac{1}{r^2}u_{\theta\theta} = 0.$$

(a) Set $u = R(r)\Theta(\theta)$ and determine the differential equations for R and Θ.

(b) Solve the differential equations of part (a). [*Hint:* For R see Section 2.7.]

43. Dirichlet's Problem for a Circular Region When the region R for Dirichlet's problem is a circle, the following initial-boundary value problem results.

$$u_{rr} + \frac{1}{r}u_r + \frac{1}{r^2}u_{\theta\theta} = 0, \qquad 0 < r < a, \quad -\infty < \theta < \infty \tag{28}$$

$$u(a, \theta) = f(\theta), \quad 0 \le \theta \le 2\pi, \tag{29}$$

$$u(r, -\pi) - u(r, \pi) = 0, \qquad 0 \le r \le a, \tag{30}$$

$$u_\theta(r, -\pi) - u_\theta(r, \pi) = 0, \qquad 0 \le r \le a, \tag{31}$$

$$u(r, \theta) \text{ is continuous at } r = 0. \tag{32}$$

f is assumed to be periodic of period 2π. Conditions (30) and (31) indicate that u is periodic as a function of θ, and condition (32) demands that u be continuous throughout the circular region. Solve the boundary value problem (28)–(32). [*Hint:* The solutions from part (b) of Exercise 42 are of the form

$$(c_1 + c_2 \log r)(b_1 + b_2\theta) + (c_3 r^{\sqrt{\lambda}} + c_4 r^{-\sqrt{\lambda}})(b_3 \cos \sqrt{\lambda}\theta + b_4 \sin \sqrt{\lambda}\theta).$$

The periodicity conditions require that $b_2 = 0$ and $\lambda = \lambda_n = n^2$, $n = 0, 1, 2, \ldots$. The continuity condition requires that $c_2 = c_4 = 0$. Thus, if

$$u_n(r, \theta) = r^n(\alpha_n \cos n\theta + \beta_n \sin n\theta),$$

we can write

$$u(r, \theta) = \sum_{n=0}^{\infty} u_n(r, \theta).$$

Complete the solution.]

44. Dirichlet's Problem for a Semicircular Region Set up the boundary value problem for the steady state temperature, u, in a semicircular region, if the temperature along the diameter is zero and if the temperature along the semicircle is a known function of θ, say $f(\theta)$. [*Hint:* Consider the center of the semicircle to be at the origin and the semicircle to lie in the upper half plane, so that $0 \le r \le a$ and $0 \le \theta \le \pi$. Note that $u(r, 0) = 0$ and $u(r, \pi) = 0$. See also Exercise 43.]

45. Solve the boundary value problem of Exercise 44 whre $f(\theta) = 100°C$, $a = 1$.

46. Set up the boundary value problem for the steady state temperature in a circular section $0 \leq r \leq a$, $0 \leq \theta \leq \alpha < \pi$ if $u = 0$ on $\theta = 0$, and $\theta = \alpha$ and $u = f(\theta)$ on the circular segment. [*Hint:* See Exercise 44.]

11.9 NONHOMOGENEOUS PARTIAL DIFFERENTIAL EQUATIONS: METHOD I

Our objective is to solve Problem 1 of Section 11.5 with $F \neq 0$. We illustrate the approach with some examples.

EXAMPLE 1 Solve the initial-boundary value problem

$$u_t - au_{xx} = F(x, t), \quad 0 < x < l, \quad t > 0, \tag{1}$$

$$u(0, t) = 0, \qquad t > 0, \tag{2}$$

$$u(l, t) = 0, \qquad t > 0, \tag{3}$$

$$u(x, 0) = f(x), \qquad 0 < x < l. \tag{4}$$

Solution If we set $F(x, t) = 0$ in Eq. (1) and leave the rest of the problem alone, then we have Example 1 of Section 11.7. The solution for this case is

$$\sum_{n=1}^{\infty} c_n e^{-(n^2\pi^2 at)/l^2} \sin \frac{n\pi x}{l},$$

where

$$c_n = \frac{2}{l} \int_0^l f(x) \sin \frac{n\pi x}{l} dx.$$

We introduce the notation $X_n(x) = \sin \dfrac{n\pi x}{l}$ and

$$\phi_n(x) = \sqrt{\frac{2}{l}} \sin \frac{n\pi x}{l}. \tag{5}$$

Applying Theorem 1 of Section 6.2, we have

$$(\phi_n, \phi_m) = \begin{cases} 0, & \text{if } n \neq m \\ 1, & \text{if } n = m \end{cases}.$$

If $\psi(x)$ is twice continuously differentiable and satisfies $\psi(0) = \psi(l) = 0$, then $\psi(x) = \sum_{n=1}^{\infty} (\psi, \phi_n)\phi_n$.

We seek a solution of (1)–(4) in the form

$$u(x, t) = \sum_{n=1}^{\infty} T_n(t)\phi_n(x),$$

where $T_n(t)$ are unknown functions to be determined. By the orthonormality of ϕ_n (Theorem 1 of Section 6.2), we can write

$$T_n(t) = (u, \phi_n) = \int_0^l u(x, t)\phi_n(x)dx. \tag{6}$$

Assuming that differentiation under the integral sign is permissible, we have

$$T_n'(t) = \int_0^l u_t(x, t)\phi_n(x)dx.$$

From Eq. (1) we know that $u_t = au_{xx} + F(x, t)$; hence,

$$\begin{aligned} T_n'(t) &= \int_0^l [au_{xx}(x, t) + F(x, t)]\phi_n(x)dx \\ &= a\int_0^l u_{xx}(x, t)\phi_n(x)dx + \int_0^l F(x, t)\phi_n(x)dx. \end{aligned} \tag{7}$$

Since $F(x, t)$ and $\phi_n(x)$ are known, the second integral in Eq. (7) is a known function of t, say $K_n(t)$. For the first integral in Eq. (7) we employ integration by parts twice to obtain

$$\begin{aligned} \int_0^l u_{xx}(x, t)\phi_n(x)dx &= u_x(x, t)\phi_n(x)\Big|_{x=0}^{x=l} - \int_0^l u_x(x, t)\phi_n'(x)dx \\ &= [u_x(x, t)\phi_n(x) - u(x, t)\phi_n'(x)]\Big|_{x=0}^{x=l} + \int_0^l u(x, t)\phi_n''(x)dx. \end{aligned} \tag{8}$$

Since

$$\phi_n(0) = \phi_n(l) = u(0, t) = u(l, t) = 0,$$

the first term disappears and we have

$$\int_0^l u_{xx}(x, t)\phi_n(x)dx = \int_0^l u(x, t)\phi_n''(x)dx.$$

Furthermore, $\phi_n'' + \mu_n\phi_n = 0$ implies that $\phi_n'' = -\mu_n\phi_n$, where $\mu_n = \dfrac{n^2\pi^2}{l^2}$ [see Eqs. (6) and (9), Section 11.7], hence

$$\int_0^l u_{xx}(x, t)\phi_n(x)dx = -\mu_n\int_0^l u(x, t)\phi_n(x)dx = -\mu_n T_n(t).$$

Substituting our results in Eq. (7), we see that $T_n(t)$ must satisfy the ordinary differential equation

$$T_n'(t) + a\mu_n T_n(t) = K_n(t). \tag{9}$$

Equation (9) is a first-order linear ordinary differential equation and can be solved by the method of Section 1.4. Thus,

$$T_n(t) = e^{-a\mu_n t}\left[\int_0^t K_n(s)e^{a\mu_n s}\,ds + c\right]. \tag{10}$$

From Eqs. (5) and (6), we note that

$$T_n(0) = \int_0^l u(x, 0)\phi_n(x)dx = \int_0^l f(x)\phi_n(x)dx.$$

Consequently, the constant c in Eq. (10) is given by

$$c = T_n(0) = \int_0^l f(x)\phi_n(x)dx. \tag{11}$$

Finally, the solution of the initial-boundary value problem (1)–(4) is given by

$$u(x, t) = \sum_{n=1}^{\infty} T_n(t)\phi_n(x),$$

where $\phi_n(x)$ is given by Eq. (5) and $T_n(t)$ is given by Eqs. (10) and (11).

The method of solution outlined in Example 1 is useful for many problems of the type encountered in Sections 11.6, 11.7, and 11.8. It is important that the associated problem obtained by setting $F(x, t) = 0$ is solvable and that the "boundary terms" in the integration by parts process [see Eq. (8)] either vanish or are known as a function of t. The latter requirement will be met if the boundary conditions for the original problem are $u(0, t) = 0$ and $u(l, t) = 0$. If this is not the case, it is sometimes possible to introduce a change of variables that makes the boundary conditions have this form (see Section 11.10).

EXAMPLE 2 Solve the initial-boundary value problem

$$u_{tt} - c^2u_{xx} = F(x, t), \quad 0 < x < l, \quad t > 0 \tag{12}$$

$$u(x, 0) = f(x), \quad 0 < x < l, \tag{13}$$

$$u_t(x, 0) = g(x), \quad 0 < x < l, \tag{14}$$

$$u(0, t) = 0, \quad t > 0, \tag{15}$$

$$u(l, t) = 0, \quad t > 0. \tag{16}$$

Solution If we set $F(x, t) = 0$ in Eq. (12), we then have the initial-boundary value problem of Section 11.6. As in Example 1, we set

$$\phi_n(x) = \sqrt{2/l}\sin\frac{n\pi x}{l},$$

and we seek a solution of problem (12)–(16) in the form

$$u(x, t) = \sum_{n=1}^{\infty} T_n(t)\phi_n(x),$$

where the functions $T_n(t)$ are to be determined. By Theorem 1 of Section 6.2, we can write

$$T_n(t) = (u, \phi_n) = \int_0^l u(x, t)\phi_n(x)dx. \tag{17}$$

Assuming that differentiation under the integral sign is permissible, we have

$$T'_n(t) = \int_0^l u_t(x, t)\phi_n(x)dx, \tag{18}$$

$$T''_n(t) = \int_0^l u_{tt}(x, t)\phi_n(x)dx. \tag{19}$$

Solving Eq. (12) for u_{tt} and substituting the result in Eq. (19) leads to

$$\begin{aligned} T''_n(t) &= \int_0^l [c^2 u_{xx}(x, t) + F(x, t)]\phi_n(x)dx \\ &= c^2 \int_0^l u_{xx}(x, t)\phi_n(x)dx + \int_0^l F(x, t)\phi_n(x)dx. \end{aligned}$$

Setting

$$K_n(t) = \int_0^l F(x, t)\phi_n(x)dx,$$

and utilizing integration by parts, we have

$$\begin{aligned} T''_n(t) &= c^2[u_x(x, t)\phi_n(x) - u(x, t)\phi'_n(x)] \Big|_{x=0}^{x=l} \\ &\quad + c^2 \int_0^l u(x, t)\phi''_n(x)dx + K_n(t). \end{aligned}$$

As in Example 1, the "boundary term" disappears and $\phi''_n = -\mu_n\phi_n$ with $\mu_n = \dfrac{n^2\pi^2}{l^2}$ [see Eq. (10), Section 11.6]. Thus,

$$T''_n = c^2 \int_0^l u(x, t)[-\mu_n\phi_n(x)]dx + K_n(t).$$

In other words, T_n must satisfy the ordinary differential equation

$$T''_n(t) + c^2\mu_n T_n(t) = K_n(t). \tag{20}$$

Equation (20) is a second-order nonhomogeneous linear ordinary differential equation and can be solved by the method of Section 2.11 or of Section 2.12, depending on the function $K_n(t)$. Coupling Eqs. (13) and (17) and Eqs. (14) and (18), we find that $T_n(t)$ must satisfy the initial conditions

$$T_n(0) = \int_0^l f(x)\phi_n(x)dx, \tag{21}$$

$$T'_n(0) = \int_0^l g(x)\phi_n(x)dx. \tag{22}$$

If the function $K_n(t)$ is continuous for $t > 0$, the initial value problem (20)–(22) has a unique solution $T_n(t)$, and the solution of the original initial-boundary value problem (12)–(16) is given by

$$u(x, t) = \sum_{n=1}^{\infty} T_n(t)\phi_n(x),$$

where $\phi_n(x)$ is given by Eq. (5) and $T_n(t)$ is the solution of (20)–(22).

EXERCISES

In Exercises 1 through 10, solve the initial-boundary value problem (1)–(4) for the given conditions.

1. $a = 1, l = 1, f(x) = x, F(x, t) = x + t$

2. $a = 1, l = 1, f(x) = x, F(x, t) = xt$

3. $a = 1, l = \pi, f(x) = \sin x, F(x, t) = \cos t$

4. $a = 1, l = \pi, f(x) = x^2, F(x, t) = xe^t$

5. $a = 5, l = 1, f(x) = e^x, F(x, t) = x - 3$

6. $a = 1, l = 1, f(x) = x(1 - x), F(x, t) = \sin t$

7. $a = 3, l = \pi, f(x) = \sin 3x, F(x, t) = \cos t - 3x$

8. $a = 2, l = 1, f(x) = \sin^2 x, F(x, t) = x^2e^{3t}$

9. $a = 2, l = \pi, f(x) = x(\pi - x)^2, F(x, t) = xt^2$

10. $a = 1, l = \pi, f(x) = x + \sin x, F(x, t) = t + \sin x$

In Exercises 11 through 20, solve the initial-boundary value problem (12)–(16) for the given conditions.

11. $c = 1, l = 1, f(x) = x(1 - x), g(x) = 0, F(x, t) = x + t$

12. $c = 1, l = 1, f(x) = x^2(1 - x), g(x) = 0, F(x, t) = xt$

13. $c = 1, l = \pi, f(x) = 0, g(x) = 3, F(x, t) = \cos \pi t$

14. $c = 1, l = \pi, f(x) = x(\pi - x), g(x) = 3, F(x, t) = xe^t$

15. $c = 1, l = 1, f(x) = x(1 - x)^2, g(x) = \pi, F(x, t) = x - 3$

16. $c = 1, l = 1, f(x) = x(1 - x), g(x) = 0, F(x, t) = \sin t$

17. $c = 1, l = \pi, f(x) = 0, g(x) = \pi, F(x, t) = \cos \pi t - 3x$

18. $c = 1, l = 1, f(x) = x^2(1 - x), g(x) = 3, F(x, t) = x^2e^{3t}$

19. $c = 1, l = \pi, f(x) = x(\pi - x)^2, g(x) = 0, F(x, t) = xt^2$

20. $c = 2, l = \pi, f(x) = 0, g(x) = \cos x, F(x, t) = t + \sin x$

21. Apply the method of this section to solve the initial-boundary value problem

$$u_{tt} - c^2 u_{xx} = F(x, t), \quad 0 < x < l, \quad t > 0,$$
$$u(x, 0) = f(x), \quad 0 < x < l,$$
$$u_t(x, 0) = g(x), \quad 0 < x < l,$$
$$u(0, t) = 0, \quad t > 0,$$
$$u_x(l, t) = 0, \quad t > 0.$$

22. Apply the method of this section to solve the initial-boundary value problem

$$u_{tt} - c^2 u_{xx} = F(x, t) \quad 0 < x < l, \quad t > 0,$$
$$u(x, 0) = f(x), \quad 0 < x < l,$$
$$u_t(x, 0) = g(x), \quad 0 < x < l,$$
$$u_x(0, t) = 0, \quad t > 0,$$
$$u(l, t) = 0, \quad t > 0.$$

23. Flexible String: Distributed External Forces If $q(x, t, u, u_t, u_x)$ denotes a distributed external force per unit mass in the positive u direction, then in the development of the equation of motion for the string of Section 11.6 one has the additional force term $\rho(\Delta x)q$. Consequently, the basic partial differential equation is

$$u_{tt} - c^2 u_{xx} = q.$$

The following special cases lend themselves to simple interpretation:

(i) $q = f(x, t)$, an applied force;
(ii) $q = -g$, gravity is acting;
(iii) $q = -ku$, k a constant, a restoring force;
(iv) $q = -ru_t$, $r(> 0)$ a constant, damping due to air resistance;
(v) $q = -ru_t - ku$, a combination of (iii) and (iv) [*Note:* this provides an alternative interpretation of the telegraph equation presented in Exercise 60, Section 11.6];
(vi) $q = -ru_t - ku + f(x, t)$, the telegraph equation with an applied force.

Find the displacement of a string π units long subjected to gravity if $f(x) = 0$, $g(x) = 3$, and $c = 1$.

24. Heat Equation: Source Terms Set up and solve the initial-boundary value problem for the temperature in a rod of length π units that is heated by an electric current so that $F(x, t) = xe^t$ (see Exercise 29, Section 11.7), if the initial temperature in the rod is given by $f(x) = x^2$ and the temperature at the ends is maintained at 0°C. Take $a = 1$.

25. A string is stretched between the points (0, 0) and (1, 0). The string is initially at rest and is subjected to the external force $\pi^2 \sin \pi x$. Find the displacement of the string as a function of the time t and the distance x. Take $c = 1$.

26. A string is stretched between the points (0, 0) and (1, 0). The string is initially at rest and is subjected to the external force $\pi^2 x$. Find the displacement of the string as a function of the time t and the distance x. Take $c = 1$.

11.10 NONHOMOGENEOUS PARTIAL DIFFERENTIAL EQUATIONS: METHOD II

We write the partial differential equation (1) in Problem 1 of Section 11.5 in the form $A[u] = F(x, y)$, where A is a linear operator. Set

$$u(x, y) = v(x, y) + w(x, y), \tag{1}$$

where $v(x, y)$ is to be a new unknown function, and $w(x, y)$ is an unknown function to be determined. Substitution of (1) into Problem 1 yields

PROBLEM 1′

$$A[v] = F(x, t) - A[w],$$

$$A_1 v(0, y) + A_2 v_x(0, y) = f_1(y) - A_1 w(0, y) - A_2 w_x(0, y)$$

$$A_3 v(l, y) + A_4 v_x(l, y) = f_2(y) - A_3 w(l, y) - A_4 w_x(l, y)$$

$$A_5 v(x, 0) = f_3(x) - A_5 w(x, 0)$$

$$A_6 v_y(x, 0) = f_4(x) - A_6 w_y(x, 0).$$

The idea is to choose a specific $w(x, t)$ so that Problem 1′ reduces to a problem that is solvable either by the method of Section 11.9 or by one of the methods in Sections 11.6, 11.7, or 11.8. For a given problem we may have considerable flexibility in the choice of w, since the only criterion is that Problem 1′ be a solvable problem. Once w is chosen and v is calculated, the desired solution u is obtained from Eq. (1); in other words, $u = w + v$.

EXAMPLE 1 Find the solution of the initial-boundary value problem

$$u_t - au_{xx} = F_1(x, t), \quad 0 < x < l, \quad t > 0,$$

$$u(0, t) = f_1(t), \qquad t > 0,$$

$$u(l, t) = 0, \qquad t > 0,$$

$$u(x, 0) = f_3(x), \qquad 0 < x < l.$$

Solution Setting $u(x, t) = v(x, t) + w(x, t)$, Problem 1′ becomes

$$\begin{aligned} v_t - av_{xx} &= F_1(x, t) + aw_{xx} - w_t, \quad 0<x<l, \quad t>0,\\ v(0, t) &= f_1(t) - w(0, t), \qquad t>0,\\ v(l, t) &= -w(l, t), \qquad t>0,\\ v(x, 0) &= f_3(x) - w(x, 0), \qquad 0<x<l. \end{aligned}$$

If we can choose $w(x, t)$ so that

$$f_1(t) - w(0, t) = 0 \tag{2}$$

and

$$-w(l, t) = 0, \tag{3}$$

then Problem 1′ will be nothing more than Example 1 of Section 11.9. Condition (3) can be satisfied in many ways, but a simple solution is

$$w(x, t) = (x - l)g(t),$$

where g is to be determined. With this choice, condition (2) becomes

$$f_1(t) + lg(t) = 0,$$

and so

$$g(t) = (-1/l)f_1(t).$$

Hence, a convenient choice for w is

$$w(x, t) = \left(\frac{l - x}{l}\right) f_1(t).$$

Then v is to be a solution of

$$\begin{aligned} v_t - av_{xx} &= F(x, t), \quad 0<x<l, \quad t>0\\ v(0, t) &= 0, \qquad t>0,\\ v(l, t) &= 0, \qquad t>0,\\ v(x, 0) &= f(x), \qquad 0<x<l, \end{aligned}$$

where

$$F(x, t) = F_1(x, t) - \left(\frac{l - x}{l}\right) f_1'(t) \qquad \text{and} \qquad f(x) = f_3(x) - \left(\frac{l - x}{l}\right) f_1(0).$$

This latter problem can be solved by the method of Section 11.9, and we can assume that v is now known. Thus, the solution of the original problem is $u = v + w$.

EXAMPLE 2 Solve the boundary value problem

$$u_{xx} + u_{yy} = x^2, \qquad 0 < x < l, \quad 0 < y < m,$$
$$u(x, 0) = f_1(x), \quad 0 < x < l,$$
$$u(x, m) = g_1(x), \quad 0 < x < l,$$
$$u(0, y) = 0, \qquad 0 < y < m,$$
$$u(l, y) = 0, \qquad 0 < y < m.$$

Solution Setting $u(x, y) = v(x, y) + w(x, y)$, Problem 1′ becomes

$$v_{xx} + v_{yy} = x^2 - w_{xx} - w_{yy}, \quad 0 < x < l, \quad 0 < y < m,$$
$$v(x, 0) = f_1(x) - w(x, 0), \quad 0 < x < l,$$
$$v(x, m) = g_1(x) - w(x, m), \quad 0 < x < l,$$
$$v(0, y) = -w(0, y), \qquad 0 < y < m,$$
$$v(l, y) = -w(l, y), \qquad 0 < y < m.$$

Choose

$$w(x, y) = \frac{1}{12} x(x^3 - l^3).$$

Then

$$w(0, y) = 0, \; w(l, y) = 0, \; w_{yy} = 0, \; w_{xx} = x^2,$$

and Problem 1′ takes the form

$$v_{xx} + v_{yy} = 0, \qquad 0 < x < l, \quad 0 < y < m,$$
$$v(x, 0) = f(x), \quad 0 < x < l,$$
$$v(x, m) = g(x), \quad 0 < x < l,$$
$$v(0, y) = 0, \qquad 0 < y < m,$$
$$v(l, y) = 0, \qquad 0 < y < m,$$

where

$$f(x) = f_1(x) - \frac{1}{12} x(x^3 - l^3) \qquad \text{and} \qquad g(x) = g_1(x) - \frac{1}{12} x(x^3 - l^3).$$

This latter problem is Example 1 of Section 11.8. Thus we can solve for v by the methods of that section, and

$$u(x, t) = v(x, t) + \tfrac{1}{12} x(x^3 - l^3).$$

REMARK 1 Note that any choice of w that has the property that $x^2 - w_{xx} - w_{yy} = 0$ will reduce Problem 1′ to a solvable problem. For example, the choice

$$w(x, y) = \tfrac{1}{12}x^4 + \alpha x + \beta y + \gamma,$$

where α, β, and γ are any constants, reduces Problem 1′ to

$$\begin{aligned} v_{xx} + v_{yy} &= 0, & 0<x<l,\quad 0<y<m,\\ v(x, 0) &= f(x), & 0<x<l,\\ v(x, m) &= g(x), & 0<x<l,\\ v(0, y) &= h(y), & 0<y<m,\\ v(l, y) &= k(y), & 0<y<m, \end{aligned}$$

with

$$\begin{aligned} f(x) &= f_1(x) - \tfrac{1}{12}x^4 - \alpha x - \gamma,\\ g(x) &= g_1(x) - \tfrac{1}{12}x^4 - \alpha x - \beta m - \gamma,\\ h(y) &= -\beta y - \gamma \end{aligned}$$

and

$$k(y) = -\tfrac{1}{12}l^4 - \alpha l - \beta y - \gamma.$$

This latter problem is solvable. (See Example 3, Section 11.8.)

REMARK 2 There is only one solution, u, of the original boundary value problem of Example 2. Therefore, different choices for w (and correspondingly different solutions v) do not give rise to different solutions, but rather to alternative versions of the same function.

REMARK 3 The utility of the method of this section hinges upon our ability to spot appropriate forms for w. Very often the question of whether this task is easy or not is dictated by the form of F, f_1, f_2, f_3, or f_4.

EXERCISES

In Exercises 1 through 5, solve the initial-boundary value problem of Example 1 for each of the given cases. Take $a = l = 1$.

1. $F_1(x, t) = 2t(1 - x)$, $f_1(t) = t^2$, $f_3(x) = x$
2. $F_1(x, t) = (x - 1)\sin t$, $f_1(t) = \cos t$, $f_3(x) = x(1 - x)$
3. $F_1(x, t) = t$, $f_1(t) = \frac{1}{2}t^2$, $f_3(x) = x$
4. $F_1(x, t) = xe^t$, $f_1(t) = -e^t$, $f_3(x) = 3x$
5. $F_1(x, t) = 5xe^{3t}$, $f_1(t) = 2t$, $f_3(x) = x$

In Exercises 6 through 10, solve the boundary value problem of Example 2 for each of the given cases.

6. $f_1(x) = \frac{1}{12}x^4$, $g_1(x) = 0$

7. $f_1(x) = \frac{1}{12}x^4$, $g_1(x) = \frac{1}{12}x(x^3 - 1)$

8. $f_1(x) = 0$, $g_1(x) = x$

9. $f_1(x) = x$, $g_1(x) = 0$

10. $f_1(x) = 1 - x$, $g_1(x) = x$

11. Solve the boundary value problem

$$\begin{aligned} &u_{xx} + u_{yy} = (2 - x^2)\sin y, && 0<x<1, \quad 0<y<\pi \\ &u(x, 0) = 0, && 0<x<1, \\ &u(x, \pi) = 0, && 0<x<1, \\ &u(0, y) = h(y), && 0<y<\pi, \\ &u(1, y) = k(y), && 0<y<\pi \end{aligned}$$

by finding an appropriate function $w(x, y)$ so that the function $v(x, y) = u(x, y) - w(x, y)$ solves a boundary value problem for which the partial differential equation is homogeneous. [*Hint:* try $w(x, y) = Ax^2 \sin y$, where A is to be determined.]

12. Solve the initial-boundary value problem

$$\begin{aligned} &u_t - 2u_{xx} = \left(\frac{\pi - x}{\pi}\right)\cos t + \frac{xt}{\pi}(2 + \pi t), && 0<x<\pi, \quad t>0, \\ &u(0, t) = \sin t, && t>0, \\ &u(\pi, t) = t^2, && t>0, \\ &u(x, 0) = x(\pi - x)^2, && 0<x<\pi \end{aligned}$$

by finding an appropriate function $w(x, t)$ so that $v(x, t) = u(x, t) - w(x, t)$ solves an initial-boundary value problem such that $v(0, t) = v(\pi, t) = 0$.

13. Solve the initial-boundary value problem

$$\begin{aligned} &u_{tt} - u_{xx} = \cos \pi t + \frac{10}{\pi}(\pi - x) + \frac{xe^t}{\pi}, && 0<x<\pi, \quad t>0, \\ &u(x, 0) = \frac{x}{\pi}, && 0<x<\pi, \\ &u_t(x, 0) = \frac{x}{\pi} + 3, && 0<x<\pi, \\ &u(0, t) = 5t^2, && t>0, \\ &u(\pi, t) = e^t, && t>0 \end{aligned}$$

by finding an appropriate function $w(x, t)$ so that $v(x, t) = u(x, t) - w(x, t)$ solves an initial-boundary value problem such that $v(0, t) = v(\pi, t) = 0$.

14. Torsion of a Beam The two-dimensional nonhomogeneous Laplace equation

$$u_{xx} + u_{yy} = f(x, y) \tag{4}$$

is known as the (two-dimensional) *Poisson equation.* If a beam of uniform cross section has its axis coincident with the z-axis, then linear elasticity theory shows that the stress function, $u(x, y)$, for the beam satisfies Eq. (4) with $f(x, y) = -2$ and $u = 0$ on the boundary of the area of intersection of the beam with the xy-plane. Find the stress function of a square beam of side π units. [*Hint:* The original rectangle is $-\pi/2 \leq x \leq \pi/2$, $-\pi/2 \leq y \leq \pi/2$. For convenience of the eigenvalue problem, set $s = x + \pi/2$, $t = y + \pi/2$, and note that Poisson's equation becomes $u_{ss} + u_{tt} = -2$. Solve the new boundary value problem as a first step.]

15. Assume a section of the earth's crust to be a rod with one end at the surface of the earth (considered to be at $x = 0$) and the other end (considered to be at $x = l$) inside the earth at such a depth that the temperature at that end is fixed (taken to be 0°C). Consider that the surface of the earth is warmed by the sun so that the temperature, u, at the surface is given by $28 \cos \dfrac{\pi}{12}(t - 12)$ (the time t is measured in hours; 1 AM corresponds to $t = 1$, and midnight corresponds to $t = 24$). If the initial temperature distribution in the earth's crust is $f(x) = 28\left(\dfrac{x - l}{l}\right)$, set up and solve the initial-boundary value problem for the temperature in the earth's crust when $a = l = 1$. (Assume that the effect of time units in hours rather than seconds has already been accounted for and that no further modifications are necessary.)

16. Suppose that a circular shaft (refer to Exercise 51, Section 11.6) has the end $x = 0$ fixed and is initially at rest in its equilibrium position. Assume that the shaft undergoes torsional vibrations due to a periodic rotation at the end $x = l$ (take $l = 1$) of the form $f(t) = \cos t$. Set up and solve the initial-boundary value problem for the angular displacement. Take $c = 1$.

17. The following initial-boundary value problem corresponds to a special case of the *telegraph equation* (Exercise 60, Section 11.6).

$$\begin{aligned}
u_{tt} + a^2u &= u_{xx}, && 0 < x < l, \quad t > 0\\
u(x, 0) &= 0, && 0 < x < l,\\
u_t(x, 0) &= 0, && 0 < x < l,\\
u(0, t) &= t^3/3!, && t > 0,\\
\rho u_x(l, t) &= u(l, t), && t > 0,
\end{aligned}$$

where a, ρ are constants. Set

$$u(x, t) = v(x, t) + \left(\frac{l - x}{l}\right)^2 \frac{t^3}{3!}$$

and determine an equation for the eigenvalues of the initial-boundary value problem for $v(x, t)$.

18. Loss of Heat through the Sides of a Pipe Suppose we have a cylindrical pipe filled with a hot fluid (of constant temperature), and we wish to investigate the loss of heat through the sides of the pipe. The pipe can be considered a hollow cylinder of inner radius r_1, outer radius r_2; therefore, the temperature *in the pipe* satisfies the initial value problem

$$u_t = \frac{a}{r}(ru_r)_r, \quad r_1 < r < r_2, \quad t > 0$$

$$u(r_1, t) = A, \qquad t > 0,$$

$$u(r_2, t) = B, \qquad t > 0,$$

$$u(r, 0) = f(r), \qquad r_1 < r < r_2,$$

where A is the temperature of the fluid, B is the temperature of the air (or medium) surrounding the pipe, and $f(r)$ is the initial distribution of temperature in the pipe. For the case $A = 100°\text{C}$, $B = 0°\text{C}$, $r_1 = 100$ cm, $r_2 = 101$ cm, and $f(r) = 100(101 - r)$, take $a = 1$ and show that the substitution

$$u(r, t) = v(r, t) + 100(101 - r)$$

produces a nonhomogeneous equation (for v) with homogeneous initial-boundary conditions (the solution of which involves Bessel functions).

REVIEW EXERCISES

In each of Exercises 1 through 4, state the order of the partial differential equation, state whether it is linear or quasilinear, and, if it is linear, indicate whether it is homogeneous or nonhomogeneous and whether it has constant coefficients or variable coefficients.

1. $u_{xxyz} - u_{xxy} + u_{zy} + 3u = 0$

2. $u_{xx} - 5y^2u_{yy} = 3xy$

3. $uu_{xx} + 5u_{yy} - 2(u_x)^2 + 2u^2u_y = 0$

4. $xu_t - 3u_{xx} + 5u_x = 0$

In Exercises 5 and 6, assume that u is a function of four variables x, y, z and t. Integrate the equation to obtain the general solution.

5. $u_{xy} = z^2 + 2xy - t$

6. $u_{xyt} + \frac{1}{t}u_{xy} = 2xy$ [*Hint:* set $v = u_{xy}$.]

7. Show that $u = 3xy + x^2$ is a particular solution of $u_x + u_y = 5x + 3y$.

8. Show that $u = x^3 + \cos(x + 3y)$ is a particular solution of $9u_{xx} - u_{yy} = 54x$.

9. Suppose that $A[u] = u_{xxy} - 5u_{xy} + 2u_y$ and that u_1 is a particular solution of the equation $A[u] = 8x - 9y$ and that u_2 is a particular solution of the equation $A[u] = 4y - 3x + \cos(xy)$. Find a particular solution to the equation $A[u] = 5(x - y) + \cos xy$.

10. Suppose that $A[u] = u_{xx} + 3u_{xy} - 7u_y + 8u$ and that u_1 is a particular solution of the equation $A[u] = e^x \cos y$ and that u_2 is a particular solution of the equation $A[u] = e^{-x} \cos y$. Find a particular solution to the equation $A[u] = \cosh x \cos y$.

In each of Exercises 11 through 14, find a particular solution in the form $e^{\lambda x + \mu y}$.

11. $u_{xx} - u_{yy} = 0$

12. $u_{xx} - u_{xy} + u_x - u_y = 0$

13. $u_{xy} - u_x + u_y - 3u = 0$

14. $u_y - u_x = 0$

In Exercises 15 and 16, assume a solution in the form $u = X(x)Y(y)$ and determine the ordinary differential equations that X and Y satisfy.

15. $3u_x + 8u_y = 2(x + y)u$

16. $u_{xy} - 3u = 0$

In Exercises 17 through 20, classify the equation as hyperbolic, parabolic, or elliptic. Find the general solution.

17. $u_{xx} + u_{yy} = 0$

18. $u_{xx} - 12u_{xy} + 36u_{yy} = 0$

19. $u_{xx} + 4u_{xy} - 5u_{yy} = 0$

20. $u_{xy} + u_{yy} = 0$

21. Solve the following initial-boundary value problem.

$$\begin{aligned} &u_t - u_{xx} = 0, && 0<x<1, \quad t>0 \\ &u(x, 0) = 3 - 3x, && 0<x<1, \\ &u(0, t) = 3, && t>0, \\ &u(1, t) = 1, && t>0. \end{aligned}$$

22. Solve the following initial-boundary value problem

$$\begin{aligned} &u_{tt} - 4u_{xx} = 0, && 0<x<1, \quad t>0, \\ &u(x, 0) = 0, && 0<x<1, \\ &u_t(x, 0) = x, && 0<x<1, \\ &u(0, t) = 1, && t>0, \\ &u(1, t) = 0, && t>0. \end{aligned}$$

23. Solve the following boundary value problem.

$$\begin{aligned} &u_{xx} + u_{yy} = 0, && 0<x<1, \quad 0<y<1, \\ &u(x, 0) = x^2, && 0<x<1, \\ &u(x, 1) = 0, && 0<x<1, \end{aligned}$$

$$u(0, y) = \cos y, \quad 0 < y < 1,$$
$$u(1, y) = 0, \qquad 0 < y < 1.$$

24. Using the method discussed in Section 11.9, solve the following initial-boundary value problem.

$$u_{tt} - u_{xx} = \frac{1}{\sqrt{2}} xt, \qquad 0 < x < 1, \quad t > 0,$$
$$u(x, 0) = \sqrt{2} \sin \pi x, \quad 0 < x < 1,$$
$$u_t(x, 0) = \sqrt{2} \sin 5\pi x, \quad 0 < x < 1,$$
$$u(0, t) = 0, \qquad t > 0,$$
$$u(1, t) = 0, \qquad t > 0.$$

25. Using the method discussed in Section 11.9, solve the following initial-boundary value problem.

$$u_t - u_{xx} = \frac{1}{\sqrt{2}} xe^{\pi^2 t}, \qquad 0 < x < 1, \quad t > 0,$$
$$u(x, 0) = \sqrt{2} \sin \pi x, \quad 0 < x < 1,$$
$$u(0, t) = 0, \qquad t > 0,$$
$$u(1, t) = 0, \qquad t > 0.$$

26. Solve the nonhomogeneous boundary value problem.

$$u_{xx} + u_{yy} = (2 - \pi^2 x^2) \cos \pi y + \left(2 - \frac{\pi^2 x^2}{4}\right) \sin \frac{\pi}{2} y, \quad 0 < x < 1, \quad 0 < y < 1,$$
$$u(x, 0) = x^2, \qquad 0 < x < 1,$$
$$u(x, 1) = 0, \qquad 0 < x < 1,$$
$$u(0, y) = y, \qquad 0 < y < 1,$$
$$u(1, y) = \sin \frac{\pi}{2} y + \cos \pi y, \qquad 0 < y < 1$$

by finding an appropriate function $w(x, y)$ so that the substitution $v(x, y) = u(x, y) - w(x, y)$ leads to a boundary value problem for v that contains a homogeneous partial differential equation. [*Hint:* try w in the form $f(x)[\cos \pi y + \sin \frac{\pi}{2} y]$.]

27. Longitudinal Vibrations in a Bar *Longitudinal* vibrations, $u(x, t)$, of a bar with one end free (see Exercise 51, Section 11.6) and the other end (taken to be $x = l$) subjected to a constant force, p, leads to the initial-boundary value problem.[13]

[13] I. N. Bronshtein and K. A. Semendyayev, *A Guidebook to Mathematics* (New York: Springer-Verlag, 1973), p. 572.

$$\begin{aligned} u_{tt} - a^2u_{xx} &= 0, & 0<x<l, \quad t>0,\\ u(x,0) &= f(x), & 0<x<l,\\ u_t(x,0) &= g(x), & 0<x<l,\\ u_x(0,t) &= 0, & t>0,\\ u_x(l,t) &= kp, & t>0, \end{aligned}$$

where k is a constant. Describe how to find u as a function of x and t.

28. A harp string[14] is plucked so that its initial velocity is zero and its initial shape is

$$u(x,0) = \begin{cases} \dfrac{20h}{9l}x, & 0<x<\dfrac{9l}{20},\\[2mm] \dfrac{20h}{l}\left(\dfrac{l}{2}-x\right), & \dfrac{9l}{20}<x<\dfrac{11l}{20},\\[2mm] \dfrac{20h}{9l}(x-l), & \dfrac{11l}{20}<x<l. \end{cases}$$

Find u as a function of x and t.

[14]This is Exercise 2 in Philip M. Morse and K. Uno Ingard, *Theoretical Acoustics* (New York: McGraw-Hill Book Co., 1968), p. 169. Used with the permission of McGraw-Hill Book Company.

APPENDIX A

Determinants and Linear Systems of Equations

Systems of linear algebraic equations occur in many places in this text. For example, to find the values of the constants of integration in an initial value problem requires us to solve a system of linear equations. Further illustrations are the method of undetermined coefficients and variation of parameters for nonhomogeneous differential equations.

In this appendix we provide tools that are sufficient to solve any linear system. However, we make no attempt at a complete discussion of the problem. The main tool presented here is Cramer's rule. For some systems other tools may be more efficient, for example, elimination or matrix methods.

DEFINITION 1

A square matrix, A, of order n is a function whose domain is the set $\{(i, j): i = 1, 2, \ldots, n; j = 1, 2, \ldots, n\}$ *and whose range is a subset of the complex numbers.*

Notation The image under A of the ordered pair (i, j) is denoted by a_{ij}. That is, $A(i, j) = a_{ij}$. The range of the square matrix A is customarily written in the form

$$\begin{bmatrix} a_{11} & a_{12} & \cdots & a_{1n} \\ a_{21} & a_{22} & \cdots & a_{2n} \\ \cdot & \cdot & & \cdot \\ \cdot & \cdot & & \cdot \\ \cdot & \cdot & & \cdot \\ a_{n1} & a_{n2} & \cdots & a_{nn} \end{bmatrix}. \tag{1}$$

It is common practice to refer to the expression (1) as *the matrix A,* and each image a_{ij} as an element of A. Thus, we speak of the matrix A as consisting of n rows and n columns. For each element a_{ij} the first subscript i refers to the row in which a_{ij} appears, and the second subscript j refers to the column in which a_{ij} appears. Thus, a_{12} appears in the first row, second column.

To every square matrix A we can associate a number (defined below) called the *determinant* of A. The determinant of A is denoted by $|A|$ and is written in the form

$$|A| = \begin{vmatrix} a_{11} & a_{12} & \cdots & a_{1n} \\ a_{21} & a_{22} & \cdots & a_{2n} \\ \cdot & \cdot & & \cdot \\ \cdot & \cdot & & \cdot \\ \cdot & \cdot & & \cdot \\ a_{n1} & a_{n2} & \cdots & a_{nn} \end{vmatrix}. \tag{2}$$

Once again we speak of $|A|$ as consisting of n rows and n columns. The expression in (2) is also referred to as a determinant of order n.

DEFINITION 2

The minor of a_{ij} is the determinant of order $n - 1$ formed by crossing out the ith *row and the* jth *column of $|A|$. The minor of a_{ij} is denoted by M_{ij}.*

EXAMPLE 1 If $|A|$ is of order 4, write the minor of a_{13} and the minor of a_{32}.

Solution We have

$$|A| = \begin{vmatrix} a_{11} & a_{12} & a_{13} & a_{14} \\ a_{21} & a_{22} & a_{23} & a_{24} \\ a_{31} & a_{32} & a_{33} & a_{34} \\ a_{41} & a_{42} & a_{43} & a_{44} \end{vmatrix}.$$

Thus,

$$M_{13} = \begin{vmatrix} a_{21} & a_{22} & a_{24} \\ a_{31} & a_{32} & a_{34} \\ a_{41} & a_{42} & a_{44} \end{vmatrix}, \qquad M_{32} = \begin{vmatrix} a_{11} & a_{13} & a_{14} \\ a_{21} & a_{23} & a_{24} \\ a_{41} & a_{43} & a_{44} \end{vmatrix}.$$

DEFINITION 3

The value of a determinant of order one is a_{11}. That is, $|A| = |a_{11}| = a_{11}$. (Be careful not to misinterpret $|\ \ |$ as being absolute value signs; they are not.)

DEFINITION 4

If $|A|$ is a determinant of order n, then

$$|A| = \sum_{i=1}^{n} a_{ij}(-1)^{i+j} M_{ij} \qquad \textit{for any fixed } j \tag{3}$$

or

$$|A| = \sum_{j=1}^{n} a_{ij}(-1)^{i+j} M_{ij} \qquad \textit{for any fixed } i. \tag{4}$$

When we evaluate $|A|$ by formula (3), we are expanding $|A|$ by the minors of the jth column. Similarly, formula (4) is spoken of as the expansion of $|A|$ by the minors of the ith row.

EXAMPLE 2 Find the value of the general determinant of order 2.

Solution We have

$$|A| = \begin{vmatrix} a_{11} & a_{12} \\ a_{21} & a_{22} \end{vmatrix}.$$

We evaluate $|A|$ by taking the expansion of $|A|$ by the minors of the first row. Hence,

$$\begin{aligned} |A| &= a_{11}(-1)^{1+1}|a_{22}| + a_{12}(-1)^{1+2}|a_{21}| \\ &= a_{11}a_{22} - a_{12}a_{21}. \end{aligned} \tag{5}$$

Equation (5) can be viewed as a formula for the value of any determinant of order 2. In fact, one can represent formula (5) by the following schematic:

$$\begin{vmatrix} a_{11} & a_{12} \\ a_{21} & a_{22} \end{vmatrix} \begin{matrix} - \\ + \end{matrix} \tag{6}$$

In this schematic the arrows indicate multiplication, and the plus and minus signs refer to the algebraic signs needed for formula (5).

EXAMPLE 3 Find the value of each of the following determinants of order 2.

$$\text{(a)} \begin{vmatrix} 3 & 4 \\ 2 & 5 \end{vmatrix} \qquad \text{(b)} \begin{vmatrix} 3 & 4 \\ -2 & -5 \end{vmatrix}$$

Solution

(a) $\begin{vmatrix} 3 & 4 \\ 2 & 5 \end{vmatrix} = +(3)(5) - (2)(4) = 15 - 8 = 7$

(b) $\begin{vmatrix} 3 & 4 \\ -2 & -5 \end{vmatrix} = +(3)(-5) - (-2)(4) = -15 + 8 = -7$

EXAMPLE 4 Find the value of the general determinant of order 3.

Solution

$$|A| = \begin{vmatrix} a_{11} & a_{12} & a_{13} \\ a_{21} & a_{22} & a_{23} \\ a_{31} & a_{32} & a_{33} \end{vmatrix}.$$

We expand by the minors of the second column.

$$\begin{aligned} |A| &= a_{12}(-1)^{1+2}\begin{vmatrix} a_{21} & a_{23} \\ a_{31} & a_{33} \end{vmatrix} + a_{22}(-1)^{2+2}\begin{vmatrix} a_{11} & a_{13} \\ a_{31} & a_{33} \end{vmatrix} \\ &\qquad + a_{32}(-1)^{3+2}\begin{vmatrix} a_{11} & a_{13} \\ a_{21} & a_{23} \end{vmatrix} \\ &= -a_{12}\begin{vmatrix} a_{21} & a_{23} \\ a_{31} & a_{33} \end{vmatrix} + a_{22}\begin{vmatrix} a_{11} & a_{13} \\ a_{31} & a_{33} \end{vmatrix} - a_{32}\begin{vmatrix} a_{11} & a_{13} \\ a_{21} & a_{23} \end{vmatrix}. \end{aligned}$$

Each of these determinants of order 2 can be evaluated by use of the schematic (6). Thus,

$$|A| = -a_{12}(a_{21}a_{33} - a_{31}a_{23}) + a_{22}(a_{11}a_{33} - a_{31}a_{13}) - a_{32}(a_{11}a_{23} - a_{21}a_{13}).$$

Finally, after some rearrangement, we can write

$$|A| = a_{11}a_{22}a_{33} + a_{12}a_{23}a_{31} + a_{13}a_{21}a_{32} - a_{31}a_{22}a_{13} - a_{32}a_{23}a_{11} - a_{33}a_{21}a_{12}. \tag{7}$$

Equation (7) can be viewed as a formula for the value of any determinant of order 3. Formula (7) can be represented by the following schematic:

$$\begin{array}{|ccc|cc} a_{11} & a_{12} & \overline{a_{13}} & \overline{a_{11}} & \overline{a_{12}} \\ a_{21} & a_{22} & a_{23} & a_{21} & a_{22} \\ a_{31} & a_{32} & a_{33} & a_{31} & a_{32} \\ & & + & + & + \end{array} \tag{8}$$

In the schematic (8), the first two columns of $|A|$ are written to the right of the determinant. The arrows indicate multiplication and the signs refer to the algebraic signs necessary to obtain formula (7).

EXAMPLE 5 Evaluate

$$\begin{vmatrix} 2 & 8 & 7 \\ 1 & 4 & -1 \\ -3 & 6 & -2 \end{vmatrix}.$$

Solution

$$\begin{aligned} \begin{array}{|ccc|cc} 2 & 8 & 7 & 2 & 8 \\ 1 & 4 & -1 & 1 & 4 \\ -3 & 6 & -2 & -3 & 6 \end{array} &= +(2)(4)(-2) + (8)(-1)(-3) + (7)(1)(6) \\ &\qquad - (-3)(4)(7) - (6)(-1)(2) - (-2)(1)(8) \\ &= -16 + 24 + 42 + 84 + 12 + 16 \\ &= 162. \end{aligned}$$

For determinants of order 4 or larger, there are *no* convenient schematics such as (6) or (8). Therefore, with such determinants we *must* evaluate them by expansion of the minors of some row or column.

EXAMPLE 6 Evaluate

$$|A| = \begin{vmatrix} 5 & 4 & -2 & 6 \\ -3 & 1 & 1 & 0 \\ 2 & -3 & 1 & 0 \\ 1 & 2 & -1 & 0 \end{vmatrix}.$$

Solution We evaluate $|A|$ by expansion of the minors of the second row.

$$|A| = (-3)(-1)^{2+1}\begin{vmatrix} 4 & -2 & 6 \\ -3 & 1 & 0 \\ 2 & -1 & 0 \end{vmatrix} + (1)(-1)^{2+2}\begin{vmatrix} 5 & -2 & 6 \\ 2 & 1 & 0 \\ 1 & -1 & 0 \end{vmatrix}$$

$$+ (1)(-1)^{2+3}\begin{vmatrix} 5 & 4 & 6 \\ 2 & -3 & 0 \\ 1 & 2 & 0 \end{vmatrix} + (0)(-1)^{2+4}\begin{vmatrix} 5 & 4 & -2 \\ 2 & -3 & 1 \\ 1 & 2 & -1 \end{vmatrix}$$

$$= (3)\begin{vmatrix} 4 & -2 & 6 \\ -3 & 1 & 0 \\ 2 & -1 & 0 \end{vmatrix} + (1)\begin{vmatrix} 5 & -2 & 6 \\ 2 & 1 & 0 \\ 1 & -1 & 0 \end{vmatrix} - (1)\begin{vmatrix} 5 & 4 & 6 \\ 2 & -3 & 0 \\ 1 & 2 & 0 \end{vmatrix}$$

$$+ (0)\begin{vmatrix} 5 & 4 & -2 \\ 2 & -3 & 1 \\ 1 & 2 & -1 \end{vmatrix}.$$

Each of these determinants of order 3 can be evaluated according to the schematic (8). Thus,

$$\begin{vmatrix} 4 & -2 & 6 \\ -3 & 1 & 0 \\ 2 & -1 & 0 \end{vmatrix}\begin{matrix} 4 & -2 \\ -3 & 1 \\ 2 & -1 \end{matrix} = +(4)(1)(0) + (-2)(0)(2) + (6)(-3)(-1)$$

$$- (2)(1)(6) - (-1)(0)(4) - (0)(-3)(-2)$$

$$= 0 + 0 + 18 - 12 - 0 - 0 = 6$$

$$\begin{vmatrix} 5 & -2 & 6 \\ 2 & 1 & 0 \\ 1 & -1 & 0 \end{vmatrix}\begin{matrix} 5 & -2 \\ 2 & 1 \\ 1 & -1 \end{matrix} = +(5)(1)(0) + (-2)(0)(1) + (6)(2)(-1))$$

$$- (1)(1)(6) - (-1)(0)(5) - (0)(2)(-2)$$

$$= 0 + 0 - 12 - 6 - 0 - 0 = -18$$

$$\begin{vmatrix} 5 & 4 & 6 \\ 2 & -3 & 0 \\ 1 & 2 & 0 \end{vmatrix}\begin{matrix} 5 & 4 \\ 2 & -3 \\ 1 & 2 \end{matrix} = +(5)(-3)(0) + (4)(0)(1) + (6)(2)(2)$$

$$- (1)(-3)(6) - (2)(0)(5) - (0)(2)(4)$$

$$= 0 + 0 + 24 + 18 - 0 - 0 = 42$$

$$\begin{vmatrix} 5 & 4 & -2 \\ 2 & -3 & 1 \\ 1 & 2 & -1 \end{vmatrix}\begin{matrix} 5 & 4 \\ 2 & -3 \\ 1 & 2 \end{matrix} = +(5)(-3)(-1) + (4)(1)(1) + (-2)(2)(2)$$

$$- (1)(-3)(-2) + (2)(1)(5) - (-1)(2)(4)$$

$$= 15 + 4 - 8 - 6 - 10 + 8 = 3. \qquad (9)$$

Thus,

$$|A| = (3)(6) + (1)(-18) - (1)(42) + (0)(3) = 18 - 18 - 42 + 0 = -42.$$

The reader no doubt observes that the calculations in (9) could have been disregarded, since the determinant involved was to be multiplied by zero. Indeed, we could have saved considerable labor if originally we had expanded $|A|$ by the minors of the fourth column. In this case we would only have to calculate the minor of the element 6 for the other products would be zero. Hence,

$$|A| = 6(-1)^{1+4}\begin{vmatrix} -3 & 1 & 1 \\ 2 & -3 & 1 \\ 1 & 2 & -1 \end{vmatrix} = (-6)\left|\begin{matrix} -3 & 1 & 1 \\ 2 & -3 & 1 \\ 1 & 2 & -1 \end{matrix}\right|\begin{matrix} -3 & 1 \\ 2 & -3 \\ 1 & 2 \end{matrix}$$

$$= (-6)[+(-3)(-3)(-1) + (1)(1)(1) + (1)(2)(2) - (1)(-3)(1) - (2)(1)(-3) - (-1)(2)(1)]$$

$$= (-6)(-9 + 1 + 4 + 3 + 6 + 2) = (-6)(7) = -42.$$

On the basis of Example 6 we may formulate a general rule for efficiency of computation: to expand by the minors of the row or column that has the most zeros. Unfortunately, some determinants do not have any zeros. In such a situation we make use of the following theorem about determinants.[1]

THEOREM 1

The value of a determinant is unchanged if all the elements of some row (or column) are multiplied by the same nonzero constant and the results added to another row (or column).

EXAMPLE 7 For the determinant

$$|A| = \begin{vmatrix} 1 & 5 & 3 & -2 \\ 7 & -2 & 6 & 1 \\ 8 & 4 & -3 & 5 \\ -3 & 1 & 4 & 2 \end{vmatrix},$$

multiply the second row by -3 and add the results to the third row.

Solution To indicate that row 2 is to be multiplied by -3 and the results added to row 3, we write $(-3)R_2 + R_3$. Thus,

$$\begin{vmatrix} 1 & 5 & 3 & -2 \\ 7 & -2 & 6 & 1 \\ 8 & 4 & -3 & 5 \\ -3 & 1 & 4 & 2 \end{vmatrix} \overset{(-3)R_2 + R_3}{=} \begin{vmatrix} 1 & 5 & 3 & -2 \\ 7 & -2 & 6 & 1 \\ -13 & 10 & -21 & 2 \\ -3 & 1 & 4 & 2 \end{vmatrix}.$$

Note that row 3 is the *only* row that changes.

[1]For a proof of this theorem, see M. Richardson, *College Algebra,* Alternate Edition (Englewood Cliffs, N.J.: Prentice-Hall, 1958).

EXAMPLE 8 For the determinant of Example 7, multiply the first column by 2 and add the results to the fourth column.

Solution.

$$\begin{vmatrix} 1 & 5 & 3 & -2 \\ 7 & -2 & 6 & 1 \\ 8 & 4 & -3 & 5 \\ -3 & 1 & 4 & 2 \end{vmatrix} \overset{2C_1 + C_4}{=} \begin{vmatrix} 1 & 5 & 3 & 0 \\ 7 & -2 & 6 & 15 \\ 8 & 4 & -3 & 21 \\ -3 & 1 & 4 & -4 \end{vmatrix}.$$

EXAMPLE 9 Evaluate the determinant of Example 7.

Solution

$$|A| = \begin{vmatrix} 1 & 5 & 3 & -2 \\ 7 & -2 & 6 & 1 \\ 8 & 4 & -3 & 5 \\ -3 & 1 & 4 & 2 \end{vmatrix} \overset{2C_1 + C_4}{=} \begin{vmatrix} 1 & 5 & 3 & 0 \\ 7 & -2 & 6 & 15 \\ 8 & 4 & -3 & 21 \\ -3 & 1 & 4 & -4 \end{vmatrix}$$

$$\overset{-5C_1 + C_2}{=} \begin{vmatrix} 1 & 0 & 3 & 0 \\ 7 & -37 & 6 & 15 \\ 8 & -36 & -3 & 21 \\ -3 & 16 & 4 & -4 \end{vmatrix}$$

$$\overset{-3C_1 + C_3}{=} \begin{vmatrix} 1 & 0 & 0 & 0 \\ 7 & -37 & -15 & 15 \\ 8 & -36 & -27 & 21 \\ -3 & 16 & 13 & -4 \end{vmatrix}.$$

Actually all three of the steps above could have been carried out simultaneously. Thus, we could have saved some time and some writing. We now expand $|A|$ by the minors of the first row, to obtain

$$|A| = (1)(-1)^{1+1} \begin{vmatrix} -37 & -15 & 15 \\ -36 & -27 & 21 \\ 16 & 13 & -4 \end{vmatrix} = \left|\begin{matrix} -37 & -15 & 15 \\ -36 & -27 & 21 \\ 16 & 13 & -4 \end{matrix}\right| \begin{matrix} -37 & -15 \\ -36 & -27 \\ 16 & 13 \end{matrix}$$

$$\begin{aligned} &= +(-37)(-27)(-4) + (-15)(21)(16) + (15)(-36)(13) \\ &\quad -(16)(-27)(15) - (13)(21)(-37) - (-4)(-36)(-15) \\ &= -3996 - 5040 - 7020 + 6480 + 10{,}101 + 2160 \\ &= -16{,}056 + 18{,}741 \\ &= 2685. \end{aligned}$$

EXAMPLE 10 Evaluate

$$|A| = \begin{vmatrix} 2 & 1 & 1 & 5 & 1 \\ -3 & 2 & -1 & -4 & 2 \\ 5 & 3 & 1 & -3 & 1 \\ 1 & -2 & 2 & 6 & -1 \\ -1 & 1 & -1 & -5 & 2 \end{vmatrix}.$$

Solution

$$|A| \begin{matrix} R_2 + R_1 \\ R_2 + R_3 \\ 2R_2 + R_4 \\ -R_2 + R_5 \\ = \end{matrix} \begin{vmatrix} -1 & 3 & 0 & 1 & 3 \\ -3 & 2 & -1 & -4 & 2 \\ 2 & 5 & 0 & -7 & 3 \\ -5 & 2 & 0 & -2 & 3 \\ 2 & -1 & 0 & -1 & 0 \end{vmatrix}.$$

Expanding by the minors of the third column, we have

$$|A| = (-1)(-1)^{2+3} \begin{vmatrix} -1 & 3 & 1 & 3 \\ 2 & 5 & -7 & 3 \\ -5 & 2 & -2 & 3 \\ 2 & -1 & -1 & 0 \end{vmatrix} \begin{matrix} -R_1 + R_2 \\ -R_1 + R_3 \\ = \end{matrix} \begin{vmatrix} -1 & 3 & 1 & 3 \\ 3 & 2 & -8 & 0 \\ -4 & -1 & -3 & 0 \\ 2 & -1 & -1 & 0 \end{vmatrix}.$$

Expanding by the minors of the fourth column, we obtain

$$|A| = (3)(-1)^{1+4} \begin{vmatrix} 3 & 2 & -8 \\ -4 & -1 & -3 \\ 2 & -1 & -1 \end{vmatrix} \begin{matrix} 2R_2 + R_1 \\ -R_2 + R_3 \\ = (-3) \end{matrix} \begin{vmatrix} -5 & 0 & -14 \\ -4 & -1 & -3 \\ 6 & 0 & 2 \end{vmatrix}.$$

Expanding by the minors of the second column, we have

$$|A| = (-3)(-1)(-1)^{2+2} \begin{vmatrix} -5 & -14 \\ 6 & 2 \end{vmatrix} = (3)(-10 + 84) = (3)(74)$$

$$= 222.$$

Determinants can be used to solve systems of linear algebraic equations. We consider the following system of n equations in n unknowns.

$$\begin{aligned} a_{11}x_1 + a_{12}x_2 + \cdots + a_{1n}x_n &= b_1 \\ a_{21}x_1 + a_{22}x_2 + \cdots + a_{2n}x_n &= b_2 \\ \vdots \qquad\qquad\qquad & \quad \vdots \\ a_{n1}x_1 + a_{n2}x_2 + \cdots + a_{nn}x_n &= b_n. \end{aligned} \tag{10}$$

In system (10) we understand that each a_{ij} and each b_i is known and that x_1, $x_2, \ldots, x_n$ are unknown. The task is to find the value(s) of $x_1, x_2, \ldots, x_n$ that satisfy system (10).

EXAMPLE 11 Show that $x_1 = 1$, $x_2 = 2$ satisfy the system

$$x_1 + x_2 = 3$$

$$3x_1 + 7x_2 = 17.$$

Solution By direct substitution, we have

$$1 + 2 = 3 \Rightarrow 3 = 3$$

$$3(1) + 7(2) = 17 \Rightarrow 17 = 17.$$

Notation The determinant of coefficients, denoted by Δ, for the system (10) is

$$\Delta = \begin{vmatrix} a_{11} & a_{12} & \cdots & a_{1n} \\ a_{21} & a_{22} & \cdots & a_{2n} \\ \cdot & \cdot & & \cdot \\ \cdot & \cdot & & \cdot \\ \cdot & \cdot & & \cdot \\ a_{n1} & a_{n2} & \cdots & a_{nn} \end{vmatrix}$$

By Δ_k we mean the determinant that resembles Δ in every respect *except* that the kth *column* of Δ_k consists of the numbers $b_1, b_2, \ldots, b_n$. That is, a_{ik} is replaced by b_i for each value of i.

We now quote a well known result for the algebraic system (10).

Cramer's Rule *If* $\Delta \neq 0$, *the system* (10) *has a unique solution given by* $x_k = \Delta_k/\Delta$, $k = 1, 2, \ldots, n$.

EXAMPLE 12 Using Cramer's rule, solve the system

$$3x_1 - x_2 + 2x_3 = 1$$

$$5x_1 + 2x_2 - x_3 = -2$$

$$2x_1 - x_2 + x_3 = 3.$$

Solution

$$\Delta = \begin{vmatrix} 3 & -1 & 2 \\ 5 & 2 & -1 \\ 2 & -1 & 1 \end{vmatrix} = \left|\begin{matrix} 3 & -1 & 2 \\ 5 & 2 & -1 \\ 2 & -1 & 1 \end{matrix}\right|\begin{matrix} 3 & -1 \\ 5 & 2 \\ 2 & -1 \end{matrix}$$

$$= +(3)(2)(1) + (-1)(-1)(2) + (2)(5)(-1) - (2)(2)(2) - (-1)(-1)(3) - (1)(5)(-1)$$

$$= 6 + 2 - 10 - 8 - 3 + 5 = -8 \neq 0$$

$$x_1 = \frac{\Delta_1}{\Delta} = \frac{\begin{vmatrix} 1 & -1 & 2 \\ -2 & 2 & -1 \\ 3 & -1 & 1 \end{vmatrix}}{-8} = \frac{\left|\begin{array}{rrr|rr} 1 & -1 & 2 & 1 & -1 \\ -2 & 2 & -1 & -2 & 2 \\ 3 & -1 & 1 & 3 & -1 \end{array}\right|}{-8}$$

$$= \frac{2 + 3 + 4 - 12 - 1 - 2}{-8} = \frac{-6}{-8} = \frac{3}{4}$$

$$x_2 = \frac{\Delta_2}{\Delta} = \frac{\begin{vmatrix} 3 & 1 & 2 \\ 5 & -2 & -1 \\ 2 & 3 & 1 \end{vmatrix}}{-8} = \frac{-6 - 2 + 30 + 8 + 9 - 5}{-8}$$

$$= \frac{34}{-8} = -\frac{17}{4}$$

$$x_3 = \frac{\Delta_3}{\Delta} = \frac{\begin{vmatrix} 3 & -1 & 1 \\ 5 & 2 & -2 \\ 2 & -1 & 3 \end{vmatrix}}{-8} = \frac{18 + 4 - 5 - 4 - 6 + 15}{-8}$$

$$= \frac{22}{-8} = -\frac{11}{4}.$$

Thus, the solution is $x_1 = \frac{3}{4}$, $x_2 = -\frac{17}{4}$, and $x_3 = -\frac{11}{4}$.

In case $b_i = 0$ for every $i = 1, 2, \ldots, n$, system (10) is called a *homogeneous* system. Otherwise, the system is called *nonhomogeneous*. The homogeneous system has an obvious solution, $x_1 = 0, x_2 = 0, \ldots, x_n = 0$. This solution is referred to as the *trivial solution*. The following theorem is important.

THEOREM 2

A homogeneous system has nontrivial solutions (in other words, solutions other than the trivial solution) if and only if $\Delta = 0$.

EXAMPLE 13 Solve the system

$$\begin{aligned} x_1 - x_2 + x_3 &= 0 \\ x_1 + x_2 + x_3 &= 0 \\ 2x_1 \qquad + 2x_3 &= 0. \end{aligned}$$

Solution

$$\Delta = \begin{vmatrix} 1 & -1 & 1 \\ 1 & 1 & 1 \\ 2 & 0 & 2 \end{vmatrix} \overset{R_2 + R_1}{=} \begin{vmatrix} 2 & 0 & 2 \\ 1 & 1 & 1 \\ 2 & 0 & 2 \end{vmatrix} = (1)(-1)^{2+2} \begin{vmatrix} 2 & 2 \\ 2 & 2 \end{vmatrix} = 4 - 4 = 0.$$

Hence, we anticipate nontrivial solutions. From the third equation we obtain

$x_3 = -x_1$. Substituting this result into either the first equation or the second yields $x_2 = 0$. Thus, we have the solution

$$x_1 = a, \qquad x_2 = 0, \qquad x_3 = -a, \text{ where } a \text{ is any number.}$$

Note that there are an infinity of solutions in this case.

Partial-Fraction Decomposition

The method of partial-fraction decomposition is useful in differential equations when we want to compute the inverse Laplace transform or the integral of functions of such forms as

$$\frac{2x+3}{x^2(x-1)}, \frac{3x}{x^2-1}, \frac{-4x^3+33x^2-58x+31}{(x-2)^2(x+1)(x-1)}, \frac{2x^4+1}{x^3(x^2+1)^2(x-2)}.$$

Such functions, being the quotient of two polynomials, are called *rational fractions.* A proper rational fraction is one for which the degree of the numerator is less than the degree of the denominator. Otherwise, the rational fraction is improper.

Using long division, improper fractions can be written as a polynomial plus a proper rational fraction. For this reason we restrict our attention to proper rational fractions.

The reader no doubt is familiar with the algebraic problem of adding the fractions

$$\frac{2}{x-2} + \frac{5}{(x-2)^2} + \frac{-7}{x+1} + \frac{1}{x-1}. \tag{1}$$

Of course, the method is to find the *least common denominator* (*L.C.D.*) of the fractions [$(x+1)(x-1)(x-2)^2$ in this illustration], express each fraction as an equivalent fraction having the L.C.D. as denominator, add the resulting numerators, and express the result over the L.C.D. For our illustration the answer would be

$$\frac{-4x^3+33x^2-58x+31}{(x-2)^2(x+1)(x-1)}. \tag{2}$$

What about the reverse problem? That is, beginning with the fraction (2), can we write this fraction in the form (1)? The process associated with this reverse problem is known as *decomposing a proper rational fraction into partial fractions.* To decide what *simple* kinds of fractions might occur in the decomposition we look at the denominator in (2). The factor $(x-2)^2$ suggests two fractions, one with denominator $(x-2)$ and the other with denominator $(x-2)^2$ [in the general case of a "repeated factor" $(x-a)^n$, there would be n fractions having, respectively, the denominators $(x-a)$, $(x-a)^2$, $(x-a)^3, \ldots, (x-a)^n$]; the factor $(x+1)$ suggests a fraction with this denominator, and the factor $(x-1)$ suggests a fraction with this denominator. Since the fraction (2) is proper, each of the partial fractions will also be proper.

Thus, the numerators for the fractions having the denominator $(x - 2)$, $(x + 1)$, or $(x - 1)$ will be constants. Since we have no way of knowing (as yet) what values these constants should have, we consider them to be unknowns and denote them by A, B, and C. Since the denominator $(x - 2)^2$ is quadratic, one might suspect that its numerator should be of degree 1, that is, of the form $Dx + E$, where D and E are unknown. Although this suspicion is correct, it is unnecessary. That a linear numerator is unnecessary can be seen as follows. The factor $(x - 2)^2$ would produce the two partial fractions

$$\frac{A}{x-2} + \frac{Dx+E}{(x-2)^2} = \frac{A}{x-2} + \frac{D(x-2)+(2D+E)}{(x-2)^2} = \frac{A}{x-2} + \frac{D(x-2)}{(x-2)^2} + \frac{2D+E}{(x-2)^2} = \frac{A+D}{x-2} + \frac{2D+E}{(x-2)^2}.$$

Since $\alpha = A + D$ and $\beta = 2D + E$ are unknown constants, it is the same as if we began with constants in both numerators.

Using similar algebraic manipulations, one can establish the following rule.

Rule 1 *For each partial fraction associated with a repeated factor, the numerator can be taken to be of one less degree than the factor that is repeated.*

For example,

$$\frac{x+2}{x^2(x-1)} = \frac{A}{x} + \frac{B}{x^2} + \frac{C}{x-1}$$

$$\frac{3}{x^2-1} = \frac{A}{x-1} + \frac{B}{x+1}$$

$$\frac{x^2-3}{x(x-1)^2(x^2+1)} = \frac{A}{x} + \frac{B}{x-1} + \frac{C}{(x-1)^2} + \frac{Dx+E}{x^2+1}$$

$$\frac{x^2+1}{x^2-1} = 1 + \frac{2}{x^2-1} = 1 + \frac{A}{x-1} + \frac{B}{x+1}$$

$$\frac{x^3-3x^2+7}{x^3(x^2+1)^2(x-3)} = \frac{A}{x} + \frac{B}{x^2} + \frac{C}{x^3} + \frac{Dx+E}{x^2+1} + \frac{Fx+G}{(x^2+1)^2} + \frac{H}{x-3}$$

For the fraction (2) we would write

$$\frac{-4x^3+33x^2-58x+31}{(x-2)^2(x+1)(x-1)} = \frac{A}{x-2} + \frac{D}{(x-2)^2} + \frac{B}{x+1} + \frac{C}{x-1}. \tag{3}$$

The problem is to find values for A, B, C, and D so that Eq. (3) holds for all values of x. If we clear fractions in Eq. (3), we obtain

$$\begin{aligned} -4x^3 + 33x^2 - 58x + 31 &= A(x-2)(x+1)(x-1) + D(x+1)(x-1) \\ &\quad + B(x-2)^2(x-1) + C(x-2)^2(x+1) \qquad (4) \\ &= A(x^3 - 2x^2 - x + 2) + D(x^2 - 1) \\ &\quad + B(x^3 - 5x^2 + 8x - 4) + C(x^3 - 3x^2 + 4) \\ &= (A + B + C)x^3 + (-2A + D - 5B - 3C)x^2 \\ &\quad + (-A + 8B)x + (2A - D - 4B + 4C). \qquad (5) \end{aligned}$$

Equation (5) is to hold for all values of x. Therefore, the coefficient of each power of x should be the same on both sides of the equation; that is,

$$\begin{aligned} A + B + C &= -4 \\ -2A - 5B - 3C + D &= 33 \\ -A + 8B &= -58 \\ 2A - 4B + 4C - D &= 31. \end{aligned} \qquad (6)$$

Equations (6) constitute a linear algebraic system of four equations in four unknowns, and so we can solve them by the methods of Appendix A.

The method outlined in Eqs. (3) through (6) is typical of partial-fraction decomposition. The method can be used successfully for the decomposition of any proper rational fraction when the factors of the denominator are known.

In some cases, however, it is possible to introduce some shortcuts. For example, if one were to substitute $x = 2$ in Eq. (4), the terms involving A, B, and C would disappear and we would have a single equation involving the single unknown D:

$$-4(2)^3 + 33(2)^2 - 58(2) + 31 = D(2 + 1)(2 - 1)$$

or

$$D = \frac{-4(2)^3 + 33(2)^2 - 58(2) + 31}{(2 + 1)(2 - 1)} = 5. \qquad (7)$$

It is convenient to write Eq. (7) in the form

$$D = \left.\frac{-4x^3 + 33x^2 - 58x + 31}{(x + 1)(x - 1)}\right|_{x=2} \qquad (8)$$

Formula (8) we note is nothing more than the left-hand side of Eq. (3) with the denominator under D deleted and then evaluated at $x = 2$ (the value that makes the denominator under D equal zero).

Similarly, we can substitute $x = -1$ in Eq. (4) and thus obtain an equation involving the single unknown, B. The solution for B can also be written in a form similar to Eq. (8). Thus,

$$B = \left.\frac{-4x^3 + 33x^2 - 58x + 31}{(x - 2)^2(x - 1)}\right|_{x=-1} = -7. \qquad (9)$$

Once again formula (9) is the left-hand side of Eq. (3) with the denominator under B deleted and evaluated at $x = -1$ (the value that makes the denominator under B equal zero).

In exactly the same way, we obtain

$$C = \left.\frac{-4x^3 + 33x^2 - 58x + 31}{(x-2)^2(x+1)}\right|_{x=1} = 1.$$

Unfortunately, we cannot obtain A by any "convenient" shortcut. [In general, for repeated factors only the numerator of $(x - a)^n$ with the highest exponent can be obtained by this shortcut method; the other numerators can be evaluated by a differentiation process but this process is often more complicated than the method already introduced.] Therefore, we must follow through with the procedure embodied in Eqs. (3) through (6). We notice, however, that since there is only one unknown left to determine, we need only one equation. We could proceed by multiplying Eq. (3) by the L.C.D., thus obtaining Eq. (4). In Eq. (4) it is easy to determine the coefficient of x^3 without doing any algebra. Thus, we could immediately write the first of Eqs. (6) and solve this equation for A. Hence,

$$A = -4 - B - C = -4 - (-7) - 1 = 2.$$

Combining our results, we have

$$\frac{-4x^3 + 33x^2 - 58x + 31}{(x-2)^2(x+1)(x-1)} = \frac{2}{x-2} + \frac{5}{(x-2)^2} + \frac{-7}{x+1} + \frac{1}{x-1}.$$

As a second illustration let us decompose the fraction

$$\frac{3x^2 - x + 3}{(x+1)(x-1)(x-3)}$$

into partial fractions. We write

$$\frac{3x^2 - x + 3}{(x+1)(x-1)(x-3)} = \frac{A}{x+1} + \frac{B}{x-1} + \frac{C}{x-3}.$$

Each of A, B, and C can be evaluated by the shortcut introduced above. Thus,

$$A = \left.\frac{3x^2 - x + 3}{(x-1)(x-3)}\right|_{x=-1} = \frac{7}{8},$$

$$B = \left.\frac{3x^2 - x + 3}{(x+1)(x-3)}\right|_{x=1} = -\frac{5}{4},$$

and

$$C = \left.\frac{3x^2 - x + 3}{(x-1)(x+1)}\right|_{x=3} = \frac{27}{8}.$$

Consequently,

$$\frac{3x^2 - x + 3}{(x+1)(x-1)(x-3)} = \frac{7/8}{x+1} + \frac{-5/4}{x-1} + \frac{27/8}{x-3}.$$

As a final illustration, let us decompose the fraction

$$\frac{18x^2 + 20x + 18}{(x+3)(x-3)(x^2+1)}$$

into partial fractions. In this case, we write

$$\frac{18x^2 + 20x + 18}{(x+3)(x-3)(x^2+1)} = \frac{A}{x+3} + \frac{B}{x-3} + \frac{Cx + D}{x^2+1}. \tag{10}$$

A and B can be obtained by the shortcut method, but C and D cannot. We first obtain A and B.

$$A = \left.\frac{18x^2 + 20x + 18}{(x-3)(x^2+1)}\right|_{x=-3} = -2$$

and

$$B = \left.\frac{18x^2 + 20x + 18}{(x+3)(x^2+1)}\right|_{x=3} = 4.$$

Clearing fractions in Eq. (10), we have

$$18x^2 + 20x + 18 = A(x-3)(x^2+1) + B(x+3)(x^2+1) + (Cx + D)(x-3)(x+3).$$

The coefficient of x^3 can be easily read off:

$$A + B + C = 0.$$

Therefore, $C = -A - B = -(-2) - 4 = -2$. Also, the constant term can be easily obtained without doing all the algebra. In this case, we have

$$-3A + 3B - 9D = 18 \Rightarrow D = \tfrac{1}{9}(-3A + 3B - 18) = \tfrac{1}{9}[-3(-2) + 3(4) - 18] = 0.$$

Finally, we can write

$$\frac{18x^2 + 20x + 18}{(x+3)(x-3)(x^2+1)} = \frac{-2}{x+3} + \frac{4}{x-3} + \frac{-2x}{x^2+1}.$$

These three illustrations incorporate all the features of partial fractions that one needs for the exercises in this text. We repeat that the method outlined in Eqs. (3) through (6) *always* works. Sometimes the amount of labor required can be reduced by using the shortcuts discussed in the illustrations.

Solutions of Polynomial Equations

In Sections 2.4 and 2.5 we encountered the problem of finding the roots of the characteristic equation

$$a_n\lambda^n + a_{n-1}\lambda^{n-1} + \cdots + a_1\lambda + a_0 = 0. \tag{1}$$

In this appendix we describe a few techniques that are useful in solving some of the exercises that appear in the text.

Consider first the problem of dividing the polynomial $4\lambda^4 + 4\lambda^3 - 25\lambda^2 - \lambda + 6$ by $\lambda - 3$. Using the usual method of long division we have

$$\begin{array}{r|l}
 & 4\lambda^3 - 16\lambda^2 + 23\lambda + 68 \\ \hline
\lambda - 3 & 4\lambda^4 + 4\lambda^3 - 25\lambda^2 - \lambda + 6 \\
 & \underline{-4\lambda^4 \pm 12\lambda^3} \\
 & \qquad 16\lambda^3 - 25\lambda^2 \\
 & \qquad \underline{-16\lambda^3 \pm 48\lambda^2} \\
 & \qquad\qquad 23\lambda^2 - \lambda \\
 & \qquad\qquad \underline{-23\lambda^2 \pm 69\lambda} \\
 & \qquad\qquad\qquad 68\lambda + 6 \\
 & \qquad\qquad\qquad \underline{-68\lambda \pm 204} \\
 & \qquad\qquad\qquad\qquad 210
\end{array}$$

Thus we note that the quotient is $4\lambda^3 + 16\lambda^2 + 23\lambda + 68$ and the remainder is 210; in other words,

$$4\lambda^4 + 4\lambda^3 - 25\lambda^2 - \lambda + 6 = (\lambda - 3)(4\lambda^3 + 16\lambda^2 + 23\lambda + 68) + 210.$$

In the special case that the divisor is of the form $\lambda - a$ ($a = 3$ in the above illustration), the long division problem can be carried out very quickly by a process known as *synthetic division*. Suppose we want to divide $a_n\lambda^n + a_{n-1}\lambda^{n-1} + \cdots + a_1\lambda + a_0$ by $\lambda - a$. Symbolically the synthetic division is

$$\begin{array}{l|cccccc}
\underline{a} & a_n & a_{n-1} & a_{n-2} & \cdots & a_1 & a_0 \\
 & & c_{n-1} & c_{n-2} & \cdots & c_1 & c_0 \, . \\ \hline
 & b_n & b_{n-1} & b_{n-2} & \cdots & b_1 & b_0
\end{array}$$

where

$$b_n = a_n$$

$$c_i = ab_{i+1}, \qquad i = n-1, n-2, \ldots, 1, 0$$

$$b_i = a_i + c_i, \qquad i = n-1, n-2, \ldots, 1, 0.$$

The quotient is $b_n\lambda^{n-1} + b_{n-1}\lambda^{n-2} + \cdots + b_2\lambda + b_1$ and the remainder is b_0.

For the above illustration we have

$$\begin{array}{r|rrrrr} 3 & 4 & 4 & -25 & -1 & 6 \\ & & 12 & 48 & 69 & 204. \\ \hline & 4 & 16 & 23 & 68 & 210 \end{array}$$

Thus the quotient is $4\lambda^3 + 16\lambda^2 + 23\lambda + 68$ and the remainder is 210.

As a second illustration let us divide $3\lambda^4 - 2\lambda^3 - 2\lambda^2 + 4$ by $\lambda+2$, using synthetic division. Note that $\lambda+2 = \lambda-(-2) \Rightarrow a = -2$. Also observe the presence of 0 in the first line of the synthetic division since the coefficient of λ is zero.

$$\begin{array}{r|rrrrr} -2 & 3 & -2 & -2 & 0 & 4 \\ & & -6 & 16 & -28 & 56 \\ \hline & 3 & -8 & 14 & -28 & 60 \end{array}$$

Thus the quotient is $3\lambda^3 - 8\lambda^2 + 14\lambda - 28$ and the remainder is 60.

Now let us divide $4\lambda^4 + 4\lambda^3 - 25\lambda^2 - \lambda + 6$ by $\lambda-2$:

$$\begin{array}{r|rrrrr} 2 & 4 & 4 & -25 & -1 & 6 \\ & & 8 & 24 & -2 & -6. \\ \hline & 4 & 12 & -1 & -3 & 0 \end{array}$$

Thus the quotient is $4\lambda^3 + 12\lambda^2 - \lambda - 3$ and the remainder is 0. In this case we can write

$$4\lambda^4 + 4\lambda^3 - 25\lambda^2 - \lambda - 6 = (\lambda-2)(4\lambda^3+12\lambda^2-\lambda-3). \tag{2}$$

From Eq. (2) we see that the problem of finding the roots of

$$4\lambda^4 + 4\lambda^3 - 25\lambda^2 - \lambda - 6 = 0 \tag{3}$$

is equivalent to finding the roots of

$$(\lambda-2)(4\lambda^3 + 12\lambda^2 - \lambda - 3) = 0.$$

Thus one root is $\lambda = 2$ and subsequent roots are found by solving the equation

$$4\lambda^3 + 12\lambda^2 - \lambda - 3 = 0. \tag{4}$$

We conclude that synthetic division provides us with a test to determine whether or not $\lambda = a$ is a root of the polynomial equation (1). Simply perform synthetic division with a and if the remainder is zero, then $\lambda = a$ is a root; if the remainder is not zero, then $\lambda = a$ is not a root.

Thus the synthetic division

$$\begin{array}{r|rrrr} -3 & 4 & 12 & -1 & -3 \\ & & -12 & 0 & 3 \\ \hline & 4 & 0 & -1 & 0 \end{array}$$

shows that $\lambda = -3$ is a root of Eq. (4) (hence a root of Eq. (3)). Subsequent roots are found by solving the equation

$$4\lambda^2 - 1 = 0,$$

in other words, $\lambda = \pm\frac{1}{2}$. Thus the roots of Eq. (3) are $\lambda = \frac{1}{2}, -\frac{1}{2}, 2, -3$.

In the previous illustration it may appear that we made lucky guesses at choices for a. This is not the case; the guesses are dictated by the following principle, which we state without proof.

Possibilities for Rational Roots If all the coefficients of Eq. (1) are integers, then $\lambda = p/q$, where p and q are integers, is a root of Eq. (1) only if p is a factor of a_0 and q is a factor of a_n.

For Eq. (3) $a_n \equiv a_4 = 4$ and $a_0 = -6$. Thus the possibilities for p are ± 1, ± 2, ± 3, ± 6, and the possibilities for q are ± 1, ± 2, ± 4. Therefore, the only rational numbers that can possibly be a root of Eq. (3) are $\pm\frac{1}{4}, \pm\frac{1}{2}, \pm\frac{3}{4}, \pm 1, \pm\frac{3}{2}, \pm 2, \pm 3, \pm 6$. We test these numbers and no others. Each time we find a root we decrease the degree of the polynomial to be solved by one. Since any root of the new polynomial must be a root of the original polynomial, there are no additional rational roots possible. In fact, the original list of possibilities will, in general, be diminished. For example, after finding $\lambda = 2$ to be a root of Eq. (3), we search for roots of Eq. (4). The only possible rational roots of Eq. (4) are $\pm\frac{1}{4}, \pm\frac{1}{2}, \pm\frac{3}{4}, \pm 1, \pm\frac{3}{2}, \pm 3$.

Proof of the Existence and Uniqueness Theorem

Here we will prove Theorem 1 of Section 1.2. For a thorough understanding of the proof, the reader should be familiar with some topics usually studied in advanced calculus. For easy reference we restate the theorem here.

THEOREM 1

Consider the initial value problem

$$y' = F(x, y) \tag{1}$$

$$y(x_0) = y_0. \tag{2}$$

Assume that the functions F and $\partial F/\partial y$ are continuous in some rectangle

$$\mathcal{R} = \left\{(x, y): \begin{matrix} |x - x_0| \leq A \\ |y - y_0| \leq B \end{matrix}\right\} \quad A > 0, B > 0$$

about the point (x_0, y_0). Then there is a positive number $h \leq A$ such that the IVP(1)–(2) has one and only one solution in the interval $|x - x_0| \leq h$.

Proof The proof is long and will be accomplished in several steps.

1. Since F is continuous in $\mathcal{R}$, it is bounded there. That is, there exists a positive number M such that $|F(x, y)| \leq M$ for every point (x, y) in $\mathcal{R}$. Let h be the smaller of the two numbers A and B/M. That is,

$$h = \min(A, B/M). \tag{3}$$

We will prove that the IVP(1)–(2) has one but no more than one solution in the interval $|x - x_0| \leq h$, where h is given by (3). Although we will not be able to find explicitly the solution of the IVP(1)–(2), we will show that it exists, is unique, and how to approximate it. Roughly speaking, the proof consists in showing that the sequence of functions

$$\begin{aligned}
y_0(x) &= y_0 \\
y_1(x) &= y_0 + \int_{x_0}^{x} F(s, y_0(s))ds \\
y_2(x) &= y_0 + \int_{x_0}^{x} F(s, y_1(s))ds \\
&\cdots\cdots\cdots\cdots \\
y_n(x) &= y_0 + \int_{x_0}^{x} F(s, y_{n-1}(s))ds \\
&\cdots\cdots\cdots\cdots
\end{aligned} \tag{4}$$

converges and the limit function is the unique solution of the IVP(1)–(2).

2. Assume that (x, y_1) and (x, y_2) are any two points in $\mathcal{R}$. Applying the mean value theorem to F, considering $F(x, y)$ as a function of y, we have

$$F(x, y_1) - F(x, y_2) = \frac{\partial F}{\partial y}(x, \tilde{y})(y_1 - y_2), \tag{5}$$

where $\tilde{y}$ is between y_1 and y_2. Since $\partial F/\partial y$ is continuous in $\mathcal{R}$, it is bounded there. That is, there exists a positive number K such that

$$\left| \frac{\partial F}{\partial y}(x, y) \right| \leq K \tag{6}$$

for every point (x, y) in $\mathcal{R}$.

From (5) and (6) it follows that for every pair of points (x, y_1) and (x, y_2) in $\mathcal{R}$, the function F satisfies the condition

$$| F(x, y_1) - F(x, y_2) | \leq K | y_1 - y_2 |. \tag{7}$$

A function F which is defined in a set $\mathcal{R}$ and satisfies the condition (7) for some positive constant K and every pair of points (x, y_1) and (x, y_2) in $\mathcal{R}$ is called *Lipschitz continuous in y over $\mathcal{R}$ with Lipschitz constant K*. Thus the hypothesis that $\partial F/\partial y$ is continuous in $\mathcal{R}$ implies that F is Lipschitz continuous in y over $\mathcal{R}$. In this theorem it is the Lipschitz continuity of F rather than the continuity of $\partial F/\partial y$ that is needed for its proof.

3. Consider the sequence of functions $y_0(x), y_1(x), \ldots, y_n(x), \ldots$ defined by (4). We will show that $y_n(x)$ for $n = 0, 1, 2, \ldots$ exists in $| x - x_0 | \leq h$, is continuous there, and satisfies the inequality

$$| y_n(x) - y_0 | \leq B \qquad \text{for} \qquad | x - x_0 | \leq h, \qquad n = 0, 1, 2, \ldots. \tag{8}$$

This will establish the fact that the sequence (4) is well defined. We will prove the above claim by induction. That is, we will first show that our claim is true for $n = 0$. Then we will assume that our claim is true for $n = k$ and show that it is true for $n = k+1$. For $n = 0$, $y_0(x) = y_0$ is continuous in $| x - x_0 | \leq h$ because a constant function is continuous everywhere. Also (8) holds because $| y_0(x) - y_0 | = 0$. Now assume that $y_k(x)$ is continuous in $| x - x_0 | \leq h$ and that (8) holds for $n = k$. Then, $F(x, y_k(x))$ is continuous in x in the interval $| x - x_0 | \leq h$ because F and y_k are. Hence, $y_{k+1}(x)$ is well defined by (4) with $n = k+1$ and is continuous in the interval $| x - x_0 | \leq h$ because the integral of a continuous function is continuous. Finally,

$$\begin{aligned} | y_{k+1}(x) - y_0 | &= \left| \int_{x_0}^{x} F(s, y_k(s))\,ds \right| \\ &\leq \left| \int_{x_0}^{x} | F(s, y_k(s)) |\,ds \right| \\ &\leq M \left| \int_{x_0}^{x} ds \right| \leq M | x - x_0 | \leq Mh \leq M \frac{B}{M} = B. \end{aligned}$$

This establishes our claim for all $n = 0, 1, 2, \ldots$. The very last inequality explains the choice of h.

4. Next we will prove, again by induction, that for $|x-x_0| \leq h$ the following inequalities hold for $n = 1, 2, \ldots$

$$|y_n(x) - y_{n-1}(x)| \leq \frac{MK^{n-1}|x-x_0|^n}{n!} \leq \frac{MK^{n-1}h^n}{n!}. \tag{9}$$

For $n = 1$, we have

$$|y_1(x) - y_0(x)| = \left| \int_{x_0}^{x} F(s, y_0(s))\, ds \right| \leq M \left| \int_{x_0}^{x} ds \right| = M|x-x_0| \leq Mh.$$

Hence (9) is true for $n = 1$. Next we assume that (9) holds for $n = m$, that is, we assume that for $x-x_0 \leq h$

$$|y_m(x) - y_{m-1}(x)| \leq \frac{MK^{m-1}|x-x_0|^m}{m!} \leq \frac{MK^{m-1}h^m}{m!} \tag{10}$$

and show that (9) holds also for $n = m+1$. In fact, using (4) and (7), we obtain

$$\begin{aligned} |y_{m+1}(x) - y_m(x)| &= \left| \int_{x_0}^{x} F(s, y_m(s))ds - \int_{x_0}^{x} F(s, y_{m-1}(s))ds \right| \\ &\leq \left| \int_{x_0}^{x} |F(s, y_m(s)) - F(s, y_{m-1}(s))|\, ds \right| \\ &\leq K \left| \int_{x_0}^{x} |y_m(s) - y_{m-1}(s)|\, ds \right|. \end{aligned}$$

Now, using (10), we find

$$\begin{aligned} |y_{m+1}(x) - y_m(s)| &\leq \frac{MK^m}{m!} \left| \int_{x_0}^{x} |s-x_0|^m ds \right| \\ &= \frac{MK^m|x-x_0|^{m+1}}{(m+1)!} \leq \frac{MK^m h^{m+1}}{(m+1)!} \end{aligned}$$

and the proof of (9) is complete.

5. The next step in the proof is to show that the sequence of functions $y_n(x)$ converges uniformly to a limit function $y(x)$ on the interval $|x-x_0| \leq h$. To this end we make the following observations: First, the nth partial sum of the series

$$y_0(x) + \sum_{n=1}^{\infty} [y_n(x) - y_{n-1}(x)] \tag{11}$$

is $y_n(x)$. Thus the series (11) and the sequence $y_n(x)$ have the same convergence properties. Second, because of (9), series (11) is dominated by the series of constant terms

$$|y_0| + \sum_{n=1}^{\infty} \frac{MK^{n-1}h^n}{n!}.$$

The latter series converges. In fact,

$$\sum_{n=1}^{\infty} \frac{MK^{n-1}h^n}{n!} = \frac{M}{K} \sum_{n=1}^{\infty} \frac{(Kh)^n}{n!} = \frac{M}{K}(e^{Kh} - 1).$$

Therefore, by the Weierstrass M-test the series (11) converges absolutely and uniformly on the interval $|x-x_0| \le h$. It follows now that the sequence $y_n(x)$ converges uniformly on the interval $|x-x_0| \le h$. Let us denote the limit function by $y(x)$. That is,

$$y(x) = \lim_{n\to\infty} y_n(x). \tag{12}$$

6. Next, we will prove that $y(x)$ is a solution of the IVP(1)–(2). First of all $y(x)$ satisfies the initial condition (2). In fact, from (4),

$$y_n(x_0) = y_0, \qquad n = 0, 1, 2, \ldots$$

and taking limits of both sides as $n \to \infty$ we find $y(x_0) = y_0$. Since $y(x)$ is the uniform limit, in the interval $|x-x_0| \le h$, of the continuous functions $y_n(x)$, $n = 0, 1, 2, \ldots$, it follows that $y(x)$ is itself continuous on the interval $|x-x_0| \le h$. Also from (8), taking limits as $n \to \infty$, we find that for $|x-x_0| \le h$,

$$|y(x)-y_0| \le B.$$

Hence, the function $F(x, y(x))$ is well defined and continuous on the interval $|x-x_0| \le h$ and the integral

$$\int_{x_0}^{x} F(s, y(s))ds$$

exists. From (7), we have for $|x-x_0| \le h$,

$$|F(x, y_n(x)) - F(x, y(x))| \le K\,|y_n(x) - y(x)|.$$

Since the sequence $y_n(x)$ converges uniformly to $y(x)$ on the interval $|x-x_0| \le h$, it follows that the sequence $F(x, y_n(x))$ also converges uniformly to $F(x, y(x))$ and consequently

$$\lim_{n\to\infty} \int_{x_0}^{x} F(s, y_n(x))ds = \int_{x_0}^{x} F(s, y(s))ds. \tag{13}$$

Taking limits on both sides of (4), as $n \to \infty$, and using (12) and (13), we obtain

$$y(x) = y_0 + \int_{x_0}^{x} F(s, y(s))ds. \tag{14}$$

Differentiating both sides of (14) with respect to x (note that the right-hand side is a differentiable function of the upper limit x) we find

$$y'(x) = F(x, y(x)),$$

and this completes the proof that $y(x)$ is a solution of the IVP(1)–(2).

7. Finally we will prove that $y(x)$ is the only solution of the IVP(1)–(2). Assume that $\tilde{y}(x)$ is another solution of the IVP(1)–(2). Then

$$y'(x) - \tilde{y}'(x) = F(x, y(x)) - F(x, \tilde{y}(x)).$$

Integrating from x_0 to x and using the fact that $y(x_0) = y_0 = \tilde{y}(x_0)$, we find

$$y(x) - \tilde{y}(x) = \int_{x_0}^{x} [F(s, y(s)) - F(s, \tilde{y}(s))]\,ds. \tag{15}$$

Now assume that $x \geq x_0$. The case $x < x_0$ is treated in a similar way. From (15), using (7), we obtain

$$| y(x) - \tilde{y}(x) | \leq K \int_{x_0}^{x} | y(s) - \tilde{y}(s) | \, ds. \tag{16}$$

Setting

$$w(x) = \int_{x_0}^{x} | y(s) - \tilde{y}(s) | \, ds \tag{17}$$

and using (16), we find

$$w'(x) = | y(x) - y(x) | \leq K \int_{x_0}^{x} | y(s) - y(s) | \, ds = Kw(x)$$

or

$$w'(x) - Kw(x) \leq 0. \tag{18}$$

Multiplying both sides of the inequality (18) by the positive integrating factor exp $(-Kx)$ we obtain

$$\frac{d}{dx}\left[w(x)e^{-Kx} \right] \leq 0. \tag{19}$$

Integrating both sides of (19) from x_0 to x we find

$$w(x)e^{-Kx} - w(x_0)e^{-Kx_0} \leq 0.$$

But from (17), $w(x_0) = 0$ and $w(x) \geq 0$. It follows then that

$$0 \leq w(x)e^{-Kx} \leq 0 \qquad \text{or} \qquad w(x) \equiv 0.$$

Hence, from (16), we conclude that $y(x) = \tilde{y}(x)$, and the proof is complete.

Answers to Odd-Numbered Exercises

Section 1.1
Page 8

1. True. **3.** False. **5.** True. **7.** False. **21.** $y' = -\frac{x}{y}$; $y^2 + x^2 = k^2$.

23. True. **25.** False. **27.** False.

29. Differentiating both sides of the first family, we obtain

$$2x + 2yy' = 2c = \frac{x^2 + y^2}{x} \Rightarrow y' = \frac{y^2 - x^2}{2xy}.$$

Differentiating both sides of the second family, we find that

$$y' = -\frac{2xy}{y^2 - x^2}.$$

Since the slopes are negative reciprocals, we conclude that every member of the family $x^2 + y^2 = 2cx$ cuts every member of the family $x^2 + y^2 = 2ky$ at right angles, and vice versa.

Section 1.2
Page 16

1. Yes. **3.** Yes. **5.** No, because $F_y(x, y)$ is not continuous for $y = 0$.

7. The solution through the point $(-3, 0)$ exists throughout the interval $\left(-\infty, -\frac{3}{2}\right)$; the solution through $(-1, 5)$ exists throughout the interval $\left(-\frac{3}{2}, 0\right)$; the solution through the point $(1, -7)$ exists throughout the interval $(0, 2)$; the solution through the point $(3, 0)$ exists throughout $(2, \infty)$.

9. Show by direct substitution that $y = 2x - 1$ is a solution of the IVP. Thus, by Example 2, $y = 2x - 1$ is the only solution of the IVP [in the interval $(0, \infty)$].

11. True. **13.** False.

15. The isoclines are the hyperbolas $xy = c$ for $c \neq 0$ and the lines $x = 0$ and $y = 0$ for $c = 0$. The heavy polygonal line shown here is a graphical approximation of the solution through $(1, 2)$.

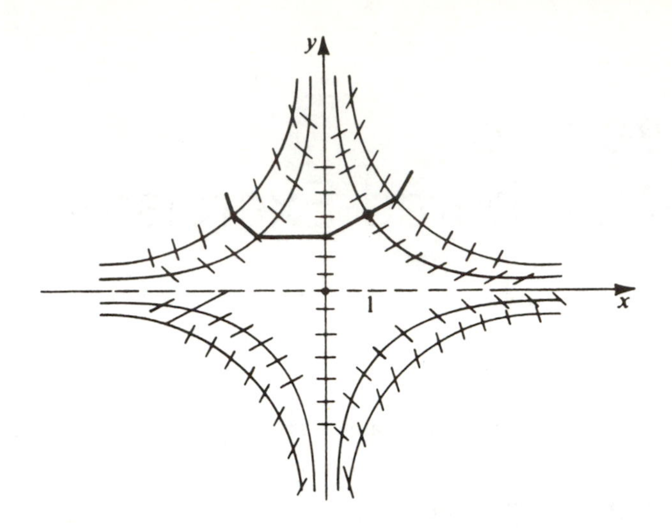

Section 1.3
Page 27

1. Separable; $y = \dfrac{c}{xe^x}$. **3.** Separable; $y = 0$ and $y = \dfrac{1}{4}(x + c)^2$.

5. It is not separable. **7.** It is not separable.

9. Separable; $y = 1 - ce^{-(1/2)x^2 - x}$.

11. $y = \cos^{-1}(ce^{-x} - 1)$ is the general solution while the curves $x = n\pi + \pi/2$, $n = 0, \pm 1, \pm 2, \ldots$ are singular solutions.

13. It is not separable. **15.** $y = e^x$.

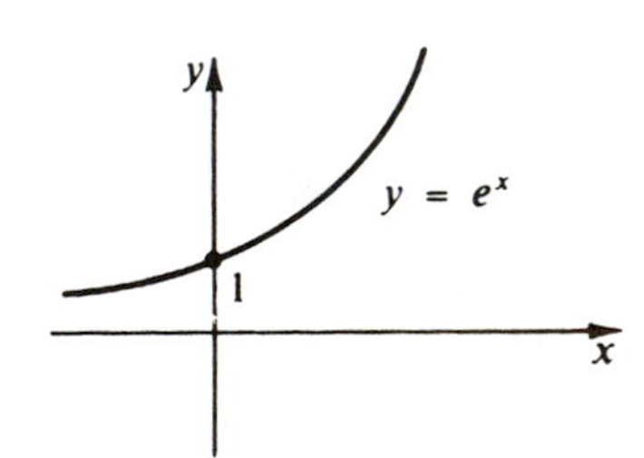

17. $y^{a_1}e^{a_2 y} = cx^{b_1}e^{b_2 x}$

19. $x(t) = \left[\dfrac{a}{3B(b - Bt)^{1/2}} + c\right]^6$. **21.** $I(w) = 2.4 + ce^{-0.088w}$. **23.** 12 hours.

25. 0.58 second. **27.** $\dfrac{100}{6}\ln 2 = 11.5$ years.

29. $y = ke^{-x/C}$, where k is an arbitrary constant. **31.** $y = k(1 - x)^{(1-A)/C}x^{A/C}$.

33. $\dfrac{1000}{55}\ln 4 = 25.2$ years. **35.** $10 \ln 3 = 10.99\%$. **37.** 86.3 years; 57.6 years.

Section 1.4
Page 37

1. Linear; $y = 3 + \dfrac{c}{x}$. **3.** It is not linear. **5.** Linear; $y = 2x^2 + cx$.

7. Linear; $y = \frac{1}{2}(x-1)^5 + c(x-1)^3$. **9.** It is not linear.

11. Linear; $y = \frac{1}{6}x^5 + \dfrac{c}{x}$. **13.** It is not linear. **15.** $y = x + \dfrac{1}{x}e^x$.

17. $y(x) = \begin{cases} e^x, & x < 0 \\ 2e^x - x - 1, & x \geq 0. \end{cases}$

21. 0.19 amperes. After a long time the current will be 1.6 amperes.

23. $I(t) = (10.2)e^{-2t} + (0.4)\sin t - (0.2)\cos t$.

25. The amount of salt in the tank after 5 minutes will be $30\left(1 - \dfrac{1}{e}\right) \simeq 18.96$ pounds. After a very long time, the amount will be 30 pounds.

29. $y = \dfrac{1}{2x^2 + cx}$. **31.** $y = \dfrac{1}{-\frac{2}{3}\ln x + c(\ln x)^{-1/2}}$. **33.** $y = x$. **35.** $y = -1$.

37. $y(x) = \frac{1}{2}(x-1)^5 - 4(x-1)^3$.

41. (a) $t = \dfrac{3\ln 3}{\ln 3 - \ln 2} = 8.13$ minutes.

(b) $40 + 30\left(\dfrac{2}{3}\right)^{5/3} = 55.26°\text{F}$.

43. $y = 1 + \dfrac{10}{t(\ln t + c)}$.

Section 1.5
Page 48

1. It is exact; $x^2 + y^2 + 4y - 2xy = c$.

3. It is exact; $x^2 - 2x + y^2 + 6y - 2xy = c$.

5. It is exact; $x^3 + 3xy^2 = c$. **7.** It is exact; $y + e^{xy} = c$.

9. It is exact; $\sin x - \cos y + xy = c$. **11.** It is not exact.

13. $-y^3 + y + 3x^2y = k$. **17.** $x^2 + y^2 + 4y - 2xy = 4$.

19. $x^3 + 3xy^2 = 4.$

25. $\mu = x^{-3}$ is an integrating factor; $\ln x + x^{-2}y = c.$

27. $\mu = x^{-2}$ is an integrating factor; $2 \ln x + yx^{-1} = c.$

29. $\mu = x^{-5}$ is an integrating factor; $\ln x - \frac{1}{4}x^{-4}y^4 = c.$

31. The solution is Eq. (4) of Section 1.4. **33.** $y = cx.$ **35.** $y = cx.$

37. $y = cx.$ **39.** $2x + y + \frac{1}{2}\ln(x^2 + y^2) = c.$

Section 1.6
Page 54

1. It is homogeneous; $y = cx^{1/3} - \frac{5}{2}x.$ **3.** It is homogeneous; $x^2 - y^2 = cx.$

5. It is homogeneous; $-\frac{x^3}{3y^3} + \ln y = c.$ **7.** It is homogeneous; $y^7(y - 10x)^3 = c.$

9. It is homogeneous; $\ln y + \frac{x^2}{2y^2} = c.$

11. $\ln\sqrt{y^2 + xy + x^2} + \sqrt{3}\tan^{-1}\frac{2\sqrt{3}}{3}\left(\frac{y}{x} + \frac{1}{2}\right) = c.$

13. $(x + y)(x - y)^5 = c.$

17. $\sqrt{2}\tan^{-1}\frac{y - 2}{\sqrt{2}(x - 1)} - \ln[2(x - 1)^2 + (y - 2)^2] = c.$

19. $\left(y + \frac{11}{3}\right)^2 + 2\left(x + \frac{1}{3}\right)\left(y + \frac{11}{3}\right) - 5\left(x + \frac{1}{3}\right)^2 = c.$

21. $[(y + 1)^2 - 2x(y + 1) - x^2][(y + 1) + (\sqrt{2} - 1)x]^{\sqrt{2}} =$
$c[(y + 1) - (\sqrt{2} + 1)x]^{\sqrt{2}}.$

23. $\frac{2}{\sqrt{7}}\tan^{-1}\frac{2(y - 1) - x}{x\sqrt{7}} = c + \ln[(y - 1)^2 - x(y - 1) + 2x^2].$

25. $y = x.$

27. $\sqrt{2}\tan^{-1}\frac{y - 2}{\sqrt{2}(x - 1)} - \ln[2(x - 1)^2 + (y - 2)^2] = -\ln 2.$

Section 1.7
Page 57

1. $y(x) = c_1 + c_2e^{-x} + 3x.$ 3. $y(x) = \dfrac{1 + (-x + 1)^2}{2}.$

9. $y(x) = c_1x^5 + c_2x^3 + c_3x^2 + c_4x + c_5.$ 11. $y(x) = c_1 \sinh (c_2 \pm x).$

13. $2c_1y + \ln \dfrac{c_1y - 1}{c_1y + 1} = 2c_1(c_1^2x + c_2)$. Also, $y =$ constant is a solution.

15. $y = c_1 + \sin^{-1} (c_2e^x).$

REVIEW EXERCISES CHAPTER 1
Page 58

1. $y(x) = \dfrac{2e}{xe^x}.$ 3. $y(x) = x(1 + x^2).$

5. $y(x) = -\left(\dfrac{5 - 2x^3}{3x}\right)^{1/2}$ 7. $y(x) = \dfrac{1}{2}x^2 + \dfrac{3}{2} - \ln x.$

9. $x - \ln |x - 2| = t + c$. Also $x = 2$ is a solution.

11. $\sin x = \sin t + c.$ 13. $y = x + \dfrac{1}{1 + ce^x}.$

15. $(y - x)x^2 = cy$ 17. $y = cx + xe^x$

19. $x^5 - y^5 + 5xy^4 = c$ 21. $x^2 + y^2 - 4xy = c$

23. $(0, \infty)$

25. $y(x) = \begin{cases} \dfrac{x^2 - 1}{2x}, & x \le 1 \\ \dfrac{x - 1}{x}, & x > 1 \end{cases}$

27. The current satisfies Eq. (7), Section 1.4.

29. $\left(\dfrac{\ln 2}{\ln 5 - \ln 2}\right)$ hours $\approx$ 45 minutes.

31. $-\dfrac{5568 (\ln 0.85)}{\ln 2} \approx 1305.5$ years.

33. (a) $y(t) = 2(30 + t) - 50\left(\dfrac{30}{30 + t}\right)^2$; (b) 51.87 pounds.

35. 2183 years. **37.** $10\,e^{-1000t}$.

39. 6.2 ft.

41. Show that the motion satisfies an IVP of the form $\dot{v} + Kv = -g$, $v(0) = v_0$, where K is a constant.

43. $(x + y)e^{[(x-y)^2]/2} = c$; $xy(x + y - 1) = c$. **45.** $I(t) = 4.5e^{-2t} + 0.5e^{-t}$.

Section 2.1
Page 71

1. It is linear, homogeneous, with variable coefficients, and of order 2.

3. It is nonlinear.

5. It is linear, nonhomogeneous, with constant coefficients, and of order 5.

7. It is linear, nonhomogeneous, with variable coefficients, and of order 1.

9. It is linear, nonhomogeneous, with variable coefficients, and of order 2.

11. It is linear, nonhomogeneous, with variable coefficients, and of order 2.

13. It is linear, nonhomogeneous, with constant coefficients, and of order 3.

15. It is linear, homogeneous, with constant coefficients, and of order 2.

17. It is linear, nonhomogeneous, with variable coefficients, and of order 3.

19. It is linear, homogeneous, with variable coefficients, and of order 2.

21. $y(t) = \frac{1}{2}\cos 16t - \frac{1}{4}\sin 16t$.

23. $y(t) = 10\sin\sqrt{5}\,t \Rightarrow$ amplitude $= 10$ cm and period $= 2\pi/\sqrt{5}$.

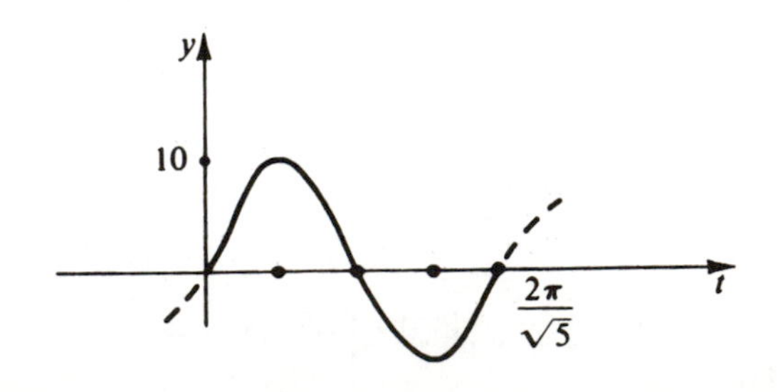

Section 2.2
Page 79

1. 2. 3. -3. 5. $e^x\left(\ln x - \frac{1}{x}\right)$. 7. $-e^{5x}$. 9. x^3. 11. e^{14x}.

13. $(\lambda_2 - \lambda_1)x^{\lambda_1 - \lambda_2 - 1}$ 15. 0 17. $y = c_1x^2 + c_2x + c_3$.

19. $y = -\sin x + c_1x + c_2$. 21. $y = c_1x^3 + c_2x^2 + c_3x + c_4$.

23. $y = c_1x^4 + c_2x^3 + c_3x^2 + c_4x + c_5$. 25. $y = c_1e^{2x} + c_2e^{-2x}$.

27. $y = \frac{c_1}{x}$ 29. $y = c_1e^{-2x} + c_2xe^{-2x}$.

31. $y = c_1(\cos x + \sin x) + c_2(\cos x - \sin x)$. 33. $y = c_1 + c_2 \cos x + c_3 \sin x$.

35. $W = ax^3$. 37. $W = c$. 43. $p(v) = c_1 + c_2 \ln v$.

Section 2.3
Page 85

1. Yes. 3. Yes.

5. No, because the leading coefficient vanishes at π and π belongs to the interval $3\pi/4 \leq x \leq 5\pi/4$.

7. Yes. 9. Yes. 11. $y(x) \equiv 0$.

13. Any interval not containing the point $x = 0$.

15. Any interval not containing any of the points $x = k\pi + \pi/2$, $k = 0, \pm 1, \pm 2, \ldots$.

17. Any interval at all.

19. A spring at rest. Its solution is $y = 0$. 21. $W(y_1, y_2; t) = -2ai \neq 0$.

23. $y = \dfrac{c_1x + c_2}{(x - a)(x - b)}$.

Section 2.4
Page 90

1. $4\lambda^3 - 2\lambda^2 + 6\lambda - 7 = 0$. 3. $\lambda^3 - \lambda + 2 = 0$.

5. $5\lambda^3 - 5\lambda^2 + \lambda - 2 = 0$. 7. $3\lambda^3 + 4\lambda^2 = 0$.

9. $3\lambda^5 - 2\lambda^4 + \lambda^2 - 2\lambda = 0$.

11. No. 13. No. 15. Yes. 17. No. 19. Yes.

21. $\lambda_1 = 4i$, $\lambda_2 = -4i$; $y(x) = c_1e^{4ix} + c_2e^{-4ix}$ or $y(x) = \alpha_1 \cos 4x + \alpha_2 \sin 4x$.

23. $\lambda_{1,2} = \dfrac{-1 \pm i\sqrt{3}}{2}$; $y(x) = c_1e^{[(-1+i\sqrt{3})/2]x} + c_2e^{[(-1-i\sqrt{3})/2]x}$ or

$$y(x) = \alpha_1 e^{-(x/2)} \cos \frac{x\sqrt{3}}{2} + \alpha_2 e^{-(x/2)} \sin \frac{x\sqrt{3}}{2}$$

25. $\lambda_{1,2} = 1 \pm \dfrac{1}{2}i$; $y(x) = c_1e^{[1-(i/2)]x} + c_2e^{[1-(i/2)]x}$ or

$$y(x) = \alpha_1 e^x \cos \frac{x}{2} + \alpha_2 e^x \sin \frac{x}{2}.$$

27. $\lambda_{1,2} = \pm 2i$; $y(x) = c_1e^{2ix} + c_2e^{-2ix}$ or $y(x) = \alpha_1 \cos 2x + \alpha_2 \sin 2x$.

29. $\lambda_{1,2} = \pm 5i$; $y(x) = c_1e^{5ix} + c_2e^{-5ix}$ or $y(x) = \alpha_1 \cos 5x + \alpha_2 \sin 5x$.

31. $y(x) = c_1e^{2x} + c_2e^{-2x} = \alpha_1 \cosh 2x + \alpha_2 \sinh 2x$.

33. $y(x) = c_1e^{5x} + c_2e^{-5x} = \alpha_1 \cosh 5x + \alpha_2 \sinh 5x$.

35. $y(x) = c_1e^{9x} + c_2e^{-9x} = \alpha_1 \cosh 9x + \alpha_2 \sinh 9x$.

37. $y(x) = c_1e^{3x} + c_2e^{-3x} = \alpha_1 \cosh 3x + \alpha_2 \sinh 3x$.

39. $\lambda_{1,2} = \pm \sqrt{\beta}$; $D(t) = c_1e^{t\sqrt{\beta}} + c_2e^{-t\sqrt{\beta}}$ or $D(t) = \alpha_1 \cosh t\sqrt{\beta} + \alpha_2 \sinh t\sqrt{\beta}$.

Section 2.5
Page 99

1. $y(x) = c_1e^{-x/2} + c_2e^{-x}$. **3.** $y(x) = c_1e^{-x/2} \cos \dfrac{x\sqrt{3}}{2} + c_2e^{-x/2} \sin \dfrac{x\sqrt{3}}{2}$.

5. $y(x) = c_1 \cos x\sqrt{6} + c_2 \sin x\sqrt{6}$. **7.** $y(x) = c_1e^{3x}$.

9. $y(x) = c_1e^{-x} + c_2e^{x} + c_3xe^{x} + c_4e^{-2x}$.

11. $y(x) = c_1e^{3x} + c_2e^{-2x} + c_3xe^{-2x} + c_4e^{x} \cos 2x + c_5e^{x} \sin 2x + c_6e^{-2x} \cos x$

$+ c_7e^{-2x} \sin x + c_8xe^{-2x} \cos x + c_9xe^{-2x} \sin x$. Order 9.

13. $y(x) = c_1e^{(3/2)x} + c_2e^{-(4/5)x} + c_3 \cos 2x + c_4 \sin 2x + c_5x \cos 2x$

$+ c_6x \sin 2x + c_7e^{[(1+\sqrt{5})/2]x} + c_8e^{[(1-\sqrt{5})/2]x}$. Order 8.

15. $y(x) = c_1 \cos 2x + c_2 \sin 2x + c_3 \cos x + c_4 \sin x + c_5\, x \cos x$

$+ c_6\, x \sin x$. Order 6.

17. $y(x) = c_1e^{7x} + c_2e^{(7/2)x} + e^{-(1/2)x}\left[c_3 \cos \frac{\sqrt{3}}{2}x + c_4 \sin \frac{\sqrt{3}}{2}x\right] + e^{-4x}[c_5 + c_6x + c_7x^2]$. Order 7.

19. $y(x) = c_1e^{-2x} + e^{5x}[c_2 + c_3x] + e^{4x}[c_4 + c_5x + c_6x^2] + \cos 2x[c_7 + c_8x + c_9x^2] + \sin 2x[c_{10} + c_{11}x + c_{12}x^2]$. Order 12.

21. $y(x) = e^{-x/2}\left[\cos \frac{x\sqrt{3}}{2} + \left(2 + \frac{1}{\sqrt{3}}\right) \sin \frac{x\sqrt{3}}{2}\right]$.

23. $y(x) = 3 \cos x\sqrt{3} - 6 \sin x\sqrt{3}$. **25.** $y(x) = -1 + x + 2e^x$.

27. $y(x) = (2 + x)e^x - 2e^{-x}$. **29.** $y(x) = 2e^x(\sin x - \cos x)$.

31. $y(x) = 3 \cos x + 2 \sin x$.

33. $y(t) = (15\sqrt{6} - 18)e^{-(2/3)t} + (12 - 15\sqrt{6})e^{-t} \to 0$ as $t \to \infty$.

35. Overdamped. **37.** Underdamped. **39.** Underdamped.

41. Critically damped. **43.** Underdamped. **45.** $f(x) = c\,(\cos x + \sin x)$.

Section 2.7
Page 106

1. $y(x) = c_1x^{3/5} + c_2x$.

3. $y(x) = (-x)^{-1/6}\left[c_1 \cos\left(\frac{\sqrt{11}}{6} \ln(-x)\right) + c_2 \sin\left(\frac{\sqrt{11}}{6} \ln(-x)\right)\right]$.

5. $y(x) = c_1(-x) + c_2(-x)^{-4}$. **7.** $y(x) = c_1(-x)^{3/2} + c_2(-x)^{-1}$.

9. $y(x) = c_1x + c_2x \ln x + c_3x^{-2}$.

11. $y(x) = (-x - 3)^{-1}[c_1 \cos 2 \ln(-x - 3) + c_2 \sin 2 \ln(-x - 3)]$.

13. $y(x) = c_1(-x)^3 + c_2(-x)^{-2}$. **15.** $y(x) = c_1x + c_2x^2 + c_3x^{-4}$.

17. $y(x) = -x + x \ln(-x)$. **19.** $y(x) = \frac{3}{7}(-x)^{1/2} + \frac{64}{7}(-x)^{-3}$.

21. (b) Apply l'Hospital's rule and use the formula

$$\frac{d}{dx}(a^x) = a^x \ln a.$$

23. $p(r) = c_1 + c_2 \ln r$. **25.** $y(x) = c_1x + c_2x^2 + c_3x^{4+\sqrt{13}} + c_4x^{4-\sqrt{13}}$.

27. $y(x) = c_1(-x)^{-1} + c_2(-x)^{(7+\sqrt{65})/4} + c_3(-x)^{(7-\sqrt{65})/4}$.

29. $y(x) = c_1x + c_2x^2 + c_3x^{(23+\sqrt{445})/14} + c_4x^{(23-\sqrt{445})/14}$.

31. Characteristic equation: $\lambda^6 - 15\lambda^5 + 73\lambda^4 - 143\lambda^3 + 142\lambda^2 - 48\lambda = 0$; characteristic roots: $\lambda_1 = 0, \lambda_2 = \lambda_3 = 1, \lambda_4 = 2, \lambda_5 = 3, \lambda_6 = 8$. $y(x) = c_1 + (c_2 + c_3 \ln x)\, x + c_4x^2 + c_5x^3 + c_6x^8$.

33. $y(x) = c_1x + c_2x^5$. **35.** $y(x) = c_1 + c_2(-x) + c_3(-x)^2 + c_4(-x)^4$.

37. $y(x) = c_1(x - 4) + c_2(x - 4)^{-4}$. **39.** $y(x) = [c_1 + c_2 \ln(1 - x)]\,(1 - x)^{-2}$.

Section 2.8
Page 114

1. $y_2(x) = x^{-1}$. **3.** $y_2(x) = (x + 1)\, e^{-3x}$. **5.** $y_2(x) = x - 2 + \frac{2}{x}$.

7. $y_2(x) = \ln x$. **9.** $y_2(x) = x^{-1/2} \cos x$. **11.** $y_2(x) = x^{-2}e^{-(3/4)x^{2/3}}$.

13. $y_2(x) = \ln x$. **15.** $y_2(x) = e^{2x}$.

17. $y_2(x) = (8x^3 - 12x) \int_{x_0}^{x} \frac{e^{t^2}\, dt}{(8t^3 - 12t)^2}$.

19. $y_2(x) = \frac{5x^3 - 3x}{2} \int_{x_0}^{x} \frac{4\, dt}{(t^2 - 1)\,(5t^3 - 3t)^2}$.

21. $y_2(x) = (2 - 4x + x^2) \int_{x_0}^{x} \frac{e^t\, dt}{t(2 - 4t + t^2)^2}$.

23. $y_2(x) = x \int_{x_0}^{x} \frac{dt}{t^2\sqrt{1 - t^2}}$.

25. $y_2(x) = (4x^3 - 3x) \int_{x_0}^{x} \frac{dt}{(4t^3 - 3t)^2\sqrt{1 - t^2}}$. **27.** $y_2(x) = e^x \ln x$.

35. $y_2(x) = -\frac{1}{4}(3x^2 - 1) \ln \frac{x - 1}{x + 1} - \frac{3}{2}x$.

Section 2.9
Page 119

1. $y(x) = 1 - \frac{1}{2}x^2 + \frac{1}{8}x^4 + 0 \cdot x^5 - \cdots$.

3. $y(x) = 2 + x - \frac{3}{2}x^2 - \frac{1}{24}x^4 + \frac{1}{24}x^5 + \cdots$.

5. $y(x) = 1 - \frac{3}{2}x^2 - \frac{1}{6}x^3 + \frac{3}{8}x^4 + \frac{1}{10}x^5 + \cdots$.

7. $y(x) = 2 + x - x^2 + \frac{1}{6}x^3 - \frac{1}{4}x^4 + \frac{1}{24}x^5 + \cdots$.

9. $y(x) = 1 + x - \frac{1}{3}x^3 + \frac{1}{3}x^4 - \frac{7}{30}x^5 + \cdots$.

11. $y(x) = a_0\left(1 - \frac{x^2}{2} + \frac{x^4}{24} - \cdots\right) + a_1\left(x - \frac{x^3}{6} + \frac{x^5}{120} - \cdots\right)$.

13. $y(x) = a_0\left(1 - \frac{1}{6}x^3 + \cdots\right) + a_1\left(x - \frac{1}{6}x^4 + \cdots\right)$.

15. $y(x) = a_0(1 - x^2 + \cdots) + a_1\left(x - \frac{1}{12}x^3 + \frac{1}{480}x^5 + \cdots\right)$.

17. $y(x) = a_0\left(1 + \frac{1}{6}x^3 - \frac{1}{8}x^4 - \frac{1}{40}x^5 + \cdots\right)$

$$+ a_1\left(x - \frac{3}{2}x^2 + \frac{3}{2}x^3 - \frac{13}{24}x^4 - \frac{13}{20}x^5 + \cdots\right).$$

19. $y(x) = a_0\left(1 + \frac{1}{12}x^4 + \cdots\right) + a_1\left(x + \frac{1}{30}x^5 + \cdots\right) + a_2(x^2 + \cdots)$.

21. For $a \neq -1$, we have $a_2 = -\frac{\lambda a_0}{1 + a}$; $a_3 = -\frac{\lambda a_1}{6(1 + a)}$; $a_4 = \frac{\lambda(\lambda - 4a)}{12(1 + a)^2}a_0$;

$a_5 = \frac{\lambda(\lambda - 12a)}{120(1 + a)^2}a_1$.

Section 2.10
Page 123

1. $y_h(x) = c_1e^{2x} + c_2e^{3x}$. **3.** $y_h(x) = c_1e^{-x} + c_2e^{2x}$.

5. $y_h(x) = c_1 \cos 3x + c_2 \sin 3x$. **7.** $y_h(x) = c_1e^{4x} + c_2e^{9x}$.

9. $y_h(x) = c_1e^{5x} + c_2xe^{5x}$. **11.** $y_h(x) = c_1e^x + c_2e^{-x} + c_3 \cos x + c_4 \sin x$.

13. $y_h(x) = c_1x^{8/3} + c_2x^{-1}$. **15.** $y_h(x) = c_1\, x \sin x + c_2\, x \cos x$.

17. True. **19.** True. **21.** False. **23.** True.

Section 2.11
Page 131

7. $y_p(x) = Ax^3 + Bx^2 + Cx + D$.

9. $y_p(x) = Ax^2 + Bx + C + Dx \cos x + Ex \sin x$.

11. $y_p(x) = Ax^6 + Bx^5 + Cx^4 + Dx^3 + Ex^2 + Fx + G \cos x + H \sin x$.

13. $y_p(x) = x + e^{-x}$. **15.** $y_p(x) = \frac{1}{2}x^2e^{2x}$. **17.** $y_p(x) = -\frac{1}{72}x^4$.

19. $y_p(x) = -\frac{1}{6}x^2e^{-2x}$. **21.** $y_p(x) = \frac{1}{2}xe^x$. **23.** $y_p(x) = -\frac{1}{2}x \sin x - \frac{1}{2} \cos x$.

25. $y_p(x) = \left(\frac{1}{2}x^2 - \frac{5}{2}x + \frac{19}{4}\right)e^x$. **27.** $y(x) = c_1 \cos x + c_2 \sin x - \frac{1}{2}x \cos x$.

29. $y(x) = c_1e^{2x} + c_2e^{-x} - \frac{1}{3}x^2e^{-x} - \frac{2}{9}xe^{-x} - \frac{1}{2}x^2 + \frac{1}{2}x - \frac{3}{4}$.

31. $y(x) = c_1e^x + c_2e^{-x} + xe^x - xe^{-x}$. **33.** $y(x) = c_1x + c_2x^{-4} + \frac{1}{14}x^3$.

35. $y(x) = c_1e^{2x} + c_2e^{3x} + \frac{1}{10}e^x(\sin x + 3 \cos x)$.

37. $y(x) = c_1e^x + c_2e^{4x} - \frac{1}{5}e^{2x}(\cos x + 2 \sin x)$.

39. $y(x) = -2 \sin x + x + e^{-x}$. **41.** $y(x) = \frac{1}{2}x^2e^{2x}$. **43.** False. **45.** True.

47. False.

49. $Q(t) = \frac{1}{104}[12e^{-3t} - 13e^{-2t} + \cos 2t + 5 \sin 2t]$.

$I(t) = \frac{1}{52}[13e^{-2t} - 18e^{-3t} - \sin 2t + 5 \cos 2t]$.

51. $u_p = \dfrac{\beta}{b\tau}$. **53.** $u(t) = \dfrac{3}{2}e^{-2t} - \dfrac{4}{3}e^{-3t} + \dfrac{5}{6}; \lim_{t\to\infty} u(t) = \dfrac{5}{6}$. **55.** $x_p = \dfrac{K}{\beta}$.

59. $u(x) = c_1 + c_2x + c_3x^2 + c_4x^3 + \cos x$.

61. $u(x) = c_1 + c_2x + c_3x^2 + c_4x^3 + \sinh x$.

63. $u(x) = c_1 + c_2x + c_3x^2 + c_4x^3 + \dfrac{1}{360}x^6$.

Section 2.12
Page 139

1. $y(x) = c_1 \cos x + c_2 \sin x - (\cos x) \ln |\sec x + \tan x|$.

3. $y(x) = c_1e^x + c_2xe^x + xe^x \ln x$. **5.** $y(x) = c_1e^{-3x} + c_2xe^{-3x} + \dfrac{1}{2x}e^{-3x}$.

7. $y(x) = c_1e^{6x} + c_2xe^{6x} + \dfrac{1}{4}x^2e^{6x}(2 \ln x - 3)$.

9. $y(x) = c_1 \cos x + c_2 \sin x + \dfrac{1}{2}\sin x \tan x$.

11. $y(x) = c_1e^{-x} + c_2xe^{-x} + e^{-x}\left[-\dfrac{1}{2}(\ln x)^2 - \ln x - 1\right]$

13. $y(x) = c_1x^{3/5} + c_2x + 4x^{1/2}$. **15.** $y(x) = c_1x^{3/2} + c_2x^{-1} + \dfrac{1}{18}x^{-3}$.

17. $y(x) = c_1x^{-2} + c_2x^{-2}\ln x + x^{-3}(\ln x + 2) + \dfrac{1}{27}x^{-5}(3 \ln x + 2)$.

19. $y(x) = \cos x + \sin x - (\cos x) \ln |\sec x + \tan x|$.

21. $y(x) = e^{-3x}\left(\dfrac{1}{2x} + \dfrac{21}{2}x - 7\right)$. **23.** $y(x) = e^{2x}\left(\dfrac{4}{3}x + \dfrac{1}{6}x^{-2} - \dfrac{3}{2}\right)$.

25. $y(x) = 24x^{3/2} - \dfrac{59}{36}x^{-1} + \dfrac{1}{18}x^{-3}$.

27. $y(x) = x[c_1e^x + c_2e^{-x} + c_3 \cos x + c_4 \sin x] + c_5x[e^x \int x^{-5}e^{-x}dx$

$$- e^{-x}\int x^{-5}e^xdx + 2\cos x \int x^{-5}\sin x\, dx + 2 \sin x \int x^{-5}\cos x\, dx].$$

29. $x(t) = c_1e^{-t} + c_2te^{-t} + e^{-t}\ln(t + 1)$.

REVIEW EXERCISES CHAPTER 2
Page 141

1. Nonhomogeneous, variable coefficients, order 3. The general solution is $y = y_h + y_p$. To determine y_h use the method associated with the Euler equation and to determine y_p use the method of undetermined coefficients.

3. Homogeneous, variable coefficients, order 2. The general solution can be obtained by the reduction-of-order method after recognizing that $y = x$ is a solution. Alternatively the general solution can be approximated by using the power-series method.

5. $W(x) = W(x_0)e^{-8(x-x_0)}$, x_0 an arbitrary value of x.

7. $y = (c_1 + c_2x + c_3x^2)e^{7x} + c_4e^{\sqrt{3}x} + c_5e^{-\sqrt{3}x} + e^{-x}(c_6 \cos x + c_7 \sin x)$. Order 7.

9. $y(x) = 0 + 3(x-1) - \frac{1}{2}(x-1)^2 + \frac{11}{18}(x-1)^3 + \frac{11}{72}(x-1)^4 + \cdots$.

11. $y = c_1e^{-x} + c_2e^{-16x} + \frac{1}{34}e^x + \frac{4}{15}xe^{-x}$.

13. $y = c_1x^3 + c_2x^{-4/3}$.

15. $y = c_1e^{2x} + c_2e^{8x} + c_3e^{-x}$.

17. $y = c_1 \tan x + c_2e^x, \quad 0 < x < 1$.

19. $y = c_1e^{-3x} + e^{-(3/2)x}(c_2 \cos \frac{3\sqrt{3}}{2}x + c_3 \sin \frac{3\sqrt{3}}{2}x) + \frac{1}{27}x$.

21. $y = c_1 \cos x + c_2 \sin x + (\sin x) \ln |\csc x - \cot x|$.

23. $y = c_1 \cos x + c_2 \sin x - \frac{1}{4}x^2 \cos x + \frac{1}{4}x \sin x$.

25. $y = x^{-4}[c_1 \cos(\ln x) + c_2 \sin(\ln x)]$.

27. (a) 2.991 sec. (b) 2392.8 ft. (c) −96.31 ft/sec.

29. Yes, the ball will clear the fence; $y = 28.1$ ft when $x = 320$ ft.

31. 6.969 mi/sec.

33. Velocity = 2576 ft/sec, height = 51,520 ft.

35. Range = 42.32 ft, maximum height = 14.42 ft, time of flight = 1.62 sec.

37. The ellipse is given by $\left(\frac{x}{a}\right)^2 + \left(\frac{\dot{x}}{a\omega}\right)^2 = 1$, and the constant energy is $ma^2\omega^2/2$.

41. $v_c = \cos 2t + \frac{5}{2}\sin 2t - t\cos 2t.$

43. Time of flight $= \left[v_0 \sin\alpha + \sqrt{(v_0\sin\alpha)^2 + 2gh}\right]/g.$

47. $y = \left(\frac{a}{4}\right)^{1/3}$ or $y = \left(\frac{a}{27c^2}\right)e^{2x} + ce^{-x}$, where c is an arbitrary constant.

Section 3.1
Page 156

1. (b) $x_1(t) = c_1e^{3t} + c_2e^{5t},\ x_2(t) = c_1e^{3t} + 3c_2e^{5t}.$ **7.** True.

9. False (they are not solutions). **11.** True. **15.** $\alpha = \beta = \gamma = \frac{1}{3}.$

17. A second linearly independent solution is $\begin{bmatrix} t\ln t + 1 \\ \ln t \end{bmatrix}$ and the general solution is

$x_1(t) = c_1t + c_2(t\ln t + 1),\ x_2(t) = c_1 + c_2\ln t.$

19. "The n solutions

$$\begin{bmatrix} x_{11}(t) \\ x_{21}(t) \\ \cdot \\ \cdot \\ \cdot \\ x_{n1}(t) \end{bmatrix}, \begin{bmatrix} x_{12}(t) \\ x_{22}(t) \\ \cdot \\ \cdot \\ \cdot \\ x_{n2}(t) \end{bmatrix}, \ldots, \begin{bmatrix} x_{1n}(t) \\ x_{2n}(t) \\ \cdot \\ \cdot \\ \cdot \\ x_{nn}(t) \end{bmatrix}$$

of the system (21) are linearly independent on an interval I if and only if the determinant

$$\begin{vmatrix} x_{11}(t) & x_{12}(t) & \cdots & x_{1n}(t) \\ x_{21}(t) & x_{22}(t) & \cdots & x_{2n}(t) \\ \cdot & \cdot & & \cdot \\ \cdot & \cdot & & \cdot \\ \cdot & \cdot & & \cdot \\ x_{n1}(t) & x_{n2}(t) & \cdots & x_{nn}(t) \end{vmatrix}$$

is never zero on I."

21. "There exist n linearly independent solutions of the system (21). Furthermore, if the n column vectors

$$\begin{bmatrix} x_{11}(t) \\ x_{21}(t) \\ \cdot \\ \cdot \\ \cdot \\ x_{n1}(t) \end{bmatrix}, \begin{bmatrix} x_{12}(t) \\ x_{22}(t) \\ \cdot \\ \cdot \\ \cdot \\ x_{n2}(t) \end{bmatrix}, \ldots, \begin{bmatrix} x_{1n}(t) \\ x_{2n}(t) \\ \cdot \\ \cdot \\ \cdot \\ x_{nn}(t) \end{bmatrix}$$

are linearly independent solutions of the system (21), then the general solution of the system (21) is given by

$$\begin{bmatrix} x_1(t) \\ x_2(t) \\ \cdot \\ \cdot \\ \cdot \\ x_n(t) \end{bmatrix} = c_1 \begin{bmatrix} x_{11}(t) \\ x_{21}(t) \\ \cdot \\ \cdot \\ \cdot \\ x_{n1}(t) \end{bmatrix} + c_2 \begin{bmatrix} x_{12}(t) \\ x_{22}(t) \\ \cdot \\ \cdot \\ \cdot \\ x_{n2}(t) \end{bmatrix} + \cdots + c_n \begin{bmatrix} x_{1n}(t) \\ x_{2n}(t) \\ \cdot \\ \cdot \\ \cdot \\ x_{nn}(t) \end{bmatrix},$$

where $c_1, c_2, \ldots, c_n$ are arbitrary constants."

23. True. **25.** True.

27. False. (The given column vector is a solution, but it fails to satisfy all the initial conditions.)

29. True.

Section 3.2
Page 168

1. $x_1(t) = c_1e^{2t} + c_2e^{3t}, x_2(t) = 2c_1e^{2t} + c_2e^{3t}$.

3. $x_1(t) = c_1e^{2t} + c_2e^{3t} - 3te^{2t} - \frac{1}{3}t - \frac{5}{18}$,

$x_2(t) = 2c_1e^{2t} + c_2e^{3t} - 6te^{2t} - \frac{4}{3}t + 6e^{2t} - \frac{7}{9}$.

5. $x(t) = c_1e^{-t} + c_2te^{-t} + 1, y(t) = c_1e^{-t} + c_2\left(te^{-t} - \frac{1}{6}e^{-t}\right) + 1$.

7. $x(t) = c_1e^{2t}\cos 3t + c_2e^{2t}\sin 3t - \frac{2}{13}t^2 + \frac{36}{169}t + \frac{534}{2197}$,

$y(t) = \frac{1}{2}c_1e^{2t}(\cos 3t + 3\sin 3t) + \frac{1}{2}c_2e^{2t}(\sin 3t - 3\cos 3t)$

$+ \frac{10}{13}t^2 + \frac{80}{169}t + \frac{567}{2197}$.

9. $I_1(t) = \frac{1}{10}e^{-3t} - \frac{1}{10}\cos t - \frac{3}{10}\sin t$,

$I_2(t) = -\frac{1}{20}e^{-3t} + \frac{1}{20}\cos t - \frac{3}{20}\sin t$.

11. $x_1(t) = 2e^{2t} - e^{3t}, x_2(t) = 4e^{2t} - e^{3t}$.

13. $x_1(t) = -3te^{2t} - \frac{1}{3}t - \frac{5}{18}, x_2(t) = -6te^{2t} - \frac{4}{3}t + 6e^{2t} - \frac{7}{9}$.

15. $x(t) = -e^{-t} + 1, y(t) = -e^{-t} + 1.$

17. $x(t) = -\frac{2}{13}t^2 + \frac{36}{169}t + \frac{534}{2197}, y(t) = \frac{10}{13}t^2 + \frac{80}{169}t + \frac{567}{2197}.$

19. $x(t) = c_1 e^{-t}, y(t) = c_2 e^{-t}.$ **21.** $x(t) = c_1 e^{-t} + c_2 e^{2t}, y(t) = 2c_1 e^{-t} + \frac{1}{2}c_2 e^{2t}.$

23. $x(t) = c_1 e^{-t}\cos t + c_2 e^{-t}\sin t,$
$y(t) = -c_1 e^{-t}\sin t + c_2 e^{-t}\cos t.$

25. $x(t) = c_1 e^t + c_2 te^t, y(t) = c_1 e^t + c_2(t + 1)e^t.$

27. $x(t) = c_1 e^{3t} + c_2 e^{5t}, y(t) = c_1 e^{3t} + 3c_2 e^{5t}.$

29. $c_1(t) = A + Be^{-[(K/V_1)+(K/V_2)]t}$
$c_2(t) = A - (V_1/V_2)\,Be^{-[(K/V_1)+(K/V_2)]t}$, A and B are arbitrary constants.

31. $y(t) = \frac{abd}{(a + d)(a + cf)} + \frac{bd}{a + d}\left[1 - \frac{a}{(a + d)(a + cf)}\right]e^{-(a-cf)t},$

$\lim_{t\to\infty} y(t) = \frac{abd}{(a + d)(a + cf)} > 0.$

Section 3.3
Page 185

1. $\lambda = 1, 2; x_1(t) = c_1 e^t + c_2 e^{2t}, x_2(t) = -\frac{3}{2}c_1 e^t - c_2 e^{2t}.$

3. $\lambda = 1, 1; x_1(t) = c_1 e^t + c_2(t + \frac{1}{2})e^t, x_2(t) = c_1 e^t + c_2 te^t.$

5. $\lambda = -2 \pm 3i; x_1(t) = c_1 e^{-2t}\cos 3t + c_2 e^{-2t}\sin 3t,$

$$x_2(t) = c_1\left(\frac{5}{2}e^{-2t}\cos 3t + \frac{3}{2}e^{-2t}\sin 3t\right) + c_2\left(\frac{5}{2}e^{-2t}\sin 3t - \frac{3}{2}e^{-2t}\cos 3t\right).$$

7. $x_1(t) = -e^{2t} + 2e^{3t}, x_2(t) = -e^{2t} + e^{3t}.$

9. $x_1(t) = -2e^{-2t}\sin 3t, x_2(t) = -e^{-2t}(5\sin 3t - 3\cos 3t).$

11. $\lambda = 2, -3; x_1(t) = c_1 e^{2t}, x_2(t) = c_2 e^{-3t}.$

13. $\lambda = 1, 3; x_1(t) = c_1 e^t + c_2 e^{3t}, x_2(t) = -c_1 e^t - 3c_2 e^{3t}.$

15. $\lambda = 1, 3; x_1(t) = c_1 e^t + 3c_2 e^{3t}, x_2(t) = -c_1 e^t - c_2 e^{3t}.$

17. $\lambda = 1, 2, 3;\ x_1(t) = c_1e^t + 2c_2e^{2t}$
$x_2(t) = -c_2e^{2t} + c_3e^{3t}$
$x_3(t) = -c_1e^t - c_3e^{3t}.$

19. $\lambda = -1, 2, 2;\ x_1(t) = c_1e^{-t} + c_2e^{2t} + c_3(t+1)e^{2t}$
$x_2(t) = -3c_1e^{-t} + c_3e^{2t}$
$x_3(t) = 4c_1e^{-t} + c_2e^{2t} + c_3(t+3)e^{2t}.$

21. $\lambda = 0, 1, -2, -3;\ x_1(t) = -2c_1 + c_2e^t + c_3e^{-2t} + 3c_4e^{-3t}$
$x_2(t) = -2c_3e^{-2t} - 4c_4e^{-3t}$
$x_3(t) = c_1 + c_3e^{-2t}$
$x_4(t) = -c_1 + 4c_4e^{-3t}.$

23. $\lambda = 1, 2, 3;\ x_1(t) = 2c_1e^t + 3c_2e^{2t} + c_3e^{3t}$
$x_2(t) = c_1e^t + c_2e^{2t} - c_3e^{3t}$
$x_3(t) = 3c_1e^t + 5c_2e^{2t} + 2c_3e^{3t}.$

25. $x_1(t) = e^t,\ x_2(t) = 0,\ x_3(t) = -e^t.$

27. $x_1(t) = (t+1)e^{2t},\ x_2(t) = e^{2t},\ x_3(t) = (t+3)e^{2t}.$

29. $x_1(t) = 2e^t + 3e^{2t} + e^{3t}$
$x_2(t) = e^t + e^{2t} - e^{3t}$
$x_3(t) = 3e^t + 5e^{2t} + 2e^{3t}.$

31. A particular solution is $\begin{bmatrix} x_{1p} \\ x_{2p} \end{bmatrix} = \begin{bmatrix} \frac{3}{2}e^{5t} \\ \frac{7}{2}e^{5t} \end{bmatrix}.$

The general solution is

$$x_1(t) = c_1e^{3t} + \left(c_2 + \frac{3}{2}\right)e^{5t}$$

$$x_2(t) = c_1e^{3t} + \left(3c_2 + \frac{7}{2}\right)e^{5t}.$$

33. A particular solution is $\begin{bmatrix} x_{1p} \\ x_{2p} \end{bmatrix} = \begin{bmatrix} 1 \\ t \end{bmatrix}.$

The general solution is

$x_1(t) = c_1e^t + c_2e^{3t} + 1$

$x_2(t) = -c_1e^t - 3c_2e^{3t} + t.$

35. A particular solution is $\begin{bmatrix} x_p \\ y_p \end{bmatrix} = \begin{bmatrix} 1 \\ 4 \end{bmatrix}$.

The general solution is

$x(t) = c_1 e^{-3t} + c_2 t e^{-3t} + 1$

$y(t) = c_1 e^{-3t} + c_2 (t+1) e^{-3t} + 4.$

37. $\lambda = 0, -\left(\frac{K}{V_1} + \frac{K}{V_2}\right); c_1(t) = A + Be^{-[(K/V_1) + (K/V_2)]t}; A, B$ constants;

$$c_2(t) = A - B\frac{V_1}{V_2}e^{-[(K/V_1) + (K/V_2)]t}.$$

REVIEW EXERCISES CHAPTER 3
Page 187

1. $x(t) = c_1 e^t + c_2 e^{2t}$, $y(t) = c_1 e^t + 2c_2 e^{2t}$.

3. $x(t) = c_1 e^t + c_2 e^{2t}$, $y(t) = c_1 e^t + 2c_2 e^{2t} - e^{3t}$.

5. $x_1(t) = c_1 e^t + c_2 e^{-t} + c_3 \cos t + c_4 \sin t$

$x_2(t) = c_1 e^t - c_2 e^{-t} - c_3 \sin t + c_4 \cos t - t^2$

$x_3(t) = c_1 e^t + c_2 e^{-t} - c_3 \cos t - c_4 \sin t - 2t$

$x_4(t) = c_1 e^t - c_2 e^{-t} + c_3 \sin t - c_4 \cos t - 2.$

7. $x(t) = 3c_1 e^{-t} + 2c_3 e^{4t}$

$y(t) = 6c_1 e^{-t} - c_3 e^{4t}$

$z(t) = -c_1 e^{-t} + c_2 e^{2t} + c_3 e^{4t}.$

9. $x_1(t) = c_1 e^t + \frac{1}{3}c_2 e^{-2t} + \frac{1}{10}c_3 e^{3t}$

$x_2(t) = c_2 e^{-2t} - \frac{1}{5}c_3 e^{3t}$

$x_3(t) = c_3 e^{3t}.$

11. $x(t) = -e^t + 2e^{2t}$, $y(t) = -e^t + 4e^{2t}$.

13. $x_1(t) = e^t$, $x_2(t) = e^t - t^2$, $x_3(t) = e^t - 2t$

$x_4(t) = e^t - 2.$

15. $x_1(t) = \frac{35}{6}e^t - \frac{14}{15}e^{-2t} + \frac{1}{10}e^{3t}$

$x_2(t) = -\frac{14}{5}e^{-2t} - \frac{1}{5}e^{3t}$

$x_3(t) = e^{3t}.$

17. Explosion.

19. $x(t) = e^{2t}(\sin t - 2\cos t) + 2$
$y(t) = e^{2t}(3\sin t - \cos t) + 1.$

Section 4.2
Page 197

13. $\cos 4t.$ **15.** $\sinh 5t.$ **17.** $\sin 2t.$

19. $\frac{5}{s} - \frac{48}{s^4}.$ **21.** $\frac{8s}{64s^2 + 9}.$ **23.** $\frac{s-3}{(s-3)^2 + 4} - \frac{5}{(s-1)^2 - 25}.$

25. $\frac{s-1}{s^2+1}.$

27. $\frac{7!}{s^8} - \frac{4!}{s^5} + \frac{5 \cdot 2!}{s^3}.$ **29.** $\frac{2s}{(s^2-1)^2}.$ **31.** $\frac{s}{(s-5)^2}.$

33. $\frac{s^3}{s^2+1} + \frac{s^2}{(s-1)^2} - s - 1.$ **35.** $\frac{s^2}{s^4 - 1}.$ **37.** $\frac{s-1}{s[(s-1)^2 + 4]}.$

39. $\frac{2}{k}\sin kt.$ **41.** $t^n e^{kt}.$ **43.** $\cosh kt.$ **45.** $\sin t - t\cos t.$

47. $2e^{-9t} - 2e^{-4t} + 1.$ **49.** $e^{-2t} + 2te^{-2t} - e^{2t} + 3te^{2t}.$ **51.** $e^{-2t} + \frac{2}{7}\sin 7t.$

53. $3e^{3t} - 2.$ **55.** $3 - 2t + \frac{1}{2}t^2 - 3e^{-t}.$ **57.** $e^{-t}\left(2\cos 4t - \frac{5}{4}\sin 4t\right).$

Section 4.3
Page 205

1. $y(t) = e^{-2t}.$ **3.** $y(t) = e^{-4t} - e^t.$ **5.** $y(t) = \frac{1}{3}t^3e^{-t} + 3e^{-t}.$

7. $y(t) = e^{-t} + e^t + 3te^t.$ **9.** $y(t) = t^2e^{-5t} - te^{-5t}.$

11. $y(t) = 3 - 5e^{3t} + 2e^{6t}.$ **13.** $y(t) = -\frac{1}{10}\cos 3t - \frac{1}{30}\sin 3t + \frac{1}{10}e^t.$

15. $y(t) = e^t + e^{-5t}\sin t.$ **17.** $y(t) = 1 - 2e^{-t} + 2e^{(1/2)t}\cos\frac{\sqrt{3}}{2}t.$

19. $y(t) = -2 + 3t + 3e^{-3t} + 6te^{-3t}.$ **21.** $y(t) = -e^{-t} - te^{-t} - \frac{5}{2}t^2e^{-t} + e^{4t}.$

23. $y(t) = 1 - te^{-t} + e^{-5t} - e^{3t}$.

25. $y(t) = -e^{-2t} + te^{-2t} + e^{-5t} + 2te^{-5t}$.

27. $y(t) = 3e^{-t} + e^{t} + 2te^{t} + 2e^{-(1/2)t} \cos \frac{\sqrt{3}}{2} t$.

29. $y(t) = -2e^{t} \cos t + 3e^{t} \sin t + e^{-2t} \cos t$.

31. $y(t) = -1 + e^{2t} - 2te^{2t} + \frac{1}{2} t^3 e^{2t}$.

33. $y(t) = -11e^{-2t} - te^{-2t} + 3e^{t} \cos t\sqrt{3}$.

35. $y(t) = -19e^{-t} + 8e^{-6t} + 13e^{t} \cos t + 9e^{t} \sin t$.

37. $x_1(t) = 2e^{2t} - e^{3t}, x_2(t) = 4e^{2t} - e^{3t}$.

39. $x_1(t) = e^{2t} + 2e^{3t}, x_2(t) = e^{3t} - e^{2t}$.

41. $x_1(t) = -2e^{-2t} \sin 3t, x_2(t) = -e^{-2t} (5 \sin 3t - 3 \cos 3t)$.

43. $x_1(t) = -3te^{2t} - \frac{1}{3} t - \frac{5}{18}, x_2(t) = 6(1 - t)e^{2t} - \frac{4}{3} t - \frac{7}{9}$.

45. $x_1(t) = (t + 1)e^{2t}, x_2(t) = e^{2t}, x_3(t) = (t + 3)e^{2t}$.

Section 4.4
Page 209

1. $y(t) = 1 - e^{-t} - 4te^{-t}$.

3. $y(t) = h_1(t - 2)[4e^{(t-2)} - 5e^{-4(t-2)}]$.

5. $y(t) = 1 + 3t + 5e^{3t} - 3e^{6t}$.

7. $y(t) = e^{-t} + e^{t}(2t - 1) + h_1(t - 3)[2e^{(t-3)} + e^{-(t-3)} - 2]$.

9. $y(t) = h_1(t - 1)[2 - 2 \cos 2(t - 1)] + h_1(t - 2)[\cos 2(t - 2) - 1] - \cos 2t$.

11. $y(t) = e^{-5t} \cos t + h_1(t - 1)[2 - 2e^{-5(t-1)} \cos (t - 1) - 10e^{-5(t-1)} \sin (t - 1)]$
$+ h_1(t - 3)[1 - e^{-5(t-3)} \cos (t - 3) - 5e^{-5(t-3)} \sin (t - 3)]$.

13. $y(t) = -5 + e^{4t} + 4e^{-t} + h_1(t - 1)[-10 + 2e^{4(t-1)} + 8e^{-(t-1)}]$
$+ h_1(t - 2)[-5 + e^{4(t-2)} + 4e^{-(t-2)}]$.

Section 4.5
Page 213

1. $y(t) = e^{4t} - e^{-t}$.

3. $y(t) = te^{-3t} + h_1(t - 2)[(t - 2)e^{-3(t-2)}]$.

5. $y(t) = 6 - e^t + e^{2t}$.

7. $y(t) = e^t - e^{-t} + h_1(t - 3)[e^{-(t-3)} - e^{-2(t-3)}]$.

9. $y(t) = -te^{-5t} + h_1(t - 1)[(t - 1)e^{-5(t-1)}] - 3h_1(t - 2)[(t - 2)e^{-5(t-2)}]$.

11. $y(t) = 9e^{3t} - e^{-3t} + h_1(t - 2)[e^{3(t-2)} - e^{-3(t-2)}]$.

13. $y(t) = e^{4t} - e^{-2t} - 9 \cos t - 2 \sin t + h_1(t - 5)[e^{4(t-5)} - e^{-2(t-5)}]$.

Section 4.6
Page 219

5. $r = 4n(0) + 6 + [4n(0) + 24]t + 12t^2$.

13. (a)

$$i(t) = \frac{(\omega^2 - \omega_0^2)v_0\omega}{L[(\omega_0^2 - \omega^2) + 2a\omega^2]} e^{-at} + \left\{\frac{v_0\omega a(\omega_0^2 - \omega^2)}{L[(\omega_0^2 - \omega^2)^2 + 2a\omega^2]} - \frac{v_0\omega\omega_0^2}{L[(\omega_0^2 - \omega^2)^2 + 2a\omega^2]}\right\} te^{-at} + \frac{v_0\omega(\omega_0^2 - \omega^2)}{L[(\omega_0^2 - \omega^2)^2 + 2a\omega^2]} \cos \omega t + \frac{v_0\omega^2}{L[(\omega_0^2 - \omega^2)^2 + 2a\omega^2]} \sin \omega t.$$

(b)

$$i(t) = \frac{v_0\omega(\omega^2 - \omega_0^2)}{L[(\omega_0^2 - \omega^2)^2 + 2a\omega^2]} e^{-at} \cos \sqrt{\omega_0^2 - a^2}t + \frac{1}{\sqrt{\omega_0^2 - a^2}}\left\{\frac{v_0\omega a(\omega_0^2 - \omega^2)}{L[(\omega_0^2 - \omega^2)^2 + 2a\omega^2]} - \frac{v_0\omega\omega_0^2}{L[(\omega_0^2 - \omega^2)^2 + 2a\omega^2]}\right\} \cdot e^{-at} \sin \sqrt{\omega_0^2 - a^2}t + \frac{v_0\omega(\omega_0^2 - \omega^2)}{L[(\omega_0^2 - \omega^2)^2 + 2a\omega^2]} \cos \omega t + \frac{v_0\omega^2}{L[(\omega_0^2 - \omega^2)^2 + 2a\omega^2]} \sin \omega t.$$

(c)

$$i(t) = \frac{-v_0\omega(\omega_0^2 - \omega^2)}{L[(\omega_0^2 - \omega^2)^2 + 2a\omega^2]} e^{-at} \cosh \sqrt{a^2 - \omega^2} t$$

$$+ \frac{1}{\sqrt{a^2 - \omega_0^2}} \left\{ \frac{v_0 \omega a(\omega_0^2 - \omega^2)}{L[(\omega_0^2 - \omega^2)^2 + 2a\omega^2]} - \frac{v_0 \omega \omega_0^2}{L[(\omega_0^2 - \omega^2)^2 + 2a\omega^2]} \right\}$$

$$\cdot e^{-at} \sinh \sqrt{a^2 - \omega_0^2} t$$

$$+ \frac{v_0 \omega(\omega_0^2 - \omega^2)}{L[(\omega_0^2 - \omega^2)^2 + 2a\omega^2]} \cos \omega t + \frac{v_0 \omega^2}{L[(\omega_0^2 - \omega^2)^2 + 2a\omega^2]} \sin \omega t.$$

REVIEW EXERCISES CHAPTER 4
Page 221

1. $\dfrac{480}{s^6} + \dfrac{6s}{(s^2+1)^2}$. **3.** $1 + \dfrac{e^{-2s}}{s}$. **5.** $\dfrac{1}{8}e^{9t} - \dfrac{1}{8}e^{t}$.

7. $h_1(t-3)e^t$. **9.** $y(t) = 91 - 48t + e^{8t} - 43e^{-t}$.

11. $y(t) = 17 - 6t - 17\cos 2t - 7\sin 2t + 5e^{-2t/5} - 5e^{-3t}$.

13. $y(t) = 3e^t - e^{-t} - 2e^{(1/2)t}\cos\dfrac{\sqrt{3}}{2}t - 2\sqrt{3}\,e^{(1/2)t}\sin\dfrac{\sqrt{3}}{2}t$.

15. $y(t) = 5h_1(t-2) - 8h_1(t-2)e^{3(t-2)} + 3h_1(t-2)e^{-2(t-2)} + 8e^{3t} - 3e^{8t}$.

17. See Exercise 16, Section 4.6, and the convolution theorem.

19. $x(t) = Ae^{-bt} + Bte^{-bt} + Ce^{\lambda_1 t} + De^{\lambda_2 t}$,

where $\lambda_1 = [-R + \sqrt{R^2 - 4Km}]/2m$, $\lambda_2 = [-R - \sqrt{R^2 - 4Km}]/2m$,

$A = Q(2mb - R)/m^2(b + \lambda_1)^2(b + \lambda_2)^2$,

$B = Q/[mb^2 - Rb + K]$,

$C = Q/[(b + \lambda_2)^2\sqrt{R^2 - 4Km}]$,

and

$D = -Q/[(b + \lambda_1)^2\sqrt{R^2 - 4Km}]$.

21. $x_1(t) = e^{-t}$, $x_2(t) = 5e^{-t}$.

23. $x_1(t) = e^{-t} - 2e^{2t}$, $x_2(t) = 2e^{-t} - e^{2t}$.

25. $x_1(t) = 7e^{-t}\cos t + 3e^{-t}\sin t$

$x_2(t) = -7e^{-t}\sin t + 3e^{-t}\cos t.$

27. $x_1(t) = e^t - 6te^t,\ x_2(t) = e^t - 6(t + 1)e^t.$

29. $x_1(t) = e^{3t} - e^{5t},\ x_2(t) = e^{3t} - 3e^{5t}.$

Section 5.2
Page 228

1. $R = \frac{1}{3}.$ **3.** $R = +\infty.$ **5.** $R = 1$

7. First derivative: $\sum_{n=1}^{\infty} 3^n n(x - 1)^{n-1}; R = \frac{1}{3}.$

Second derivative: $\sum_{n=2}^{\infty} 3^n n(n - 1)(x - 1)^{n-2}; R = \frac{1}{3}.$

9. First derivative: $\sum_{n=1}^{\infty} \frac{(-1)^n}{(2n-1)!} x^{2n-1}; R = +\infty.$

Second derivative: $\sum_{n=1}^{\infty} \frac{(-1)^n}{(2n-2)!} x^{2n-2}; R = +\infty.$

11. First derivative: $\sum_{n=1}^{\infty} \frac{n}{n+1}(x + 1)^{n-1}; R = 1.$

Second derivative: $\sum_{n=2}^{\infty} \frac{n(n-1)}{n+1}(x + 1)^{n-2}; R = 1.$

15. Solution: $y(x) = a_0 \sum_{n=0}^{\infty} \frac{3^n(x-1)^n}{n!} = a_0 e^{3(x-1)}.$

17. $\cos x = \sum_{n=0}^{\infty} \frac{x^{2n}}{(2n)!}$; $R = +\infty$; converges for all x.

19. $\sin x = \sum_{n=0}^{\infty} \frac{(-1)^n(x-\pi)^{2n-1}}{(2n+1)!}$; $R = +\infty$; converges for all x.

21. $\frac{1}{x+3} = \sum_{n=0}^{\infty} \frac{(-1)^n(x-1)^n}{4^{n+1}}$; $R = 4$; converges for $-3 < x < 5$.

23. $\frac{1}{x} = \sum_{n=0}^{\infty} -\frac{(x+3)^n}{3^{n+1}}$; $R = 3$; converges for $-6 < x < 0$.

Section 5.3
Page 232

1. $x = 0$ is a regular singular point. Every other finite value of x is an ordinary point.

3. $x = 1$ is a regular singular point. Every other finite value of x is an ordinary point.

5. $x = 1$ is a regular singular point. Every other finite value of x is an ordinary point.

7. $x = 0$ is an irregular singular point. $x = 1$ and $x = -1$ are regular singular points. Every other finite value of x is an ordinary point.

9. There are no singular points.

11. $x = 1$ and $x = -1$ are regular singular points.

13. There are no singular points.

15. $x = 1$ and $x = -1$ are regular singular points.

17. False.

19. True.

21. True.

Section 5.4
Page 244

1. $y(x) = 1 - 2x^2$.

3. $y(x) = -7 - 8x - 2x^2$.

5. $y(x) = x$.

7. $y(x) = x$.

9. $y(x) = -x$.

11. $a_0 = 0, a_1 = 1, a_2 = 0, a_3 = 0$.

13. $a_0 = -20, a_1 = -2, a_2 = -1, a_3 = 1$.

15. $a_0 = 0, a_1 = 1, a_2 = 0, a_3 = 0$.

17. $a_0 = 0, a_1 = 1, a_2 = 0, a_3 = -\frac{2}{3}$.

19. $a_0 = 1, a_1 = 0, a_2 = -1, a_3 = 0$.

21. Apply Theorem 1.

23. Apply Theorem 1.

25. $P_1(x) = x$.

27. $P_3(x) = \frac{5}{2}x^3 - \frac{3}{2}x$.

31. $T_1(x) = x$.

33. $T_3(x) = 4x^3 - 3x$.

35. $H_2(x) = 4x^2 - 2$.

37. $2a_2 + 3a_1 + 2a_0 = 0$;
$$a_{n+2} = -\frac{3(n+1)a_{n+1} + (n+2)a_n}{(n+2)(n+1)},$$
$n = 1, 2, 3, \ldots$.

Section 5.5
Page 264

1. $y_1(x) = x^{1/3} \sum_{n=0}^{\infty} a_n x^n$; $y_2(x) = x^{-1/3} \sum_{n=0}^{\infty} b_n x^n$.

3. $y_1(x) = \sum_{n=0}^{\infty} a_n x^n$; $y_2(x) = y_1(x) \ln|x| + \sum_{n=0}^{\infty} b_n x^n$.

5. $y_1(x) = \sum_{n=0}^{\infty} a_n x^n; y_2(x) = y_1(x) \ln|x| + \sum_{n=0}^{\infty} b_n x^n.$

7. $y_1(x) = |x|^{1/4} \sum_{n=0}^{\infty} a_n x^n; y_2(x) = |x|^{-1/4} \sum_{n=0}^{\infty} b_n x^n.$

9. $y_1(x) = \sum_{n=0}^{\infty} a_n x^n; y_2(x) = y_1(x) \ln|x| + \sum_{n=0}^{\infty} b_n x^n.$

11. $y_1(x) = \sum_{n=0}^{\infty} a_n x^n; y_2(x) = Cy_1(x) \ln|x| + x^{-2} \sum_{n=0}^{\infty} b_n x^n.$

13. $y_1(x) = x^2\left[1 + \sum_{n=1}^{\infty} \frac{(-1)^n}{2^{2n} n! 3\cdot 4 \cdots (2+n)} x^{2n}\right].$

15. $y_1(x) = 1 + 2x + \frac{8}{3}x^2 + \cdots.$ **17.** $y_1(x) = 1 - 2x + \frac{1}{2}x^2.$

19. $y_1(x) = 1 - x; y_2(x) = (1 - x) \ln|x| + (1 + 2x + \cdots).$

21. True. **23.** False. **25.** True. **27.** False.

29. $y_1(x) = 1 - \frac{1}{2}x^4 + \frac{1}{24}x^8 - \cdots,$

$y_2(x) = x^2 - \frac{1}{6}x^6 + \frac{1}{120}x^{10} - \cdots.$

31. $a_n = \frac{1}{n(n+3)} a_{n-1}, n = 1, 2, 3, \ldots.$

33. $a_n = \frac{1}{2n+3} a_{n-1}, n = 1, 2, 3, \ldots.$ **37.** $y_1(x) = \sum_{n=0}^{\infty} \frac{(-1)^n}{(n!)^2} \left(\frac{x}{2}\right)^{2n}.$

39. $L_1(x) = 1 - x.$ **41.** $L_3(x) = 6 - 18x + 9x^2 - x^3.$

43. $F\left(1, 2, \frac{3}{4}; x\right) = 1 + \sum_{n=1}^{\infty} \frac{4^n(n+1)!}{3\cdot 7\cdot 11 \cdots (4n-1)} x^n.$

45. $t(1 - t)\ddot{y} + \left(\frac{1}{2} - t\right)\dot{y} - 2y = 0.$

REVIEW EXERCISES CHAPTER 5

Page 266

1. $y(x) = -1 - \sum_{n=1}^{\infty} \frac{1\cdot 4 \cdots (3n-2)}{(3n)!} x^{3n}.$

3. $y(x) = -3x + 4x^3$

5. $a_0 = 1, a_1 = 0, a_2 = -\frac{1}{2}, a_3 = \frac{1}{6}.$

7. $a_0 = 4, a_1 = 1, a_2 = \frac{1}{7}, a_3 = \frac{6}{245}.$

9. It has two linearly independent solutions, each of the form
$$y(x) = \sum_{n=0}^{\infty} a_n x^n; \ R = \infty.$$

11. It has two linearly independent solutions, each of the form
$$y(x) = \sum_{n=0}^{\infty} a_n (x-1)^n; \ R \geq 1.$$

13. $y_1(x) = |x+1|^{1/2} \sum_{n=0}^{\infty} a_n (x+1)^n$

$y_2(x) = \sum_{n=0}^{\infty} b_n (x+1)^n; \ R \geq 2.$

15. $y(x) = a_0 x^3 \left[1 + \sum_{n=1}^{\infty} \frac{(-1)^n}{2^{2n} \cdot n! \cdot 4 \cdot 5 \cdot 6 \cdots (3+n)} x^{2n}\right].$

17. $y(x) = a_0(6 - 18x + 9x^2 - x^3).$

19. $\cos x = 1 - \frac{x^2}{2!} + \frac{x^4}{4!} - \frac{x^6}{6!} + \cdots.$

Section 6.1
Page 274

1. True. 3. False (compare Exercise 3 with Exercises 1 and 2).

5. False $[y(\pi/2) = +1, \text{ not } -1]$. 7. $y(x) = \cos 3x + c_2 \sin 3x$.

9. $y(x) = \cos 3x + \sin 3x$. 11. $y(x) = 0$. 13. $y(x) = xe^x$.

15. $y_1(x) = x$ and $y_2(x) = x^3$ are linearly independent solutions of the homogeneous differential equation. Also,
$$\begin{vmatrix} y_1(1) & y_2(1) \\ y_1(2) & y_2(2) \end{vmatrix} = \begin{vmatrix} 1 & 1 \\ 2 & 8 \end{vmatrix} = 8 - 2 = 6 \neq 0.$$
Therefore, by Theorem 1, part (a), there exists one and only one solution.

17. $y(x) = c_1 + \frac{c_2 - c_1}{l} x$. Note that $y(0) = c_1$ and $y(l) = c_2$.

Section 6.2
Page 283

1. $\lambda_n = 4\pi^2 n^2, \ y_n = \sin 2\pi n x; \ n = 1, 2, 3, \ldots.$

3. $\lambda_n = n^2\pi^2, \ n = 1, 2, 3, \ldots, \ y_n(x) = \begin{cases} \sin n\pi x & \text{if } n \text{ is even} \\ \cos n\pi x & \text{if } n \text{ is odd} \end{cases} \quad n = 1, 2, 3, \ldots.$

5. $\lambda_n = -2 + \frac{n^2\pi^2}{4}, n = 1, 2, 3, \ldots, y_n(x) = \sin\frac{n\pi x}{2}, n = 1, 2, 3, \ldots.$

7. $\lambda_n = -2 + n^2\pi^2, \ n = 1, 2, 3, \ldots, \ y_n = \sin n\pi x, \ n = 1, 2, 3, \ldots.$

11. Step 1. Solve the EVP $X'' - \lambda X = 0$

$$X(0) = X(l) = 0$$

to obtain $\lambda_n = -\frac{n^2\pi^2}{l^2}, n = 1, 2, 3, \ldots$

$$X_n(x) = \sin\frac{n\pi x}{l}, \; n = 1, 2, 3, \ldots.$$

Step 2. Solve $T'' + \frac{c^2n^2\pi^2}{l^2}T = 0$ to obtain

$$T_n(t) = c_1 \cos\frac{cn\pi}{l}t + c_2 \sin\frac{cn\pi}{l}t.$$

Step 3. $u_n(x, t) = X_n(x)T_n(t)$ and $\sum_{n=1}^{\infty} \alpha_n u_n(x, t)$ are solutions of the DE.

Step 4. Choose β_n and γ_n, where $\beta_n = \alpha_n c_1$, $\gamma_n = \alpha_n c_2$ such that

$$\sum_{n=1}^{\infty} \beta_n \sin\frac{n\pi x}{l} = f(x)$$

and

$$\sum_{n=1}^{\infty} \gamma_n \sin\frac{n\pi x}{l} = g(x).$$

REVIEW EXERCISES CHAPTER 6
Page 285

1. It has no solution.

3. It has exactly one solution: $y(x) = 2x - \frac{\pi}{4}\sin 4x$.

5. No solution if $A \neq B - 2\pi$; infinitely many solutions if $A = B - 2\pi$. It never has exactly one solution.

7. $\lambda_n = -n^2, \quad n = 1, 2, \ldots; y_n(x) = \sin nx, \quad n = 1, 2, \ldots.$

9. $\lambda_n = n^2, n = 0, 1, 2, \ldots; y_n(x) = \cos nx, n = 0, 1, 2, \ldots.$

Section 7.2
Page 293

1. {(0, 3), (0.2, 1.8), (0.4, 1.12), (0.6, 0.752), (0.8, 0.5712), (1, 0.5027)}.

3. {(0, 0), (0.2, 0), (0.4, 0.024), (0.6, 0.1104), (0.8, 0.2822), (1, 0.5533)}.

5. {(0, 0), (0.2, 0.2), (0.4, 0.392), (0.6, 0.5587), (0.8, 0.6772), (1, 0.7129)}.

7. {(0, 3), (0.1, 2.4), (0.2, 1.93), (0.3, 1.564), (0.4, 1.2812), (0.5, 1.0650), (0.6, 0.9020), (0.7, 0.7816), (0.8, 0.6953), (0.9, 0.6362), (1, 0.5990)}.

9. $\{(0, 0), (0.1, 0.1), (0.2, 0.199), (0.3, 0.2950), (0.4, 0.3857), (0.5, 0.4685), (0.6, 0.5405), (0.7, 0.5977), (0.8, 0.6354), (0.9, 0.6478), (1, 0.6278)\}$.

11. $y(0.85) \approx 0.2344$; actual value $= 0.2544$.

13. $\{(0, 0), (0.1, 0), (0.2, 0.01), (0.3, 0.031), (0.4, 0.0641), (0.5, 0.1105), (0.6, 0.1716), (0.7, 0.2487), (0.8, 0.3436), (0.9, 0.4579), (1, 0.5937)\}$; actual values $= \{(0, 0), (0.1, 0.0052), (0.2, 0.0214), (0.3, 0.0499), (0.4, 0.0918), (0.5, 0.1487), (0.6, 0.2221), (0.7, 0.3137), (0.8, 0.4255), (0.9, 0.5596), (1, 0.7183)\}$.

Section 7.3
Page 297

1. $\{(0, 0), (0.2, 0), (0.4\ 0.0168), (0.6, 0.0717), (0.8, 0.1907), (1, 0.4054)\}$.

3. $\{(0, 1), (0.2, 0.84), (0.4, 0.7448), (0.6, 0.7027), (0.8, 0.7042), (1, 0.7415)\}$.

5. $\{(0, 1), (0.2, 1.02), (0.4, 1.0820), (0.6, 1.1937), (0.8, 1.3694), (1, 1.6334)\}$.

7. $\{(0, 2), (0.2, 2), (0.4, 2.0321), (0.6, 2.1306), (0.8, 2.3407), (1, 2.7344)\}$.

9. $\{(0, 0), (0.2, 0.02), (0.4, 0.0884), (0.6, 0.2158), (0.8, 0.4153), (1, 0.7027)\}$; actual values—see the answer to Exercise 13, Section 7.2.

11. $\{(0, 0), (0.1, 0.005), (0.2, 0.0210), (0.3, 0.0492), (0.4, 0.0909), (0.5, 0.1474), (0.6, 0.2204), (0.7, 0.3116), (0.8, 0.4228), (0.9, 0.5562), (1, 0.7141)\}$; actual values—see the answer to Exercise 13, Section 7.2.

13. $\{(0, 0), (0.1, 0), (0.2, 0.0021), (0.3, 0.0085), (0.4, 0.0218), (0.5, 0.0449), (0.6, 0.0809), (0.7, 0.1331), (0.8, 0.2056), (0.9, 0.3024), (1, 0.4282)\}$.

15. $\{(0, -1), (0.1, -1.205), (0.2, -1.4221), (0.3, -1.6540), (0.4, -1.9038), (0.5, -2.1757), (0.6, -2.4743), (0.7, -2.8057), (0.8, -3.1769), (0.9, -3.5969), (1, -4.0762)\}$.

Section 7.4
Page 301

1. $\{(0, 1), (0.2, 1.5587), (0.4, 2.5387), (0.6, 4.1467), (0.8, 6.6909), (1, 10.6307)\}$.

3. $\{(0, 0), (0.2, 0.0147), (0.4, 0.0493), (0.6, 0.0926), (0.8, 0.1397), (1, 0.1885)\}$.

5. $\{(0, 0), (0.2, -0.0398), (0.4, -0.1576), (0.6, -0.3446), (0.8, -0.5810), (1, -0.8391)\}$.

7. $\{(0, 0), (0.2, -0.0179), (0.4, -0.0638), (0.6, -0.1170), (0.8, -0.1401), (1, -0.0649)\}$.

9. $\{(0, 1), (0.2, 1.5605), (0.4, 2.5442), (0.6, 4.1591), (0.8, 6.7157), (1, 10.6769)\}$.

11. {(0, 0), (0.2, −0.0399), (0.4, −0.1580), (0.6, −0.3452), (0.8, −0.5819), (1, −0.8398)}.

13. {(0, 0), (0.2, 0.0213), (0.4, 0.0917), (0.6, 0.2218), (0.8, 0.4250), (1, 0.7175)}; actual values—see the answer to Exercise 13, Section 7.2.

15. {(0, 0), (0.1, 0.0052), (0.2, 0.0214), (0.3, 0.0498), (0.4, 0.0918), (0.5, 0.1487), (0.6, 0.2221), (0.7, 0.3137), (0.8, 0.4255), (0.9, 0.5595), (1, 0.7182)}; actual values—see the answer to Exercise 13, Section 7.2.

17. {(0, 1), (0.1, 1.2373), (0.2, 1.5604), (0.3, 1.9882), (0.4, 2.5438), (0.5, 3.2556), (0.6, 4.1582), (0.7, 5.2938), (0.8, 6.7139), (0.9, 8.4815), (1, 10.6735)}.

Section 7.5
Page 307

1. {(0, 0, 0), (0.2, 0.2, 0), (0.4, 0.44, 0.008), (0.6, 0.7216, 0.0483), (0.8, 1.0513, 0.1621), (1, 1.4437, 0.4156)}.

3. {(0, 0, 0), (0.2, 0.2201, 0.0034), (0.4, 0.4828, 0.0316), (0.6, 0.7969, 0.1281), (0.8, 1.1829, 0.3696), (1, 1.6820, 0.8961)}.

5. {(0, 1, 0), (0.2, 1.2, 0.2), (0.4, 1.44, 0.52), (0.6, 1.704, 1.016), (0.8, 1.9616, 1.7632), (1, 2.1613, 2.8608)}.

7. {(0, 1, 1), (0.2, 1.4, 1), (0.4, 1.88, 1.04), (0.6, 2.4723, 1.128), (0.8, 3.2213, 1.2769), (1, 4.1916, 1.5057)}.

9. {(0, 0, 1), (0.2, −0.12, 1), (0.4, −0.3984, 1.04), (0.6, −0.6019, 1.1232), (0.8, −0.8110, 1.2580), (1, −1.0251, 1.4599)}.

11. From Example 1, Section 3.2 $x(1) \approx 7.8666$, $y(1) \approx 13.2448$; actual values: $x(1) = 10.1073$, $y(1) = 17.4964$.

13. {(0, 2, 3), (0.2, 2.712, 4.2027), (0.4, 3.7137, 5.9358), (0.6, 5.1342, 8.4466), (0.8, 7.1627, 12.1004), (1, 10.0779, 17.4384)}; actual values—see the answer to Exercise 11.

Section 7.6
Page 309

1. {(0, 22.3), (0.2, 23.1675), (0.4, 24.1069), (0.6, 25.1273), (0.8, 26.2395), (1, 27.4563)}.

3. {(0, 22.3), (0.2, 23.2042), (0.4, 24.1883), (0.6, 25.2633), (0.8, 26.4426), (1, 27.7418)}.

5. {(0, 9.45), (0.2, 9.4071), (0.4, 9.3646), (0.6, 9.3224), (0.8, 9.2807), (1, 9.2393)}.

7. $\{(0, 9.45), (0.2, 9.4073), (0.4, 9.3650), (0.6, 9.3230), (0.8, 9.2814), (1, 9.2402)\}$.

9. $\{(0, 30), (0.2, 29.52), (0.4, 29.0704), (0.6, 28.6479), (0.8, 28.2497), (1, 27.8736)\}$.

11. $\{(0, 30), (0.2, 29.5200), (0.4, 29.0704), (0.6, 28.6479), (0.8, 28.2497), (1, 27.8736)\}$.

13. (a) $\frac{1}{4T_1^3} \ln \left| \frac{T - T_1}{T + T_1} \right| - \frac{1}{2T_1^3} \tan^{-1}\left(\frac{T}{T_1}\right) = -\frac{A}{M}t + C$, where C is a constant of integration.

15. $n = \frac{e^{2\sqrt{Akt}} - 1}{e^{2\sqrt{Akt}} + 1}\sqrt{\frac{A}{k}}$. From Exercise 14, we have $\{(0, 0), (0.2, 20{,}000), (0.4, 39{,}600), (0.6, 58{,}032), (0.8, 74{,}664), (1, 89{,}089)\}$; actual values (using the formula above) $= \{(0, 0), (0.2, 19{,}867.7248), (0.4, 38{,}966.3962), (0.6, 56{,}641.5992), (0.8, 72{,}433.6982), (1, 86{,}105.172)\}$.

17. $\{(1, 1), (1.2, 0.9822), (1.4, 0.9227), (1.6, 0.8124), (1.8, 0.6250), (2, 0.1793)\}$.

19. $c_1 = a/(b + ace^{a\xi})$, where c is a constant of integration.

REVIEW EXERCISES CHAPTER 7
Page 312

1. $\{(1, 0), (1.2, 0.2), (1.4, 0.488), (1.6, 0.9132), (1.8, 1.5827), (2, 2.7602)\}$.

3. $\{(1, 0), (1.2, 0.2452), (1.4, 0.5564), (1.6, 1.056), (1.8, 1.916), (2, 3.8013)\}$.

5. $\{(1, 0), (1.2, 0.2466), (1.4, 0.6294), (1.6, 1.2936), (1.8, 2.7854), (2, 9.9533)\}$.

7. $\{(1, 1, -1), (1.2, 1.3745, -0.5829), (1.4, 1.7924, -0.0720), (1.6, 1.6219, 0.5609), (1.8, 1.1283, 1.3056), (2, 0.6285, 2.0004)\}$.

9. $\{(1, 1, 1), (1.2, 1.3748, -0.5803), (1.4, 1.5792, -0.1035), (1.6, 1.4895, 0.3972), (1.8, 1.1607, 0.8374), (2, 0.7659, 1.2441)\}$.

11. The equation $\dot{y} = x^2 + y^2$ indicates that y is an increasing function of t. Thus the relationship of Exercise 10 cannot be valid for all $t \geq 1$ since in each of Exercises 6 through 9 (see the answers for Exercises 7 and 9) we note that $y(t) > 0$ for all $t > t_0$, where $1.4 < t_0 < 1.6$ and positive values for y cannot be obtained from the relation given in Exercise 10.

13. $\{(1, 4.5), (1.2, 4.3812), (1.4, 4.3211), (1.6, 4.2795), (1.8, 4.2436), (2, 4.2076)\}$. Actual values (see Example 2, Section 1.5): $\{(1, 4.5), (1.2, 4.4112), (1.4, 4.3553), (1.6, 4.3124), (1.8, 4.2734), (2, 4.2333)\}$.

15. $\{(1, 4.5), (1.2, 4.4113), (1.4, 4.3562), (1.6, 4.3132), (1.8, 4.2741), (2, 4.2339)\}$. Actual values—see the answer to Exercise 13.

Section 8.2
Page 317

1. In the region $-1 \leq t \leq 1$, $0 \leq x_1 \leq 1$, $0 \leq x_2 \leq 1$, choose $x_1 = 1/n^2$, $x_2 = y_2$, $y_1 = 1/(n+1)^2$. In this case, the assumption that the Lipschitz condition is satisfied leads to the impossible requirement that the constant L_1 satisfy the inequality

$$\frac{1}{n(n+1)} \leq L_1 \frac{2n+1}{n^2(n+1)^2}.$$

In the region $-1 \leq t \leq 1$, $3 \leq x_1 \leq 4$, $0 \leq x_2 \leq 1$, we can write

$$|f(t, x_1, x_2) - f(t, y_1, y_2)| \leq |x_1^{1/2} - y_1^{1/2}| + |x_2 - y_2|$$

$$= \frac{|x_1 - y_1|}{\sqrt{x_1} + \sqrt{y_1}} + |x_2 - y_2| \leq \frac{|x_1 - y_1|}{3 + 0} + |x_2 - y_2|.$$

Thus, we can choose $L_1 = \dfrac{1}{3}$ and $L_2 = 1$.

3. The general solution of the IVP is $y(t) = A\cos t + B\sin t$. If $\bar{y}(t) = \bar{A}\cos t + \bar{B}\sin t$ is the true value of the solution, then

$$|\bar{y}(t) - y(t)| \leq |\bar{A} - A| + |\bar{B} - B| \leq 2 \times 10^{-3}.$$

Section 8.3
Page 320

1. (0, 0) is the only critical point. For $xy \neq 0$, the trajectories are the half-lines $y = cx$, $c \neq 0$. The direction for increasing t is away from the origin.

3. (0, 0) is the only critical point. For $xy \neq 0$, the trajectories are the hyperbolas $x^2 - y^2 = c$, $c \neq 0$. For $c = 0$, there are four half-lines $y = x$, $x > 0$; $y = x$, $x < 0$; $y = -x$, $x > 0$; $y = -x$, $x < 0$.

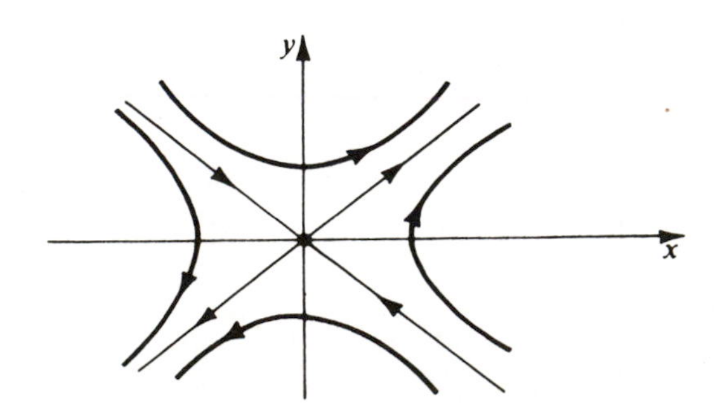

5. $x = e^t$, $y = -e^{2t}$. The trajectory is $y = -x^2$, $x > 0$.

7. The trajectories are given by $y = x \ln |x|$, $x \geq 1$.

9. $x(t) = e^{-t}(-\cos t + \sin t)$, $y(t) = e^{-t}(\sin t + \cos t)$. This curve is a logarithmic spiral approaching the origin as $t \to \infty$.

11. The only critical point is (0, 0). The trajectories are given by $y(y^2 - 3x^2) = C$.

13. The critical points are at $(n\pi, 0)$, $n = 0, \pm 1, \pm 2, \ldots$. The trajectories are given by $\frac{1}{2}y^2 - 4\cos x = C$.

15. Set $\dot{x} = y$; then $\dot{y} = -x$. The trajectories are the circles $x^2 + y^2 = C$.

17. Set $\dot{x} = y$; then $\dot{y} = x$. The trajectories are the hyperbolas $y^2 - x^2 = C$.

19. (a) $ax + by = 0$, $cx + dy = 0$. From Appendix A, the solution of this homogeneous system is (0, 0) if and only if the determinant of coefficients is not zero (if $ad - bc \neq 0$).

(b) If $ad - bc = 0$, the two equations coincide and every point on the line $ax + by = 0$ is a critical point.

Section 8.4
Page 326

1. Stable. **3.** Asymptotically stable. **5.** Unstable. **7.** Unstable.

9. Asymptotically stable. **11.** (0, 0) is asymptotically stable.

13. Asymptotically stable. **15.** Unstable. **17.** Asymptotically stable.

19. Asymptotically stable. **21.** Asymptotically stable.

23. Asymptotically stable. **25.** Stable.

Section 8.5
Page 336

1. $\lambda = 1, 3$. Unstable node. **3.** $\lambda = -1, 1$. Saddle point.

5. $\lambda = 2, 2$. Unstable node. **7.** $\lambda = -1, -1$. Stable node.

9. $\lambda = -2 \pm i$. Stable focus. **11.** $\lambda = -1 \pm i$. Stable focus.

13. $\lambda = -2, -3$. Stable node. **15.** $\lambda = \pm\sqrt{7}$. Saddle point.

17. $\lambda = \pm 1$. Saddle point. **19.** $\lambda = \pm 1$. Saddle point. **21.** Stable node.

23. Stable node. **25.** Stable focus.

27. $s = -2, p = -4, s^2 - 4p > 0$. Saddle point.

29. $s = 0, p = -13, s^2 - 4p > 0$. Saddle point.

31. $s = 2, p = 1, s^2 - 4p = 0$. Unstable node.

33. $s = -2, p = 1, s^2 - 4p = 0$. Asymptotically stable node.

Section 8.6
Page 345

1. The linearized system for X, Y is $\dot{X} = -(bc/d)\,Y$, $\dot{Y} = (ad/b)\,X$. The point $(X, Y) = (0, 0)$ is a center.

3. The linearized system is $\dot{x} = y$, $\dot{y} = -w^2x - ky$.

(a) Stable focus.
(b) Stable node.
(c) Stable node.

REVIEW EXERCISES CHAPTER 8
Page 346

1. The hypotheses of Theorem 1, Section 8.1 are satisfied.

3. The line $y = 2x$.

5. $(0, 0)$ is the only critical point; the equation of trajectories is $y(x) = x^2(c - \ln x)$; the trajectory through $(1, 0)$ is $y(x) = -x^2 \ln x$.

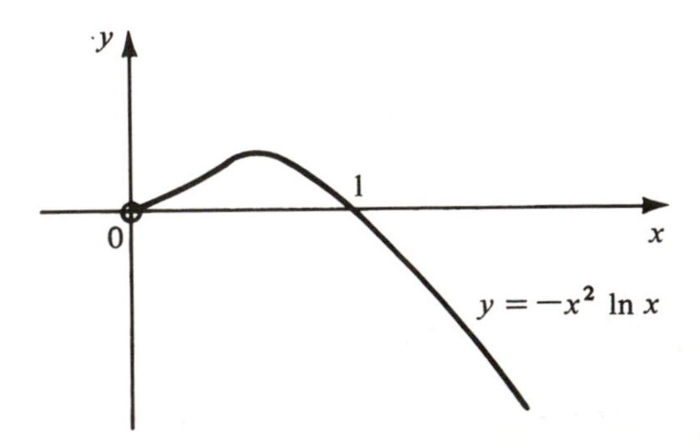

7. Stable.

9. Asymptotically stable.

11. Unstable.

13. Saddle point.

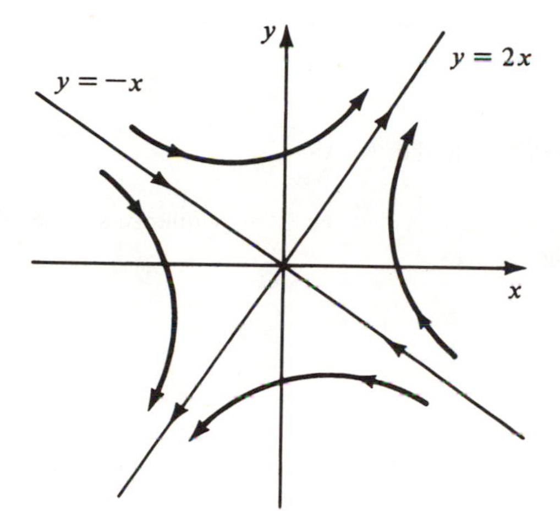

15. Center.

17. Node.

19. Asymptotically stable focus.

Section 9.1
Page 354

1. Linear, homogeneous, variable coefficients, order two.

3. Nonlinear.

5. Linear, nonhomogeneous, constant coefficients, order two.

7. Linear, nonhomogeneous, variable coefficients, order one.

9. Linear, nonhomogeneous, variable coefficients, order one.

11. Linear, nonhomogeneous, variable coefficients, order two.

13. Linear, nonhomogeneous, constant coefficients, order three.

15. Linear, nonhomogeneous, constant coefficients, order two.

17. True. **19.** False.

21. False. **23.** True.

Section 9.2
Page 357

1. Yes.

3. No, the coefficient q_k vanishes for $k = 2$.

5. Verify by direct substitution that the given sequence satisfies the IVP. Uniqueness follows from Theorem 1.

7. See Exercise 5.

9. See Exercise 5.

11. $y_k = k \cdot 2^k$. **13.** $y_k = (\frac{1}{2})^k$.

15. THEOREM *The initial value problem*

$$a_n(k)y_{k+n} + a_{n-1}(k)y_{k+n-1} + \cdots + a_1(k)y_{k+1} + a_0(k)y_k = f(k)$$
$$y_0 = A_0, y_1 = A_1, \ldots, y_{n-1} = A_{n-1}$$

where $a_n, a_{n-1}, \ldots, a_0$ and f are given functions of k, with $a_n(k) \cdot a_0(k) \neq 0$ for $k = 0, 1, 2, \ldots$ and $A_0, \ldots, A_{n-1}$ given constants, has exactly one solution.

Section 9.3
Page 364

1. $y_k = c_1 3^k + c_2 5^k$.

3. The solutions are linearly dependent.

5. The solutions are linearly dependent.

7. $y_k = c_1 + c_2(-1)^k + 2k^3 - 6k^2 + 4k$.

11. The Casoratian of the sequences $y_k^{(1)}, y_k^{(2)}, \ldots, y_k^{(n)}$ is denoted by $C(y_k^{(1)}, \ldots, y_k^{(n)})$ and is defined to be the determinant

$$\begin{vmatrix} y_k^{(1)} & y_k^{(2)} & \cdots & y_k^{(n)} \\ y_{k+1}^{(1)} & y_{k+1}^{(2)} & \cdots & y_{k+1}^{(n)} \\ \cdots & \cdots & \cdots & \cdots \\ y_{k+n-1}^{(1)} & y_{k+n-1}^{(2)} & \cdots & y_{k+n-1}^{(n)} \end{vmatrix}.$$

13. THEOREM

Let $y_k^{(1)}, y_k^{(2)}, \ldots, y_k^{(n)}$ be n linearly independent solutions of the linear homogeneous difference equation

$$a_n(k)y_{k+n} + a_{n-1}(k)y_{k+n-1} + \cdots + a_1(k)y_{k+1} + a_0(k)y_k = 0,$$

where $a_n(k)a_0(k) \neq 0$ for $k = 0, 1, 2, \ldots$. Then the general solution of the equation is $c_1y_k^{(1)} + c_2y_k^{(2)} + \cdots + c_ny_k^{(n)}$, where $c_1, c_2, \ldots, c_n$ are arbitrary constants.

15. True. **17.** False. **19.** False.

21. $y_k = y_0 + \sum_{n=0}^{k-1} r_n$, where y_0 is arbitrary.

23. $y_k = y_0 + \frac{1}{2}k(k-1)$, where y_0 is arbitrary.

25. $y_k = c(-3)^k - \frac{1}{4}$, where c is an arbitrary constant.

27. $y_k = c(2k+1)$, where c is an arbitrary constant.

Section 9.4
Page 370

1. $y_k = c_1 2^k + c_2 5^k$. **3.** $y_k = c_1(-3)^k + c_2 k(-3)^k$.

5. $y_k = c_1(3\sqrt{2})^k \cos\frac{3k\pi}{4} + c_2(3\sqrt{2})^k \sin\frac{3k\pi}{4}$.

7. $y_k = c(\frac{2}{3})^k$. **9.** $y_k = c_1 2^k + c_2 k 2^k$.

11. $y_k = c_1(-\frac{4}{3})^k + c_2(\frac{4}{3})^k$.

13. $y_k = c_1 + c_2(-1)^k$.

15. $y_k = c(-0.1)^k$.

17. $$y_k = c_1(\sqrt{2})^k \cos\frac{k\pi}{4} + c_2(\sqrt{2})^k \sin\frac{k\pi}{4} + c_3(\sqrt{2})^k \cos\frac{3k\pi}{4} + c_4(\sqrt{2})^k \sin\frac{3k\pi}{4}.$$

19. $y_k = c_1 + c_2 \cos\frac{k\pi}{2} + c_3 \sin\frac{k\pi}{2}$.

21. $y_k = c_1(-\frac{1}{3})^k + c_2(-\frac{1}{2})^k$.

23. $y_k = c_1(-1)^k + c_2 \cos\frac{k\pi}{3} + c_3 \sin\frac{k\pi}{3}$.

25. $y_k = c_1 2^k + c_2 k 2^k + c_3 k^2 2^k$.

27. $y_k = c_1 + c_2 k + c_3(-1)^k + c_4(-2)^k$.

29. $y_k = 3 \cdot 5^k$. **31.** $y_k = \frac{1}{3}(3\sqrt{2})^k \sin \frac{3k\pi}{4}$.

33. $y_k = \frac{1}{\sqrt{5}}\left[\left(\frac{1+\sqrt{5}}{2}\right)^k - \left(\frac{1-\sqrt{5}}{2}\right)^k\right]$.

35. $y_k = 1 + \frac{1}{2}\sin \frac{2k\pi}{3}$. **37.** $y_k = 1 - \cos \frac{k\pi}{2}$.

39. The general solution has one of the forms described by Theorem 1. Now use the fact that $|a| < 1$ if and only if $\lim_{k\to\infty} a^k = 0$ and $\lim_{k\to\infty} ka^k = 0$.

41. Eventually the ray will get out of the lenses.

43. 11.974.

45. $$V_k = \frac{V_0}{\left(\frac{3+\sqrt{5}}{2}\right)^n - \left(\frac{3-\sqrt{5}}{2}\right)^n}\left[\left(\frac{3+\sqrt{5}}{2}\right)^n\left(\frac{3-\sqrt{5}}{2}\right)^k - \left(\frac{3-\sqrt{5}}{2}\right)^n\left(\frac{3+\sqrt{5}}{2}\right)^k\right]$$

47. $y_k = c_1 \cdot 1 + c_2 k^2$.

Section 9.5
Page 383

1. $y_k^p = -4 \cdot 3^k$. **3.** $y_k^p = \frac{1}{8} - 3^{k-2}$.

5. $y_k^p = \frac{3}{2}k$. **7.** $y_k^p = -\frac{3}{25}k2^k + \frac{96}{625}2^k$.

9. $y_k^p = \frac{1}{2}\sin \frac{k\pi}{4} + \frac{1-\sqrt{2}}{2}\cos \frac{k\pi}{4}$.

11. $y_k^p = k^3 - 3k^2$. **13.** $y_k^p = \frac{1}{2}k\cos \frac{k\pi}{3} - \frac{3}{4}\cos \frac{k\pi}{3}$.

15. $y_k^p = -\frac{1}{12}2^k \cos \frac{3k\pi}{4}$.

17. $y_k^g = c_1 2^k + c_2 5^k - 4 \cdot 3^k$.

19. $y_k^g = c_1(-3)^k + c_2k(-3)^k + \frac{1}{8} - 3^{k-2}$.

21. $y_k^g = c_1 + c_2(-1)^k + \frac{3}{2}k$.

23. $y_k^g = c_1(\frac{3}{2})^k \cos\frac{k\pi}{2} + c_2(\frac{3}{2})^k \sin\frac{k\pi}{2} - \frac{3}{25}k \cdot 2^k + \frac{96}{625}2^k$.

25. $y_k^g = c(-1)^k + \frac{1}{2}\sin\frac{k\pi}{4} + \frac{1-\sqrt{2}}{2}\cos\frac{k\pi}{4}$.

27. $y_k^g = c_1 + c_2k + k^3 - 3k^2$.

29. $y_k^g = c_1(-\frac{1}{2})^k + c_2(\frac{1}{2})^k - \frac{3^k}{1297}\left(\sin\frac{k\pi}{4} + 36\cos\frac{k\pi}{4}\right)$.

31. $y_k^g = c_1 + c_2\cos\frac{2k\pi}{3} + c_3\sin\frac{2k\pi}{3} + (\frac{1}{2}k - \frac{3}{4})\cos\frac{k\pi}{3}$.

33. $y_k^g = (\sqrt{2})^k\left(c_1\cos\frac{k\pi}{4} + c_2\sin\frac{k\pi}{4} + c_3\cos\frac{3k\pi}{4} + c_4\sin\frac{3k\pi}{4}\right) - \frac{2^k}{12}\cos\frac{3k\pi}{4}$.

35. $y_k = 3 \cdot 2^k - 4 \cdot 3^k$. **37.** $y_k = 1 + \frac{3}{2}k$.

39. $y_k = \frac{3-\sqrt{2}}{2}(-1)^k + \frac{1}{2}\sin\frac{k\pi}{4} + \frac{1-\sqrt{2}}{2}\cos\frac{k\pi}{4}$.

41. $y_k = \frac{1}{1297}\left[\frac{36-111\sqrt{2}}{2}\left(-\frac{1}{2}\right)^k + \frac{36+111\sqrt{2}}{2}\left(\frac{1}{2}\right)^k - 3^k\left(\sin\frac{k\pi}{4} + 36\cos\frac{k\pi}{4}\right)\right]$.

43. $y_k = (-1)^k - \cos\frac{k\pi}{3} + 1$.

45. $y_k^p = Ak5^k$.

47. $y_k^p = A\sin\frac{k\pi}{2} + B\cos\frac{k\pi}{2} + Ck(\frac{3}{2})^k\cos\frac{k\pi}{2} + Dk(\frac{3}{2})^k\sin\frac{k\pi}{2}$.

49. $y_k^p = Ak + (Bk^2 + Ck + D)\sin\frac{k\pi}{3} + (Ek^2 + Fk + G)\cos\frac{k\pi}{3}$.

51. $y_k^p = -4 \cdot 3^k$. **53.** $y_k^p = k^3 - 3k^2$.

55. $y_k^p = \frac{1}{4}(3k^4 - 2k^3 - k^2)$.

57. $x_k = c_1 + c_2(-1)^k, \quad y_k = c_1 + \frac{1}{2}c_2(-1)^k$.

59. $x_k = c_1(-1)^k + c_2k(-1)^k + \frac{1}{2}$,
$y_k = c_1(-1)^k + c_2(k + \frac{1}{6})(-1)^k + \frac{1}{2}$.

61. Observe that E is a particular solution, so $y_k^g = y_k^h + E$. Now use Exercise 40 of Section 9.4.

63. 117 years. **65.** 36,85.

REVIEW EXERCISES CHAPTER 9
Page 387

1. $y_k = c_1 2^k + c_2 5^k - 4 \cdot 3^k - \frac{1}{2}k2^k$.

3. $y_k = c_1(-3)^k + c_2k(-3)^k + 9 - 8 \cdot 3^k$.

5. $y_k = \frac{3}{2} + c_1(-1)^k + c_2 \cos\dfrac{k\pi}{3} + c_3 \sin\dfrac{k\pi}{3}$.

7. $y_k = 2 - k\cos\dfrac{k\pi}{2} - 2\cos\dfrac{k\pi}{2}$.

9. $y_k = 1 - (-\frac{2}{3})^k$.

13. $y_k = 1 + \dfrac{4k^3 - 9k^2 - k}{6}$.

15. $x_k = \frac{1}{2} + (-1)^k$, $y_k = -\frac{1}{2} - (-1)^k$.

17. 0.31 volts. **19.** \$215.67.

21. From $y_{k+2} = y_{k+1} + y_k$, we find $y_{k+1} = y_{k+2} - y_k$. And for $k = 2, 4, 6, \ldots$, we obtain

$$y_3 = y_4 - y_2$$
$$y_5 = y_6 - y_4$$
$$y_7 = y_8 - y_6$$
$$\cdots\cdots$$
$$y_{2k-1} = y_{2k} - y_{2k-2}.$$

Adding them and also adding $y_1 = 1$ to both sides, we find

$$y_1 + y_3 + \cdots + y_{2k-1} = y_{2k}.$$

23. Subtracting the identities in Exercises 21 and 22, we find

$$\begin{aligned} y_1 - y_2 + y_3 - y_4 + \cdots + y_{2k-1} - y_{2k} &= y_{2k} - y_{2k+1} + 1 \\ &= y_{2k} - (y_{2k} + y_{2k-1}) + 1 \\ &= 1 - y_{2k-1}. \end{aligned}$$

Adding y_{2k+1} to both sides, we also find

$$\begin{aligned} y_1 - y_2 + \cdots + y_{2k-1} - y_{2k} + y_{2k+1} &= 1 - y_{2k-1} + y_{2k+1} \\ &= 1 - y_{2k-1} + y_{2k} + y_{2k-1} \\ &= 1 + y_{2k}. \end{aligned}$$

From these two results the desired identity follows for all k (even or odd).

Section 10.2
Page 393

1. True. **3.** False. **5.** True. **7.** False. **9.** True. **11.** True.

13. See Figure 10.1(a). **15.** See Figure 10.1(b).

17. Compute $(af + bg)(x + T)$ and $(fg)(x + T)$.

19. $p > 0$; the period is $2\pi/\sqrt{p}$.

Section 10.3
Page 400

1. $f(x) \sim -\dfrac{2}{\pi}\displaystyle\sum_{n=1}^{\infty}\frac{(-1)^n}{n}\sin n\pi x.$

3. $f(x) \sim \dfrac{1}{2} + \dfrac{1}{\pi}\displaystyle\sum_{n=1}^{\infty}\frac{1-(-1)^n}{n}\sin nx.$

5. $f(x) \sim \pi - 2\displaystyle\sum_{n=1}^{\infty}\frac{1}{n}\sin nx.$

7. $f(x) \sim \dfrac{4\pi^2}{3} + \displaystyle\sum_{n=1}^{\infty}\left(\frac{4}{n^2}\cos nx - \frac{4\pi}{n}\sin nx\right).$

9. $f(x) \sim 1 + \cos 2x.$

11. $f(x) \sim \sin 2x.$

13. $f(x) \sim 2\displaystyle\sum_{n=1}^{\infty}(-1)^n\frac{6 - n^2\pi^2}{n^3}\sin nx.$

15. $f(x) \sim \dfrac{2}{\pi} - \dfrac{4}{\pi}\displaystyle\sum_{n=1}^{\infty}\frac{(-1)^n}{4n^2 - 1}\cos nx.$

17. $f(x) \sim \dfrac{1}{2} + \dfrac{2}{\pi} \sum_{n=1}^{\infty} \dfrac{1}{n} \sin \dfrac{n\pi}{2} \cos \dfrac{n\pi x}{2}$.

19. $f(x) \sim \dfrac{1}{2} + \dfrac{1}{\pi} \sum_{n=1}^{\infty} \dfrac{1-(-1)^n}{n} \sin \dfrac{n\pi x}{l}$.

Section 10.4
Page 407

1. $f(x) \sim \sum_{n=1}^{\infty} \dfrac{2}{n} (-1)^{n+1} \sin nx.$

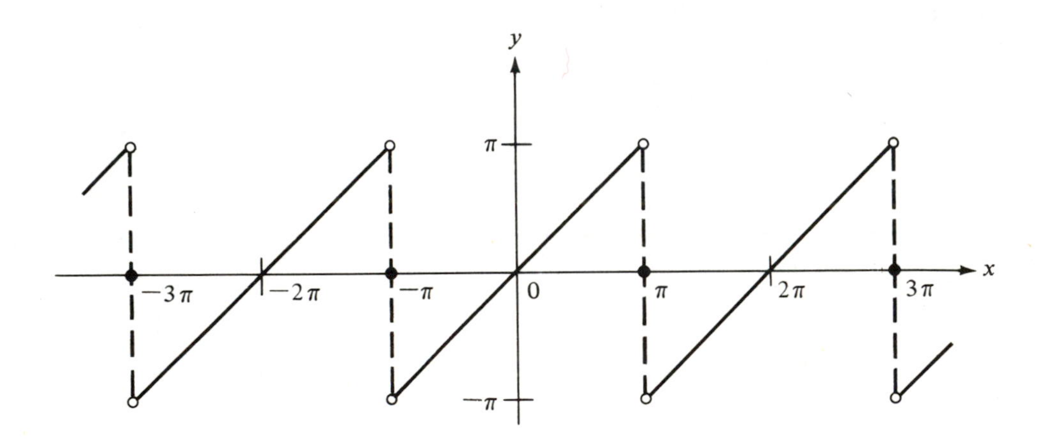

3. 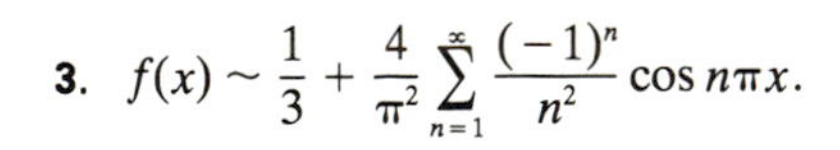
$f(x) \sim \dfrac{1}{3} + \dfrac{4}{\pi^2} \sum_{n=1}^{\infty} \dfrac{(-1)^n}{n^2} \cos n\pi x.$

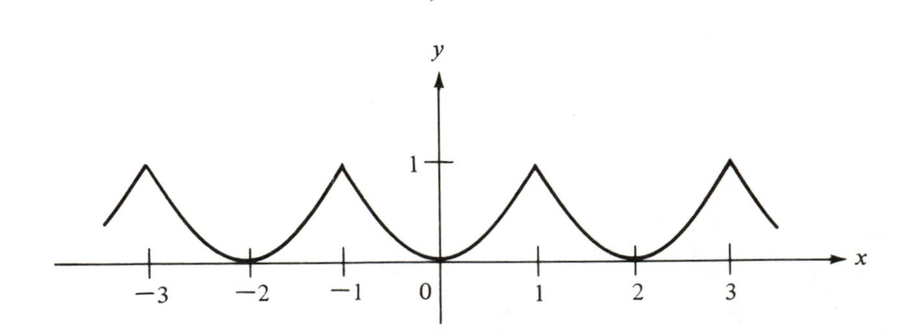

5. $f(x) \sim \dfrac{4}{\pi^2} \sum_{n=1}^{\infty} \dfrac{1 - (-1)^n}{n^2} \cos \dfrac{n\pi x}{2}$.

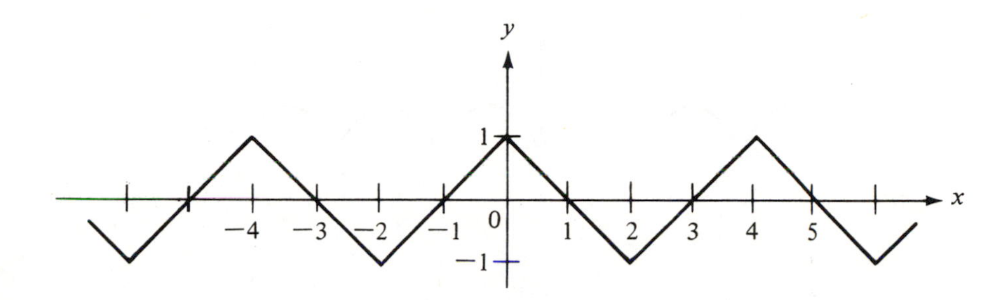

7. $f(x) \sim \dfrac{1}{\pi} + \dfrac{1}{2} \sin x + \dfrac{1}{\pi} \sum_{n=2}^{\infty} \dfrac{1 + (-1)^n}{1 - n^2} \cos nx.$

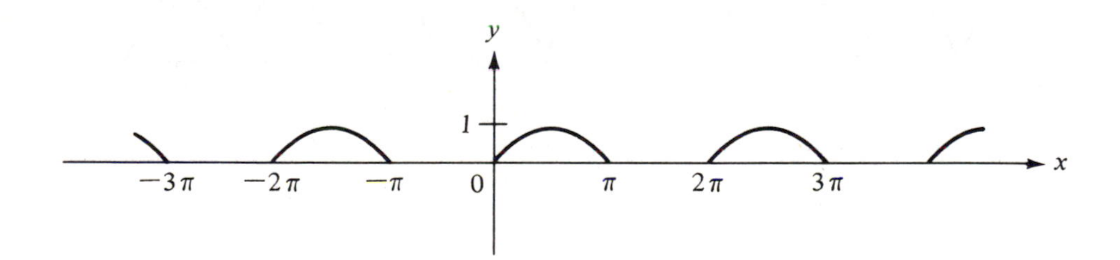

9. $f(x) \sim \dfrac{1}{2} + \dfrac{2}{\pi} \sum_{n=1}^{\infty} \dfrac{1}{n} \sin \dfrac{n\pi}{2} \cos \dfrac{n\pi x}{2}$.

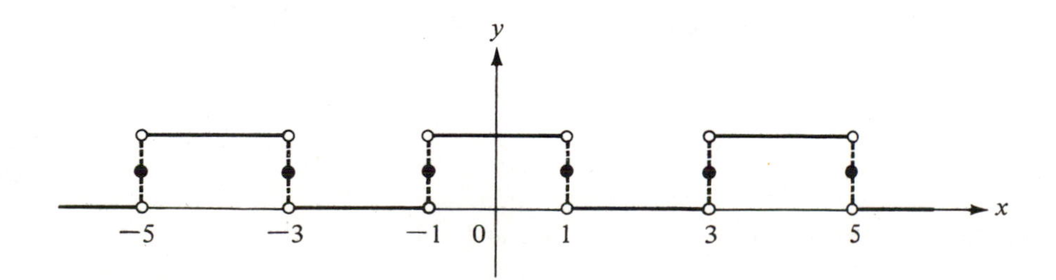

11. $f(x) = \frac{1}{2} - \frac{1}{2}\cos 2x.$

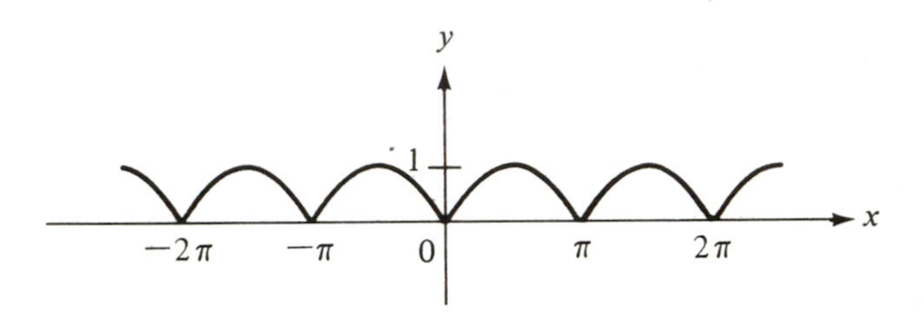

13. $f(x) = \frac{1}{2} + \frac{1}{2}\cos 2x.$

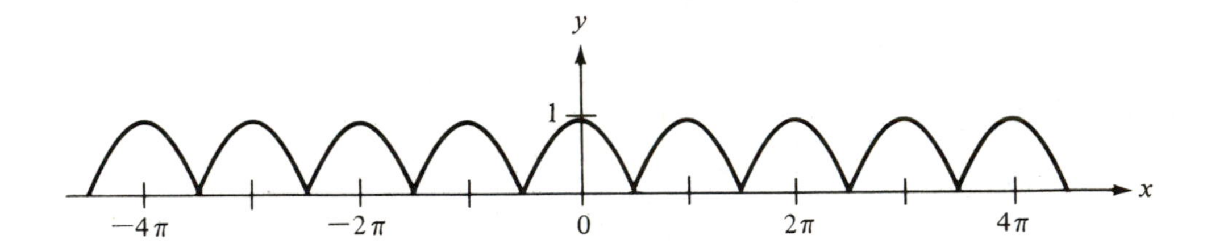

15. $f(x) = \cos 2x.$

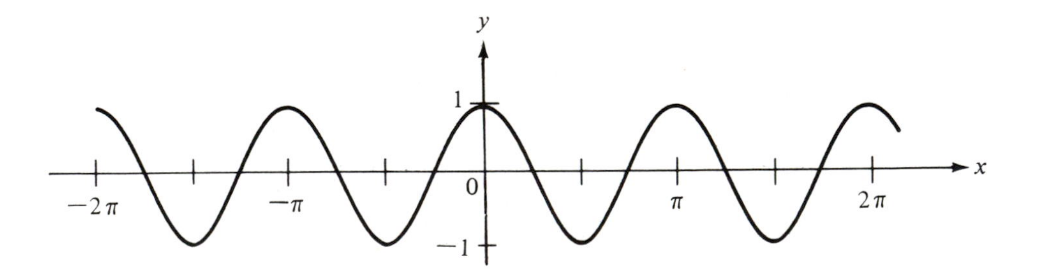

17. Use Exercise 1.

19. In the identity of Exercise 18, set $x = 1$ and $x = 0$.

21. False. **23.** True.

25. False. **27.** False.

29. True.

Section 10.5
Page 416

1. True. **3.** False. **5.** False. **7.** True. **9.** False.

11. Even; $f(x) \sim \dfrac{1}{2} + \dfrac{2}{\pi^2}\sum_{n=1}^{\infty}\dfrac{(-1)^n - 1}{n^2}\cos n\pi x.$

13. Odd; $f(x) \sim \dfrac{2}{\pi}\sum_{n=1}^{\infty}\dfrac{1-(-1)^n}{n}\sin nx.$

15. Odd; $f(x) \sim \dfrac{2}{\pi^3}\sum_{n=1}^{\infty}(-1)^n\dfrac{6 - n^2\pi^2}{n^3}\sin n\pi x.$

17. Odd; $f(x) \sim \sin 2x.$

19. $f(x) \sim \dfrac{4}{\pi}(\sin x + \frac{1}{3}\sin 3x + \frac{1}{5}\sin 5x + \cdots).$

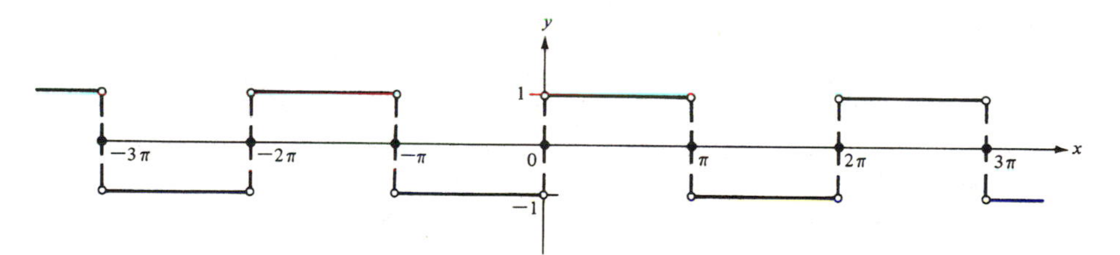

21. $f(x) \sim 2(\sin x - \frac{1}{2}\sin 2x + \frac{1}{3}\sin 3x - \cdots)$. For the graph of the function to which the series converges, see Exercise 1, Section 10.4.

23. $f(x) \sim -2\sum_{n=1}^{\infty}\left[\dfrac{(-1)^n}{n\pi} + \dfrac{2[1-(-1)^n]}{n^3\pi^3}\right]\sin n\pi x.$

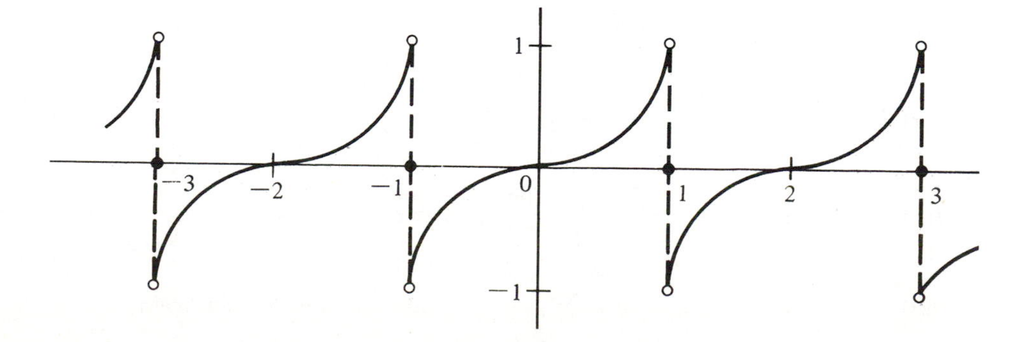

25. $f(x) \sim \dfrac{2}{\pi} \sum_{n=1}^{\infty} \dfrac{1}{n}\left(1 - \cos\dfrac{n\pi}{2}\right) \sin nx.$

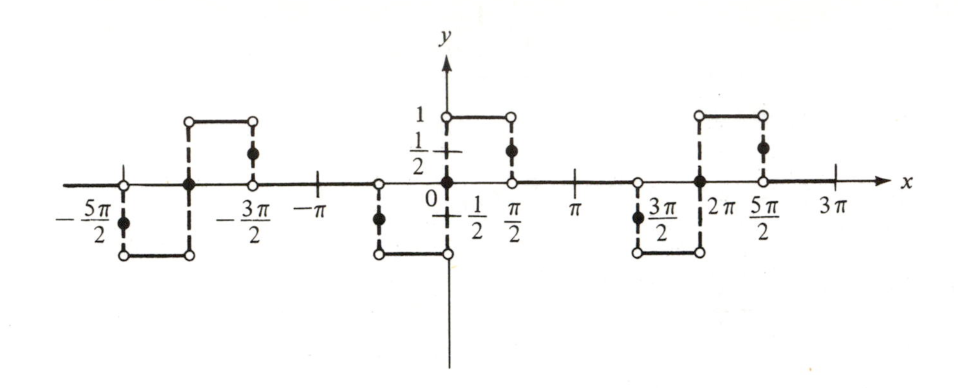

27. $f(x) \sim \sum_{n=1}^{\infty} \left(\dfrac{2}{n^2\pi^2} \sin\dfrac{n\pi}{2} - \dfrac{1}{n\pi} \cos\dfrac{n\pi}{2}\right) \sin n\pi x.$

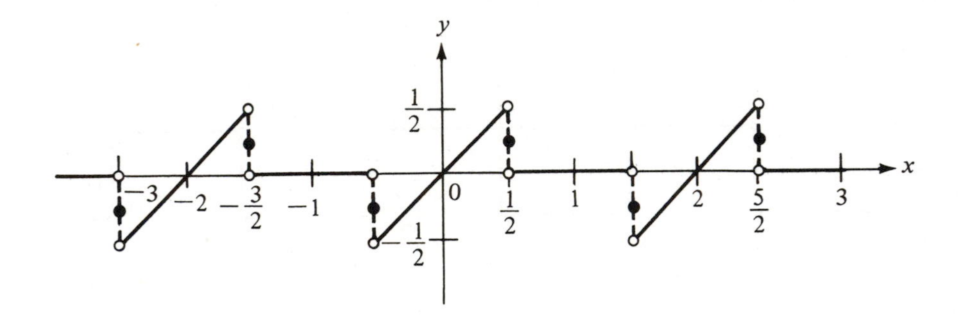

REVIEW EXERCISES CHAPTER 10
Page 418

1. $f(x) \sim -3 - \dfrac{36}{\pi^2} \sum_{n=1}^{\infty} \dfrac{(-1)^n}{n^2} \cos\dfrac{n\pi x}{3}.$

3. $f(x) \sim \dfrac{3\sqrt{3}}{2\pi} + \dfrac{3\sqrt{3}}{\pi} \sum_{n=1}^{\infty} \dfrac{(-1)^n}{1 - 9n^2} \cos nx.$

5. $f(x) \sim 1 + \dfrac{e - e^{-1}}{2} + (e - e^{-1}) \sum_{n=1}^{\infty} \dfrac{(-1)^n}{1 + n^2\pi^2} (\cos n\pi x - n\pi \sin n\pi x).$

7. $P(t) \sim P_0(1 - e^{-1}) + 2P_0(1 - e^{-1}) \sum_{n=1}^{\infty} \dfrac{1}{1 + 4n^2\pi^2} (\cos 200n\pi t + 2n\pi \sin 200n\pi t).$

9. $f(x) \sim \frac{4}{\pi}\left(\sin x + \frac{1}{3}\sin 3x + \frac{1}{5}\sin 5x + \cdots\right).$

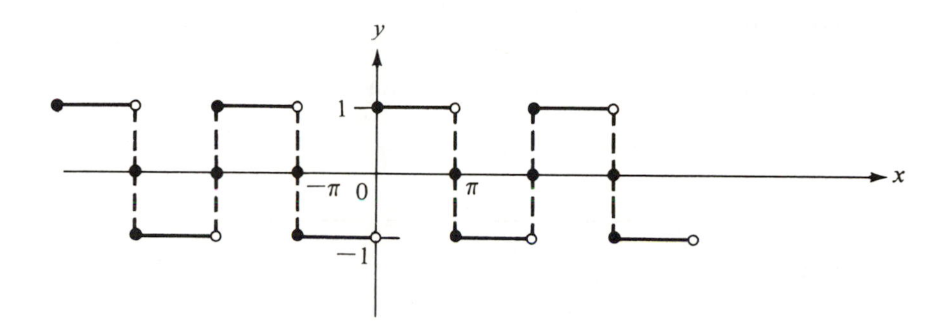

11. $f(x) \sim 1 + \cos x.$

13. Find the Fourier series of the function $f(x) = |x|, -1 \leq x < 1.$

15. $f(x) \sim \left(1 + \frac{\pi}{2}\right) + \frac{2}{\pi}\sum_{n=1}^{\infty}\frac{(-1)^n - 1}{n^2}\cos nx.$

17. $f(x) \sim \frac{1}{\pi} - \frac{2}{\pi}\sum_{n=1}^{\infty}\frac{(-1)^n}{4n^2 - 1}\cos 2nt.$

Section 11.2
Page 426

1. third order, homogeneous, variable coefficients.

2. fourth order, homogeneous, constant coefficients.

5. second order, homogeneous, constant coefficients.

7. first order, nonhomogeneous, variable coefficients.

9. first order, nonhomogeneous, constant coefficients.

11. $u(x, y) = f(x).$ 13. $u(x, y) = x^3 + 4xy + f(y).$

15. $u(x, y) = f(x) + g(y).$ 17. $u(x, y) = yf(x) + g(x).$

19. $u(x, y) = x \sin y + ye^x + f(x) + g(y).$

21. $u(x, y, z) = xyz + f(y, z) + g(x, y).$

23. $u(x, y, z) = f(x, z) + xg(y, z) + h(y, z)$.

25. $u(x, y, z) = f(x, y) + g(x, z) + xh(y, z) + k(y, z)$.

27. $u(x, y, z) = yz + f(x, y)$.

29. $u(x, y, z) = \frac{1}{2}yz^2 + \frac{3}{2}xz^2 + zf_1(x, y) + f_2(x, y)$.

Section 11.3
Page 429

25. $u(x, y) = \frac{1}{2}xy^2 + f(x)$.

27. $u(x, y) = xf(y) + g(y)$.

29. $u(x, y) = \frac{1}{12}x^2y^3 + yf(x) + g(x) + h(y)$.

31. $u(x, y) = y \sin x - e^y + f(x)$.

33. $u_p = e^{\lambda x + (1-3\lambda)y}$, λ arbitrary.

35. $u_p = e^{[4-(3/2)\mu]x + \mu y}$, μ arbitrary.

37. $u_p = e^{[(3/5)\mu + 2/5]x + \mu y}$, μ arbitrary.

39. $u_p = e^{(-\mu-2)x + \mu y}$ or $u_p = e^{(-\mu-1)x + \mu y}$, μ arbitrary.

41. $u_p = e^{(\mu-4)x + \mu y}$ or $u_p = e^{(\mu-1)x + \mu y}$, μ arbitrary.

43. $u_p = e^{\lambda x + (2\lambda+3)y}$ or $u_p = e^{\lambda x + (2\lambda-1)y}$, λ arbitrary.

45. $u_p = e^{\lambda x + [1-(2/3)\lambda]y}$ or $u_p = e^{\lambda x + [-2/3-(2/3)\lambda]y}$, λ arbitrary.

47. $u_p = e^{x+y}$, corresponding to $\lambda = \mu = 1$.

49. (a) $u_p = -3x^2 - 6x - 6$. (b) $u_p = \cos y + \sin y$.
(c) $u_p = 2x^2 - 4x - 3y^2 - 6y$. (d) $u_p = \frac{10}{53}e^{3x+4y}$

Section 11.4
Page 436

1. $X' - \lambda X = 0$, $3Y' - \lambda Y = 0$.

3. $2X' - \lambda X = 0$, $5Y' + \lambda Y = 0$.

5. $X' + (1 - \lambda)X = 0$, $Y' + \lambda Y = 0$.

7. $X'' - \lambda X = 0,\ Y' - \lambda Y = 0.$

9. $X'' - \lambda X = 0,\ Y'' - \lambda Y = 0.$

11. $X'' + X' - \lambda X = 0,\ Y'' + Y' + \lambda Y = 0.$

13. $X'' - X' - \lambda X = 0,\ Y'' - Y' - \lambda Y = 0.$

15. $X'' - X' - \lambda X = 0,\ Y' - \lambda Y = 0.$

17. $5X'' + 6X' - \lambda X = 0,\ 3Y' - \lambda Y = 0.$

19. $2X'' + (1 - \lambda)X = 0,\ 3Y'' + \lambda Y = 0.$

21. $aX'' + dX' + (f - \lambda)X = 0,\ cY'' + eY' + \lambda Y = 0.$

23. $X' - \lambda X = 0,\ Y' - \mu Y = 0,\ Z' + (\lambda - \mu)Z = 0.$

25. $X' - \lambda X = 0,\ 2Y' - \mu Y = 0,\ 2Z' - (\lambda + \mu)Z = 0.$

27. $X'' - \lambda X = 0,\ Y'' + \mu Y = 0,\ Z'' + (\lambda - \mu)Z = 0.$

29. $X'' - \lambda X = 0,\ Y'' - \mu Y = 0,\ Z'' + (\lambda - \mu)Z = 0.$

31. $2X'' + (1 - \lambda)X = 0,\ Y'' - \mu Y = 0,\ Z'' + (\lambda - \mu)Z = 0.$

41. Elliptic. $u(x, y) = f\left(x + \left[-\frac{1}{4} + i\frac{\sqrt{3}}{12}\right]y\right) + g\left(x + \left[-\frac{1}{4} - i\frac{\sqrt{3}}{12}\right]y\right).$

43. Elliptic. $u(x, y) = f\left(x + \left[-\frac{1}{4} + i\frac{\sqrt{3}}{4}\right]y\right) + g\left(x + \left[-\frac{1}{4} - i\frac{\sqrt{3}}{4}\right]y\right).$

45. Hyperbolic. $u(x, y) = f\left(x + \frac{-9 + \sqrt{65}}{8}y\right) + g\left(x + \frac{-9 - \sqrt{65}}{8}y\right).$

47. Hyperbolic. $u(x, y) = f\left(x + \frac{-5 + \sqrt{21}}{2}y\right) + g\left(x + \frac{-5 - \sqrt{21}}{2}y\right).$

49. Parabolic. $u(x, y) = f(x - \frac{1}{4}y) + (\alpha x + \beta y)g(x - \frac{1}{4}y)$, α, β arbitrary.

51. Hyperbolic. $u(x, y) = f(x + 2y) + g(x - 3y).$

53. Hyperbolic. $u(x, y) = f(x + 3y) + g(x + 7y).$

55. Elliptic. $u(x, y) = f\left(x + \frac{-2 + i}{5}y\right) + g\left(x + \frac{-2 - i}{5}y\right).$

57. Hyperbolic. $u(x, y) = f(x + \frac{1}{4}y) + g(x - y)$.

59. Parabolic. $u(x, y) = f(x - \frac{1}{3}y) + (\alpha x + \beta y)g(x - \frac{1}{3}y)$, α, β arbitrary.

61. (b) $X'' + \left[\dfrac{8m\pi^2 E}{h^2} - \lambda\right]X = 0,\ Y'' + (\lambda - \mu)Y = 0,\ Z'' + \mu Z = 0.$

63. $(M^2 - 1)X'' - \lambda X = 0,\ Y'' - \lambda Y = 0.$

65. $T'' + (a^2 - \lambda)T = 0,\ X'' - \lambda X = 0.$

$\lambda < 0$: $T(t) = c_1 \cos\sqrt{a^2 - \lambda}\,t + c_2 \sin\sqrt{a^2 - \lambda}\,t$

$X(x) = d_1 \cos\sqrt{-\lambda}\,x + d_2 \sin\sqrt{-\lambda}\,x.$

$\lambda = 0$: $T(t) = c_1 \cos at + c_2 \sin at$

$X(x) = d_1 + d_2 x.$

$0 < \lambda < a^2$: $T(t) = c_1 \cos\sqrt{a^2 - \lambda}\,t + c_2 \sin\sqrt{a^2 - \lambda}\,t$

$X(x) = d_1 e^{\sqrt{\lambda}x} + d_2 e^{-\sqrt{\lambda}x}$

$\lambda = a^2$: $T(t) = c_1 + c_2 t$

$X(x) = d_1 e^{ax} + d_2 e^{-ax}$

$\lambda > a^2$: $T(t) = c_1 e^{\sqrt{\lambda - a^2}t} + c_2 e^{-\sqrt{\lambda - a^2}t}$

$X(x) = d_1 e^{\sqrt{\lambda}x} + d_2 e^{-\sqrt{\lambda}x}$

67. $u(x, y) = \sqrt{\lambda}\,x + \sqrt{1 - \lambda}\,y + c, \quad (0 < \lambda < 1).$

Section 11.6
Page 448

1. $u(x, t) = \displaystyle\sum_{n=1}^{\infty} \frac{8}{\pi(2n - 1)^3} \cos(2n - 1)t \sin(2n - 1)x.$

3. $u(x, t) = \displaystyle\sum_{n=1}^{\infty} \left(\frac{-4}{n^3}\right)[1 + 2(-1)^n] \cos nt \sin nx.$

5. $u(x, t) = \displaystyle\sum_{n=1}^{\infty} \frac{12}{(2n - 1)^2\pi} \sin(2n - 1)t \sin(2n - 1)x.$

7. $u(x, t) = \displaystyle\sum_{n=1}^{\infty} \frac{4A}{(2n - 1)^2\pi^2} \sin(2n - 1)\pi t \sin(2n - 1)\pi x.$

9. $u(x, t) = \displaystyle\sum_{n=1}^{\infty} \left[\frac{8}{(2n - 1)^3\pi} \cos(2n - 1)t + \frac{12}{(2n - 1)^2\pi} \sin(2n - 1)t\right] \sin(2n - 1)x.$

11. $u(x,t) = \sum_{n=1}^{\infty} \frac{3}{2n^3} \cos 2nt \sin 2nx$

$$+ \sum_{n=1}^{\infty} \left[\frac{4}{(2n-1)^3} \cos (2n-1)t + \frac{4}{(2n-1)^2} \sin (2n-1)t \right] \sin (2n-1)x.$$

13. $u(x,t) = \cos 20t \sin 5t + \sum_{n=1}^{\infty} \frac{1}{\pi[(2n-1)^2 - 4]} \sin 4(2n-1)t \sin (2n-1)x.$

15. $u(x,t) = \sum_{n=1}^{\infty} \frac{-1}{n^2} \cos 2nt \cos 2nx.$

17. $u(x,t) = \sum_{n=1}^{\infty} \frac{-\pi}{2n^2} \cos 2nt \cos 2nx$

$$+ \sum_{n=1}^{\infty} \left[\frac{24}{(2n-1)^4 \pi} + \frac{2\pi}{(2n-1)^2} \right] \cos (2n-1)t \cos (2n-1)x.$$

19. $u(x,t) = 0.$ **21.** $u(x,t) = 0.$

23. $u(x,t) = \sum_{n=1}^{\infty} \frac{-1}{n^2} \cos 2nt \cos 2nx.$

25. $u(x,t) = \sum_{n=1}^{\infty} \frac{-\pi}{2n^2} \cos 2nt \cos 2nx$

$$+ \sum_{n=1}^{\infty} \left[\frac{24}{(2n-1)^4 \pi} - \frac{2\pi}{(2n-1)^2} \right] \cos (2n-1)t \cos (2n-1)x.$$

27. $u(x,t) = \frac{1}{8} \sin 8t \cos 2t + \sum_{n=1}^{\infty} \frac{-20}{[4n^2 - 25]\pi} \cos 4nt \cos 2nx.$

29. $u(x,t) = \sum_{n=1}^{\infty} \left[\alpha_n \cos \frac{(2n-1)\pi ct}{2l} + \beta_n \sin \frac{(2n-1)ct}{2l} \right] \sin \frac{(2n-1)\pi x}{2l},$

where

$$\alpha_n = \frac{2}{l} \int_0^l f(x) \sin \frac{(2n-1)\pi x}{2l} \, dx$$

and

$$\beta_n = \frac{4}{(2n-1)\pi c} \int_0^l g(x) \sin \frac{(2n-1)\pi x}{2l} \, dx.$$

33. $u(x,t) = \frac{3x}{\pi} + \sum_{n=1}^{\infty} \frac{3}{n\pi} \cos 2nt \sin 2nx$

$$+ \sum_{n=1}^{\infty} \left[\frac{8}{(2n-1)^3 \pi} - \frac{6}{(2n-1)\pi} \right] \cos (2n-1)t \sin (2n-1)x.$$

35. $$u(x, t) = 2 + \sum_{n=1}^{\infty} \frac{-3}{2n^3} \cos 2nt \sin 2nx + \sum_{n=1}^{\infty} \left[\frac{8}{(2n-1)^3} - \frac{8}{(2n-1)\pi}\right] \cos (2n-1)t \sin (2n-1)x.$$

37. $$u(x, t) = x + \frac{10(\pi - x)}{\pi} + \sum_{n=1}^{\infty} \frac{\pi - 10}{n\pi} \cos 2nt \sin 2nx + \sum_{n=1}^{\infty} \left\{\left[\frac{2(10-\pi)}{(2n-1)\pi} - \frac{40}{\pi}\right] \cos (2n-1)t + \frac{12}{(2n-1)^2\pi} \sin (2n-1)t\right\} \sin (2n-1)x.$$

39. $$u(x, t) = 7 - 5x + \sum_{n=1}^{\infty} \frac{-5}{n\pi} \cos 2n\pi t \sin 2n\pi x + \sum_{n=1}^{\infty} \left[\frac{-18}{(2n-1)\pi} \cos (2n-1)t + \frac{4A}{(2n-1)^2\pi^2} \sin (2n-1)t\right] \sin (2n-1)x.$$

41. $$u(x, t) = \frac{5x}{\pi} + \sum_{n=1}^{\infty} \frac{5}{n\pi} \cos 2nt \sin 2nx + \sum_{n=1}^{\infty} \left\{\left[\frac{8}{(2n-1)^3\pi} - \frac{10}{(2n-1)\pi}\right] \cos (2n-1)t + \frac{12}{(2n-1)^2\pi} \sin (2n-1)t\right\} \sin (2n-1)x.$$

43. $$u(x, t) = \frac{8x}{\pi} + \sum_{n=1}^{\infty} \left[\frac{3}{2n^3} + \frac{8}{n\pi}\right] \cos 2nt \sin 2nx + \sum_{n=1}^{\infty} \left\{\left[\frac{4}{(2n-1)^3} - \frac{16}{(2n-1)\pi}\right] \cos (2n-1)t + \frac{4}{(2n-1)^2} \sin (2n-1)t\right\} \sin (2n-1)x.$$

45. $$u(x, t) = 6 - \frac{6x}{\pi} + \sum_{n=1}^{\infty} \frac{-6}{n\pi} \cos 8nt \sin 2nx + \left[\left(1 - \frac{12}{5\pi}\right) \cos 20t + \frac{1}{21\pi} \sin 20t\right] \sin 5x$$
$$+ \sum_{n=1}^{2} \left\{\frac{-12}{(2n-1)\pi} \cos 4(2n-1)t + \frac{1}{\pi[(2n-1)^2 - 4]} \sin 4(2n-1)t\right\} \sin (2n-1)x$$
$$+ \sum_{n=4}^{\infty} \left\{\frac{-12}{(2n-1)\pi} \cos 4(2n-1)t + \frac{1}{\pi[(2n-1)^2 - 4]} \sin 4(2n-1)t\right\} \sin (2n-1)x.$$

47. The initial shape of the string is that of the curve $\sin 3x$ with $0 < x < \pi$, and is released with an initial velocity of 4 units per second.

49. (e) u_1 and u_2 are damped waves travelling to the right with speed $\beta/\sqrt{\gamma}$.

51. $\theta(x, t) = \sum_{n=1}^{\infty} \frac{24(-1)^{n+1}}{(2n-1)^2\pi^2} \cos\frac{(2n-1)\pi t}{2} \sin\frac{(2n-1)\pi x}{2}$.

53. $u(x, t) = \sum_{n=1}^{\infty} \frac{(-1)^{n+1}}{(2n-1)^2\pi^2} \cos(2n-1)\pi t \sin(2n-1)\pi x$.

55. $u(x, t) = \sum_{n=1}^{\infty} \frac{72}{n^2\pi^2} \sin\frac{n\pi}{3} \sin\frac{n\pi}{6} \sin\frac{n\pi t}{6} \sin\frac{n\pi x}{6}$.

57. $u(x, t) = \sum_{n=1}^{\infty} \frac{120}{n^2\pi^2} \sin\frac{7n\pi}{20} \sin\frac{n\pi}{20} \sin\frac{n\pi t}{10} \sin\frac{n\pi x}{10}$.

59. $\rho S T_n'' + \left[EI\frac{n^4\pi^4}{l^4} + \frac{n^2\pi^2}{l^2}F(t)\right]T_n = 0.$

63. 3.932 Newtons/m.

Section 11.7
Page 458

1. $u(x, t) = \sum_{n=1}^{\infty} \frac{2(-1)^{n+1}}{n\pi} e^{-n^2\pi^2 t} \sin n\pi x.$

3. $u(x, t) = \sum_{n=1}^{\infty} \frac{-\pi}{n} e^{-16n^2 t} \sin 2nx$

$+ \sum_{n=1}^{\infty} \left[\frac{-2\pi}{2n-1} - \frac{8}{\pi(2n-1)^3}\right] e^{-4(2n-1)^2 t} \sin(2n-1)x.$

5. $u(x, t) = \sum_{n=1}^{\infty} \left\{\frac{2n\pi}{n^2\pi^2 - 1}[(-1)^{n+1}\cos 1 + 1] + \frac{6(-1)^{n+1}}{n\pi}\right\} e^{-2n^2\pi^2 t} \sin n\pi x.$

7. $u(x, t) = \sum_{n=1}^{\infty} \frac{2n\pi}{n^2\pi^2 + 1}[1 + e(-1)^{n+1}] e^{-5n^2\pi^2 t} \sin n\pi x.$

9. $u(x, t) = 3e^{-t}\sin x + \sum_{n=2}^{\infty} \frac{2(-1)^{n+1}}{n} e^{-n^2 t} \sin nx.$

11. $u(x,t) = \sum_{n=1}^{\infty} \frac{4}{\pi^3(2n-1)^3} e^{-(2n-1)^2\pi^2 t} \sin(2n-1)\pi x.$

13. $u(x,t) = \sum_{n=1}^{\infty} \frac{3}{2n^3} e^{-8n^2 t} \sin 2nx$

$+ \sum_{n=1}^{\infty} \frac{4}{(2n-1)^3}[2\pi^2(2n-1)^2 + 1]e^{-2(2n-1)^2 t} \sin(2n-1)x.$

15. $u(x,t) = \sum_{n=1}^{\infty} 2\left\{\frac{(-1)^{n+1}}{n\pi} + \frac{n\pi}{n^2\pi^2+1}[1 + e(-1)^{n+1}]\right\} e^{-n^2\pi^2 t} \sin n\pi x.$

17. $u(x,t) = \frac{10(\pi - x)}{\pi} + \sum_{n=1}^{\infty} \frac{2}{n\pi}[10 + (-1)^{n+1}\pi]e^{-5n^2 t} \sin nx.$

19. $u(x,t) = 7 - 4x + \sum_{n=1}^{\infty} \left\{\frac{2}{n\pi}[3(-1)^n - 7] + \frac{2}{n^3\pi^3}[1 - (-1)^n]\right\} e^{-n^2\pi^2 t} \sin n\pi x.$

21. $u(x,t) = 7 - 5x + \sum_{n=1}^{\infty} \left\{\frac{-10}{n\pi} + \frac{4n\pi}{n^2\pi^2 - 4}[1 - (-1)^n \cos 2]\right\} e^{-n^2\pi^2 t} \sin n\pi x.$

23. $u(x,t) = 7 - 4x + \sum_{n=1}^{\infty} \left\{\frac{2n\pi}{n^2\pi^2+1}[1 + e(-1)^{n+1}] + \frac{2}{n\pi}[3(-1)^n - 7\right\} e^{-3n^2\pi^2 t} \sin n\pi x.$

25. $u(x,t) = 10 + \sum_{n=1}^{\infty} \left\{\frac{4}{n^3\pi^3}[(-1)^n + 2] + \frac{20}{n\pi}[(-1)^n - 1]\right\} e^{-5n^2\pi^2 t} \sin n\pi x.$

27. $u(x,t) = e^{-\alpha t} \sum_{n=1}^{\infty} \frac{4}{(2n-1)^3\pi^3} e^{-(2n-1)^2\pi^2 t} \sin(2n-1)\pi x.$

29. $u(x,t) = \sum_{n=1}^{\infty} \frac{2A}{n\pi}\left[1 - \cos\frac{n\pi}{2}\right] e^{-(n^2\pi^2 at)/l^2} \sin\frac{n\pi x}{l}.$

31. $u(15,t) = \sum_{n=1}^{\infty} \frac{200}{n\pi}\left[\cos\frac{2n\pi}{3} - (-1)^n\right] \sin\frac{n\pi}{2} e^{-[n^2\pi^2(1.02)t]/30^2}.$

33. (a) $v(r,t) = \sum_{n=1}^{\infty} c_n e^{-(n^2\pi^2 at)/R^2} \sin\frac{n\pi r}{R}$, with $c_n = \frac{2}{R}\int_0^R rg(r) \sin\frac{n\pi r}{R}\,dr.$

(b) $u(r,t) = \frac{1}{r}v(r,t)$ with v given in part (a). Note that $\lim_{r\to 0+} u(r,t)$ makes sense.

35. (a) $v(r, t) = \sum_{n=1}^{\infty} \frac{12(-1)^{n+1}R^3}{n^3\pi^3} e^{-(n^2\pi^2 at)/R^2} \sin \frac{n\pi r}{R}$.

(b) $u(r, t) = \frac{1}{r} v(r, t)$ with v given in part (a). Note that $\lim_{r \to 0+} u(r, t)$ makes sense.

---

1. $u(x, y) = \sum_{n=1}^{\infty} \left[\frac{2(-1)^{n+1}}{n\pi} \cosh n\pi y + \frac{2(-1)^n}{n\pi} \coth n\pi \sinh n\pi y\right] \sin n\pi x.$

3. $u(x, y) = \sum_{n=1}^{\infty} 2\left[\frac{2 - n^2\pi^2}{n^3\pi^3}(-1)^n - \frac{2}{n^3\pi^3}\right][\cosh n\pi y - \coth n\pi \sinh n\pi y] \sin n\pi x.$

5. $u(x, y) = [\cosh \pi y - \coth \pi \sinh \pi y] \sin \pi x.$

7. $u(x, y) = \sum_{n=1}^{\infty} \frac{2n\pi}{(\sinh n\pi)(n^2\pi^2 - 1)}[1 + (-1)^{n+1} \cos 1] \sinh n\pi y \sin n\pi x.$

9. $u(x, y) = \sum_{n=1}^{\infty} \frac{2n\pi}{(n^2\pi^2 + 4) \sinh (n\pi/2)}[1 - (-1)^n e^2] \sinh \frac{n\pi y}{2} \sin \frac{n\pi x}{2}.$

11. $u(x, y) = \left[\frac{-2}{\pi} \cosh \pi y + \left(1 - \frac{\cosh \pi}{\pi}\right)\frac{\sinh \pi y}{\sinh \pi}\right] \sin \pi x$

$+ \sum_{n=2}^{\infty} \frac{2(-1)^{n+1}}{n\pi}[\cosh n\pi y - \coth n\pi \sinh n\pi y] \sin n\pi x.$

13. $u(x, y) = \sum_{n=1}^{\infty} \left\{\frac{2n\pi}{n^2\pi^2 + 4}[1 - (-1)^n e^2] \cosh \frac{n\pi y}{2}\right.$

$+ \left[\frac{2n\pi(1 - (-1)^n \cos 2)}{(n^2\pi^2 - 4) \sinh n\pi} - \frac{2n\pi (\coth n\pi)(1 - (-1)^n e^2)}{n^2\pi^2 + 4}\right].$

$\left.\cdot \sinh \frac{n\pi y}{2}\right\} \sin \frac{n\pi x}{2}.$

15. $u(x, y) = \sum_{n=1}^{\infty} \left\{2\left[\frac{2 - n^2\pi^2}{n^3\pi^3}(-1)^n - \frac{2}{n^3\pi^3}\right] \cosh n\pi y\right.$

$+ \left[\frac{2n\pi(1 + (-1)^{n+1}e)}{(n^2\pi^2 + 1) \sinh 2n\pi} - 2 \coth 2n\pi \left(\frac{2 - n^2\pi^2}{n^3\pi^3}(-1)^n - \frac{2}{n^3\pi^3}\right)\right].$

$\left.\cdot \sinh n\pi y\right\} \sin n\pi x.$

17. $u(x, y) = \sum_{n=1}^{\infty} \left\{ \frac{(-1)^{n+1} 2n\pi \sin 1}{n^2\pi^2 - 1} [\cosh n\pi x - \coth n\pi \sinh n\pi x] \right\} \sin n\pi y.$

19. $u(x, y) = \sum_{n=1}^{\infty} \frac{2(-1)^{n+1}}{n\pi \sinh n\pi} \sinh n\pi x \sin n\pi y.$

21. $u(x, y) = \sum_{n=1}^{\infty} \frac{2n\pi[1 + (-1)^{n+1} \cos 1]}{n^2\pi^2 - 1} [\cosh n\pi x - \coth n\pi \sinh n\pi x] \sin n\pi y.$

23. $u(x, y) = \sum_{n=1}^{\infty} \frac{-8[2 + (-1)^n(n^2\pi^2 - 2)]}{n^3\pi^3 \sinh \frac{n\pi}{2}} \sinh \frac{n\pi x}{2} \sin \frac{n\pi y}{2}.$

25. $u(x, y) = \sum_{n=1}^{\infty} \frac{2n\pi[1 + (-1)^{n+1}e]}{n^2\pi^2 + 1} [\cosh n\pi x - \coth 2n\pi \sinh n\pi x] \sin n\pi y.$

27. $$u(x, y) = \sum_{n=1}^{\infty} \left\{ \frac{2n\pi[1 + (-1)^{n+1}e^3]}{n^2\pi^2 + 9} \cosh n\pi x + \left[\frac{6(-1)^{n+1}}{n\pi \sinh n\pi} - \frac{2n\pi(\coth n\pi)(1 + (-1)^{n+1}e^3)}{n^2\pi^2 + 9} \right] \sinh n\pi x \right\} \sin n\pi y .$$

29. $u(x, y) = \sum_{n=1}^{\infty} \frac{4}{n\pi} [1 - 2(-1)^n] \left[\cosh \frac{n\pi x}{2} - \coth n\pi \sinh \frac{n\pi x}{2} \right] \sin \frac{n\pi y}{2} .$

31. $$u(x, y) = \sum_{n=1}^{\infty} \frac{4}{(2n - 1)\pi} \left[\cosh \frac{(2n - 1)\pi x}{2} + \frac{3 - \cosh \frac{(2n - 1)\, 3\pi}{2}}{\sinh \frac{(2n - 1)\, 3\pi}{2}} \sinh \frac{(2n - 1)\pi x}{2} \right] \sin \frac{(2n - 1)\pi y}{2} .$$

33. $u(x, y) = u_1(x, y) + u_2(x, y)$, where u_1 is the solution in Exercise 1, and u_2 is the solution in Exercise 17.

35. $u(x, y) = u_1(x, y) + u_2(x, y)$, where u_1 is the solution in Exercise 9, and u_2 is the solution in Exercise 25.

37. $u(x, y) = u_1(x, y) + u_2(x, y)$, where u_1 is the solution in Exercise 17, and

$$u_2(x, y) = \left[\cosh \pi y + \left(\frac{2}{\pi} - \frac{8}{\pi^3} - \cosh \pi \right) \frac{\sinh \pi y}{\sinh \pi} \right] \sin \pi x + \sum_{n=2}^{\infty} \left[\frac{2 - n^2\pi^2}{n^3\pi^3} (-1)^n - \frac{2}{n^3\pi^3} \right] \frac{2 \sinh n\pi y}{\sinh n\pi} \sin n\pi x.$$

39. $u(x, y) = u_1(x, y) + u_2(x, y)$, where u_1 is the solution in Exercise 13, and

$$u_2(x, y) = \sum_{n=1}^{\infty} \left\{ \frac{2n\pi[1 - (-1)^n e^2]}{n^2\pi^2 + 4} \cosh \frac{n\pi x}{2} + \left[\frac{2n\pi(1 - (-1)^n \cos 2)}{(n^2\pi^2 - 4) \sinh 2n\pi} - \frac{2n\pi \coth 2n\pi\,(1 - (-1)^n e^2)}{n^2\pi^2 + 4} \right] \cdot \sinh \frac{n\pi x}{2} \right\} \sin \frac{n\pi y}{2}.$$

41. $$u(x, y) = \sum_{n=1}^{\infty} \frac{200}{n\pi} [1 - (-1)^n] \left[\cosh \frac{n\pi y}{10} - \coth \frac{3n\pi}{2} \sinh \frac{n\pi y}{10} \right] \sin \frac{n\pi x}{10}.$$

43. $u(r, \theta) = \sum_{n=0}^{\infty} r^n (\alpha_n \cos n\theta + \beta_n \sin n\theta)$, with $\alpha_n = \frac{1}{a^n\pi} \int_0^{2\pi} f(\theta) \cos n\theta \, d\theta$,

$\beta_0 = 0, \quad \beta_n = \frac{1}{a^n\pi} \int_0^{2\pi} f(\theta) \sin n\theta \, d\theta.$

45. $$u(r, \theta) = \sum_{n=1}^{\infty} \frac{200}{n\pi} [1 - (-1)^n] r^n \sin n\theta.$$

Section 11.9

Page 473

1. $$u(x, t) = \sum_{n=1}^{\infty} \frac{2}{n^5\pi^5} \{[e^{-n^2\pi^2 t} - 1][1 - (-1)^n + (-1)^n n^2\pi^2] + (-1)^{n+1} n^4\pi^4 e^{-n^2\pi^2 t} - [(-1)^n - 1] n^2\pi^2 t\} \sin n\pi x.$$

3. $$u(x, t) = \frac{2 \sin x}{\pi} (\sin t + \cos t) + \sum_{n=1}^{\infty} \frac{4 \sin (2n + 1)x}{\pi(2n + 1)[(2n + 1)^4 + 1]} [\sin t + (2n + 1)^2 \cos t - (2n + 1)^2 e^{-(2n+1)^2 t}].$$

5. $$u(x, t) = \sum_{n=1}^{\infty} \left\{ \frac{2[2(-1)^n - 3]}{5\, n^3\pi^3} (1 - e^{5n^2\pi^2 t}) + \frac{2n\pi}{n^2\pi^2 + 1} [1 + e(-1)^{n-1}] \right\} e^{-5n^2\pi^2 t} \sin n\pi x.$$

7. $u(x, t) = \left\{\frac{1}{5}[-3 + (3\cos t + \sin t)e^{3t} - \pi(e^{3t} - 1)\right\}\frac{2e^{-3t}\sin x}{\pi}$

$+ \frac{1}{4}(1 - e^{-12t})\sin 2x + \left\{\frac{1}{1095}[-27 + (27\cos t + \sin t)e^{27t}] - \frac{\pi}{27}(e^{27t} - 1) + \frac{\pi}{2}\right\}\cdot$

$\cdot\frac{2e^{-27t}\sin 3x}{\pi} + \sum_{n=4}^{\infty}\left\{\frac{1-(-1)^n}{n(9n^4+1)}[-3n^2 + (3n^2\cos t + \sin t)e^{3n^2t}] + \frac{(-1)^n\pi}{n^3}(e^{3n^2t} - 1)\right\}\cdot$

$\cdot\frac{2e^{-3n^2t}\sin nx}{\pi}.$

9. $u(x, t) = \sum_{n=1}^{\infty}\left\{\frac{(-1)^{n+1}t^2e^{2n^2t}}{2n^2} + \frac{(2n^2t - 1)(-1)^ne^{2n^2t}}{4n^6} + \frac{(-1)^n}{4n^6} + 2\pi^2[1 - (-1)^n]\right.$

$\left. + \frac{1}{n^2}[4 + 2(-1)^n]\right\}\frac{e^{-2n^2t}\sin nx}{n}.$

11. $u(x, t) = \sum_{n=1}^{\infty}\left\{\left[\frac{2 - 3(-1)^n}{n^3\pi^3}\right]\cos n\pi t - \left[\frac{1-(-1)^n}{n^4\pi^4}\right]\sin n\pi t\right.$

$\left. + \frac{(-1)^n}{n^3\pi^3} - \left[\frac{(-1)^n - 1}{n^3\pi^3}\right]t\right\}2\sin n\pi x.$

13. $u(x, t) = \sum_{n=1}^{\infty}\left\{\frac{-\cos(2n-1)t}{(2n-1)[(2n-1)^2 - \pi^2]} + \frac{3\sin(2n-1)t}{(2n-1)^2}\right.$

$\left. + \frac{\cos \pi t}{(2n-1)[(2n-1)^2 - \pi^2]}\right\}\frac{4\sin(2n-1)x}{\pi}.$

15. $u(x, t) = \sum_{n=1}^{\infty}\left\{\left[\frac{7}{n^3\pi^3}\right]\cos n\pi t + \left[\frac{1-(-1)^n}{n^2\pi}\right]\sin n\pi t\right.$

$\left. - \frac{\sqrt{2}}{n^3\pi^3}[3 - 2(-1)^n]\right\}2\sin n\pi x.$

17. $u(x, t) = \sum_{n=1}^{\infty}\left\{\left[\frac{-1 + (-1)^n}{n(n^2 - \pi^2)} - \frac{3(-1)^n\pi}{n^3}\right]\cos nt + \frac{\pi}{n^2}[1 - (-1)^n]\sin nt\right.$

$\left. - \left[\frac{(-1)^n - 1}{n(n^2 - \pi^2)}\right]\cos \pi t + \frac{3\pi(-1)^n}{n^3}\right\}\frac{2\sin nx}{\pi}.$

19. $u(x, t) = \sum_{n=1}^{\infty}\left\{\left[\frac{2\pi^3}{n}(1 - (-1)^n) + \frac{\pi}{n^3}(4 + 2(-1)^n) - \frac{2\pi(-1)^n}{n^5}\right]\cos nt\right.$

$\left. - \frac{\pi(-1)^nt^2}{n^3} + \frac{2\pi(-1)^n}{n^5}\right\}\frac{2\sin nx}{\pi}.$

21. $u(x, t) = \sum_{n=1}^{\infty} T_n(t)\phi_n(x)$, where $\phi_n(x) = \sqrt{2/l}\sin\frac{(2n-1)\pi x}{2l}$ and $T_n(t)$ is the solution of the initial value problem

$$T_n'' + c^2\mu_n T_n = K_n(t),$$

$$T_n(0) = \int_0^l f(x)\phi_n(x)dx, \quad T_n'(0) = \int_0^l g(x)\phi_n(x)dx,$$

with

$$K_n(t) = \int_0^l F(x, t)\phi_n(x)dx,$$

and

$$\mu_n = \frac{(2n-1)^2\pi^2}{4l^2}.$$

23. $u(x, t) = \sum_{n=1}^{\infty} \frac{[1-(-1)^n]l}{n^3\pi^3}\left\{gl\left(\cos\frac{n\pi t}{l} - 1\right) + 3n\pi\sin\frac{n\pi t}{l}\right\}\sin\frac{n\pi x}{l}.$

25. $u(x, t) = (1 - \cos \pi t)\sin \pi x.$

Section 11.10
Page 478

1. $u(x, t) = (1 - x)t^2 + \sum_{n=1}^{\infty} \frac{2(-1)^{n+1}}{n\pi} e^{-n^2\pi^2 t}\sin n\pi x.$

3. $u(x, t) = \frac{1}{2}(1 - x)t^2 + \sum_{n=1}^{\infty} \frac{2(-1)^{n+1}e^{-n^2\pi^2 t}}{n^5\pi^5}[1 + n^4\pi^4 - (1 - n^2\pi^2 t)e^{n^2\pi^2 t}] \cdot \sin n\pi x.$

5. $u(x, t) = (1 - x)2t + \sum_{n=1}^{\infty}\left\{\frac{5(-1)^{n+1}}{3 + n^2\pi^2}[-1 + e^{(3+n^2\pi^2)t}] + \frac{2}{n^2\pi^2}[1 - e^{n^2\pi^2 t}] + (-1)^{n+1}\right\}\frac{2e^{-n^2\pi^2 t}\sin n\pi x}{n\pi}.$

7. $u(x, y) = \frac{1}{12}x(x^3 - 1) + \sum_{n=1}^{\infty} \frac{(-1)^{n+1}}{6n\pi}[\cosh n\pi y - \coth n\pi \sinh n\pi y]\sin n\pi x.$

9. $$u(x, y) = \frac{1}{12}x(x^3 - 1) + \sum_{n=1}^{\infty}\left\{\left(\frac{2(-1)^{n+1}}{n\pi} + \frac{2}{n^5\pi^5}[2 + (n^2\pi^2 - 2)(-1)^n][\cosh n\pi - 2]\right)\cosh n\pi y - \left(\frac{2(-1)^n \cosh n\pi}{n\pi} + \frac{2}{n^5\pi^5}[2 + (n^2\pi^2 - 2)(-1)^n] - \frac{2\cosh n\pi}{n^5\pi^5}[2 + (n^2\pi^2 - 2)(-1)^n][\cosh n\pi - 2]\right)\frac{\sinh n\pi y}{\sinh n\pi}\right\}\sin n\pi x.$$

11. $$u(x, y) = \sum_{n=1}^{\infty}[\alpha_n \cosh nx + \beta_n \sinh nx]\sin ny + x^2 \sin y,$$

where

$$\alpha_n = \frac{2}{\pi}\int_0^{\pi}[h(y) - 2\sin y]\sin ny\, dy$$

and

$$\beta_n = \frac{1}{\sinh n}\left\{\frac{2}{\pi}\int_0^{\pi}[k(y) - \sin y]\sin ny\, dy - (\cosh n)(\alpha_n)\right\}.$$

13. $$u(x, t) = \frac{5}{\pi}(\pi - x)t^2 + \frac{1}{\pi}xe^t + \sum_{n=1}^{\infty}\left\{\frac{-\cos(2n-1)t}{(2n-1)[(2n-1)^2 - \pi^2]} + \frac{3\sin(2n-1)t}{(2n-1)^2} + \frac{\cos \pi t}{(2n-1)[(2n-1)^2 - \pi^2]}\right\}\frac{4\sin(2n-1)x}{\pi}.$$

15. $$u(x, t) = \sum_{n=1}^{\infty}\frac{-56}{\pi[144n^4\pi^2 + 1]}\left[12n^2\pi \sin\frac{\pi t}{12} - \cos\frac{\pi t}{12} + e^{-n^2\pi^2 t}\right]\sin n\pi x + 28(1 - x)\cos\frac{\pi}{12}(t - 12).$$

17. The eigenvalues are the solutions of the equation

$$\tan\sqrt{-\lambda}l = \rho\sqrt{-\lambda}, \qquad \lambda < 0.$$

REVIEW EXERCISES CHAPTER 11
Page 481

1. Order 4, linear, homogeneous, constant coefficients.

3. Order 2, quasilinear.

5. $u = (z^2 - t)xy + \frac{1}{2}x^2y^2 + f(x, z, t) + g(y, z, t)$.

9. $u_p = u_1 + u_2$. **11.** $u_p = e^{\mu(y-x)}$, μ arbitrary.

13. $u_p = e^{\lambda x + [(\lambda+3)/(\lambda+1)]y}$, λ arbitrary.

15. $3X' - (2x + \lambda)X = 0$, $8Y' + (2y - \lambda)Y = 0$.

17. Elliptic. $u(x, y) = f(x + iy) + g(x - iy)$.

19. Hyperbolic. $u(x, y) = f(x + y) + g(x - \frac{1}{5}y)$.

21. $u = 3 - 2x + \sum_{n=1}^{\infty} \frac{2(-1)^n}{n\pi} e^{-n^2\pi^2 t} \sin n\pi x$.

23.
$$u = \sum_{n=1}^{\infty} 2\left[\frac{2 - n^2\pi^2}{n^3\pi^3}(-1)^n - \frac{2}{n^3\pi^3}\right][\cosh n\pi y - \coth n\pi \sinh n\pi y] \sin n\pi x$$
$$+ \sum_{n=1}^{\infty} \frac{2n\pi[1 + (-1)^{n+1}\cos 1]}{n^2\pi^2 - 1}[\cosh n\pi x - \coth n\pi \sinh n\pi x] \sin n\pi y.$$

25.
$$u = \sqrt{2}\left[\frac{e^{2\pi^2 t} - 1}{2\pi^3} + 1\right] e^{-\pi^2 t} \sin \pi x + \sum_{n=2}^{\infty} \frac{\sqrt{2}(-1)^n}{n(n^2+1)\pi^3}[e^{-n^2\pi^2 t} - e^{\pi^2 t}] \sin n\pi x.$$

27. Set $v(x, t) = u(x, t) - \dfrac{kpx^2}{2l}$, then $v(x, t)$ is a solution of the initial-boundary value problem

$$\begin{aligned}
v_{tt} - a^2 v_{xx} &= \frac{a^2kp}{l}, && 0<x<l, \quad t>0 \\
v(x, 0) &= f(x) - \frac{kpx^2}{2l}, && 0<x<l, \\
v_t(x, 0) &= g(x) - \frac{kpx^2}{2l}, && 0<x<l, \\
v_x(0, t) &= 0, && t>0, \\
v_x(l, t) &= 0, && t>0.
\end{aligned}$$

The solution corresponding to this initial-boundary value problem with homogeneous partial differential equation is given in Example 3, Section 11.6. Thus $v(x, t)$ can be determined by using the method of Section 11.9.

Index

31. $$\int \frac{x\,dx}{ax^2 + bx + c} = \frac{1}{2c} \ln |ax^2 + bx + c| - \frac{b}{2c} \int \frac{dx}{ax^2 + bx + c}$$

32. $$\int \frac{dx}{(ax^2 + bx + c)^{n+1}} = \frac{2ax + b}{n(4ac - b^2)(ax^2 + bx + c)^n} + \frac{2(2n - 1)a}{n(4ac - b^2)} \int \frac{dx}{(ax^2 + bx + c)^n}, \quad b^2 \neq 4ac \quad \text{and} \quad n > 0.$$

33. $$\int \cos^n ax\,dx = \frac{\cos^{n-1} ax \sin ax}{na} + \frac{n - 1}{n} \int \cos^{n-2} ax\,dx.$$

34. $$\int \sin^n ax\,dx = \frac{-\sin^{n-1} ax \cos ax}{na} + \frac{n - 1}{n} \int \sin^{n-2} ax\,dx.$$

35. $$\int \sec^n ax\,dx = \frac{\sec^{n-2} ax \tan ax}{a(n - 1)} + \frac{n - 2}{n - 1} \int \sec^{n-2} ax\,dx, \quad n \neq 1.$$

36. $$\int \csc^n x\,dx = -\frac{\csc^{n-2} ax \cot ax}{a(n - 1)} + \frac{n - 2}{n - 1} \int \csc^{n-2} ax\,dx, \quad n \neq 1.$$

37. $$\int \sec^n ax \tan ax\,dx = \frac{\sec^n ax}{na} + c, \quad n \neq 0.$$

38. $$\int \csc^n ax \cot ax\,dx = -\frac{\csc^n ax}{na} + c, \quad n \neq 0.$$

39. $$\int \cos ax \cos bx\,dx = \frac{\sin (a - b)x}{2(a - b)} + \frac{\sin (a + b)x}{2(a + b)} + c, \quad a^2 \neq b^2.$$

40. $$\int \sin ax \sin bx\,dx = \frac{\sin (a - b)}{2(a - b)} x - \frac{\sin (a + b)x}{2(a + b)} + c, \quad a^2 \neq b^2.$$

41. $$\int \sin ax \cos bx\,dx = -\frac{\cos (a - b)x}{2(a - b)} - \frac{\cos (a + b)x}{2(a + b)} + c, \quad a^2 \neq b^2.$$

42. $$\int \cos^2 ax\,dx = \frac{x}{2} + \frac{\sin 2ax}{4a} + c.$$

43. $$\int \sin^2 ax\,dx = \frac{x}{2} - \frac{\sin 2ax}{4a} + c.$$

44. $$\int \sin ax \cos ax\,dx = -\frac{\cos 2ax}{4a} + c.$$

45. $$\int \tan^2 ax\,dx = \frac{1}{2} \tan ax - x + c.$$

46. $$\int \cot^2 ax\,dx = -\frac{1}{a} \cot ax - x + c.$$

47. $$\int \sec^2 ax\,dx = \frac{1}{a} \tan ax + c.$$

48. $$\int \csc^2 ax\,dx = -\frac{1}{a} \cot ax + c.$$